Advances in Nanotechnology for Smart Agriculture

The yield of major agricultural crops can be severely decreased due to the inappropriate application of commonly used harmful chemicals. Excessive agrochemicals in field application can negatively affect microbial populations and their diversity, which in turn ultimately affects plant growth. Thus, it is necessary to turn toward more eco-friendly approaches which equally protect crops as well as the desirable microbial populations of complex soil systems. Nanoparticles are considered as potential agents for the production and development of sustainable agriculture. Green synthesis of nanoparticles has gained attention as a useful measure to diminish the harmful effects associated with the old methods of nanoparticle synthesis. *Advances in Nanotechnology for Smart Agriculture: Techniques and Applications* illustrates the science and practical applications of nanoparticles for sustainable agriculture. The book also examines the role of nanotechnology in agricultural best practices, including sustainable development, precision farming, and long-term soil health.

Microbial Biotechnology for Food, Health, and the Environment Series

Series Editor: Ashok Kumar Nadda

Plant-Microbial Interactions and Smart Agricultural Biotechnology
Edited by Swati Tyagi, Robin Kumar, Baljeet Saharan, and Ashok Kumar Nadda

Microbial Products
Applications and Translational Trends
Edited by Mamtesh Singh, Gajendra Pratap Singh, and Shivani Tyagi

For more information about this series, please visit: www.routledge.com
https://www.routledge.com/Microbial-Biotechnology-for-Food-Health-and-the-Environment/book-series/CRCMBFHE

Advances in Nanotechnology for Smart Agriculture
Techniques and Applications

Edited by
Parul Chaudhary, Anuj Chaudhary,
Ashok Kumar Nadda, and Priyanka Khati

CRC Press
Taylor & Francis Group
Boca Raton London New York

CRC Press is an imprint of the
Taylor & Francis Group, an **informa** business

Cover Image Credit: Cover image supplied by Anuj Chaudhary

First edition published 2023
by CRC Press
6000 Broken Sound Parkway NW, Suite 300, Boca Raton, FL 33487-2742

and by CRC Press
4 Park Square, Milton Park, Abingdon, Oxon, OX14 4RN

CRC Press is an imprint of Taylor & Francis Group, LLC

ISBN: 978-1-032-38547-1 (hbk)
ISBN: 978-1-032-38548-8 (pbk)
ISBN: 978-1-003-34556-5 (ebk)

DOI: 10.1201/9781003345565

Typeset in Times
by codeMantra

Contents

Editors

Dr. Parul Chaudhary is working as an Assistant Professor in the School of Agriculture, Graphic Era Hill University, Dehradun, Uttarakhand, India. She has teaching and research experience of more than 5 years. She received her PhD in Microbiology at the Govind Ballabh Pant University of Agriculture & Technology from Pantnagar, U.S. Nagar, India. Her doctoral research focused on nanotechnology and microbial inoculants utilized for maintenance of plant and soil health for sustainable agriculture. Her areas of interests in the agricultural field include microbiology, nanotechnology, bioremediation and soil metagenomics. She has published more than 36 research and review articles in reputed journals. She co- authored 13 book chapters with Elsevier and Springer. She has reviewed publications for various national and international journals. She is also a member of the guest editorial board in Frontiers in Environmental Science. She has received the Young Scientist Award from Bioved Research Institute, Prayagraj, India.

Mr. Anuj Chaudhary worked as an Assistant Professor at Shobhit University, Gangoh, Saharanpur, Uttar Pradesh, India. He has completed his Bachelor of Technology in Biotechnology and Master of Technology in Agricultural Processing & Food Engineering from Sardar Vallabhbhai Patel University of Agriculture & Technology, Modipuram, Meerut, India. His area of interest is in agriculture microbiology, biotechnology, agricultural processing, and nanotechnology. He has published 12 research articles in highly reputed international journals and authored six book chapters.

Dr. Ashok Kumar Nadda is working as an Assistant Professor in the Department of Biotechnology and Bioinformatics, Jaypee University of Information Technology, Waknaghat, Solan, Himachal Pradesh, India. He holds an extensive Research and Teaching experience of more than 8 years in the field of microbial biotechnology, with research expertise focusing on various issues pertaining to nano-biocatalysis, microbial enzymes, biomass, bioenergy, and climate change. Dr. Ashok teaches Enzymology and Enzyme Technology, Microbiology, Environmental Biotechnology, Bioresources, and Industrial Products to undergraduate, graduate, and PhD students. He holds international work experiences in South Korea, India, Malaysia, and People's Republic of China. He worked as a post-doctoral fellow in the State Key Laboratory of Agricultural Microbiology, Huazhong Agricultural University, Wuhan, China. He also worked as a Brain Pool Researcher/Assistant Professor at Konkuk University, Seoul, South Korea. Dr. Ashok has a keen interest in microbial enzymes, biocatalysis, CO_2 conversion, biomass degradation, biofuel synthesis, and bioremediation. His work has been published in various internationally reputed journals, namely *Chemical Engineering Journal, Bioresource Technology, Scientific Reports, Energy, International Journal of Biological Macromolecules, Science of the Total Environment*, and *Journal of Cleaner Production*. Dr. Ashok has published more than 185 scientific contributions in the form of research, review, books, book chapters, and others at several platforms in various journals of international repute. The research output includes 120 research articles, 43 book chapters, and 22 books. He is the main series editor of *Microbial Biotechnology for Food, Health, and the Environment* that publishes books under Taylor & Francis/CRC Press, USA. He is also a member of the editorial board and reviewer committee of various journals of international repute. He has presented his research findings in more than 40 national/ international conferences. Dr. Ashok is also an active reviewer for many high-impact journals published by Elsevier, Springer Nature, ASC, RSC, and Nature Publishers. His research works have gained broad interest through his highly cited research publications, book chapters, conference presentations, and invited lectures.

Dr. Priyanka Khati is working as a Scientist at ICAR-Vivekananda Parvatiya Krishi Anusandhan Sansthan, Almora, Uttarakhand, India. She has completed her MSc and PhD in Microbiology from Govind Ballabh Pant University of Agriculture & Technology, Pantnagar, India. Her research interests are in microbiology, nanotechnology, soil metagenomics, soil health, and biofertilizers in the agriculture field. She has published more than 43 research papers in highly esteemed international journals and authored 15 book chapters. She has reviewed publications for various national and international journals.

Contributors

Bartholomew Saanu Adeleke
Department of Biological Sciences
Microbiology Unit, Faculty of Science
Olusegun Agagu University of Science
and Technology
Okitipupa, Nigeria
and
Department of Microbiology
North-West University
Mafikeng, South Africa

Modupe Stella Ayilara
Food Security and Safety Focus Area,
Faculty of Natural and Agricultural
Sciences
North-West University
Mafikeng, South Africa

Geeta Bhandari
Department of Biosciences
Swami Rama Himalayan University
Dehradun, Uttarakhand, India

Mahendar Singh Bhinda
Crop Improvement Division
ICAR-Vivekananda Parvatiya Krishi
Anusandhan Sansthan
Almora, India

Pritom Biswas
Department of Biotechnology
Central University of South Bihar
Gaya, India

Payal Chakraborty
Department of Plant Physiology
Banaras Hindu University
Varanasi, India

Ajay Kumar Chandra
Division of Genetics
ICAR-Indian Agricultural Research
Institute
New Delhi, India

Anuj Chaudhary
School of Agriculture & Environmental
Sciences
Shobhit University
Gangoh, India

Parul Chaudhary
School of Agriculture, Graphic Era Hill
University
Dehradun, Uttarakhand
India

Shalu Chaudhary
Department of Biosciences
Swami Rama Himalayan
University
Dehradun, India

Sadhna Chauhan
Department of Plant Pathology
Govind Ballabh Pant University of
Agriculture & Technology
Pantnagar, India

Shaohua Chen
State Key Laboratory for Conservation
and Utilization of Subtropical
Agro-bioresources, Integrative
Microbiology Research
South China Agricultural
University
Guangzhou, China

Padma Chorol
Department of Zoology
Maya Group of Colleges
Dehradun, India

Hemant Dasila
Department of Microbiology
Akal College of Basic Sciences
Eternal University
Baru Sahib, India

Vaibhav Dhaigude
Animal Nutrition Division
ICAR-NDRI
Karnal, India

Sami Abou Fayssal
Department of Agronomy
University of Forestry
Sofia, Bulgaria
and
Department of Plant Production
Lebanese University
Beirut, Lebanon

Saurabh Gangola
School of Agriculture
Graphic Era Hill University
Bhimtal, India

Saipayan Ghosh
Department of Horticulture
Dr. Rajendra Prasad Central
    Agricultural University
Samastipur, India

Sanjay Gupta
Department of Biosciences
Swami Rama Himalayan
    University
Dehradun, India

Naliyapara HB
Animal Nutrition Division
ICAR-NDRI
Karnal, India

Priyanka Khati
Crop Production Division
ICAR-Vivekananda Parvatiya Krishi
    Anusandhan Sansthan
Almora, India

Ashish Kumar
Department of Microbiology
Govind Ballabh Pant University of
    Agriculture & Technology
Pantnagar, India

Govind Kumar
Crop Production Division
ICAR-Central Institute for Subtropical
    Horticulture
Lucknow, India

Pankaj Kumar
Department of Zoology and
    Environmental Science
Gurukul Kangri University
Haridwar, India

Tarun Kumar
Department of Agriculture Engineering
Dr. Rajendra Prasad Central
    Agricultural University
Samastipur, India

Ankita Kumari
Department of Biotechnology
Central University of South Bihar
Gaya, India

Asha Kumari
School of Crop Sciences
ICAR-Indian Agricultural Research
    Institute
Hazaribagh, India

Swati Lohani
Crop Production Division
ICAR-Vivekananda Parvatiya Krishi
    Anusandhan Sansthan
Almora, India

Chandan Maharana
Plant Protection Division
ICAR-Central Potato
Research Institute
Shimla, India

Damini Maithani
School of Biotechnology
Department of Microbiology
IFTM University
Moradabad, India

Sandeep Manuja
Department of Agronomy
Chaudhary Sarwan Kumar Himachal
Pradesh Krishi Vishvavidyalaya
Palampur, India

Manas Mathur
School of Agriculture
Suresh Gyan Vihar University
Jaipur, India

Jyotsana Maura
Defence Institute of Bio-Energy
Research
DRDO
Haldwani, India

Tatiana Minkina
Academy of Biology and
Biotechnology
Southern Federal University
Rostov-on-Don, Russia

Ashok Kumar Nadda
Department of Biotechnology and
Bioinformatics
Jaypee University of Information
Technology
Waknaghat, India

Mayur Giridhar Naitam
Department of Microbiology
Dr. Rajendra Prasad Central
Agricultural University
Samastipur, India

Heena Parveen
Dairy Microbiology Division
National Dairy Research Institute
Karnal, India

Vivek Kumar Patel
Department of Plant Pathology
Dr. Rajendra Prasad Central
Agricultural University
Samastipur, India

Shalu Priya
Department of Microbiology
Govind Ballabh Pant University of
Agriculture & Technology
Pantnagar, India

Vishnu D Rajput
Academy of Biology and
Biotechnology
Southern Federal University
Rostov-on-Don, Russia

Alka Rani
Department of Microbiology
Gurukul Kangri University
Haridwar, India

Vijaya Rani
Crop Protection Division
ICAR-Indian Institute of Vegetable
Research
Varanasi, India

Pratima Raypa
Department of Biochemistry
Govind Ballabh Pant University of
Agriculture & Technology
Pantnagar, India

Sanjay Kumar Sanadya
Department of Genetics and Plant
Breeding
Chaudhary Sarwan Kumar Himachal
Pradesh Krishi Vishvavidyalaya
Palampur, India

Adita Sharma
College of Fisheries
Dr. Rajendra Prasad Central
    Agricultural University
Samastipur, India

Anita Sharma
Department of Microbiology
Govind Ballabh Pant University of
    Agriculture & Technology
Pantnagar, India

Anuj Sharma
Department of Biotechnology
CSIR-Central Institute of Medicinal and
    Aromatic Plants
Bangalore, India

Barkha Sharma
Department of Microbiology
Govind Ballabh Pant University of
    Agriculture & Technology
Pantnagar, India

Deepak Sharma
School of Agricultural Sciences
Jaipur National University
Jaipur, India

Rahul Sharma
Department of Agronomy
Chaudhary Sarwan Kumar Himachal
    Pradesh Krishi Vishvavidyalaya
Palampur, India

Akashdeep Singh
Department of Agronomy
Chaudhary Sarwan Kumar Himachal
    Pradesh Krishi Vishvavidyalaya
Palampur, India

Harshita Singh
Department of Microbiology
Govind Ballabh Pant University of
    Agriculture & Technology
Pantnagar, India

Neha Singh
Department of Plant Pathology
Dr. Yashwant Singh Parmar University
    of Horticulture and Forestry
Nauni, India

Pragati Srivastava
Department of Microbiology
Govind Ballabh Pant University of
    Agriculture & Technology
Pantnagar, India

Rabiya Sultana
Department of Biotechnology
Gauhati University
Gauhati, India

Gohar Taj
Department of Molecular Biology &
    Genetic Engineering
Govind Ballabh Pant University of
    Agriculture & Technology
Pantnagar, India

Anjali Tiwari
G.B. Pant National Institute of
    Himalayan Environment
Almora, India

Shalini Tiwari
Department of Microbiology
Govind Ballabh Pant
    University of Agriculture &
    Technology
Pantnagar, India

Bhavya Trivedi
Department of Microbiology
Maya Group of Colleges
Dehradun, India

Viabhav Kumar Upadhayay
Department of Microbiology
Dr. Rajendra Prasad Central
    Agricultural University
Samastipur, India

1 Role of Nanoparticles in Agriculture

Heena Parveen and Parul Chaudhary
National Dairy Research Institute and
Graphic Era Hill University

Pragati Srivastava and Shalu Priya
Govind Ballabh Pant University of Agriculture & Technology

Geeta Bhandari
Swami Rama Himalayan University

Anuj Chaudhary
Shobhit University

CONTENTS

1.1 INTRODUCTION

Both food and the essential ingredients required for a better life are produced by agriculture, which serves as a crucial foundation of the global economy. However, in the current situation, agriculture is facing a number of difficulties such as reduction in arable land, insufficiency of water, effects of climate change and low efficacy of agrochemicals, which exacerbate abiotic and biotic stresses on crops as well as rising food demand of growing population, which results in lower yields (Pouratashi & Iravani, 2012; Tyagi et al., 2022). It is predicted that there will be 9.6 billion people on the planet by 2050,

DOI: 10.1201/9781003345565-1

which will require a 70%–100% increase in agricultural production to meet the demand for food (United Nations, 2019). Consequently, it is imperative to boost food production by more than 50% in order to fulfil the demands of expanding population (Mittal et al., 2020). However, agricultural land expansion may result in increased deforestation, endangering sustainability because it will have an adverse effect on greenhouse gas emissions and biodiversity (Ioannou et al., 2020). Additionally, the development of stress-tolerant crop varieties is still not commercially available (de Lange et al., 2018). Newer methods and approaches are continuously being developed to address these problems. Priming is a new technique for plants to defend themselves against environmental challenges. Chemical priming holds great promise for increasing plant tolerance to various stress stimuli by activating defence systems that have evolved without the need for genetic modifications (Savvides et al., 2016). As a result, one step has been taken: the usage of nanomaterial (NM)-based solution for developing current agricultural practices.

Nanomaterials are particles with a size of at least 1 nm in at least one dimension. Due to their nano-dimensions, NMs are generally employed in different sectors such as biological sciences, aircraft, electronics, agriculture and chemical manufacturing (Jeevanandam et al., 2018). These materials have a high reactivity with their huge surface area and new physicochemical features, allowing for easy change in response to rising demand. The NMs including nanoclays, nanowires and nanotubes have unique surface chemistry analogies and optical properties that boost sensitivity, improve detection limits and provide a quicker response (Kumbhakar et al., 2014; Yaqoob et al., 2020). Numerous inorganic nanoparticles (NPs) including carbon nanotubes, iron and silica NPs have drawn research interest because of their large surface area. Copper, gold and silver nanoparticles have also been studied in other investigations (Anandhi, 2020).

Agriculture is now being studied as a potential method for plant development and boosting agricultural output, among its many other uses. The National Planning Workshop of America promoted the usage of NPs in agriculture, particularly for food packing and pesticide residue detection (Scott & Chen, 2013). Furthermore, nanoparticles such as nanoenzymes have an important function in protecting plants from a variety of abiotic/biotic stress. Subsequent research on NMs has shown their role in scavenging reactive oxygen species (ROS) as well as protecting photosynthetic apparatus and simultaneously improving photosynthetic rate by controlling stress factors (Gohari et al., 2020; Chaudhary et al., 2021a, b). Several scientists have discussed the potential role of NMs including titanium, nanochitosan, cerium and zinc oxide (TiO_2, CeO_2 and ZnO) towards plant growth promotion by improving stress-tolerating abilities in plants (Ioannou et al., 2020; Chaudhary et al., 2021c, d), although the structural assets of NMs such as small size and large surface area with numerous binding sites also lead to the delivery of genetic materials (plasmid DNA or RNA) by acting as nanocarriers. Kwak et al. (2019) described how single-walled carbon nanotubes (SWCNTs) work to carry genetic material into the nucleus and chloroplast. These findings concluded that NPs can serve as a chassis for transferring useful genetic material to a variety of plant species (Demirer et al., 2019). MP-based nano-agrochemical transfer has a lot of potential to improve the agricultural input efficacy, reducing environmental pollution and lowering labour costs, all of which contribute to agricultural system sustainability and food security.

However, the application of NPs via seed and foliar spray is still in its infancy due to numerous aspects of sustainable agriculture and crop productivity development (Figure 1.1; Table 1.1). Surprisingly, some NPs, especially when used in excessive concentrations, can produce negative effects. Some NPs cause oxidative stress, which lowers agricultural yields and germination rates (Begum & Fugetsu, 2012; Da Casta & Sharma, 2016). Therefore, to develop green technologies, it is necessary to understand the functionality of NPs in relation to their interactions with plants at molecular and cellular levels. In order to improve process design, risk valuation of NMs and laws for the commercialization of nanotechnology are also necessary. In this chapter, we highlight the different types of NPs in agriculture for crop productivity and plant protection.

1.2 TYPES OF NANOMATERIALS

On the basis of their agricultural applications, NMs are generally classified into three major groups, *viz.*, inorganic, organic and combined nanoparticles (He et al., 2019) (Figure 1.2).

1.2.1 INORGANIC-BASED NANOMATERIALS

The inorganic nanoparticles might be semiconductor nanoparticles, magnetic (gold) or noble NPs (zinc oxide). Due to their better material qualities and wide range of applications, inorganic nanoparticles have garnered increased attention. There have also been reports of NPs that cause the overexpression of genes linked to improved

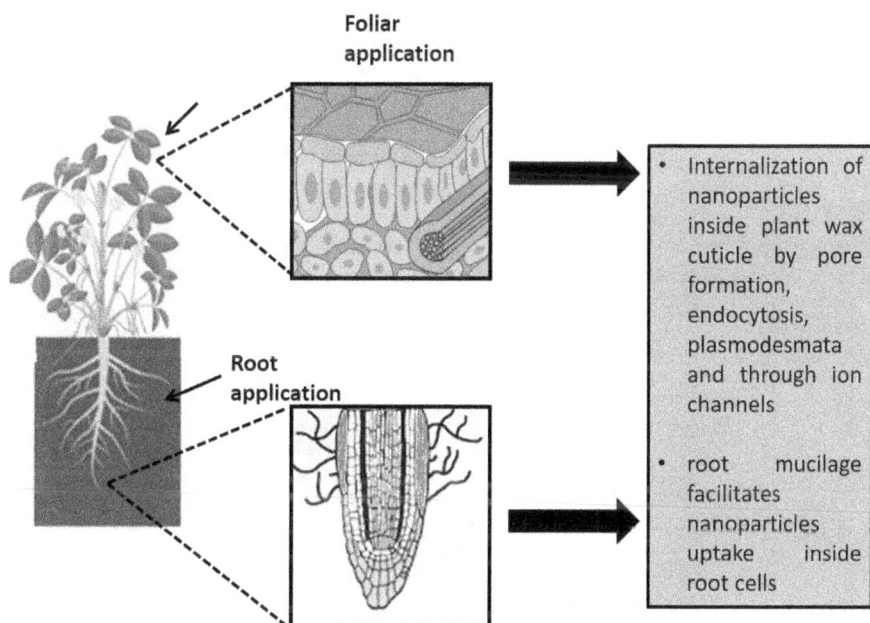

FIGURE 1.1 Application of nanomaterials via root and foliar spray.

TABLE 1.1

Seed and Foliar Application of Nanomaterials to Improve Plant Health Parameters

Nanoparticles	Plant Species	Application	Effects on Plants	References
TiO_2	*Linum usitatissimum*	Leaf treatment	Improved carotenoids and chlorophyll content, decreased MDA levels	Aghdam et al. (2016)
ZnO	*Glycine max*	Petri dish exposure	Increased germination rate	Sedghi et al. (2013)
ZnO	*Mangifera indica*	Foliar spray	Improved plant growth, nutrients uptake and carbon assimilation	Elsheery et al. (2020)
SiO_2	*Solanum lycopersicum*	Exposure *in vitro*	Up- and down-regulation of salt stress genes to mitigate salt stress	Almutairi (2016)
Fe	*Dracocephalum moldavica*	Foliar application	Increased leaf area, phenolic flavonoid content, guaiacol peroxidase, catalase, and glutathione reductase activities	Moradbeygi et al. (2020)
Ag	*Triticum aestivum*	Potting soil, seed priming	Improved root and shoot length, root number, increased morphological growth of the plant-like increased leaf area and number	Abou-Zeid and Ismail (2018); Iqbal et al. (2019)
Cu	*Zea mays*	Plant priming	Higher leaf water content and plant biomass, increased anthocyanin, chlorophyll and carotenoid contents	Van Nguyen et al. (2020)
Si	*Triticum aestivum*	Nutrient solution	Enhanced antioxidants to protect against UV-B-generated oxidative stress	Tripathi et al. (2017)
Mn	*Capsicum annuum*	Nanopriming	Role in plant salt stress management in order to promote sustainable agriculture	Ye et al. (2020)
Chitosan	*Hordeum vulgare*	Foliar application	Increased the relative water content, grain weight and grain protein content	Behboudi et al. (2018)

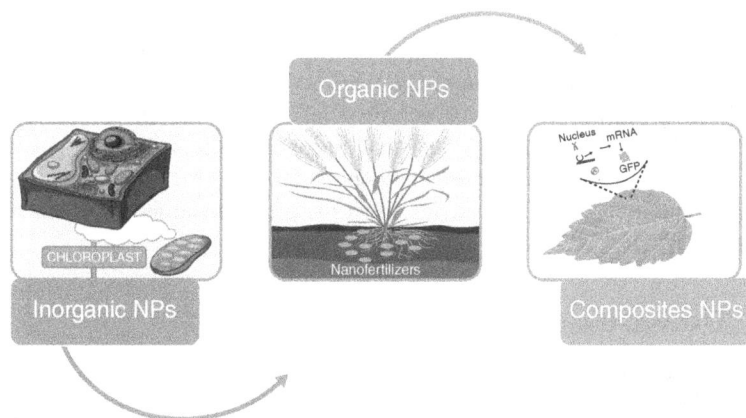

FIGURE 1.2 Types of nanomaterials used in agriculture.

photosynthesis rate, stress tolerance and antioxidant defence (Ahmad et al., 2018). Abdal-Latef et al. (2018) reported that TiO_2 NPs (0.01%) enhanced plant growth and chlorophyll content. It is also widely known that titanium dioxide has antibacterial properties. According to several studies, crops exposed to titanium dioxide may be able to reduce bacterial and fungal diseases. Trials conducted in greenhouses and on fields showed that TiO_2 NPs can significantly lessen the severity of bacterial spots (Paret et al., 2013). Iron NPs may be beneficial in the remediation of organic pollutants from soil by reducing the toxicity of contaminants and improving the growth of Chinese cabbage (Wu et al., 2016). Bharti et al. (2018) also studied the impact and interaction of TiO_2 NPs with garlic using laser-induced fluorescence and UV visible spectral techniques. The leaves of garlic plants had a better photosynthetic activity over control.

Besides its plant-promoting role, sometimes NMs are toxic to plants when entered into the soil. The prevalence of metal-based NPs in the soil matrix is influenced by the amount of dissolved organic matter (DOM), which is the most reactive and dynamic kind of organic matter in agricultural soil. The fate of NPs in the soil microenvironment depends on their interaction with soil particles, where it reveals their bioavailability to plants and also reveals their phytotoxicity (Qiu & Smolders, 2017). Due to the frequent release of nano-based agricultural products, it is also essential to examine how NPs interact with soil bacteria (Duhan et al., 2017). ZnO NPs exhibit bactericidal effects on *Sinorhizobium meliloti*, which fixes nitrogen. TiO_2 and ZnO NPs also harm the order Rhizobiales by reducing bacterial diversity and adversely influencing N_2 fixation directly (Bandyopadhyay et al., 2015). Silicon dioxide and nanochitosan NPs improved maize growth and soil enzyme activities such as dehydrogenase (Kukreti et al., 2020; Agri et al., 2022). Silicon NPs inhibit the germination of pumpkin seeds and decrease plant development (Hawthorne et al., 2012). Silver NPs (Ag-NPs) are the most prominent NPs and are employed in a wide variety of applications. It makes for around 25% of all products made with nanomaterials (Vance et al., 2015). To this, higher concentration of NMs also leads to genotoxicity in plants rather than only phytotoxicity. Haliloglu et al. (2020) also reported geno-toxicity causing DNA damage induced by the higher dosage of NPs. According to a study on *Allium cepa*, Ag-NPs can be cytotoxic and genotoxic even at low concentrations, affecting morpho-anatomical and metabolic alterations (Souza et al., 2020). The use of NPs to supply vital nutrients and boost agricultural productivity is gaining popularity; nevertheless, further study on the toxicity of recently produced NMs should be done to allay public worries about the problems allied with nanotoxicity (Agri et al., 2021).

1.2.2 ORGANIC-BASED NANOMATERIALS

These are of commercial interest with versatile activities including electrical conductivity, electron affinity and structural high strength that directly promotes plant growth. These mainly include carbon nanoparticles (carbon nanotubes (CNTs), fullerenes, carbon dots, graphene) where CNTs and fullerenes are the major representative class (Khan et al., 2019). Few reports have shown the role of carbon dots (CDs) and CNTs in plant-promoting activity, where CDs have self-fluorescence, which makes them useful for tracking functions, such as stress monitoring. Carbon nanotubes have been designed for use as sensors in plants to detect signalling chemicals such as H_2O_2, nitric oxide and calcium (Giraldo et al., 2015). Furthermore, because carbon nanotubes (CNTs) exhibit competitive mechanical, thermal and chemical possessions, they might be used as nutrient transporters for macro- and micronutrients to reduce their applied quantities, resulting in positive agricultural outcomes (Taha, 2016). Also, SW (single-walled) CNTs function as nano-transporters, allowing DNA and dye molecules to be transported into plant cells. In addition to altering seed germination and plant growth, MW (multi-walled) CNTs have a comparable transferring ability via inducing water, Ca and Fe absorption efficiency, which results in increased root and stem growth as well as peroxidase and dehydrogenase activities, leading to water retention capacity and increased biomass, flowering and fruit production (Khodakovskaya et al., 2012; Siddiqui et al., 2015). Lahiani et al. (2013) reported the growth promotion applications of MWCNTs in various crops including corn, soybeans and barley, which has been further demonstrated through TEM imaging and Raman spectroscopy showing their aggregation inside the seed endosperm. The internalization of the CNTs has resulted in increased water uptake efficiency that directly results in plant growth promotion.

There have been several reports of adverse effects over plant growth that would be largely dependent on CNT concentrations, soil minerals, plant species and exposure conditions (Mukherjee et al., 2016). The effect of MWCNT (1,000 mg/L) exposure on hydroponically grown *Zucchini* plant reduced biomass loss by 60% as compared to control and bulk carbon (Stampoulis et al., 2009). Furthermore, Anjum et al. (2013) observed the impact of GO (graphene oxide) on glutathione redox system in *fava bean*, a key predictor of cellular redox homeostasis. Reduced oxidative enzyme activity and a concentration-dependent stress response were noted. It is important to consider how NMs interact with pollutants (heavy metals). Several of them can adsorb heavy metals to help clean up soil and protect plants (Kolluru et al., 2021). The nanomaterial produced cellular damage in wheat plants treated with graphene oxide, which increased arsenic absorption. Carbohydrates were inhibited and fatty acids were disturbed, while GO improved amino acid metabolism, it also reduced As(V) to As(III), making it more toxic (Hu et al., 2014). Despite the fact that some research indicates that carbon-based NMs might be used in agriculture, certain investigations have revealed harmful effects in a variety of plant species.

1.2.3 COMPOSITE-BASED NANOMATERIALS

When a composite material has a phase with nanoscale morphology, such as nanoparticles, nanotubes or lamellar nanostructure, the material is said to be a nanocomposite. They are multiphase materials because they have many phases, and at least one of those phases should have diameters between 10 and 100 nm. Nanocomposites have evolved as useful alternatives to engineering materials that currently have limitations in order to overcome them. Crop productivity has been increased by the engineering of a number of micro-/macronutrients, nutrient-loaded and plant growth-promoting nanoparticles. The nutrients in these ENCs (engineered nanocomposites) can be readily released as soluble ions, directly promoting plant growth (Guha et al., 2020). There are several reports that have mentioned the application of these nanocomposites in pant growth promotion. Olad et al. (2018) have reported NPK-loaded unique nanocomposites which showed enhanced water absorption capacity that is chiefly dependent upon acrylamide and montmorillonite particles. Phosphorus-rich hydroxyapatite nanoparticles are seen to be exceptionally effective at increasing the yield of soybean (*Glycine max*) plants. The pace of growth and height of the plants in the hydroxyapatite nanoparticle-treated plants were noticeably higher than in the plants that were treated with phosphate fertilizers such as triple superphosphate (Liu & Lal, 2012). Nano-biochar has also come up as a suitable nanocomposite bearing effective physical and chemical properties supporting plant growth. Additionally, nano-biochar-based nanocomposite demonstrated superior absorption capacity for different pollutants like heavy metals and herbicides, because of their higher alkalinity and pH; they could also be used to mitigate soil acidity (Gamiz et al., 2017). Liu et al. (2020) reported that nanobiochar improved seed germination, plant biomass, cadmium stability and *Actinobacteria* diversity in soil under Cd-contaminated conditions.

Even though the capacity of selected and non-selected nanocompositions to bind molecules allows for the identification of their distribution, bioavailability, level of toxicity and exclusion from plant cells, the protein-nanopesticide combination can lead to protein denaturation and slight structural disarray. Similar to this, cytogenetic aberrations were also brought on by the mobility and structural alterations of genomic DNA, which were mediated by NPs (Deka et al., 2021). The homeostasis between essential aspects like biodegradability, concentration and size of incorporated active ingredients determines the toxicity of composite nano-agrochemicals on plant growth (Kumar et al., 2021). However, nanocomposites have emerged as a crucial component of nanomaterials in promoting plant life cycle. The development of nanoparticle-based products, including efficient agrochemical delivery systems with nanocomposites as main components, is ongoing and still required to enhance working design, regulate commercialization and examine the risks of nano-pesticides, nano-herbicides and nano-fertilizers in order to accomplish the practical benefits of nano-agrochemicals.

1.3 NANOMATERIALS AS PLANT REGULATORS

1.3.1 Nanoenzymes

The antioxidant enzyme-imitating properties of iron oxide NPs (Fe$_3$O$_4$) were disclosed by Gao (2022). As a result, various inorganic nanomaterials such as fullerene C60, platinum and caesium oxide NPs were shown to have comparable characteristics. For instance, MoS$_2$ nanosheets exhibiting peroxidase, catalase and superoxide dismutase-like activities were developed by Chen et al. (2018). Boghossian et al. (2013) showed that CeO$_2$NPs at relatively low concentrations (5 M) may significantly lower reactive oxygen species level and safeguard chloroplasts. Wu et al. (2018) later showed that CeO$_2$NPs with poly(acrylic acid) coatings with SOD and CAT activities preserved *Arabidopsis* plants' capacity to photosynthesize under high salinity. Wu et al. (2017) demonstrated that spherical CeO$_2$NPs coated with poly(acrylic acid) may penetrate chloroplasts, scavenge ROS and improve photosynthesis in plants. The quantum yield of PSII, carbon absorption and Rubisco carboxylation rate in *Arabidopsis* plants altered with CeO$_2$NPs were increased by 19%, 67% and 61%, respectively, under stress conditions.

It has also been demonstrated that enzyme-mimicking NPs (CeO$_2$, C60) can increase plant growth and stress tolerance through mechanisms other than ROS scavenging. For instance, Borisev et al. (2016) demonstrated that fullerenol NPs act as an extra intercellular water source to reduce the oxidative stress of sugar beet during drought stress. In another promising potential nanoenzyme with agricultural uses, Mn$_3$O$_4$NPs showed more ROS-scavenging activity than CeNPs. This finding and Mn role as a micronutrient for plants point to the potential application of Mn$_3$O$_4$ nanoenzymes in agriculture to increase plant stress tolerance (Yao et al., 2018). Some NPs may enhance stress-related and antioxidant pathways, furthermore to directly scavenging ROS and promoting stress tolerance.

1.3.2 Nano-Seed Priming

In relation to plant growth development from infections and insects, seeds have frequently been treated throughout time with fungicides and/or insecticides. Plant growth and production rates of crops may be greatly increased by using growth promoters such as agrochemicals (Chaudhary et al., 2022b). However, their extensive use in plant growth-promoting functionality has resulted in their uncontrolled release into the environment that showed significant negative effects on both the environment and the economy and also harmful effects on human health (Nicolopoulou-Stamati et al., 2016). In addition, polymer coatings have garnered a lot of interest in seed priming methodology during the past few years. Biodegradable polymers have been used as seed coating to allow for the encapsulation and persistent release of nutrients and growth regulators in order to promote seed germination, root and shoot development (Ioannou et al., 2020). Additionally, the application of polymer coatings in seed priming may increase the ability of plants and crops to withstand abiotic stresses including drought, salt, heat and heavy metals, among others (Wheeler & von Braun, 2013). There are several studies that demonstrate the possible use of synthetic polymers as seed-coating agents. A polymer coating and a biocontrol agent (*Bacillus amyloliquefaciens*) were tested on *Capsicum annuum* seedling. Their findings showed that in contrast to control (uncoated) seedlings, treated seedlings' root exudates generated a significant amount of VOCs such as hydroxylamine and hexadecenoic acid, and are regarded to be important carbon sources in soil (Sathya et al., 2016). The impact of a polymer-based film coating is used as a delivery mechanism for protective chemical agents on onion seeds in terms of growth and yield parameters (Yalamalle et al., 2019).

Polystyrene, polyamides, amino resins and polyacrylamide are the synthetic polymers that are employed most frequently. However, due to their inability to degrade, the synthetic polymers discussed above cannot be entirely eliminated from the environment. As a result, any hazardous residues left behind might be extremely harmful to the environment. Several products based on organically produced, environmentally friendly polymeric film coatings have been developed in order to effectively apply, controlled release of growth promoters, and nutrients in seed priming (Changcheng et al., 2022). The polysaccharides such as alginates, cellulose, chitosan and starch are the most often utilized natural polymers (Neri-Batang & Chakraborty, 2019). It was interestingly found that chitosan-treated seeds improved seed germination efficiency, seedling vigour and growth traits (Rakesh et al., 2017) (Table 1.2).

1.3.3 Nano-Stress Sensors

Crop plants have the ability to endure stressful situations by utilizing various systems (stress sensing and signalling), and H_2O_2 is one of these molecules. Other than H_2O_2, signalling molecules associated with plant stress responses include sugars (mainly glucose/sucrose), gases (H_2S, carbon monoxide and nitric oxide), plant hormones (ethylene and jasmonic acid) and volatile organic acids (Choi et al., 2014). When plants are under stress, calcium signalling events are engaged in the signal transduction from root to the shoot. Plant nanobiotechnology has been shown to be a successful way for tracking signalling molecules in plant species, *viz.*, SWCNT coated with hemin-complexed DNA aptamer for H_2O_2 and carbon nanotubes coated with AT15 for nitric oxide (Wu et al., 2020; Wu & Zhaohu, 2022). Additionally, stomatal activity, humidity and temperature are all tracked by nanosensors (Figure 1.3).

Researchers used NPs to construct conductive ink to print an electrical conductometric sensor that is controlled by the stomatal pore, enabling real-time measurement of the length of stomatal opening and closing (Koman et al., 2017). Fluorescence resonance energy transfer (FRET) nanosensors have been extensively used in protein dynamics and cellular concentrations of small molecules (Kwak et al., 2017). Fluorescence proteins (FPs) are FRET-based nanosensors that are either genetically determined inside the plant itself or introduced exogenously (NPs). Proteins, NPs and chemical ligand that report conformational changes can modify the distance between the donor and acceptor fluorescence domains of two fluorophores. DNA, metal ions and chemical molecules have all been detected using nanosensors based

TABLE 1.2

Role of Nanomaterials in Agricultural Field

Nanoparticles (NPs)	Plant	Potential Effect	References
Ag-NPs	*Trigonella foenum-graecum*	Increases seed germination, root and shoot length, promoting plant growth	Sharma et al. (2012)
Cerium oxide NPs	*Oryza sativa*	Improves seedling growth	Rico et al. (2013)
Carbon NPs	*Triticum*	Improve seed germination and shoot length	Saxena et al. (2014)
Manganese NPs	*Vigna radiata*	Enhanced nitrogen metabolism and photosynthesis rate	Pradhan et al. (2014)
Graphene quantum dots	*Allium sativum*	Promotes plant growth	Chakravarty et al. (2015)
Carbon dots	*Lactuca sativa L. var. longifolia*	Increase lettuce biomass	Zheng et al. (2017)
Nanochitosan, Nanozeolite NPs	*Zea mays*	Improves plant and soil enzyme activities	Khati et al. (2017, 2018)
ZnO NPs	*Oryza sativa*	Bio-remediation of soil and fortification of rice plant	Bala et al. (2019)
Titanium dioxide NPs	*Zea mays*	Improves soil enzymes and plant health	Kumari et al. (2021)
Carbon nano-horns	*Arabidopsis thaliana*	NP content increases secondary metabolites	Sun et al. (2020)
Nanozeolite, Nanogypsum	*Zea mays, Trigonella foenum-graecum*	Improves growth of beneficial microbes	Chaudhary and Sharma (2019); Khati et al. (2019a, b); Kumari et al. (2020)
MWCNTs	Thymus daenensis	Maximizes phenolic content, antioxidant activity, shoot length and seedling biomass	Samadi et al. (2020)
Iron oxide and halloysite NPs	*Lolium perenne* L.	Increased total C/N content of soil directly promoting nutritional pattern and plant growth	Medina et al. (2021)
Nanophos, Nanochitosan	*Zea mays*	Improves seed germination, plant health parameters and soil bacterial population	Chaudhary et al. (2021e, 2022a)

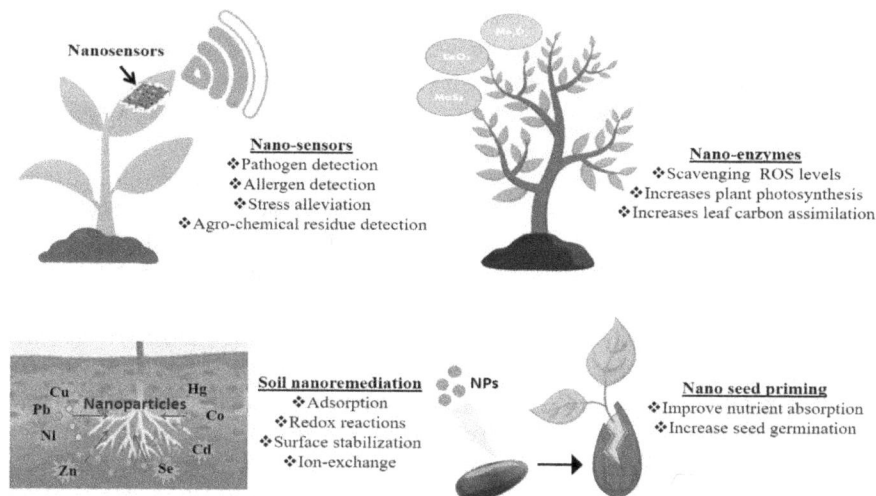

FIGURE 1.3 Role of nanoparticles in agriculture field.

on FRET (Maxwell et al., 2002). Additionally, electrochemical nanosensors have been used to detect heavy metals and other redox active species inside the plants and to signal its growth by detecting biological molecules like vitamin C, H_2O_2, oxygen and root auxin. These nanosensors are highly sensitive, able to analyse data instantly and are reasonably inexpensive (Wujcik et al., 2014).

Further research, particularly in outdoor conditions, is required for the proposed smart plant sensors. Heat and dryness in the field may complicate chemical signalling processes, necessitating greater precision in the decoded signals provided by nanosensors in response to specific signalling molecules (Wu & Li, 2022). The usage of nanosensors in agriculture could be increased by enhancing their sensitivity, specificity and consistency to allow greater decoded signal resolution under field conditions.

1.4 ACCESSIBILITY OF NANOMATERIAL

An accurate assessment of the interactions between nanoparticles and plants is required for the use of nanotechnology in plant applications. However, NP uptake by plants is scarcely observable as it depends on a number of parameters including nanoparticle size, chemical composition, surface function, routes of application, communications with environmental factors (soil quality, microbial diversity and availability of water), and the physiology and anatomy of specific plant species (Khan et al., 2020). Plant roots or vegetative portion can both be treated with nanomaterials. Stomata, stigma and hydathodes are examples of natural plant apertures with nano- or microscale exclusion sizes that can passively absorb NPs at the shoot surface. Cutin and related waxes are frequently found in the cuticle on the surface of roots. The main above-ground plant organs are shielded by this cuticle, which acts as a lipophilic barrier. Although it has been demonstrated that TiO_2 NPs can create holes in the cuticles, this barrier appears to be an impermeable layer to NPs (Sanzari et al., 2019). The effect of TiO$_2$ on canola (*Brassica napus*) also stimulates growth, which led to enhanced seed vigour and seed germination. Despite the greatest effect being seen at concentration between 1,200 and 1,500 mg/L, the treated plant had a larger radicle and improved plumule development (Mittal et al., 2020). TiO$_2$ NPs are acknowledged in improving the plant hydration rate with the accumulation of soil nutrients that directly augment several enzyme activities, *viz.*, nitrate reductase, glutamate hydrogenase and glutamine synthase, which in turn promotes plant growth (Khan et al., 2020). In relation to the plant aerial component, the dynamics of NP absorption generally seem to be more complicated in the soil. The availability of NPs may be inclined by different variables like the presence of muci-lage, root exudates, soil microbiota and organic matter. Moreover, rhizodermis lateral root connections at the root level may make it simple for NMs to enter into the root cells (Chichiricco & Poma, 2015). It was found that positively charged NPs are seques-tered by *Arabidopsis thaliana* mucilage root cap and enter tissues of root, although negatively charged NPs are not sequestered by mucilage and enter root tissue directly. However, regardless of the particle charge, root mucilage tends to trap gold NPs and make its way to the root cells (Avellan et al., 2017). Metal NPs translocate into seed-lings in various crops including maize and cabbage without affecting seed viability and its germination and development of shoot (Pokhrel & Dubey, 2013). These findings support the potential value of functional NPs for boosting plant development and seed priming under adverse environmental conditions.

1.5 CONCLUSION

This chapter emphasizes on the various types of nanoparticles and their role in agri-cultural field. The application of nanotechnology in modern agriculture supports and develops the global economy in a variety of ways. Nanomaterials support plant stress resistance via nanoenzymes, aid in seed priming and help in detection and removal of metal toxicity with the help of nanosensors. In comparison to conventional resources, the effectiveness and agronomic efficiency of NPs have greatly increased. An important factor to take into account when attempting to comprehend the physiological and bio-chemical reactions imposed by plants is the study of NP-plant interaction. Furthermore, the lack of guidelines and regulatory bodies create barriers to the commercialization of novel NP-based fertilizers. In order to advance and build large trials for future studies based on identified knowledge shortages, it is necessary to carry out extensive trials.

REFERENCES

Abdal-Latef, A., Srivastava, A. K., El-sadek, M. S. A., Kordrostami, M., & Tran, L.-S. P. (2018). Titanium dioxide nanoparticles improve growth and enhance tolerance of broad bean plants under saline soil conditions. *Land Degrad Dev*. 29, 1065–1073.

Abou-Zeid, H., & Ismail, G. (2018). The role of priming with biosynthesized silver nanopar-ticles in the response of *Triticum aestivum* L to salt stress. *Egypt J Bot*. 58, 73–85.

Aghdam, M.T., Mohammadi, H., & Ghorbanpour, M. (2016). Effects of nanoparticulate ana-tase titanium dioxide on physiological and biochemical performance of *Linum usitatis-simum* (Linaceae) under well-watered and drought stress conditions. *Braz J Bot*. 39, 139–146.

Agri, U., Chaudhary, P., & Sharma, A. (2021). In vitro compatibility evaluation of agriusable nanochitosan on beneficial plant growth-promoting rhizobacteria and maize plant. *Nat Acad Sci Lett.* 44, 555–559.

Agri, U., Chaudhary, P., Sharma, A., & Kukreti, B. (2022). Physiological response of maize plants and its rhizospheric microbiome under the influence of potential bioinoculants and nanochitosan. *Plant Soil.* 474, 451–468.

Ahmad, B., Shabbir, A., Jaleel, H., Khan, M.M.A., & Sadiq, Y. (2018). Efficacy of titanium dioxide nanoparticles in modulating photosynthesis, peltate glandular trichomes and essential oil production and quality in *Mentha piperita* L. *Cur Plant Biol.* 13, 6–15.

Almutairi, Z.M. (2016). Effect of nano-silicon application on the expression of salt tolerance genes in germinating tomato ('*Solanum lycopersicum*' L.) seedlings under salt stress. *Plant Omics.* 9, 106.

Anandhi, S (2020). Nano-pesticides in pest management. *J Entomol Zool Stud.* 8, 685–90.

Anjum, N.A., Singh, N., Singh, M. K., Shah, Z.A., Duarte, A.C., & Pereira, E. (2013). Single-bilayer graphene oxide sheet tolerance and glutathione redox system significance assessment in faba bean (*Vicia faba* L.). *J Nanop Res.* 15, 1770.

Avellan, A., Schwab, F., Masion, A., Chaurand, P., Borschneck, D., & Vidal, V. (2017). Nanoparticle uptake in plants: gold nanomaterial localized in roots of *Arabidopsis thaliana* by X-ray computed nanotomography and hyperspectral imaging. *Environ Sci Technol.* 51, 8682–8691.

Bala, R., Kalia, A., & Dhaliwal, S.S. (2019). Evaluation of efficacy of ZnO nanoparticles as remedial zinc nanofertilizer for rice. *J Soil Sci Plant Nutr.* 19, 379–389.

Bandyopadhyay, S., Plascencia-Villa, G., Mukherjee, A., Rico, C.M., Jose-Yacaman, M., Peralta-Videa, J.R., & Gardea-Torresdey, J.L. (2015). Comparative phytotoxicity of ZnO NPs bulk ZnO, and ionic zinc onto the alfalfa plants symbiotically associated with *Sinorhizobium meliloti* in soil. *Sci Total Environ.* 515, 60–69.

Begum, P., & Fugetsu, B. (2012). Phytotoxicity of multi-walled carbon nanotubes on red spinach (*Amaranthus tricolor* L) and the role of ascorbic acid as an antioxidant. *J Hazard Mater.* 243, e212–e222.

Behboudi, F., Tahmasebi Sarvestani, Z., Kassaee, M.Z., Modares Sanavi, S.A.M., Sorooshzadeh, A., & Ahmadi, S.B. (2018). Evaluation of chitosan nanoparticles effects on yield and yield components of barley (*Hordeum vulgare* L.) under late season drought stress. *J Water Environ Nanotechnol.* 3, 22–39.

Bharti, A.S., Sharma, S., Shukla, N., & Uttam, K.N. (2018). Steady state and time resolved laser-induced fluorescence of garlic plants treated with titanium dioxide nanoparticles. *Spect Lett.* 51, 45–54.

Boghossian, A.A., Sen, F., Gibbons, B.M., Sen, S., Faltermeier, S.M., Giraldo, J.P., Zhang, C.T., Zhang, J., Heller, D.A., & Strano, M.S. (2013). Application of nanoparticle antioxidants to enable hyperstable chloroplasts for solar energy harvesting. *Adv Energy Mat.* 3, 881–893.

Borisev, M., Borisev, I., Zupunski, M., Arsenov, D., Pajevic, S., Curcic, Z., Vasin, J., & Djordjevic, A. (2016). Drought impact is alleviated in sugar beets (*Beta vulgaris* L.) by foliar application of fullerenol nanoparticles. *PLoS One.* 11, e0166248.

Chakravarty, D., Erande, M.B., & Late, D.J. (2015). Graphene quantum dots as enhanced plant growth regulators: effects on coriander and garlic plants. *J Sci Food Agric.* 95, 2772–2778.

Changcheng, A., Changjiao, S., Ningjun, L., Huang, B., Jiang, J., Shen, Y., Wang, C., Zhao, X., Cui, B., Wang, C., Li, X., Zhan, S., Gao, F., Zeng, Z., Cui, H., & Wang, Y (2022). Nanomaterials and nanotechnology for the delivery of agrochemicals: strategies towards sustainable agriculture. *J Nanobiotech* 20, 11.

Chaudhary, P., Chaudhary, A., Bhatt, P., Kumar, G., Khatoon, H., Rani, A., Kumar, S., & Sharma, A. (2022a). Assessment of soil health indicators under the influence of nano-compounds and *Bacillus* spp. in field condition. *Front Environ Sci.* 9, 769871.

Chaudhary, P., Chaudhary, A., Parveen, H., Rani, A., Kumar, G., Kumar, A., & Sharma, A. (2021e). Impact of nanophos in agriculture to improve functional bacterial community and crop productivity. *BMC Plant Biol.* 21, 519.

Chaudhary, P., Khati, P., Chaudhary, A., Gangola, S., Kumar, R., & Sharma, A. (2021a). Bioinoculation using indigenous *Bacillus* spp. improves growth and yield of *Zea mays* under the influence of nanozeolite. *3 Biotech.* 11, 11.

Chaudhary, P., Khati, P., Chaudhary, A., Maithani, D., Kumar, G., & Sharma, A. (2021d). Cultivable and metagenomic approach to study the combined impact of nanogypsum and *Pseudomonas taiwanensis* on maize plant health and its rhizospheric microbiome. *PLoS One.* 16, e0250574.

Chaudhary, P., Khati, P., Gangola, S., Kumar, A., Kumar, R., & Sharma, A. (2021c). Impact of nanochitosan and *Bacillus* spp. on health, productivity and defence response in *Zea mays* under field condition. *3 Biotech.* 11, 237.

Chaudhary, P., & Sharma, A. (2019). Response of nanogypsum on the performance of plant growth promotory bacteria recovered from nanocompound infested agriculture field. *Environ Ecol.* 37, 363–372.

Chaudhary, P., Sharma, A., Chaudhary, A., Khati, P., Gangola, S., & Maithani, D. (2021b). Illumina based high throughput analysis of microbial diversity of rhizospheric soil of maize infested with nanocompounds and *Bacillus* sp. *Appl Soil Ecol.* 159, 103836.

Chaudhary, P., Singh, S., Chaudhary, A., Sharma, A., & Kumar, G. (2022b). Overview of bio-fertilizers in crop production and stress management for sustainable agriculture. *Front Plant Sci.* 13, 930340.

Chen, T., Zou, H., Wu, X., Liu, C., Situ, B., Zheng, L., & Yang, G. (2018). Nanozymatic anti-oxidant system based on MoS2 nanosheets. *ACS Appl Mater Interf.* 10, 12453–12462.

Chichiriccò, G., & Poma, A. (2015). Penetration and toxicity of nanomaterials in higher plants. *Nanomaterials.* 5, 851–873.

Choi, W.G., Toyota, M., Kim, S.H., Hilleary, R., & Gilroy, S. (2014). Salt stress-induced Ca^{2+} waves are associated with rapid, long-distance root-to-shoot signaling in plants. *Proc Natl Acad Sci U S A.* 111, 6497–6502.

Da Costa, M.V.J., & Sharma, P.K. (2016). Effect of copper oxide nanoparticles on growth, morphology, photosynthesis, and antioxidant response in *Oryza sativa*. *Photosynthetica.* 54, e110–e119.

de Lange, O., Klavins, E., & Nemhauser, J. (2018). Synthetic genetic circuits in crop plants. *Curr Opin Biotechnol.* 49, 16–22.

Deka, B., Babu, A., Baruah, C., & Barthakur, M. (2021). Nanopesticides: a systematic review of their prospects with special reference to tea pest management. *Front Nutr.* 8, 686131.

Demirer, G.S., Zhang, H., Matos, J.L., Goh, N.S., Cunningham, F.J., Sung, Y., Chang, R., Aditham, A.J., Chio, L., Cho, M.-J., Staskawicz, B., & Landry, M.P. (2019). High aspect ratio nanomaterials enable delivery of functional genetic material without DNA integration in mature plants. *Nat Nanotechnol.* 14, 456.

Duhan, J.S., Kumar, R., Kumar, N., Kaur, P., Nehra, K., & Duhan, S. (2017). Nanotechnology: the new perspective in precision agriculture. *Biotechnol Rep.* 15, 11–23.

Elsheery, N.I., Helaly, M.N., El-Hoseiny, H.M., & Alam-Eldein, S.M. (2020). Zinc oxide and silicone nanoparticles to improve the resistance mechanism and annual productivity of salt-stressed mango trees. *Agronomy.* 10, 558.

Gamiz, B., Cox, L., Hermosin, M.C., Spokas, K., & Celis, R. (2017). Assessing the effect of organoclays and biochar on the fate of abscisic acid in soil. *J Agric Food Chem.* 65. 29–38.

Gao, L. (2022). Enzyme-like property (nanozyme) of iron oxide nanoparticles. *Iron Oxide Nanopart.* doi: 10.5772/intechopen.102958

Giraldo, J.P., Landry, M.P., Kwak, S.-Y., Jain, R.M., Wong, M.H., Iverson, N.M., Ben-Naim, M., & Strano, M.S.A. (2015). Ratiometric Sensor using single chirality near-infrared fluorescent carbon nanotubes: application to in vivo monitoring. *Small.* 11, 3973–3984.

Gohari, G., Safai, F., Panahirad, S., Akbari, A., Rasouli, F., Dadpour, M.R., & Fotopoulos, V. (2020). Modified multiwall carbon nanotubes display either phytotoxic or growth promoting and stress protecting activity in *Ocimum basilicum* L. In a concentration dependent manner. *Chemosphere* 13, 126171.

Guha, T., Gopal, G., Kundu, R., & Mukherjee, A. (2020). Nanocomposites for delivering agrochemicals: a comprehensive review. *J Agric Food Chem*. 68, 3691–3702.

Haliloglu, K., Hosseinpour, A., Cinisli, K.T., Ozturk, H.I., Ozkan, G., Pour-Aboughadareh, A., & Poczai, P. (2020). Investigation of the protective roles of zinc oxide nanoparticles and plant growth promoting bacteria on DNA damage and methylation in tomato (*Solanum lycopersicum* L.) under salinity stress. *Hort Environ Biotechnol*. 10, 16.

Hawthorne, J., Musante, C., Sinha, S.K., & White, J.C. (2012). Accumulation and phytotoxicity of engineered nanoparticles to *Cucurbita pepo*. *Int J Phytoremediation*. 14, 429–442.

He, X., Deng, H., & Hwang, H. (2019). The current application of nanotechnology in food and agriculture. *J Food Drug Anal*. 27, 1–21.

Hu, X., Kang, J., Lu, K., Zhou, R., Mu, L., & Zhou, Q. (2014). Graphene oxide amplifies the phytotoxicity of arsenic in wheat. *Sci Rep*. 4, 6122.

Ioannou, A., Gohari, G., Papaphilippou, P., Panahirad, S., Akbari, A., Dadpour, M.R., Krasia-Christoforou, T., & Fotopoulos, V. (2020). Advanced nanomaterials in agriculture under a changing climate: the way to the future? *Environ Exp Bot*. 176, 104048.

Iqbal, M., Raja, N.I., Hussain, M., Ejaz, M., & Yasmeen, F. (2019). Effect of silver nanoparticles on growth of wheat under heat stress. *Iran J Sci Technol Trans A Sci*. 43, 387–395.

Jeevanandam, J., Barhoum, A., Chan, Y.S., Dufresne, A., & Danquah, M.K. (2018). Review on nanoparticles and nanostructured materials: history, sources, toxicity and regulations. *Beilstein J Nanotechnol*. 9, 1050–1074.

Khan, I., Saeed, K., & Khan, I. (2019). Nanoparticles: properties, applications and toxicities. *Arab J Chem*. 12, 908–931.

Khan, M.N., AlSolami, M.A., Basahi, R.A., Siddiqui, M.H., Al-Huqail, A A , & Abbas, Z. K. (2020). Nitric oxide is involved in nano-titanium dioxide-induced activation of antioxidant defense system and accumulation of osmolytes under water-deficit stress in *Vicia faba* L. *Ecotoxicol Environ Saf*. 190, 110152.

Khati, P., Bhatt, P., Kumar, R., & Sharma, A. (2018). Effect of nanozeolite and plant growth promoting rhizobacteria on maize. *3Biotech*. 8, 141.

Khati, P., Chaudhary, P., Gangola, S., & Sharma, P. (2019a). Influence of nanozeolite on plant growth promotory bacterial isolates recovered from nanocompound infested agriculture field. *Environ Ecol*. 37, 521–527.

Khati, P., Chaudhary, P., Gangola, S., Bhatt, P., & Sharma, A. (2017). Nanochitosan supports growth of *Zea mays* and also maintains soil health following growth. *3 Biotech*. 7, 81.

Khati, P., Sharma, A., Chaudhary, P., Singh, A.K., Gangola, S., & Kumar R. (2019b). High-throughput sequencing approach to access the impact of nanozeolite treatment on species richness and evens of soil metagenome. *Biocatal Agric Biotechnol*. 20, 101249.

Khodakovskaya, M.V., de Silva, K., Biris, A.S., Dervishi, E., & Villagarcia, H. (2012). Carbon nanotubes induce growth enhancement of tobacco cells. *ACS Nano*. 27, 2128–2135.

Kolluru, S.S., Agarwal, S., Sireesha, S., Sreedhar, I., & Kale, S.R. (2021). Heavy metal removal from wastewater using nanomaterials-process and engineering aspects. *Process Saf Environ Prot*. 150, 323–355.

Koman, V.B., Lew, T.T.S., Wong, M.H., Kwak, S.Y., Giraldo, J.P., & Strano, M.S. (2017). Persistent drought monitoring using a microfluidic-printed electromechanical sensor of stomata: in planta. *Lab Chip*. 17, 4015–4024

Kukreti, B., Sharma, A., Chaudhary, P., Agri, U., & Maithani, D. (2020). Influence of nanosilicon dioxide along with bioinoculants on *Zea mays* and its rhizospheric soil. *3 Biotech*. 10, 345.

Kumar, A., Choudhary, A., Kaur, H., Mehta, S., & Husen, A. (2021). Smart nanomaterial and nanocomposite with advanced agrochemical activities. *Nanoscale Res Lett*. 16, 156.

Kumari, H., Khati, P., Gangola, S., Chaudhary, P., & Sharma, A. (2021). Performance of plant growth promotory rhizobacteria on Maize and soil characteristics under the influence of TiO$_2$ nanoparticles. *Pantnagar J Res*. 19, 28–39.

Kumari, S., Sharma, A., Chaudhary, P., & Khati, P. (2020). Management of plant vigor and soil health using two agriusable nanocompounds and plant growth promotory rhizobacteria in Fenugreek. *3 Biotech*. 10, 461.

Kumbhakar, P., Ray, S. S., & Stepanov, A. L. (2014). Optical properties of nanoparticles and nanocomposites. *J Nanomater*. 2014, 181365.

Kwak, S.Y., Lew, T.T.S., Sweeney, C.J., Koman, V.B., Wong, M. H., Bohmert-Tatarev, K., Snell, K.D., Seo, J.S., Chua, N.-H., & Strano, M.S. (2019). Chloroplast-selective gene delivery and expression in planta using chitosan-complexed single-walled carbon nanotube carriers. *Nat Nanotechnol*. 14, 447.

Kwak, S.Y., Wong, M.H., Lew, T.T.S., Bisker, G., Lee, M.A., Kaplan, A., Dong, J., Liu, A.T., Koman, V.B., Sinclair, R., Hamann, C., & Strano, M.S. (2017). Nanosensor technology applied to living plant systems. *Annu Rev Anal Chem*. 10, 113–140.

Lahiani, M. H., Dervishi, E., Chen, J., Nima, Z., Gaume, A., & Biris, A. S. (2013). Impact of carbon nanotube exposure to seeds of valuable crops. *ACS Appl Mat Interf*. 5, 7965–7973.

Liu, R., & Lal, R. (2012). Synthetic apatite nanoparticles as a phosphorous fertilizer for soyabean (*Gycine max*). *Sci Rep*. 4, 5686.

Liu, W., Li, Y., Qiao, J., Zhao, H., Xie, J., Fang, Y., Shen, S., & Liang S. (2020). The effectiveness of nanobiochar for reducing phytotoxicity and improving soil remediation in cadmium-contaminated soil. *Sci Rep*. 10, 858.

Maxwell, D., Taylor, J.R., & Nie, S. (2002). Self-assembled nanoparticle probes for recognition and detection of biomolecules. *J Am Chem Soc*. 124, 9606–9612.

Medina, J., Calabi-Floody, M., Aponte, H., Santander, C., Paneque, M., Meier, S., Panettieri, M., Cornejo, P., Borie, F., & Knicker, H. (2021). Utilization of inorganic nanoparticles and biochar as additives of agricultural waste composting: effects of end-products on plant growth, C and nutrient stock in soils from a Mediterranean region. *Agronomy*. 11, 767.

Mittal, D., Kaur, G., Singh, P., Yadav, K., & Ali, S.A. (2020) Nanoparticle-based sustainable agriculture and food science: recent advances and future outlook. *Front Nanotechnol*. 2, 579954.

Moradbeygi, H., Jamei, R., Heidari, R., & Darvishzadeh, R. (2020). Investigating the enzymatic and non-enzymatic antioxidant defense by applying iron oxide nanoparticles in *Dracocephalum moldavica* L. plant under salinity stress. *Sci Hort*. 272, 109537.

Mukherjee, A., Majumdar, S., Servin, A.D., Pagano, L., Dhankher, O.P., & White, J.C. (2016). Carbon nanomaterials in agriculture: a critical review. *Front Plant Sci*. 7, 172.

Neri-Badang, M.C. & Soma Chakraborty (2019). Carbohydrate polymers as controlled release devices for pesticides. *J. of Carbohydrate Chem*. 38(1), 67–85.

Nicolopoulou-Stamati, P., Maipas, S., Kotampasi, C., Stamatis, P., & Hens, L. (2016). Chemical pesticides and human health: the urgent need for a new concept in agriculture. *Front Public Health*. 4, 148.

Olad, A., Gharekhani, H., Mirmohseni, A., & Bybordi, A. (2018). Superabsorbent nanocomposite based on maize bran with integration of water-retaining and slow-release NPK fertilizer. *Adv Polym Technol*. 37, 1682–1694.

Paret, M.L., Palmateer, A.J., & Knox, G.W. (2013). Evaluation of a light-activated nanoparticle formulation of titanium dioxide with zinc for management of bacterial leaf spot on rosa 'Noare'. *HortScience*. 48, 189–192.

Pokhrel, L. R., & Dubey, B. (2013). Evaluation of developmental responses of two crop plants exposed to silver and zinc oxide nanoparticles. *Sci. Total Environ*. 452–453, 321–332.

Pouratashi, M., & Iravani, H. (2012). Farmers knowledge of integrated pest management and learning style preferences: implications for information delivery. *Int J Pest Manag*. 58, 347–353.

Pradhan, S., Patra, P., Mitra, S., Dey, K.K., Jain, S., Sarkar, S., Roy, S., Palit, P., & Goswami, A. (2014). Manganese nanoparticles: impact on non-nodulated plant as a potent enhancer in nitrogen metabolism and toxicity study both in vivo and in vitro. *Agric Food Chem.* 62, 8777.

Qiu, H., & Smolders, E. (2017). Nanospecific phytotoxicity of CuO nanoparticles in soils disappeared when bioavailability factors were considered. *Environ Sci Technol.* 51, 11976–11985.

Rakesh, P., Prasad, R.D., Uma Devi, G., & Bhat, B.N. (2017). Effect of biopolymers and synthetic seed coating polymers on castor and groundnut seed. *Int J Pure App Biosci.* 5, 2043–2048.

Rico, C.M., Hong, J., Morales, M.I., Zhao, L., Barrios, A.C., Zhang, J.Y., Peralta-Videa, J.R., & Gardea-Torresdey, J.L. (2013). Effect of cerium oxide nanoparticles on rice: a study involving the antioxidant defense system and *in vivo* fluorescence imaging. *Environ Sci Technol.* 47, e5635–e5642.

Samadi, S., Saharkhiz, M.J., Azizi, M., Samiei, L., & Ghorbanpour, M. (2020). Multi-walled carbon nanotubes stimulate growth, redox reactions and biosynthesis of antioxidant metabolites in *Thymus daenensis* celak. in vitro. *Chemosphere.* 249, 126069.

Sanzari, I., Leone, A., & Ambrosone, A. (2019). Nanotechnology in plant science: To make a long story short. *Front. Bioeng. Biotechnol.* 7, 120.

Sathya, S., Lakshmi, S., & Nakkeeran, S. (2016). Combined effect of biopriming and polymer coating on chemical constituents of root exudation in chili (*Capsicum annuum* L.) cv. K 2 seedlings. *J Nat Appl Sci.* 8, 2141–2154.

Savvides, A., Ali, S., Tester, M., & Fotopoulos, V. (2016). Chemical priming of plants against multiple abiotic stresses: mission possible? *Trends Plant Sci.* 21, 329–340.

Saxena, M., Maity, S., & Sarkar, S. (2014). Carbon nanoparticles in 'biochar' boost wheat (*Triticum aestivum*) plant growth. *RSC Adv.* 4, 39948–39954.

Scott, N., & Chen, H (2013). Nanoscale science and engineering for agriculture and food systems. *Ind Biotechnol.* 9, 17–8.

Sedghi, M., Hadi, M., & Toluie, S.G. (2013). Effect of nano zinc oxide on the germination parameters of soybean seeds under drought stress. *Ann West Univ Timiş Ser Biol.* 16, 73–78.

Sharma, P., Bhatt, D., Zaidi, M.G., Saradhi, P.P., Khanna, P.K., & Arora, S. (2012). Silver nanoparticle-mediated enhancement in growth and antioxidant status of *Brassica juncea*. *Appl Biochem Biotechnol.* 167, 2225–2233.

Siddiqui, M.H., Al-Whaibi, M.H., Firoz, M., & Al-Khaishany, M.Y. (2015). Role of nanoparticles in plants. In: Siddiqui, M.H., Al-Whaibi, M.H., & Firoz, M. (Eds.), *Nanotechnology and Plant Sciences*. Springer International Publishing, Switzerland, pp. 19–35.

Souza, I.R., Silva, L.R., Fernandes, L.S.P., Salgado, L.D., de Assis, H.C.S., Firak, D.S., Bach, L., Santos-Filho, R., Voigt, C.L., & Barros, A.C. (2020). Visible-light reduced silver nanoparticles' toxicity in *Allium cepa* test system. *Environ Poll.* 257, 113551.

Stampoulis, D., Sinha, S. K., & White, J. C. (2009). Assay-dependent phytotoxicity of nanoparticles to plants. *Environ Sci Technol.* 43, 9473–9479.

Sun, L., Wang, R., Ju, Q., & Xu, J. (2020). Physiological, metabolic, and transcriptomic analyses reveal the responses of arabidopsis seedlings to carbon nanohorns. *Environ Sci Technol.* 54, 4409–4420.

Taha, R. (2016). Nano carbon applications for plant. *Adv Plants Agric Res.* 5, 00172.

Tripathi, D.K., Singh, S., Singh, V.P., Prasad, S.M., Dubey, N.K., & Chauhan, D.K. (2017). Silicon nanoparticles more effectively alleviated UV-B stress than silicon in wheat (Triticum aestivum) seedlings. *Plant Physiol Biochem.* 110, 70–81.

Tyagi, J., Chaudhary, P., Mishra, A., Khatwani, M., Dey, S., & Varma, A. (2022). Role of endophytes in abiotic stress tolerance: with special emphasis on *Serendipita indica*. *Int J Environ Res.* 16, 62.

United Nations (2019). Department of economic and social affairs, population division. World Population Prospects 2019: highlights (ST/ESA/SER.A/423).

Vance, M.E., Kuiken, T., Vejerano, E.P., McGinnis, S.P., Hochella Jr, M.F., Rejeski, D., & Hull, M.S. (2015). Nanotechnology in the real world: redeveloping the nanomaterial consumer products inventory. *Beilstein J Nanotechnol.* 6, 1769–1780.

Van Nguyen, D., Nguyen, H.M., Le, N.T., Nguyen, H.T., Le, H.M., Nguyen, A.T., Dinh, N.T.T., Hoang, S.A., & Van Ha, C. (2020). Copper nanoparticle application enhances plant growth and grain yield in maize under drought stress conditions. *J Plant Growth Regul.* 41(1), 364–375.

Wheeler, T., & von Braun, J. (2013). Climate change impacts on global food security. *Science.* 341, 508–513.

Wu, H & Li, Z. (2022). Recent advances in nano-enabled agriculture for improving plant performance, *Crop J.* 10, 1–12.

Wu, H., Nißler R., Morris, V., Herrmann, N., Hu, P., Jeon, S.J., Kruss, S., & Giraldo, J.P. (2020). Monitoring plant health with near-infrared fluorescent H_2O_2 nanosensors, *Nano Lett.* 20, 2432–2442.

Wu, H., Shabala, L., Shabala, S., & Giraldo, J.P. (2018). Hydroxyl radical scavenging by cerium oxide nanoparticles improves *Arabidopsis* salinity tolerance by enhancing leaf mesophyll potassium retention. *Environ Sci Nano.* 5, 1567.

Wu, H., Tito, N., & Giraldo, J.P. (2017). Anionic cerium oxide nanoparticles protect plant photosynthesis from abiotic stress by scavenging reactive oxygen species. *ACS Nanomat.* 11, 11283−11297.

Wu, H., & Zhaohu, L. (2022). Recent advances in nano-enabled agriculture for improving plant performance. *Crop J.* 10, 1–12.

Wu, J., Xie, Y., Fang, Z., Cheng, W., & Tsang, P.E. (2016). Effects of Ni/Fe bimetallic nanoparticles on phytotoxicity and translocation of polybrominated diphenyl ethers in contaminated soil. *Chemosphere.* 162, 235–242.

Wujcik, E.K., Wei, H., Zhang, X., Guo, J., & Yan, X. (2014). Antibody nanosensors: a detailed review. *RSC Adv.* 4, 43725–43745.

Yalamalle, V.R., Tomar, B.S., Kumar, A., & Ahammed, S.T.P. (2019). Polymer coating for higher pesticide use efficiency, seed yield and quality in onion (*Allium cepa*). *Indian J Agric Sci.* 89, 135–139.

Yao, J., Cheng, Y., Zhou, M., Zhao, S., Lin, S., Wang, X., Wu, J., Li, S., & Wei, H. (2018). ROS scavenging Mn_3O_4 nanozymes for in vivo anti-inflammation. *Chem Sci.* 9, 2927−2933.

Yaqoob, A. A., Parveen, T., Umar, K., & Mohamad, M. N. (2020). Role of nanomaterials in the treatment of wastewater: a review. *Water.* 12, 495.

Ye, Y., Cota-Ruiz, K., Hernández-Viezcas, J.A., Valdés, C., Medina-Velo, I. A., & Turley, R.S. (2020). Manganese nanoparticles control salinity-modulated molecular responses in *Capsicum annuum* L. through priming: a sustainable approach for agriculture. *ACS Sustain Chem Eng.* 8, 1427–1436.

Zheng, Y., Xie, G., Zhang, X., Chen, Z., Cai, Y., Yu, W., Liu, H., Shan, J., Li, Y., & Lei, B. (2017). Bioimaging application and growth-promoting behavior of carbon dots from pollen on hydroponically cultivated Rome lettuce. *ACS Omega.* 2, 3958–3965.

2 Techniques for Integrating Nanotechnology in Agriculture

Viabhav Kumar Upadhayay,
Mayur Giridhar Naitam, Saipayan Ghosh,
Vivek Kumar Patel, and Adita Sharma
Dr. Rajendra Prasad Central Agricultural University

Ajay Kumar Chandra
ICAR-Indian Agricultural Research Institute

Gohar Taj
Govind Ballabh Pant University of Agriculture & Technology

CONTENTS

DOI: 10.1201/9781003345565-2

2.1 INTRODUCTION

Agriculture is a very important sector as it provides raw materials for the food and feed industries and is also an important source of the agriculture-based economy in many countries. Agriculture serves as a source of livelihood for 60% of the Indian people; consequently, agriculture productivity must be maintained to sustain the overall national growth (Pramanik et al., 2020). Recent agricultural practices related to the Green Revolution have greatly expanded the global food supply. In addition, they have negative impacts on the environment and ecosystem services, underscoring the need for more sustainable agricultural practices (Mukhopadhyay, 2014). Fertilizer application is essential for increasing agricultural output; nevertheless, excessive fertilizer application permanently alters the chemical ecology of soil, hence reducing available space for cultivation (Savci, 2012). Sustainable agriculture requires the use of as few agro-chemicals as possible in order to protect the environment (An et al., 2022). To resolve these concerns, it is necessary to investigate one of the cutting-edge technologies, such as "nanotechnology" in agriculture, which enhances efficiency while ensuring environmental safety and better usage efficacy (Shang et al., 2019; Chaudhary & Sharma, 2019; Kumari et al., 2020; Kukreti et al., 2020).

This promising technology has been recognized to revitalize the agriculture and food sectors, thereby improving the livelihoods of economically deprived people (Bai et al., 2020; Pramanik et al., 2020). Norio Taniguchi used the word "nanotechnology" for the first time in 1974 (Zarzycki, 2014). The field of nanotechnology refers to the understanding of matter at the nanoscale (1–100 nm) (Bayda et al., 2019). Nanotechnology provides new nano-based agro-chemicals and new ways to deliver them to improve crop yields and soil health and reduce the amount of pesticides used (Chaudhary et al., 2022a). The concept of nanotechnological implication in agriculture may encompass: (1) nano-based formulations of agro-chemicals for the improvement of crops; (2) detecting pathogens and chemical residues by nanosensors; (3) the use of nanodevices in the genetic modification of plants; and (4) post-harvest management (Ghidan & Antary, 2019). Agriculture is a dynamic sector of research and development (R&D) where a huge number of nanotechnology-based inventions and advancements have been recorded (Cheng et al., 2016). In plants, for instance, nanotechnology methods have been discovered to transfer DNA to plant cells (Yan et al., 2022), improve nutrient absorption (Shang et al., 2019; Chaudhary et al., 2021a, b, c, d), detect plant pathogen (Cardoso et al., 2021) and enhance photosynthetic rate (Swift et al., 2018; Chaudhary et al., 2021e; Kumari et al., 2021), among many other applications (Beyth et al., 2015; Shang et al., 2019; Zahedi et al., 2020; Khati et al., 2019a, b). This chapter outlines the advanced applications of nanotechnology in agriculture, including genetic engineering, tissue culture, plant pathogen detection, photosynthesis rate enhancement, chloroplast engineering and improvement in shelf life of fruits/vegetables (Figure 2.1).

2.2 CONCEPT OF NANOTECHNOLOGY IN AGRICULTURE

Chemical fertilizers and insecticides are widely employed in agriculture to increase agricultural output, but they have unfavourable impacts on the environment and

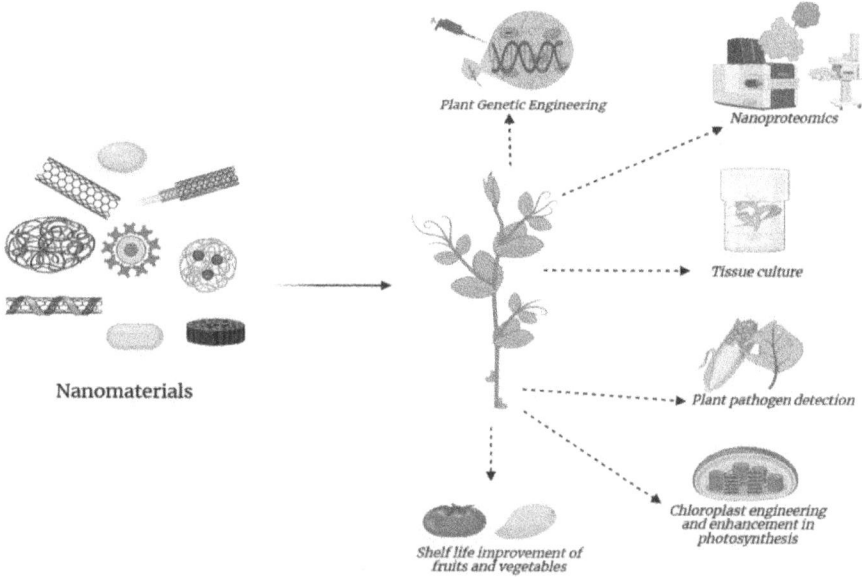

FIGURE 2.1 Advanced nanotechnological application in agriculture.

present a threat to human life. This requires the need to modernize agricultural practices using the safest and most effective technology, with a focus on increasing agricultural output with minimal adverse effects on the environment or on people (Chaudhary et al., 2022b). In recent decades, nanotechnology has evolved as the most revolutionary and cutting-edge scientific topic dealing with matter in the nanoscale range (1–100 nm). These nanoscale materials, which are more often referred to as nanoparticles (NPs), exhibit peculiar physical, chemical and biological characteristics (Mishra et al., 2014). The promising applications of nanotechnology to speed up the development of sustainable agriculture have been described in many studies (Agri et al., 2021, 2022; Bhatt et al., 2022). These include exploring the efficacy of nanoparticles as a pesticide alternative and developing nano-formulations of agrochemicals for enhanced crop protection and productivity (Khati et al., 2017, 2018; Saranya et al., 2019).

Nano-inputs allow for the targeted distribution of active ingredients while also reducing the need for fertilizers and pesticides. Consequently, these nanotools have no effect on non-target organisms, and environmental safety may be maintained. Nanosensors also offer prompt and precise details concerning the conditions of the soil or the detection of pathogens, which helped in the mitigation of financial losses for farmers and enhanced their economic situation (Chhipa, 2019). In addition to the positive benefits that nanotechnology is anticipated to have in agriculture, there are some biosafety concerns that will need to be tackled in the course of future research (Mishra et al., 2014). The prolific contribution of nano-inputs such as nano-fertilizers and nano-pesticides is shown in Figure 2.2.

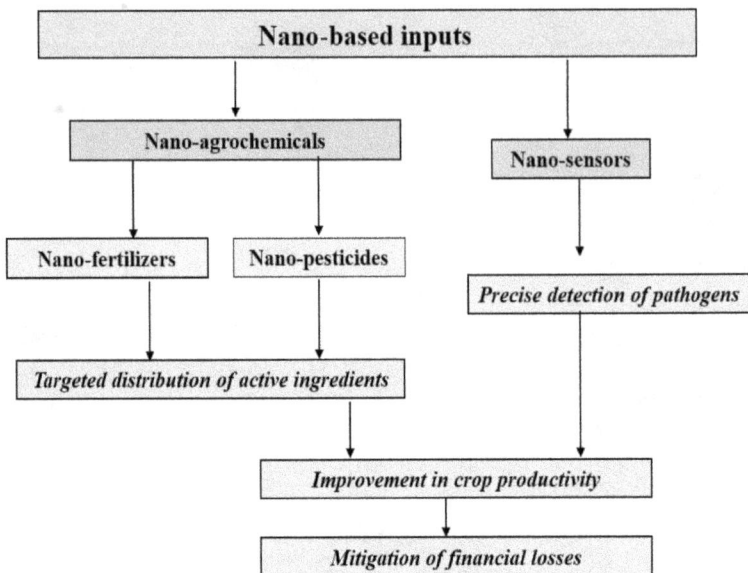

FIGURE 2.2   Prolific role of using nano-inputs in agriculture.

2.3   CUTTING-EDGE NONTECHNOLOGICAL APPLICATIONS IN AGRICULTURE

2.3.1   NANOTECHNOLOGY IN PLANT GENETIC ENGINEERING

The vectors and techniques applied in biotransformation of plants have been improved every year with the objective of enhancement of the effectiveness of the process. Currently, there are two biotransformation approaches, i.e., "indirect" and "direct" methods. The indirect strategy necessitates the transport of a plasmid-based gene construct into the particular location, i.e., targeted cells (Cha et al., 2011). The transformation system driven by *Agrobacterium tumefaciens* and other related species of *Agrobacterium* has been developed in economically valuable crops such as wheat (Cheng et al., 1997), banana (Namuddu et al., 2013) and sugar cane (Enriquez-Obregon et al., 1997). The prevalent biotransformation approach in plants causes genetic instabilities in the environment, ineffective penetrability and repeatability issues. Therefore, the present nanotechnology technique has few advantages including faster gene transfer compared to the transformation of plants mediated by *Agrobacterium* and DNA protection due to the encapsulation process (Chen et al., 2011).

Using nano-based carriers, many bioactive compounds can be transported into cells. Nanomaterials have been used extensively in animals for efficient drug delivery, cancer therapy and the cure of hereditary ailments. Despite increasing attention, the use of nanomaterials in plant systems is still in its infancy (Wang et al., 2016). Several studies have shown the successful delivery of pDNA, proteins, phytohormones, RNA and siRNA into cells and/or protoplasts. A particularly notable achievement was the use of a biolistic system to construct mesoporous silica nanoparticles (MSNs) as carriers for delivery of biomolecules (such as pDNA and protein) into cells (Martin-Ortigosa et al., 2014). This tactic involves the simultaneous integration of pDNA and its chemical inducer into MSNs. Additionally, gold nanoparticles were used as gate-keeper by capping the terminal ends of the MSNs. After the MSNs are transported inside the cells, the gold (Au) nanoparticles are unsealed by a gate-opening trigger, a flow switch which opens in response to fluctuations in the chemical/physiological environment, and the biogenic molecules are released in a more regulated manner.

Other nanomaterials, in addition to MSNs, have been developed as delivery systems. These nanomaterials include DNA nanostructures, carbon nanotubes, magnetic nanoparticles and layered double hydroxide clay nanosheets (Kwak et al., 2019). Despite recent advancements in this field, it is still unclear how these nanoscale materials interact with cells and overcome the multiple obstacles at the cellular and subcellular levels (Gu et al., 2011). Numerous plant species are able to passively integrate functionalized CNTs into their intact and walled cells. These technologies allow for the introduction of exogenous genes into cells in the form of either pDNA or siRNA, resulting in either robust expression of the foreign gene or efficient gene silencing (Demirer et al., 2019). Carbon nanotubes can shield polynucleotides from nuclease-mediated destruction for a certain time period. These methods are able to target not only tissues but also subcellular components. CNTs, for instance, were constructed to preferentially transfer pDNA into plastids via a special mechanism, i.e., lipid exchange envelope penetration (Kwak et al., 2019). This advanced technology makes it possible to release loaded pDNA in a manner that is dependent on pH, which in turn makes it possible for release in the chloroplast stroma. It was also shown that LDH nanosheets can facilitate the transport of dsRNA in *Nicotiana tabacum* cells. This paves the way for the silencing of homologous RNA and serves to defend against plant viruses. The strategy of using magnetic nanoparticles can be implemented to introduce foreign genes into pollen for genetic transformation of crops. Therefore, this strategy is very useful for plant species that show difficulties in genetic modification using traditional techniques (Zhao et al., 2017).

Unlike traditional plant genetic modification approaches, a delivery system based on nanomaterials can passively cross cell walls to release a broad range of biomolecules. This delivery technique provides massive advantages for genetic manipulation of plants. Primarily, it circumvents host-range limits, and consequently, it has the capability to be utilized in a wide variety of plants, including those that are challenging to regenerate via tissue culture.

Second, it is simple to modify through the functionalization of nanomaterials, which makes it possible to enable the delivery of the cargo in a manner that is either regulated or target-specific. This can be accomplished through the use of an exogenous gate-opening trigger or through the construction of a condition-responsive gatekeeper. Both of these approaches are decent alternatives. Because of these benefits, this technique may be "savvier" for regulated and selective cargo release when conditions are fulfilled (Kwak et al., 2019). Third, it enables delivery of "exogenote" or siRNA/dsRNA into plants, which results in the production of genetically manipulated plants without the incorporation of transgenes into the genome of the plant (also known as an "engineered but non-transgenic plant"). In this scenario, changed traits that are the consequence of either transient expression of genes or silencing of genes are not able to be passed to the progeny. Ultimately, the delivery efficacy of the system is increased. It is capable of simultaneously delivering several biomolecules to the target locations and can load a substantial amount of biomolecules (Demirer et al., 2019).

More advanced plant genetic manipulation based on nanomaterials will allow wide-ranging applications in plant biotechnology. For example, due to the strict rules and regulations governing GMOs, there is significant interest in developing plants that improve through genetic modification, in which the expression pattern of genes is regulated but external DNA is not incorporated into the genome. The stringent rules that govern the usage of GMOs may therefore not apply to this strategy. Moreover, the continuing advancement in nanomaterial-assisted plant genetic modification can be applied to a variety of plant functional research. This is mainly helpful for plant species that are difficult to modify through genetic manipulation using traditional techniques. Nanotechnology could be used as a highly efficient and species-independent delivery system to identify genotypes that can lead to desirable phenotypes via temporal expression of genes or gene silencing (Ahmar et al., 2021).

Likewise, the strategy could be extended to nuclease-based genome editing techniques such as CRISPR/Cas9 for the production of stable transgenic plants. Future research will require collaboration between nanomaterial specialists and plant biologists to produce functionalized nanomaterials and improve their techniques for broad application.

However, the nanomaterials employed for this purpose are not optimized to benefit from the mechanisms by which they interact with plants. As a result, it would be ideal to generate a new class of nanomaterials that are both unique and biologically distinct and have improved efficiency. Finally, and most crucially, additional research will be required to study the hazard related to the use of these nanomaterials in plants, and more specifically in agricultural systems, to ensure safe application in transgenic possibilities.

2.3.2 Nanoproteomics Approach

Nanoproteomics represents the quantitative proteome profiling of smaller cellular populations (<5,000 cells) which may be used to provide vital information on rare cell populations, difficult-to-obtain biological specimens and the cellular variability of pathological tissues (Zhu et al., 2018). The auspicious concept of nanoproteomics provides numerous benefits in terms of ultra-low detection, high-throughput efficiency, quick assay time and less consumption of sample (Ray et al., 2011). Nanomaterials such as gold NPs, quantum dots and nanowires, among others, have revealed the ability to circumvent the sensitivity challenges encountered by proteomics for biomarker detection and identification (Agrawal et al., 2013). Contemporary proteomics methods, such as gel-based and/or gel-free techniques coupled with high-throughput analysis, are widely employed to decipher the molecular processes behind plant stress responses.

Analysing the impact of NPs on proteomics profiling and gene expression of crop plants, especially under stressful conditions, is an effective strategy for developing stress-resistant crop varieties (Mustafa et al., 2021). Katz et al. (2007) carried out a proteome investigation of *Dunaliella salina* (halotolerant alga) in order to elucidate the molecular mechanism for the notable resistance against high salt stress. This was accomplished by evaluating the migrating patterns of native protein complexes on blue native gels and their identification by nano-LC-MS/MS. Nanoscale proteomics of rice phloem and xylem sap showed that both contain >100 proteins, including regulatory proteins, and discovered three members of the "TERMINAL FLOWER 1 (TFL1)/FLOWERING LOCUS T (FT)" protein family in phloem sap (Aki et al., 2008). Recent investigations on nano- and peptide-carriers to transfer biomolecules (DNA, RNA, proteins, ribonucleoproteins, etc.) in plants have demonstrated the ability to overcome the limits of conventional methods and accelerate the genetic transformation process (Arya et al., 2021). Nevertheless, nanoscale proteomics may be valuable for finding novel proteins that are specifically localized in target tissues, cells or organelles. In the future, it would also be much simpler to explore a vast array of protein-level biological processes in plants (Aki et al., 2008).

2.3.3 NANOTECHNOLOGY IN PLANT TISSUE CULTURE

Plant tissue culture (PTC) refers to the culture of different types of plant parts or explants under aseptic conditions. The use of nanotechnology in plant biotechnology exhibited an intriguing influence on in vitro plant growth and development, as evidenced by the beneficial effect of nanoparticles on micropropagation, somatic embryogenesis, cell suspension culture and plant decontamination (Al-Qudah et al., 2022). The imperative role of nanotechnology in PTC is represented in Figure 2.3

Aghdaei et al. (2013) reported that silver nanoparticles in combination with indole acetic acid (IAA) and 6-benzylaminopurine (6-BAP) in MS media enhanced the number of shoots, the proportion of explants and plant vigour. Zafar et al. (2016) demonstrated that zinc oxide nanoparticles can promote root development from explants cultured on appropriate media that can be used to produce significant secondary metabolites. AgNPs have shown remarkably beneficial effects on the in vitro

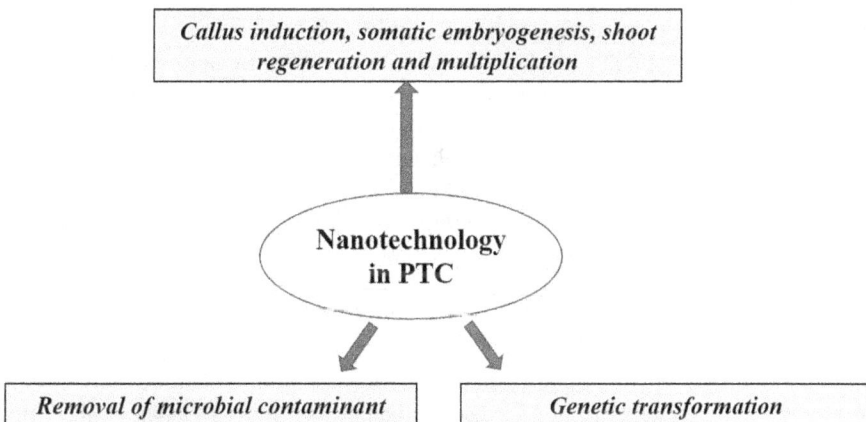

FIGURE 2.3 Multifarious role of nanotechnology in plant tissue culture (PTC).

morphogenesis of King banana, with the best callus formation occurring in culture media supplemented with 8.0 ppm of AgNPs, while the maximum shoot regeneration and shoot multiplication occurred at 4 and 6 ppm of AgNPs, respectively (Huong et al., 2021). The use of nanoparticles has been well recognized for seed germination, enhancement of plant development patterns, genetic modification of plants, protection of plants from diseases and pests, etc. (Wang et al., 2010; Ruttkay-Nedecky et al., 2017). The seeds of *Solanum lycopersicum* treated with SiO_2 (8 g/L) improved numerous metrics such as seed germination, seed vigour and weight of seedlings (Siddiqui & Al-Whaibi, 2014).

Ma et al. (2013) illustrated that CeO_2 nanoparticle (NP) (250 ppm) exposure on *Arabidopsis thaliana* enhanced plant biomass, while indium oxide (In_2O_3) NP exposure increased glutathione synthetase (GS) transcript synthesis by 3.8–4.6 fold. The application of iron oxide nanoparticles and zinc oxide nanoparticles led to maximum callus formation and regeneration of plants (tomato cultivars) under salinity stress (Aazami et al., 2021).

Strictly aseptic conditions are required for successful tissue culture. Therefore, the elimination of both external and internal infectious microorganisms is imperative (Abdi et al., 2008). Explants, such as shoot portions or zygotic embryos, are sterilized with H_2O_2, mercury chloride, calcium hypochlorite, 70% ethanol, sodium hypochlorite and silver nitrate (Álvarez et al., 2019). Metal and metal oxide nanoparticles eliminate surface-contaminating microorganisms.

A number of nanoparticles, including gold (Au), aluminium oxide, cupric oxide, iron oxide, magnesium oxide, nickel, silver, titanium dioxide and zinc oxide, have been reported to have antibacterial property against a wide range of microorganisms (Upadhayay et al., 2019). Some nanoparticles have been shown to be particularly effective in removing microbial contaminants from plant tissue cultures (Tariq et al., 2020).

2.3.4 NANODIAGNOSTIC APPROACH FOR PLANT PATHOGEN DETECTION

The most common types of nanoscale approaches being developed for agricultural diagnostics are wearable sensors for plants, microneedle patches, nanopore sequencing systems and nanoparticle-based sensors. There are numerous techniques to monitor diseases in agriculture, such as enzyme-linked immunosorbent assays (ELISAs), polymerase chain reactions (PCRs), loop-mediated isothermal amplifications (LAMPs) (Zhang & Gleason, 2019; Ristaino et al., 2020) and recombinase polymerase amplifications (RPAs) (Lau et al., 2016). Biomolecules are routinely analysed to identify specific bacterial strains and pathogens through these measurements. Some of the nanotech-based methods for diagnosis of plant diseases are illustrated below:

2.3.4.1 Microneedle Patches

Transdermal drug delivery and subcutaneous biosensing have been extensively performed in nanomedicine using microneedle patches (Van der Maaden et al., 2015; Miller et al., 2016). A small number of studies on point-of-care (POC) diagnosis of plant diseases have been performed. An oomycete that causes late blight on potatoes and tomatoes can be detected quickly and sensitively using the microneedle patch technique in combination with real-time PCR (or quantitative PCR).

2.3.4.2 Nanopore Sequencing Platform

It generates current signals dependent on nucleotide to enable facile analysis of pathogenic polynucleotides using a protein or solid-state nanopore that transports a single-stranded DNA or RNA through either a protein or solid-state nanopore (Eisenstein, 2017). The researchers have developed a standard protocol using Oxford Nanopore Technologies' handheld sequencing system (named "MinION") to detect different bacteria, viruses, fungi and phytoplasmas in plants, including *Penicillium digitatum* in lemon and *S. lycopersicum* in tomato.

2.3.4.3 Nanoparticle-Based Sensors

It has been reported that nanoparticles might be useful for probing the interaction between nanoparticles and bioanalytes associated with plant diseases in vitro or in living plants (Kwak et al., 2017). Silver nanorods have rarely been used as chemical sensor substrates for identifying pathogens or toxins using surface-enhanced Raman spectroscopy (SERS) (Yüksel et al., 2015; Wu et al., 2012). A variety of QD-based biosensors have been developed for the imaging of plants and the detection of diseases (Kwak et al., 2017). QD was reported to be CdSe-ZnS core-shell with 3-mercaptopropionic acid coating, which was readily absorbed by fungal hyphae (Rispail et al., 2014). Electronic noses (e-noses) merge electronic transducers into array rather than chemical sensors and are referred to as array-based sensors. Röck et al. (2008) and Li (2018) reported a FRET-based complex sensor for probing *Citrus tristeza virus* (CTV) (Safarnejad et al., 2017). Table 2.1 shows some nano-based sensors with their detection mechanisms.

2.4 ROLE OF NANOPARTICLES IN IMPROVING PHOTOSYNTHESIS RATE

The process through which plants transform light energy into chemical energy is called photosynthesis. This process converts carbon dioxide (CO_2) into organic molecules (sugars) using sunlight (Bryant & Frigaard, 2006). The process of photosynthesis is the result of the interaction of two different kinds of chemical reactions: light-dependent processes and light-independent processes (CO_2 fixation reactions). During light-dependent reactions, the plant absorbs energy from the sun and converts it into ATP and NADPH, two molecules that are capable of storing energy for later use. In the dark reaction, the end product of the light reaction combines with CO_2 to produce glucose as the end result of the dark reaction. Photosynthetic apparatus consumed less than 10% of the available solar radiation energy. However, the efficiency of solar energy conversion in photosynthetic organisms can be enhanced by manipulating techniques and gene modification methods (Khatri & Rathore, 2018; Kataria et al., 2019). Recently, functionally advanced nanomaterials with enhanced solar energy harvesting have been developed (Khatri & Rathore, 2018; Li et al., 2018). These nanoparticles enabled the kinetic entrapment of nanomaterials inside chloroplast, a photosynthetic organelle, leading to the development of nanobionic

TABLE 2.1

Nano-Based Sensors for Plant Pathogen Detection

Target Pathogen	Detection Mechanism	References
Phytophthora infestans	Integrating universal primer mediated with gold NP-based lateral flow biosensor	Zhan et al. (2018)
Begomovirus (CLCuKoV-Bur)	Carbon nanotube-based copper nanoparticle composite	Tahir et al. (2018)
Ralstonia solanacearum	Colorimetric probe based on gold nanoparticles (AuNPs)	Aoko et al. (2021)
Capsicum chlorosis virus (CaCV)	Nano-Au/C-MWCNT-based label-free amperometric immunosensor	Sharma et al. (2017)
Pseudomonas syringae	Coupling recombinase polymerase amplification (RPA) with AuNPs as electrochemical probes	Lau et al. (2017)
Xanthomonas axonopodis pv. *vesicatoria*	Silica nanoparticles conjugated with the secondary antibody of goat anti-rabbit immunoglobulin G (IgG)	Yao et al. (2009)
Ralstonia solanacearum	Gold nanoparticles functionalized with single-stranded oligonucleotides	Khaledian et al. (2017)
Sclerotinia sclerotiorum	Gold electrode modified with copper nanoparticles	Wang et al. (2010)
Phytophthora species	A helicase-dependent isothermal amplification (HDA) in combination with on-chip hybridization and subsequent silver nanoparticle deposition	Schwenkbier et al. (2015)
Aspergillus niger	Biosensors based on copper oxide (CuO) nanoparticles and nanolayers	Etefagh et al. (2013)

plants (Kataria et al., 2019). Trapping nanomaterials/nanotubes enhanced chloroplast carbon uptake, or photosynthesis, by enhancing chloroplast solar energy capture and electron transport rate. In addition to enhancing photosynthesis, nanotubes such as poly(acrylic acid) nanoceria (PAA-NC) and single-walled nanotube-nanoceria (SWNT-NC) reduce the level of reactive oxygen species (ROS) within extracted chloroplast and alter the sensing process in plants, which is advantageous for a number of physiological activities (Giraldo et al., 2014; Kataria et al., 2019; Velikova et al., 2021).

Carbon dot-induced photosynthesis enhancement in living plants, as reported by Li et al. (2018), resulted in a maximum 25% increase in electron transport rates, establishing a successful example of "nanobionics engineering" of plant performance in vivo. Metal nanoparticles have the potential to substantially improve the efficiency of chemical energy production in photosynthetic systems (Ranjan et al., 2022). Chlorophyll can produce ten times more excited electrons when bound to Au and Ag nanoparticles because of plasmon resonance and quick electron-hole separation (Govorov & Carmeli, 2007). After treatment with SWNTs, Giraldo et al.

(2014) observed a 49% increase in the electron transfer rate under ex vivo settings (in extracted chloroplast from spinach leaves). In a similar fashion, the carbon nanotubes in the thylakoid of spinach improved photoelectrochemical activity (Calkins et al., 2013). Besides protecting chloroplasts from ageing, TiO_2 nanoparticles have been studied to enhance chlorophyll biosynthesis and stimulate Rubisco activity, thereby enhancing the rate of photosynthesis or photosynthetic carbon assimilation (Hong et al., 2005; Khatri & Rathore, 2018).

Qi et al. (2013) observed that the exogenous application of TiO_2 led to an improvement in the net photosynthetic rate, water conductance and transpiration rate. Tighe-Neira et al. (2020) showed a transitory enhancement in net photosynthetic rate (approximately 50%) and stomatal conductance in *Raphanus sativus* treated with TiO_2 nanoparticles and microparticles (µPs). The application of AgNPs to plants resulted in an increase in leaf surface of *S. lycopersicum* area by +41 cm^2/plant, which led to a rise in photosynthetic efficiency of +1.6 g/m^2 per day (Zakharova et al., 2017). Exogenous water-soluble fullerene (FLN) derivatives eliminated cobalt toxicity in maize chloroplasts by enhancing the expression of genes encoding reaction centre proteins of photosystems, elevating the level of defence-related enzymes and enhancing N assimilation (Ozfidan-Konakci et al., 2022). Reddy Pullagurala et al. (2018) observed that zinc oxide nanoparticle treatment increased photosynthetic pigment production in cilantro (*Coriandrum sativum*). As a prospective seed priming agent, ZnO nanoparticles enhanced photosynthetic pigments as well as the number of active reaction centres per chlorophyll molecule in wheat (Rai-Kalal & Jajoo, 2021). The treatment provided in the form of seed priming with iron oxide (Fe_3O_4)-based nanoparticles enhanced the rate of photosynthesis performance and also augmented the content of iron and potassium in wheat plant (Feng et al., 2022).

2.5 ADVANCES IN CHLOROPLAST ENGINEERING USING NANOTECHNOLOGY

Photosynthetic organisms are thought to be sources of foods, biofuels, new biopharmaceuticals and next-generation biomaterials that are important for modern society (Newkirk et al., 2021). Chloroplasts are essential cellular components that serve as a platform for a wide variety of intricate biochemical and molecular processes that are required for plant growth (De-la-Peña et al., 2022). Significant efforts are being made in the field of plant biotechnology to improve both the efficiency of photosynthesis and the production of a wide variety of important compounds that are synthesized in this organelle (Singhal et al., 2022). These compounds are of importance in the fields of biology, medicine and industry (Khan et al., 2019a). Chloroplast biotechnology developments have larger effects on medicine, energy, food, bioplastics and chemicals. This approach may also offer new avenues for crop development (Maliga & Bock, 2011). Current standard approaches for monitoring plant activity, stress and photosynthesis depend on remote sensing techniques for measuring chlorophyll fluorescence or gas analysers for assessing carbon dioxide assimilation (Pérez-Bueno et al., 2019; Newkirk et al., 2021).

Carbon nanotubes were recently modified to detect H_2O_2, a critical signalling chemical produced by chloroplasts and related to plant stress (Wu et al., 2020;

Newkirk et al., 2021). Setting up a uniform and reliable way to deliver biomolecules into chloroplasts could speed up synthetic biology research into novel photosynthetic organisms and their molecular pathways and make it easier to make high-value bio-molecules (Newkirk et al., 2021). In most plants, the plastid genome is inherited maternally, requiring the development of organelle-specific and selective nanocarri-ers. Kwak et al. (2019) created chitosan-complexed single-walled carbon nanotubes that deliver plasmid DNA to chloroplasts of various plant species.

Quantum dots coated with biorecognition moiety have been used to specifically transport chemicals to chloroplasts, and this strategy may be applied to the trans-portation of genetic elements, nanosensors, nutrients or pesticides to a wide range of plant species using nanomaterials (Santana et al., 2021). Thagun et al. (2019) cre-ated peptide–DNA complexes by fusing a cell-penetrating peptide and a chloroplast-targeting peptide with DNA. These complexes were able to transport DNA across the cellular membrane and efficiently deliver the DNA to plastids. Without the need for stable plant transformation, this technique can be utilized to modify plastids in plants and to temporally regulate gene expression in a plant (Lv et al., 2020).

Boghossian et al. (2013) explored the impact of dextran-wrapped nanoceria (dNC) (along with cerium ions, fullerenol and DNA-wrapped single-walled carbon nanotubes) on the ROS generation of isolated chloroplasts from *Spinacia oleracea*. They observed that dNC offered a promising mechanism for preserving restorative chloroplast photoactivity in light-harvesting applications, as it was the most effec-tive for lowering oxidizing species and superoxide concentrations while retaining chloroplast photoactivity. Present understanding gaps of nanotechnology-enabled chloroplast bioengineering include a) precise mechanisms for entry into plant cellu-lar components, b) inadequate knowledge of nanoparticle-based plastid transforma-tions and c) translation of laboratory nanotechnology tools to crop plants in the field. Future research in chloroplast biotechnology can develop tools for efficient monitor-ing and controlling chloroplast activity in vivo and ex vivo across varied plant spe-cies, increasing plant growth and productivity (Newkirk et al., 2021).

2.6 NANOTECHNOLOGY IN THE SHELF LIFE IMPROVEMENT OF PERISHABLE FRUITS AND VEGETABLES

Although agricultural production has greatly increased recently, post-harvest losses and shelf life extension remain problems. Particularly, high moisture, high sugar con-tent, slightly low to neutral pH, microbial attack and damage from pre-harvest and post-harvest handling and transport cause fruits and vegetables to spoil quickly (Liu et al., 2020). The pre- and post-harvest losses, as well as damage, should be kept to a minimum in order to extend the shelf life and guarantee delivery to desired custom-ers (Sridhar et al., 2021). There are many post-harvest processes in use, including thermal, freezing, chilling, ultrasound, pulsed electric field, chemical and gaseous ones, but they are not without drawbacks (Dumont et al., 2016). Due to character-istics such as high surface area, slow-release action, precise active site action and target-specific nature, nanotechnology can be extremely important in addressing current fruit and vegetable losses and shelf life improvement (Khan et al., 2019a). Food processing and packaging are two crucial post-harvest technology areas where nanotechnology has the potential to improve. The idea of nanotechnology has covered the way for the processing and formulation of sensors, flavours, additives and food supplements in both animal- and plant-based products (nanoencapsulation and nanoemulsion) (Khan et al., 2019b). Magnesium oxide and silicon dioxide nanoparticles can be used as food colour and flavouring agents and titanium oxide as additive. Moreover, copper, zinc and iron oxides have been categorized as GRAS materials by the European Food Safety Authority (EFSA) (Liu et al., 2020). Because of their optical and electrochemical characteristics, nanomaterials can be used as nanosensors in fruit and vegetable products such as juice, sauces and beverages.

Antioxidants in juices can be found using graphene nanoribbons with improved surface and electrochemical properties (Iderawumi & Yusuff, 2021). Chilli sauce adulterants can be found using Pb/Au nanocrystals with high surface area and enhanced catalytic activity. Silver nanoparticles with a higher surface area for thiophenol modified species can detect heavy metals, and TiO_2/Pb nanostructure with improved electrochemical properties and conductivity can detect residual pesticides in onion, cabbage and potato (Iderawumi & Yusuff, 2021; Sridhar et al., 2021). Pathogenic infection in fruits can also be detected using nanobiosensors in fruits and vegetables. Fruits and vegetables can be effectively packaged and coated with composites of nanoparticles and biopolymers. In order to stop dehydration and limit the growth of microorganisms, edible nano-coatings have been developed that act as a barrier to the exchange of gas and moisture (Poonia & Mishra, 2022). Nanobased packaging systems are more effective than traditional packaging processes because of nanomaterials' inherent antimicrobial properties. Silicon oxide-chitosan composite was found to prevent deterioration of Fuji apple for longer periods than conventional preservation techniques (Zahedi et al., 2020). On a similar note, a zinc oxide-containing packing material was reported to improve the shelf life of orange juice up to 28 days (Zahedi et al., 2020). A silver nanoparticle-cellulose composite film with antibacterial properties was suggested to improve shelf life in tomatoes (Zahedi et al., 2020).

Likewise, a titanium dioxide complexed with chitosan and polyacrylonitrile, having antimicrobial and ethylene scavenging properties, respectively, were reported to prevent early deterioration in grapes and tomatoes (Zahedi et al., 2020). Zinc oxide-chitosan composite polymer possessing antibacterial and antioxidant properties can be employed for enhancing the shelf life of black grape, apple, mango and tomato (Zahedi et al., 2020; Neme et al., 2021). Despite being in its early stages, the use of NMs in smart packaging has advanced quickly because it provides a secure and sustainable method. Recent nanotechnology-based approaches for extending the shelf life of fruits and vegetables are presented in Table 2.2.

2.7 CONCLUSION

Nanotechnology is used as a highly skilled technology in agriculture to reap the benefits of higher crop yields while reducing reliance on more expensive fertilizers. Several different nanomaterials have been shown to be useful in many areas of agricultural research, including nano-delivery of DNA, proteomic profiling, tissue culture, enhancing photosynthesis, diagnosing plant diseases and reducing post-harvest

TABLE 2.2

Nano-Based Approaches for Extending Shelf Life of Fruits and Vegetables

Nanomaterial(s)/ Nano-Formulation	Fruit/Vegetable	Characteristics Indicating Shelf Life Improvement	References
ZnO nanoparticles	*Solanum lycopersicum* (tomato), *Cucumis sativus* (cucumber)	Reduced weight loss, delayed disease development	Javad et al. (2022)
Chitosan + nano-ZnO	Orange fruits (*Citrus sinensis* L.)	Reduced in rate of the pH change, total acidity, least weight loss	Dulta et al. (2022)
Chitosan and proline-coated chitosan nanoparticles	*Strawberry* fruit	Less decay and weight loss, level of ascorbic and phenolic acid increased and antioxidant activity of fruit	Bahmani et al. (2022)
Herbal extract-based nano-formulations created via sodium alginate and $CaCl_2$	Guava (*Psidium guajava* L.)	Least changes in colour and physiological weight	Rani et al. (2022)
Starch-silver nanoparticle	Strawberry	Low weight loss and decay	Taha et al. (2022)
Cellulose nanofibers (CNFs)	Apple fruits	Inhibition of respiration rate, ethylene production, and cell permeability of membrane to preserve the inflexibility and fruit weight	Wang et al. (2022)
Iron oxide NPs	Apple fruits	Maintenance in the firmness, elevated percentage of ascorbic acid, soluble solids and sugars	Akbar et al. (2022)
Nano-chitosan	Date palm fruits (Barhi cultivar)	Enhancement in the fruit quality, increased content of total soluble solids and total sugars, reduction in weight loss	El-Gioushy et al. (2022)
Amla essential oil-based nano-coatings	Amla fruits	Lower decay percentage, low weight loss, higher content of ascorbic acid, delayed browning, higher retaining of antioxidant activity and bioactive components	Braich et al. (2022)

losses. The controlled release and targeted distribution of nanomaterials make them valuable components in smart agricultural systems.

The application of nanomaterials in agriculture is still relatively new and requires further investigation. However, as with the implementation of any new technology, a reliable risk-benefit analysis and a comprehensive cost accounting assessment are required. For nanotechnology applications, this requires the establishment of credible approaches to characterize and quantify nanomaterials in different matrices and to assess their impact on the environment and human health. Additionally, engaging all stakeholders, including NGOs and consumer groups, in an open discourse is crucial to achieving customer satisfaction and public support for this technology.

REFERENCES

Aazami, M.A., Rasouli, F., & Ebrahimzadeh, A. (2021). Oxidative damage, antioxidant mechanism and gene expression in tomato responding to salinity stress under in vitro conditions and application of iron and zinc oxide nanoparticles on callus induction and plant regeneration. *BMC Plant Biol.* 21, 597.

Abdi, G., Salehi, H., & Khosh-Khui, M. (2008). Nano silver: a novel nanomaterial for removal of bacterial contaminants in valerian (*Valeriana officinalis* L.) tissue culture. *Acta Physiol Plant.* 30, 709–714.

Aghdaei, M., Salehi, H., & Sarmast, M. (2013). Effects of silver nanoparticles on *Tecomella undulate* (Roxh.) Seem. micropropagation. *Adv Hortic Sci.* 26, 21–24.

Agrawal, G.K., Timperio, A.M., Zolla, L., Bansal, V., Shukla, R., & Rakwal, R. (2013). Biomarker discovery and applications for foods and beverages: proteomics to nanoproteomics. *J Proteom.* 93, 74–92.

Agri, U., Chaudhary, P., & Sharma, A. (2021). In vitro compatibility evaluation of agriusable nanochitosan on beneficial plant growth-promoting rhizobacteria and maize plant. *Natl Acad Sci Lett.* 44, 555–559.

Agri, U., Chaudhary, P., Sharma, A., & Kukreti, B. (2022). Physiological response of maize plants and its rhizospheric microbiome under the influence of potential bioinoculants and nanochitosan. *Plant Soil.* 474, 451–468.

Ahmar, S., Mahmood, T., Fiaz, S., Mora-Poblete, F., Shafique, M.S., Chattha, M.S., Jung, K-H. (2021). Advantage of nanotechnology-based genome editing system and its application in crop improvement. *Front Plant Sci.* 12, 663849.

Akbar, M., Haroon, U., Ali, M., Tahir, K., Chaudhary, H.J., & Munis, M.F.H. (2022). Mycosynthesized Fe_2O_3 nanoparticles diminish brown rot of apple whilst maintaining composition and pertinent organoleptic properties. *J Appl Microbiol.* 132(5), 3735–3745.

Aki, T., Shigyo, M., Nakano, R., Yoneyama, T., & Yanagisawa, S. (2008). Nano scale proteomics revealed the presence of regulatory proteins including three FT-Like proteins in phloem and xylem saps from rice. *Plant Cell Physiol.* 49, 767–790.

Al-Qudah, T., Mahmood, S. H., Abu-Zurayk, R., Shibli, R., Khalaf, A., Lambat, T.L., & Chaudhary, R.G. (2022). Nanotechnology applications in plant tissue culture and molecular genetics: a holistic approach. *Curr Nanosci.* 18, 442–464.

Álvarez, S.P., Tapia, M.A.M., Vega, M.E.G., Ardisana, E.F.H., Medina, J.A.C., Zamora, G.L.F., & Bustamante, D.V. (2019). Nanotechnology and plant tissue culture. In *Plant Nanobionics* (pp. 333–370). Springer, Cham.

An, C., Sun, C., Li, N., Huang, B., Jiang, J., Shen, Y., Wang, C., Zhao, X., Cui, B., Wang, C., Li, X., Zhan, S., Gao, F., Zeng, Z., Cui, H., & Wang, Y. (2022). Nanomaterials and nanotechnology for the delivery of agrochemicals: strategies towards sustainable agriculture. *J Nanobiotechnol.* 20, 11.

Aoko, I.L., Ondigo, D., Kavoo, A.M., Wainaina, C., & Kiirika, L. (2021). A gold nanoparticle-based colorimetric probe for detection of gibberellic acid exuded by *Ralstonia solanacearum* pathogen in tomato (*Solanum lycopersicum* L.). *Int J Hort Sci Technol.* 8, 203–214.

Arya, S.S., Tanwar, N., & Lenka, S. K. (2021). Prospects of nano- and peptide-carriers to deliver CRISPR cargos in plants to edit across and beyond central dogma. *Nanotechnol Environ Eng.* 6, 1–13.

Bahmani, R., Razavi, F., Mortazavi, S.N., Gohari, G., & Juárez-Maldonado, A. (2022). Evaluation of proline-coated chitosan nanoparticles on decay control and quality preservation of strawberry fruit (cv. *Camarosa*) during cold storage. *Horticulturae* 8(7), 648.

Bai, C., Dallasega, P., Orzes, G., & Sarkis, J. (2020). Industry 4.0 technologies assessment: a sustainability perspective. *Int J Prod Econ.* 229, 107776.

Bayda, S., Adeel, M., Tuccinardi, T., Cordani, M., & Rizzolio, F. (2019). The history of nano-science and nanotechnology: from chemical-physical applications to nanomedicine. *Molecules*. 25, 112.

Beyth, N., Houri-Haddad, Y., Domb, A., Khan, W., & Hazan, R. (2015). Alternative antimicrobial approach: nano-antimicrobial materials. *Evid Based Complement Alternat Med*. 2015, 246012.

Bhatt, P., Pandey, S.C., Joshi, S., Chaudhary, P., Pathak, V.M., Huang, Y., Wu, X., Zhou, Z., & Chen, S. (2022). Nanobioremediation: a sustainable approach for the removal of toxic pollutants from the environment. *J Hazard Mater*. 427, 128033.

Boghossian, A.A., Sen, F., Gibbons, B.M., Sen, S., Faltermeier, S.M., Giraldo, J.P., Zhang, C.T., Zhang, J., Heller, D.A., & Strano, M.S. (2013). Application of nanoparticle antioxidants to enable hyperstable chloroplasts for solar energy harvesting. *Adv Energy Mater*. 3. 881–893.

Braich, A.K., Kaur, G., Singh, A., & Dar, B.N. (2022). *Amla* essential oil-based nano-coatings of Amla fruit: analysis of morphological, physiochemical, enzymatic parameters, and shelf-life extension. *J Food Process Preserv*. 46(6), e16498.

Bryant, D.A., & Frigaard, N.U. (2006). Prokaryotic photosynthesis and phototrophy illuminated. *Trends Microbiol*. 14, 488–496.

Calkins, J.O., Umasankar, Y., O'Neill, H., & Ramasamy, R.P. (2013). High photo-electrochemical activity of thylakoid–carbon nanotube composites for photosynthetic energy conversion. *Energy Environ Sci*. 6, 1891–1900.

Cardoso, R.M., Pereira, T.S., Facure, M.H., dos Santos, D.M., Mercante, L.A., Mattoso, L. H., & Correa, D. S. (2021). Current progress in plant pathogen detection enabled by nanomaterials-based (bio) sensors. *Sensors Actuators Rep*. 4, 100068.

Cha, T.S., Chen, C.F., Yee, W., Aziz, A., & Loh, S.H. (2011). Cinnamic acid, coumarin and vanillin: alternative phenolic compounds for efficient agrobacterium-mediated transformation of the unicellular green alga, *Nannochloropsis* sp. *J Microbiol Met*. 84, 430–434.

Chaudhary, P., Chaudhary, A., Bhatt, P., Kumar, G., Khatoon, H., Rani, A., Kumar, S., & Sharma, A. (2022a). Assessment of soil health indicators under the influence of nano-compounds and *Bacillus* spp. in field condition. *Front Environ Sci*. 9, 769871.

Chaudhary, P., Chaudhary, A., Parveen, H., Rani, A., Kumar, G., Kumar, A., & Sharma, A. (2021e). Impact of nanophos in agriculture to improve functional bacterial community and crop productivity. *BMC Plant Biol*. 21, 519.

Chaudhary, P., Khati, P., Chaudhary, A., Gangola, S., Kumar, R., & Sharma, A. (2021a). Bioinoculation using indigenous *Bacillus* spp. improves growth and yield of *Zea mays* under the influence of nanozeolite. *3 Biotech*. 11, 11.

Chaudhary, P., Khati, P., Chaudhary, A., Maithani, D., Kumar, G., & Sharma, A. (2021d). Cultivable and metagenomic approach to study the combined impact of nanogypsum and *Pseudomonas taiwanensis* on maize plant health and its rhizospheric microbiome. *PLoS One*. 16, e0250574.

Chaudhary, P., Khati, P., Gangola, S., Kumar, A., Kumar, R., & Sharma, A. (2021c). Impact of nanochitosan and *Bacillus* spp. on health, productivity and defence response in *Zea mays* under field condition. *3 Biotech*. 11, 237.

Chaudhary, P., & Sharma, A. (2019). Response of nanogypsum on the performance of plant growth promotory bacteria recovered from nanocompound infested agriculture field. *Environ Ecol*. 37, 363–372.

Chaudhary, P., Sharma, A., Chaudhary, A., Khati, P., Gangola, S., & Maithani, D. (2021b). Illumina based high throughput analysis of microbial diversity of rhizospheric soil of maize infested with nanocompounds and *Bacillus* sp. *Appl Soil Ecol*. 159, 103836.

Chaudhary, P., Singh, S., Chaudhary, A., Sharma, A., & Kumar, G. (2022b). Overview of bio-fertilizers in crop production and stress management for sustainable agriculture. *Front Plant Sci*. 13, 930340.

Chen, Z.Y., Liang, K., Qiu, R.X., & Luo, L.P. (2011). Ultrasound-and liposome microbubble-mediated targeted gene transfer to cardiomyocytes in vivo accompanied by polyethyleni-mine. *J Ultrasound Med.* 30, 1247–1258.

Cheng, H.N., Klasson, K.T., Asakura, T., & Wu, Q. (2016). Nanotechnology in agriculture. In *ACS Symposium Series* (pp. 233–242). American Chemical Society.

Cheng, M., Fry, J.E., Pang, S., Zhou, H., Hironaka, C.M., Duncan, D.R., & Wan, Y. (1997). Genetic transformation of wheat mediated by *Agrobacterium tumefaciens*. *Plant Physiol.* 115, 971–980.

Chhipa, H. (2019). Applications of nanotechnology in agriculture. In *Methods in Microbiology* (Vol. 46, pp. 115–142). Academic Press.

De-la-Peña, C., León, P., & Sharkey, T.D. (2022). Editorial: chloroplast biotechnology for crop improvement. *Front Plant Sci.* 13, 848034.

Demirer, G.S., Zhang, H., Matos, J.L., Goh, N.S., Cunningham, F.J., Sung, Y., Chang, R., Aditham, A. J., Chio, L., Cho, M.-J., Staskawicz, B., & Landry, M.P. (2019). High aspect ratio nanomaterials enable delivery of functional genetic material without DNA integra-tion in mature plants. *Nature Nanotechnol.* 14, 456–464.

Dulta, K., Koşarsoy Ağçeli, G., Thakur, A., Singh, S., Chauhan, P., & Chauhan, P. K. (2022). Development of alginate-chitosan based coating enriched with ZnO nanoparticles for increasing the shelf life of orange fruits (*Citrus sinensis* L.). *J Poly Environ.* 30, 3293–3306.

Dumont, M.-J., Orsat, V., & Raghavan, V. (2016). Reducing postharvest losses. In C. Madramootoo (Ed.) *Emerging Technologies for Promoting Food Security* (pp. 135–156). A volume in Woodhead publishing series in Food Science, Technology and Nutrition, Elsevier. https://doi.org/10.1016/C2014-0-02820-1

Eisenstein, M. (2017). An ace in the hole for DNA sequencing. *Nature.* 550, 285–288.

El-Gioushy, S.F., El-Masry, A.M., Fikry, M., El-Kholy, M.F., Shaban, A.E., Sami, R., Algarni, E., Alshehry, G., Aljumayi, H., Benajiba, N., Al-Mushhin, A A M., Algheshairy, R.M., & El-Badawy, H.E. (2022). Utilization of active edible films (chitosan, chitosan nanopar-ticle, and CaCl$_2$) for enhancing the quality properties and the shelf life of date palm fruits (Barhi cultivar) during cold storage. *Coatings.* 12(2), 255.

Enriquez-Obregon, G.A., Vazquez-Padron, R.I., Prieto-Samsonov, D.L., Perez, M., & Selman-Housein, G. (1997). Genetic transformation of sugarcane by *Agrobacterium tumefaciens* using antioxidant compounds. *Biotecnol Apl.* 14, 169–174.

Etefagh, R., Azhir, E., & Shahtahmasebi, N. (2013). Synthesis of CuO nanoparticles and fabri-cation of nanostructural layer biosensors for detecting *Aspergillus niger* fungi. *Sci Iran.* 20, 1055–1058.

Feng, Y., Kreslavski, V.D., Shmarev, A.N., Ivanov, A.A., Zharmukhamedov, S.K., Kosobryukhov, A., Yu, M., Allakhverdiev, S.I., & Shabala, S. (2022). Effects of iron oxide nanoparticles (Fe$_3$O$_4$) on growth, photosynthesis, antioxidant activity and distribu-tion of mineral elements in wheat (*Triticum aestivum*) plants. *Plants.* 11, 1894.

Ghidan, A. Y., & Antary, T. M. A. (2019). Applications of nanotechnology in agriculture. In M. Stoytcheva, & R. Zlatev (Eds.), *Applications of Nanobiotechnology*. IntechOpen. doi: 10.5772/intechopen.82976

Giraldo, J.P., Landry, M.P., Faltermeier, S.M., McNicholas, T.P., Iverson, N.M., Boghossian, A.A., Reuel, N.F., Hilmer, A.J., Sen, F., Brew, J.A., & Strano, M.S. (2014). Plant nano-bionics approach to augment photosynthesis and biochemical sensing. *Nature Mat.* 13, 400–408.

Govorov, A.O., & Carmeli, I. (2007). Hybrid structures composed of photosynthetic system and metal nanoparticles: plasmon enhancement effect. *Nano Lett.* 7, 620–625.

Gu, Z., Biswas, A., Zhao, M., & Tang, Y. (2011). Tailoring nanocarriers for intracellular protein delivery. *Chem Soc Rev.* 40, 3638–3655.

Hong, F., Yang, F., Liu, C., Gao, Q., Wan, Z., Gu, F., Wu, C., Ma, Z., Zhou, J., & Yang, P. (2005). Influences of nano-TiO_2 on the chloroplast aging of spinach under light. *Biol Trace Element Res.* 104, 249–260.

Huong, B.T.T., Xuan, T.D., Trung, K.H., Ha, T.T.T., Duong, V.X., Khanh, T.D., & Gioi, D.H. (2021). Influences of silver nanoparticles in vitro morphogenesis of specialty king banana (Musa ssp.) in Vietnam. *Plant Cell Biotechnol Mol Biol.* 22, 163–175.

Iderawumi, A.M., & Yusuff, M.A. (2021). Effects of nanoparticles on improvement in quality and shelf life of fruits and vegetables. *J Plant Biology Crop Res.* 4, 1042.

Javad, S., Azam, N., Ghaffar, N., Jabeen, K., & Ahmad, A. (2022). Effect of zinc oxide nanoparticles on shelf life of commonly used vegetables. *Proc Bulgarian Acad Sci.* 75(6), 933–942.

Kataria, S., Jain, M., Rastogi, A., Živčák, M., Brestic, M., Liu, S., & Tripathi, D.K. (2019). Role of nanoparticles on photosynthesis: avenues and applications. In D. K. Tripathi, P. Ahmad, S. Sharma, D. K. Chauhan, & N. K. Dubey (Eds.) *Nanomaterials in Plants, Algae and Microorganisms* (pp. 103–127). Academic Press. ISBN 9780128114889, https://doi.org/10.1016/B978-0-12-811488-9.00006-8.

Katz, A., Waridel, P., Shevchenko, A., & Pick, U. (2007). Salt-induced changes in the plasma membrane proteome of the halotolerant alga *Dunaliella salina* as revealed by blue native gel electrophoresis and nano-LC-MS/MS analysis. *Mol Cell Proteom.* 6, 1459–1472.

Khaledian, S., Nikkhah, M., Shams-bakhsh, M., & Hoseinzadeh, S. (2017). A sensitive biosensor based on gold nanoparticles to detect *Ralstonia solanacearum* in soil. *J Gen Plant Pathol.* 83, 231–239.

Khan, I., Saeed, K., & Khan, I. (2019a). Nanoparticles: properties, applications and toxicities. *Arab J Chem.* 12, 908–931.

Khan, M.S., Mustafa, G., & Joyia, F.A. (2019b). Technical advances in chloroplast biotechnology. In M. S. Khan, & K. A. Malik (Eds.) *Transgenic Crops - Emerging Trends and Future Perspectives*. IntechOpen, doi: 10.5772/intechopen.81240

Khati, P., Bhatt, P., Kumar, R., & Sharma, A. (2018). Effect of nanozeolite and plant growth promoting rhizobacteria on maize. *3Biotech.* 8, 141.

Khati, P., Chaudhary, P., Gangola, S., Bhatt, P., & Sharma, A. (2017). Nanochitosan supports growth of *Zea mays* and also maintains soil health following growth. *3 Biotech.* 7, 81.

Khati, P., Chaudhary, P., Gangola, S., & Sharma, P. (2019a). Influence of nanozeolite on plant growth promotory bacterial isolates recovered from nanocompound infested agriculture field. *Environ Ecol.* 37, 521–527.

Khati, P., Sharma, A., Chaudhary, P., Singh, A.K., Gangola, S., & Kumar R. (2019b). High-throughput sequencing approach to access the impact of nanozeolite treatment on species richness and evens of soil metagenome. *Biocatalysis Agric Biotechnol.* 20, 101249.

Khatri, K., & Rathore, M. S. (2018). Plant nanobionics and its applications for developing plants with improved photosynthetic capacity. In J. C. G. Cañedo, & G. L. L. Lizárraga (Eds.) *Photosynthesis - From Its Evolution to Future Improvements in Photosynthetic Efficiency Using Nanomaterials*. IntechOpen, doi: 10.5772/intechopen.71794.

Kukreti, B., Sharma, A., Chaudhary, P., Agri, U., & Maithani, D. (2020). Influence of nanosilicon dioxide along with bioinoculants on *Zea mays* and its rhizospheric soil. *3 Biotech.* 10, 345.

Kumari, H., Khati, P., Gangola, S., Chaudhary, P., & Sharma, A. (2021). Performance of plant growth promotory rhizobacteria on maize and soil characteristics under the influence of TiO_2 nanoparticles. *Pantnagar J Res.* 19, 28–39.

Kumari, S., Sharma, A., Chaudhary, P., & Khati, P. (2020). Management of plant vigor and soil health using two agriusable nanocompounds and plant growth promotory rhizobacteria in Fenugreek. *3 Biotech.* 10, 461.

Kwak, S.Y., Lew, T.T.S., Sweeney, C.J., Koman, V.B., Wong, M.H., Bohmert-Tatarev, K., & Strano, M. S. (2019). Chloroplast-selective gene delivery and expression in planta using chitosan-complexed single-walled carbon nanotube carriers. *Nat Nanotechnol.* 14, 447–455.

Kwak, S.Y., Wong, M. H., Lew, T. T. S., Bisker, G., Lee, M. A., Kaplan, A., & Strano, M. S. (2017). Nanosensor technology applied to living plant systems. *Arab J Chem.* 10, 113–140.

Lau, H.Y., Wang, Y., Wee, E.J., Botella, J.R., & Trau, M. (2016). Field demonstration of a multiplexed point-of-care diagnostic platform for plant pathogens. *Anal Chem.* 88, 8074–8081.

Lau, H.Y., Wu, H., Wee, E.J.H., Trau, M., Wang, Y., & Botella, J.R. (2017). Specific and sensitive isothermal electrochemical biosensor for plant pathogen DNA detection with colloidal gold nanoparticles as probes. *Sci Rep.* 7, 38896.

Li, L.L. (2018). Self-assembled nanomaterials for bacterial infection diagnosis and therapy. In H. Wang & L. L. Li (Eds.). *In Vivo Self-Assembly Nanotechnology for Biomedical Applications* (pp. 57–88). Springer, Singapore.

Li, W., Wu, S., Zhang, H., Zhang, X., Zhuang, J., Hu, C., Liu, Y., Lei, B., Ma, L., & Wang, X. (2018). Enhanced biological photosynthetic efficiency using light-harvesting engineering with dual-emissive carbon dots. *Adv Funct Mater.* 28, 1804004.

Liu, W., Zhang, M., & Bhandari, B. (2020). Nanotechnology - a shelf life extension strategy for fruits and vegetables. *Crit Rev Food Sci Nutr.* 60, 1706–1721.

Lv, Z., Jiang, R., Chen, J., & Chen, W. (2020). Nanoparticle-mediated gene transformation strategies for plant genetic engineering. *Plant J.* 104, 880–891.

Ma, C., Chhikara, S., Xing, B., Musante, C., White, J.C., Dhankher, O.P. (2013). Physiological and molecular response of *Arabidopsis thaliana* (L.) to nanoparticle cerium and indium oxide exposure. *ACS Sustain. Chem. Eng.* 1, 768–778.

Maliga, P., & Bock, R. (2011). Plastid biotechnology: food, fuel, and medicine for the 21st century. *Plant Phys.* 155, 1501–1510.

Martin-Ortigosa, S., Peterson, D.J., Valenstein, J.S., Lin, V.S.-Y., Trewyn, B.G., Lyznik, L. A., & Wang, K. (2014). Mesoporous silica nanoparticle-mediated intracellular Cre protein delivery for maize genome editing via loxP site excision. *Plant Physiol.* 164, 537–547.

Miller, P.R., Narayan, R.J., & Polsky, R. (2016). Microneedle-based sensors for medical diagnosis. *J Mater Chem B.* 4, 1379–1383.

Mishra, S., Singh, A., Keswani, C., & Singh, H.B. (2014). Nanotechnology: exploring potential application in agriculture and its opportunities and constraints. *Biotech Today.* 4, 9.

Mukhopadhyay, S.S. (2014). Nanotechnology in agriculture: prospects and constraints. *Nanotechnol Sci Appl.* 7, 63–71.

Mustafa, G., Farooq, A., Riaz, Z., & Hasan, M., (2021). Nano-proteomics of stress tolerance in crop plants. In Faizan, M., Hayat, S., & Yu, F. (Eds.) *Sustainable Agriculture Reviews 53. Sustainable Agriculture Reviews* (vol. 53). Springer, Cham.

Namuddu, A., Kiggundu, A., Mukasa, S.B., Kurnet, K., Karamura, E., & Tushemereirwe, W. (2013). *Agrobacterium* mediated transformation of banana (*Musa* sp.) cv. Sukali Ndiizi (ABB) with a modified *Carica papaya* cystatin (CpCYS) gene. *Afr J Biotechnol.* 12, 1811–1819.

Neme, K., Nafady, A., Uddin, S., & Tola, Y.B. (2021). Application of nanotechnology in agriculture, postharvest loss reduction and food processing: food security implication and challenges. *Heliyon.* 7, e08539.

Newkirk, G.M., de Allende, P., Jinkerson, R.E., & Giraldo, J.P. (2021). Nanotechnology approaches for chloroplast biotechnology advancements. *Front Plant Sci.* 12, 691295.

Ozfidan-Konakci, C., Alp, F. N., Arikan, B., Balci, M., Parmaksizoglu, Z., Yildiztugay, E., & Cavusoglu, H. (2022). The effects of fullerene on photosynthetic apparatus, chloroplast-encoded gene expression, and nitrogen assimilation in *Zea mays* under cobalt stress. *Physiologia Plant.* 174, e13720.

Pérez-Bueno, M.L., Pineda, M., & Barón, M. (2019). Phenotyping plant responses to biotic stress by chlorophyll fluorescence imaging. *Front in Plant Sci.* 10, 1135.

Poonia, A., & Mishra, A. (2022). Edible nanocoatings: potential food applications, challenges and safety regulations. *Nutr Food Sci.* 52, 497–514.

Pramanik, P., Krishnan, P., Maity, A., Mridha, N., Mukherjee, A., & Rai, V. (2020). Application of nanotechnology in agriculture. In N. Dasgupta, S. Ranjan, & E. Lichtfouse (Eds.) *Environmental Nanotechnology Volume 4* (pp. 317–348). Springer, Cham.

Qi, M., Liu, Y., & Li, T. (2013). Nano-TiO_2 improve the photosynthesis of tomato leaves under mild heat stress. *Biol Trace Elem Res*. 156, 323–328.

Rai-Kalal, P., & Jajoo, A. (2021). Priming with zinc oxide nanoparticles improve germination and photosynthetic performance in wheat. *Plant Physiol Biochem*. 160, 341–351.

Rani, A., Tokas, J., Gupta, P., Punia, H., & Baloda, S. (2022). Efficacy of herbal extracts-based nano-formulations in extending guava fruit shelf-life. *Appl Sci*. 12(17), 8630.

Ranjan, A., Rajput, V.D., Kumari, A., Mandzhieva, S.S., Sushkova, S., Prazdnova, E. V., Zargar, S.M., Raza, A., Minkina, T., & Chung, G. (2022). Nanobionics in crop production: an emerging approach to modulate plant functionalities. *Plants*. 11, 692.

Ray, S., Reddy, P. J., Choudhary, S., Raghu, D., & Srivastava, S. (2011). Emerging nanoproteomics approaches for disease biomarker detection: a current perspective. *J Proteom*. 74, 2660–2681.

Reddy Pullagurala, V.L., Adisa, I.O., Rawat, S., Kalagara, S., Hernandez-Viezcas, J.A., Peralta-Videa, J. R., & Gardea-Torresdey, J. L. (2018). ZnO nanoparticles increase photosynthetic pigments and decrease lipid peroxidation in soil grown cilantro (*Coriandrum sativum*). *Plant Physiol Biochem*. 132, 120–127.

Rispail, N., De Matteis, L., Santos, R., Miguel, A.S., Custardoy, L., Testillano, P.S., & Prats, E. (2014). Quantum dot and superparamagnetic nanoparticle interaction with pathogenic fungi: internalization and toxicity profile. *ACS Appl Mater Interfaces* 6, 9100–9110.

Ristaino, J.B., Saville, A.C., Paul, R., Cooper, D.C., & Wei, Q. (2020). Detection of phytophthora infestans by loop-mediated isothermal amplification, real-time LAMP, and droplet digital PCR. *Plant Dis*. 104, 708–716.

Röck, F., Barsan, N., & Weimar, U. (2008). Electronic nose: current status and future trends. *Chem Rev*. 108, 705–725.

Ruttkay-Nedecky, B., Krystofova, O., Nejdl, L., & Adam, V. (2017). Nanoparticles based on essential metals and their phytotoxicity. *J Nanobiotechnol*. 15, 1–19.

Safarnejad, M. R., Samiee, F., Tabatabie, M., & Mohsenifar, A. (2017). Development of quantum dot-based nanobiosensors against *citrus tristeza* virus (CTV). *Sens Transducers*. 213, 54.

Santana, I., Hu, P., Jeon, S.-J., Castillo, C., Tu, H., & Giraldo, J. P. (2021). Peptide-mediated targeting of nanoparticles with chemical cargoes to chloroplasts in arabidopsis plants. *Bio-Protocol*. 11, e4060.

Saranya, S., Aswani, R., Remakanthan, A., Radhakrishnan, E.K. (2019). Nanotechnology in agriculture. In Panpatte, D., & Jhala, Y. (Eds.) *Nanotechnology for Agriculture*. Springer, Singapore.

Savci, S. (2012). An agricultural pollutant: chemical fertilizer. *Int J Environ Sci Dev*. 3, 73.

Schwenkbier, L., Pollok, S., König, S., Urban, M., Werres, S., Cialla-May, D., Weber, K., & Popp, J. (2015). Towards on-site testing of phytophthora species. *Anal Methods*. 7, 211–217.

Shang, Y., Hasan, M.K., Ahammed, G.J., Li, M., Yin, H., & Zhou, J. (2019). Applications of nanotechnology in plant growth and crop protection: a review. *Molecules*. 24, 2558.

Sharma, A., Kaushal, A., & Kulshrestha, S. (2017). A Nano-Au/C-MWCNT based label free amperometric immunosensor for the detection of capsicum chlorosis virus in bell pepper. *Arch Virol*. 162, 2047–2052.

Siddiqui, M.H., & Al-Whaibi, M.H. (2014). Role of nano-SiO_2 in germination of tomato (*Lycopersicum esculentum* seeds Mill.). *Saudi J Biol Sci*. 21, 13–17.

Singhal, R., Pal, R., & Dutta, S. (2022). Chloroplast engineering: fundamental insights and its application in amelioration of environmental stress. *Appl Biochem Biotechnol*.

Sridhar, A., Ponnuchamy, M., Kumar, P. S., & Kapoor, A. (2021). Food preservation techniques and nanotechnology for increased shelf life of fruits, vegetables, beverages and spices: a review. *Environ Chem Lett.* 19, 1715–1735.

Swift, T. A., Oliver, T. A. A., Galan, M. C., & Whitney, H. M. (2018). Functional nanomaterials to augment photosynthesis: evidence and considerations for their responsible use in agricultural applications. *Interface Focus.* 9, 20180048.

Taha, I.M., Zaghlool, A., Nasr, A., Nagib, A., El Azab, I. H., Mersal, G.A.M., Ibrahim, M. M., & Fahmy, A. (2022). Impact of starch coating embedded with silver nanoparticles on strawberry storage time. *Polymers.* 14(7), 1439.

Tahir, M. A., Bajwa, S. Z., Mansoor, S., Briddon, R. W., Khan, W. S., Scheffler, B. E., & Amin, I. (2018). Evaluation of carbon nanotube based copper nanoparticle composite for the efficient detection of agroviruses. *J Hazard Mater.* 346, 27–35.

Tariq, A., Ilyas, S., Naz, S. (2020). Nanotechnology and plant tissue culture. In Javad, S. (Ed.) *Nanoagronomy.* Springer, Cham.

Thagun, C., Chuah, J.-A., & Numata, K. (2019). Targeted gene delivery into various plastids mediated by clustered cell-penetrating and chloroplast-targeting peptides. *Adv Sci.* 6, 1902064.

Tighe-Neira, R., Reyes-Díaz, M., Nunes-Nesi, A., Recio, G., Carmona, E., Corgne, A., Rengel, Z., & Inostroza-Blancheteau, C. (2020). Titanium dioxide nanoparticles provoke transient increase in photosynthetic performance and differential response in antioxidant system in *Raphanus sativus* L. *Sci Hortic.* 269, 109418.

Upadhayay, V. K., Khan, A., Singh, J., & Singh, A. V. (2019). Splendid role of nanoparticles as antimicrobial agents in wastewater treatment. In R. Bharagava (Ed) *Microorganisms for Sustainability* (pp. 119–136). Springer, Singapore.

Van der Maaden, K., Luttge, R., Vos, P. J., Bouwstra, J., Kersten, G., & Ploemen, I. (2015). Microneedle-based drug and vaccine delivery via nanoporous microneedle arrays. *Drug Deliv Transl Res.* 5, 397–406.

Velikova, V., Petrova, N., Kovács, L., Petrova, A., Koleva, D., Tsonev, T., Taneva, S., Petrov, P., & Krumova, S. (2021). Single-walled carbon nanotubes modify leaf micromorphology, chloroplast ultrastructure and photosynthetic activity of pea plants. *Int J Mol Sci.* 22, 4878.

Wang, P., Lombi, E., Zhao, F.-J., & Kopittke, P. M. (2016). Nanotechnology: a new opportunity in plant sciences. *Trends Plant Sci.* 21, 699–712.

Wang, Y., Zhang, J., Wang, X., Zhang, T., Zhang, F., Zhang, S., Li, Y., Gao, W., You, C., Wang, X., & Yu, K. (2022). Cellulose nanofibers extracted from natural wood improve the post-harvest appearance quality of apples. *Front Nutr.* 9, 881783.

Wang, Z., Wei, F., Liu, S.-Y., Xu, Q., Huang, J.-Y., Dong, X.-Y., Yu, J.-H., Yang, Q., Zhao, Y.-D., & Chen, H. (2010). Electrocatalytic oxidation of phytohormone salicylic acid at copper nanoparticles-modified gold electrode and its detection in oilseed rape infected with fungal pathogen *Sclerotinia sclerotiorum*. *Talanta.* 80, 1277–1281.

Wu, H., Nißler, R., Morris, V., Herrmann, N., Hu, P., Jeon, S.-J., Kruss, S., & Giraldo, J.P. (2020). Monitoring plant health with near-infrared fluorescent H_2O_2 nanosensors. *Nano Lett.* 20, 2432–2442.

Wu, X., Gao, S., Wang, J. S., Wang, H., Huang, Y. W., & Zhao, Y. (2012). The surface-enhanced Raman spectra of aflatoxins: spectral analysis, density functional theory calculation, detection and differentiation. *Analyst.* 137, 4226–4234.

Yan, Y., Zhu, X., Yu, Y., Li, C., Zhang, Z., & Wang, F. (2022). Nanotechnology strategies for plant genetic engineering. *Adv Mater.* 34, e2106945.

Yao, K.S., Li, S.J., Tzeng, K.C., Cheng, T.C., Chang, C.Y., Chiu, C.Y., Liao, C.Y., Hsu, J.J., & Lin, Z.P. (2009). Fluorescence silica nanoprobe as a biomarker for rapid detection of plant pathogens. *Adv Mater Res.* 79, 513–516.

Yüksel, S., Schwenkbier, L., Pollok, S., Weber, K., Cialla-May, D., & Popp, J. (2015). Label-free detection of *Phytophthora ramorum* using surface-enhanced Raman spectroscopy. *Analyst*. 140, 7254–7262.

Zafar, H., Ali, A., Ali, J.S., Haq, I.U., & Zia, M. (2016). Effect of ZnO nanoparticles on *Brassica nigra* seedlings and stem explants: growth dynamics and antioxidative response. *Front Plant Sci*. 7, 535.

Zahedi, S.M., Karimi, M., & Teixeira da Silva, J.A. (2020). The use of nanotechnology to increase quality and yield of fruit crops. *J Sci Food Agric*. 100, 25–31.

Zakharova, O.V., Gusev, A.A., Zherebin, P.M., Skripnikova, E.V., Skripnikova, M.K., Ryzhikh, V.E., Lisichkin, G.V., Shapoval, O.A., Bukovskii, M.E., & Krutyakov, Y.A. (2017). Sodium tallow amphopolycarboxyglycinate-stabilized silver nanoparticles suppress early and late blight of *Solanum lycopersicum* and stimulate the growth of tomato plants. *BioNano Sci*. 7, 692–702.

Zarzycki, A. (2014). At source of nanotechnology. *TecnoLógicas*. 17, 09–10.

Zhan, F., Wang, T., Iradukunda, L., & Zhan, J. (2018). A gold nanoparticle-based lateral flow biosensor for sensitive visual detection of the potato late blight pathogen, *Phytophthora infestans*. *Anal Chim Acta*. 1036, 153–161.

Zhang, L., & Gleason, C. (2019). Loop-mediated isothermal amplification for the diagnostic detection of *Meloidogyne chitwoodi* and *M. fallax*. *Plant Dis*. 103, 12–18.

Zhao, X., Meng, Z., Wang, Y., Chen, W., Sun, C., Cui, B., Cui, J., Yu, M., Zeng, Z., Guo, S., Luo, D., Cheng, J.Q., Zhang, R., & Cui, H. (2017). Pollen magnetofection for genetic modification with magnetic nanoparticles as gene carriers. *Nat Plant*. 3, 956–964.

Zhu, Y., Piehowski, P.D., Kelly, R.T., & Qian, W.-J. (2018). Nanoproteomics comes of age. *Expert Rev Proteomics*. 15, 865–871.

3 Microbe-Mediated Synthesis of Nanoparticles

Bhavya Trivedi and Padma Chorol
Maya Group of Colleges

Pankaj Kumar and Alka Rani
Gurukul Kangri University

Sami Abou Fayssal
University of Forestry

Anuj Chaudhary
Shobhit University

CONTENTS

DOI: 10.1201/9781003345565-3

3.1 INTRODUCTION

Nanotechnology (nano – one billionth/dwarf) is the branch of science that deals with matter having the scale of one billionth of a meter ($1\ m = 10^9\,nm$) with the study of material manipulation at molecular and atomic levels. NPs are microscopic substances that can be shaped as 0D, 1D, 2D and 3D. Nowadays, NPs are being used in various fields such as biomedical, electronics, remediation purposes as well as in agriculture to improve plant health and productivity (Kumari et al., 2020; Kukreti et al., 2020; Chaudhary et al., 2021a, b; Bhatt et al., 2022). Nanoparticles are widely recognized in the agricultural sector for their significant contribution to crop protection against pathogens and maintenance of plant and soil health (Chaudhary et al., 2021c, d, e; Agri et al., 2021, 2022). Due to their size, NPs interact with complex biomolecules with a distinctive approach; this allows them to model and develop a versatile characteristic letting them to be employed in agriculture to improve nutrient efficacy, water holding capacity and microbial diversity (Khati et al., 2018, 2019; Chaudhary & Sharma, 2019; Chaudhary et al., 2022). Drug efficiency is also increased through the multidimensional characteristics of NPs. On the other hand, toxicity of drugs can also be decreased by the development of various NPs. The physicochemical properties of biomolecules present within and over the cell are not amended while interacting with these NPs (Gao et al., 2020; Stillman et al., 2020).

Several methods have been developed for physical and chemical synthesis of nanoparticles to attain the goal of increased crop production and protection against phytopathogenic fungi and bacteria. However, they are proving to be very costly and include the application of toxic, poisonous and hazardous chemicals responsible for unusual biological threats (Moghaddam et al., 2015). Several plants, fungi and bacteria or the byproducts of their metabolism can be used for the biological synthesis of inorganic nanoparticles, which act as stabilizing and reducing materials. Biological synthesis is comparatively sustainable, economical, clean, non-toxic, biocompatible and simple, and offers wide adaptability of the nanoparticles (Gholami-Shabani et al., 2014).

Bacteria were among the first organisms utilized for the synthesis of NPs due to their ease of isolation, quick manipulation and existence of built-in mechanisms for metal ion detoxification. *Lactobacillus plantarum* was used for the synthesis of zinc oxide NPs that ranged in size from 7 to 19 nm as reported by Selvarajan and Mohanasrinivasan (2013). There are different bacterial species involved in NP synthesis such as *Bacillus megaterium, Rhodococcus* and *Bacillus licheniformis* (Gomaa, 2022). Consequently, in recent years fungi are smart agents for biological synthesis of metallic nanoparticles, due to their easy handling, high tolerance to metals, efficiency of biomimetic mineralization, high secretion of intracellular or extracellular enzymes, and effective reduction and stabilization of nanoparticles (Netala et al., 2016). Currently several species of fungi such as *Aspergillus, Alternaria, Agaricus, Acremonium, Amylomyces, Sclerotium, Trichoderma* and *Verticillium* have been successfully exploited for the production of biologically effective metallic nanoparticles (Khan et al., 2018). This chapter focuses on the microbial-assisted synthesis of NPs and variables influencing the process; their potential applications in agricultural field have also been investigated.

Algae is a plentiful and widely dispersed organism, their capacity to thrive in a lab setting is an additional benefit and offers low-cost assistance in large-scale production. Silver NP synthesis using algae is interesting due to its strong capacity to absorb metals and decrease metal ions. In order to create NP synthesis, the most popular types of algae are *Phaeophyceae, Rhodophyceae* and *Chlorophyceae* (Chugh et al., 2021). Silver NPs were synthesized using *Nannochloropsis oculata* and *Chlorella vulgaris* algae species only at a dosage of 1 mM as reported by Dahoumane et al. (2017). Plant viral capsid proteins can self-assemble and generate well-organized icosahedral viruses with variable coat protein components and can also help produce unique nanoparticles. Virus-like NPs have found use in a variety of sectors such as optics, nano catalysis and biosensing (Ma et al., 2012). This chapter discusses the microbially mediated synthesis of nanoparticles, the methods used to characterize them, the variables that affect them and the uses for them.

NPs are prepared via two routes: top-down and bottom-up modes. Microbial synthesis is included in bottom-up method while chemical synthesis follows the top-down method. Microbial synthesis is more convenient than the chemical one due to the former's sustainability and cost-effectiveness. Furthermore, the size and shape of nanoparticles can be controlled via the biological method used for NPs' production (Senapati et al., 2004). As it is well acknowledged, NPs can be produced by microbes intra- and extracellularly. In case of extracellular production, the product's recovery is easier than in the intracellular route avoiding downstream processing. NPs are predominantly classified into various categories on the basis of their characteristics such as morphology, size and chemical properties.

3.2 CLASSIFICATION OF NANOPARTICLES ON THE BASIS OF SYNTHESIS

There are two general approaches for the synthesis of nanoparticles (Figure 3.1):

1. Top-down method
2. Bottom-up method

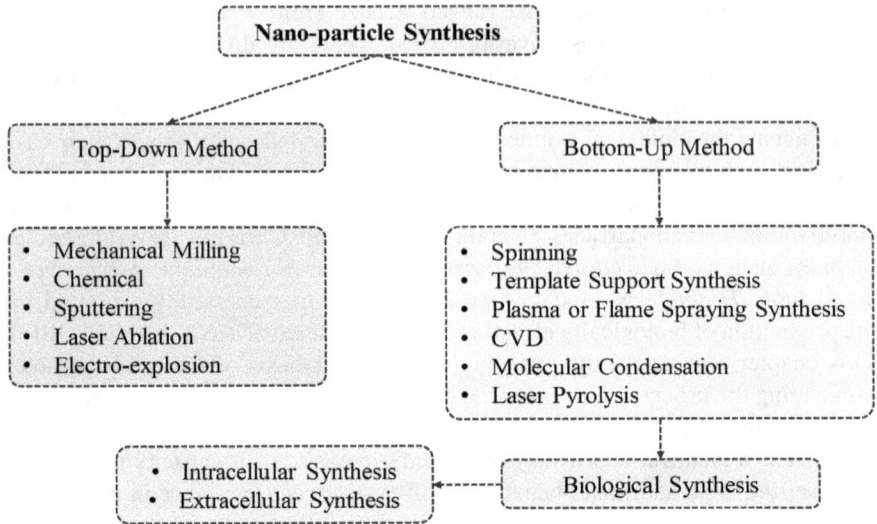

FIGURE 3.1 Methods used for the fabrication of NPs.

3.2.1 TOP-DOWN METHOD

Solid substances could be broken down into nano-sized particles by using this method. This method could be further classified as (1) dry and (2) wet grinding approaches.

3.2.1.1 Dry Grinding

In the process of grain refining, the accumulation of particles increases due to the rise in interfacial energy of the particles. In this method, the solid substance breaks due to shock, compression or friction following fluid energy mill, hammer mill, roller mill, shock shearing mill and tumbling mill. However, through this method it is difficult to obtain particles smaller than 3 nm in size due to simultaneous pulverization and condensation of small particles.

3.2.1.2 Wet Grinding

Unlike dry grinding method, greatly scattered NPs could be obtained since grinding of a substrate is executed through centrifugal fluid mill, agitating bead mill, annular gap bead mill, flow conduit bead mill, wet jet mill, vibratory ball mill, tumbling ball mill or planetary ball mill. The wet grinding approach suitably prevents the condensation of the resulting NPs.

3.2.2 BOTTOM-UP METHOD

Bottom-up method involves the development of material from the bottom: atom by atom, molecule by molecule or cluster by cluster. Through this method, the synthesis of NPs involves a miniaturization of material to an atomic level with further lead to a self-assembly. In such manner, the starting material is either in gaseous or liquid state, on the basis of which this method can be further classified as gaseous-phase approach or liquid-phase one.

The gaseous-phase approach uses the chemical vapor deposition (CVP) method, which involves chemical reactions, and physical vapor deposition (PVD) method, which requires cooling of vaporized materials. In comparison with the liquid-phase approach, the gaseous-phase approach minimizes the appearance of organic adulterants in the particles, although it involves the use of complex vacuum equipment that has low productivity and high cost. As a result of chemical reactions occurring in the gaseous phase, particles of size less than 1 nm can be produced by the CVD method. By careful control of the reactions, 10- to 100-nm-sized NPs can also be synthesized. Heat sources such as chemical flame, electric furnace, laser or plasma processes are required to enable high-temperature chemical reactions via the CVD method. On the other hand, in PVD method, liquid or solid material is evaporated using an arc discharge method; then, the vapor is cooled down, producing the desired NPs (Horikoshi & Serpone, 2017). In the bottom-up synthesis process, nanoparticles are synthesized using biological systems, i.e., microorganisms, plant extracts and even human cells. Gold nanoparticles (AuNPs) have been reportedly synthesized using plant extracts and microorganisms as reducing agents and also from biomass of oat and wheat crops (Azmath et al., 2016; Parveen et al., 2016).

Among the top-down and bottom-up methods, the bottom-up process is more accepted in NP synthesis due to numerous advantages, i.e., lesser defects, more homogeneous chemical composition, potency of creating less waste and hence more economically desired.

3.3 SCHEME FOR BIOPRODUCTION OF NPs

The bioproduction of NPs is much more advantageous in comparison with chemical and physical methods. Bacteria are considered more convenient compared to other microorganisms for the production of NPs as they can easily adopt the artificial environment with a rapid growth rate and can be easily genetically engineered (Saravanan et al., 2018a; Salem & Fouda, 2021). Microorganisms can survive in extreme metal concentrations and have the potency to reduce inorganic materials into NPs via two strategies: extracellular and intracellular. A microorganism absorbs metal ions from the artificial environment and then converts them into elemental ions with the help of enzymatic reduction (Li et al., 2011).

In extracellular mode, microorganisms are grown in suitable medium. After incubation, the broth is centrifuged and divided into two phases: supernatant and pellets. Microbial enzymes are present in supernatant and are subsequently used for NP production. Reductase enzymes are transferred to the vessel containing metal ions, which are then converted into NPs (Jeevan et al., 2012). In intracellular mode, the cellular machinery of microorganisms is used for the production of NPs. Via this method, a microorganism is inoculated in liquid media and after incubation, the microbial mass is washed with distilled water followed by centrifugation to obtain the pellets (Fernández-Llamosas et al., 2017).

Microbial mass is then allowed to react with the aqueous solution of metals. The biomass along with metal solution is allowed to incubate at optimum environmental

conditions until a specific chromatic change is observed. The appearance of a specific color indicates the production of specific NPs, i.e., if pale yellow to pinkish color
is observed, the formation of gold NPs is conclusively confirmed (Waghmare et al.,
2015). As acknowledged, the bacterial cell wall is negatively charged due to lipopoly-
saccharides (LPS), resulting in the immobilization of the positively charged metals
within the cell wall. The reduction of these immobilized metal ions occurs with the
help of microbial enzymes leading to the formation of nano clusters; thus, NPs diffuse
from the cell wall into the solution (Marooufpour et al., 2019). The characteristics
of NPs are quantified by X-Ray Diffraction (XRD), Scanning Electron Microscopy
(SEM), Transmission Electron Microscopy (TEM) and Fourier Transform Infrared
Spectroscopy (FTIR) methods.

3.4 CHARACTERIZATION OF NANOPARTICLES

Various techniques are used for the determination of NPs' physicochemical properties (Figure 3.2). These are TEM, SEM, XRD and so on. The physicochemical
characterization of NPs affects most NP characteristics. A vast variety of characterization techniques are used for morphological research on NPs. In comparison
with the study on bulk material, the perceptible physicochemical characterization
of nanoparticles, i.e., size, shape, composition, surface properties, stability and

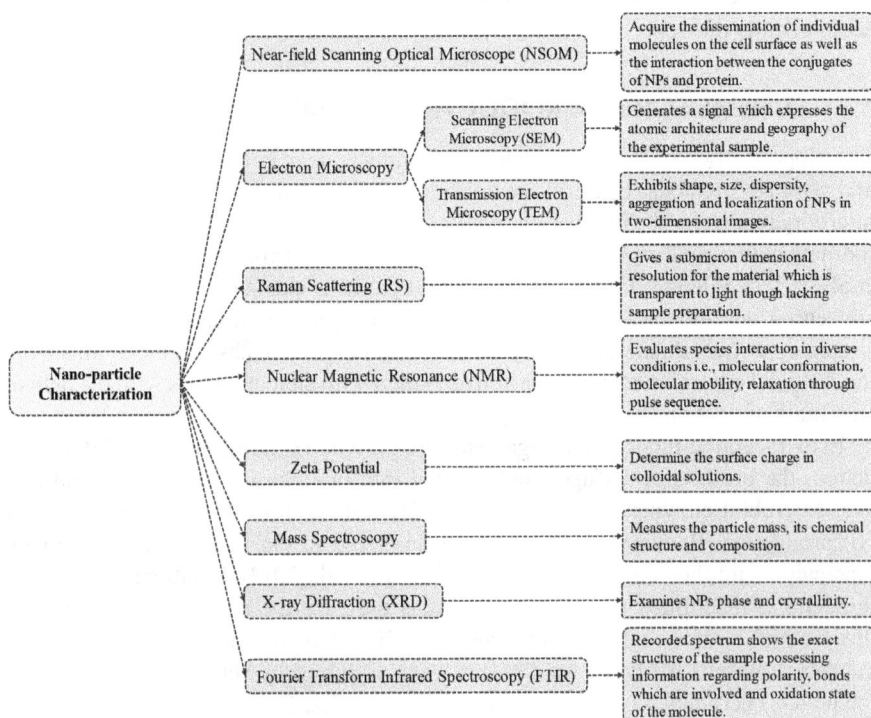

FIGURE 3.2 NP characterization methods.

solubility, is specifically significant to peculiar physiological interaction (Patri et al., 2006). These reactions help in understanding the improvement of various processes, i.e., improving drug efficiency, reduction in side effects of drugs, interaction and medication (Hall et al., 2007).

The influence of NPs on their physiological behavior controls the therapeutic efficiency of drugs. Therefore, it is prime to perceive how the various physicochemical properties of NPs influence their behavior and distribution. Various trustworthy and potent methods are required for the study of NPs; they are selected on the basis of NP features. The following techniques are used to determine the peculiar physicochemical characteristics of NPs:

3.4.1 Near-Field Scanning Optical Microscope (NSOM)

NSOM is a technique of characterization that works on the principle of optical microscopy and Scanning Probe Microscopy (SPM) to exceed the limit of far-field resolution (Hayazawa et al., 2012). In NSOM, laser light passed through an optical fiber emerges via the tip fissure at the vicinity of an object. The spatial resolution is a function of the fissure size due to the disappearance of the distance to the object light emitting from the fissure; i.e., light field is extremely compacted and limited at the fissure (Hayazawa et al., 2012). NSOM can learn about the distribution of individual molecules on the cell surface as well as the interaction between the conjugates of NPs and protein (Ianoul & Johnston, 2007; Song et al., 2011; Hinterdorfer et al., 2012).

3.4.2 Electron Microscopy

Electron microscopy is used to study the ultracellular components. This method uses a focused electron beam as the light source for examining an object. Because of its high resolution power (0.001 μm), TEM and SEM are routinely used for the evaluation of NP shape, size and surface.

3.4.2.1 Scanning Electron Microscopy (SEM)

SEM is a surface illustrating technique in which the sample's surface is scanned by incident electron beam. After interaction with the sample, it generates a signal that expresses the atomic architecture and geography of the experimental sample (Ratner et al., 2004). Two types of electrons are emitted by incident electrons: backscattered electrons, which are the elastic scattering electrons, and inelastic scattering electrons, which are also known as low-energy secondary electrons. The shape and size of NPs can be explicitly obtained by SEM. Nevertheless, contrast and drying are responsible for the contraction of specimen and change in NP properties (Hall et al., 2007). An ultrathin layer of electrically conducting material is necessary to avoid any false reflection of the sample's image (Suzuki, 2002).

3.4.2.2 Transmission Electron Microscopy (TEM)

TEM exhibits information regarding the shape, size, dispersity, aggregation and localization of NPs in two-dimensional images. An electromagnetic lens is used for focusing an electron beam into a thin part of the sample. Then, the electron beam is passed through the specimen where the electron may scatter or permeate the sample and hit a fluorescent screen at the bottom of the microscope. Both TEM and SEM can show the shape and size of NPs, although the former provides a better resolution than the latter. However, TEM also has some drawbacks because of its time-consuming preparation process besides the intensive high-voltage electron beam that can damage the spectrum (Chen & Wen, 2012). The morphological identification of silver NPs produced by *Corynebacterium glutamicum* has been studied by Gowramma et al. (2015). TEM analysis showed that the size of spherical-shaped AgNPs was 15 nm. In another study, the production of CuONPs was a result of *Streptomyces* sp. activity. The particle's shape was spherical with a 1.72–13.49 nm diameter determined by TEM (Bukhari et al., 2021). A change in the solution color was also expressed by the formation of NPs, which has been reported in various biological systems. The size of gold NPs produced by *Stenotrophomonas maltophilia* has been identified as around 40 nm by TEM analysis (Nangia et al., 2009).

3.4.3   RAMAN SCATTERING (RS)

RS helps in NPs' characterization by giving a submicron dimensional resolution for the material, which is transparent to light though lacking sample preparation (Jones et al., 2019). RS is based on the quantification principle of inelastic photon scattering, having discrete frequencies emitted from the previously emerged incident light after the reaction with electric dipoles of molecules (Popovic et al., 2011).

3.4.4   NUCLEAR MAGNETIC RESONANCE (NMR)

NMR is used to evaluate species interaction in diverse conditions, i.e., molecular conformation, molecular mobility and relaxation through pulse sequence (Tiede et al., 2008; Sapsford et al., 2011). Various physicochemical characteristics can be identified by NMR such as purity and functionality in dendrimer structure. Any change after interaction between NPs and ligand can be determined via NMR (Patri et al., 2006).

3.4.5   ZETA POTENTIAL

Zeta potential is used to determine the surface charge in colloidal solutions. A thin (stern) layer is produced when NP surface charge attracts the opposite charge. At the end of the stern layer, NP zeta potential analyzers are obtained. NP stability is generally associated with increasing zeta potential (Sikora et al., 2015; Kumar & Dixit, 2017). At −200 to +200 mV, zeta potential analysis corresponds to a scattered angle of 173°. A negative zeta potential was observed when silver NP produced by *Streptomyces albidoflavus* was characterized (Sadowski et al., 2008; Shetty et al., 2014).

3.4.6   MASS SPECTROMETRY

A particle's mass, its chemical structure and composition are estimated by mass spectrometry. On the basis of discrete masses, NP mass over charge ratio is used to differentiate the charged particles. Diverse structural and compositional proper-ties of NPs are examined via MS, leading to their differentiation based on the ion source, separation technique and detector system (McNaught & Wilkinson, 1997; Gmoshinski et al., 2013).

3.4.7 X-RAY DIFFRACTION (XRD)

XRD is used to examine NP phase and crystallinity. It consists of the irradiation of the material through X-ray and its scattering angle; then the intensity is measured. The three-dimensional structure of crystalline material is determined via XRD (Sapsford et al., 2011). The crystalline structure of gold NP produced by *Klebsiella pneumoniae* was detected using this technique. The sample was already prepared using a reduced AuNP solution, which is first drop-coated on a glass surface and then the equipment was processed at a 20 mA voltage (Bennur et al., 2016). In another study, *Bacillus cereus* was used for the production of AgNPs. X-ray diffraction revealed four peaks in a spectrum of 2θ value ranging between 20 and 80 (Ibrahim et al., 2021).

3.4.8 FOURIER TRANSFORM INFRARED SPECTROSCOPY (FTIR)

FTIR is an approach that resolves the surface chemical composition of NPs. The basic principle of this technique is illuminating the particle with infrared light; then the transmitted or absorbed radiation is documented. Recorded spectrum shows the exact structure of the sample possessing information regarding polar-ity, bonds that are involved and oxidation state of the molecule. Contaminants are also identified by FTIR (Manor et al., 2012; Chevali & Kandare, 2016; Deepty et al., 2019). In the biological synthesis of NPs, FTIR is used for the recogni-tion of functional groups situated on the surface of NPs, which are responsible for the reduction and stabilization of the latter (Rodrigues et al., 2020). Menon et al. (2017) characterized gold NPs via FTIR. They identified the biomolecules that can be used as natural capping for AuNP stabilization. Jayaseelan et al. (2013) also applied FTIR for the characterization of AuNP produced by the aqueous extract of *Abelmoschus esculentus* seed.

3.5 BACTERIA-MEDIATED FABRICATION OF NPs

Bacteria are more advantageous in comparison with other organisms used for the synthesis of NPs due to their special characteristics, i.e., high growth rate, easy manipulation, abundant availability and more efficient adaptation to the environment. Bacteria have a tremendous ability to reduce heavy metal concentrations and produce NPs. They can produce NPs both extracellularly and intracellularly, although the former is applied more conveniently than the latter as there is no need for DSP (Singh et al., 2013). This process is summarized in Figure 3.3. Extensive researches have been performed regarding the fabrication of NPs through bacteria. These NPs are used in various biomedical fields including biosensor devices (Marooufpour et al., 2019) besides bio-tagging and cell imaging (Pal et al., 2018).

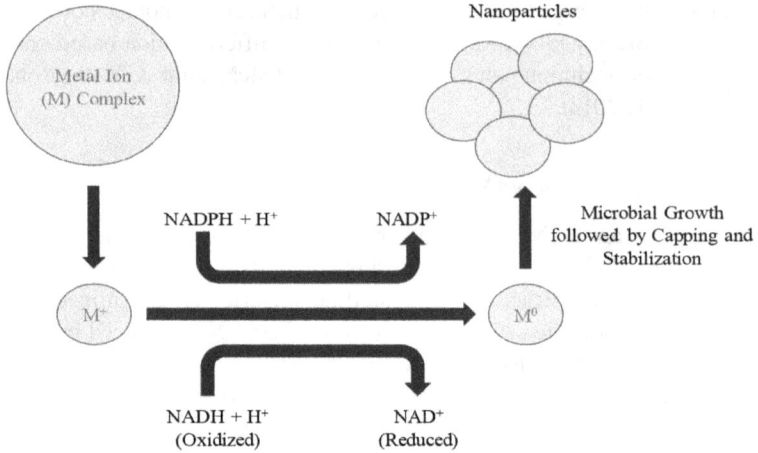

FIGURE 3.3   Bacteria-mediated fabrication of NPs.

3.5.1   GOLD NANOPARTICLES (AuNPs)

To the best of our knowledge, AuNPs have immense applications in electronics, bio-technology and biomedical fields. Bacteria can produce a diversity of NPs having different sizes, shapes and morphologies. Bacterial heterogeneity was found closely related to the fabrication of AuNPs. It was reported that AuNPs can be isolated from marine organisms. That study found that marine bacteria produced AuNPs sized less than 10 nm owing to different shapes that are changeable in each short period. Anticancer and antioxidant properties of AuNPs were also determined (Sharma et al., 2012; Shunmugam et al., 2021). Another report described the use of thermo-philic bacteria *Geobacillus* as a potent bacterium for synthesis of AuNPs. The micro-bial synthesis of these AuNPs was accomplished enzymatically and using NADH as a co-factor (Correa-Llantén et al., 2013). Srinath et al. (2018) explored the biological fabrication of 20- to 25-nm-sized AuNPs with the help of *Bacillus subtilis*. They disclosed the application of AuNPs in bioremediation of toxic dye.

3.5.2   SILVER NANOPARTICLES (AgNPs)

Silver NPs have a broad-spectrum use. A vast diversity of bacteria can produce AgNPs, i.e., *Escherichia coli* (Arshad et al., 2022), *Lactobacillus* (Suba et al., 2022), *B. cereus* (Kumar et al., 2020), *Pseudomonas* (Singh et al., 2018), *Corynebacterium* (Gowramma et al., 2015) and *Klebsiella* (Sayyid & Zghair, 2021). The first report on the production of AgNPs returns to Klaus et al. (1999). *Pseudomonas stutzeri* was used for the fabrication of AgNPs and it was reported that the periplasmic space of this Gram-negative bacteria was the site accumulator of silver atom responsible for the synthesis of different AgNP sizes. The environmental condition would also be responsible for the formation of different-sized NPs (Klaus et al., 1999). Gowramma et al. (2015) used *C. glutamicum* to produce AgNPs and characterized them by UV-vis spectroscopy, SEM, TEM and energy-dispersive X-ray analyses. They also determined an antimicrobial potential of AgNPs against various bacteria. Sayyid and Zghair (2021) produced AgNPs from *K. pneumonia* originating from human and sheep. A chromatic change was detected after NP synthesis. The fabrication of AgNPs through *Lactobacillus* sp. was confirmed by UV spectrum, and their size and shape were determined by SEM. Antimicrobial, anticancer and sensing catalysis potential of AgNPs were also checked (Suba et al., 2022).

3.6 FUNGUS-MEDIATED BIOSYNTHESIS OF NANOPARTICLES

In comparison with bacterial cell, a more viable and greater number of NPs can be produced from fungi, due to their greater potency to tolerate metal concentrations, potency of bioreduction, cell wall binding with metal ions and greater potential to accumulate metals and hydrolyze metal ions. The extracellular synthesis of NPs from fungal species has attracted researchers as the resultant NPs are devoid of any cellular attachment, which makes their purification easier. Fungi produce large numbers of enzymes and proteins, from which more sustainable and reliable NPs can be synthesized. Hence, they are more advantageous for NP biosynthesis compared to other microorganisms (Vahabi et al., 2011; Alghuthaymi et al., 2015). Various fungal species have been used for the synthesis of AuNPs, AgNPs, and other metal and metal oxide NPs.

3.6.1 AuNP Synthesis from Fungi

According to the literature, AuNP fabrication is significantly affected by tempera ture, pH, incubation time and salt concentration. The synthesis of AuNPs extracellularly from two fungal species, i.e., *Hypocrea lixii* and *Trichoderma viride*, besides temperature and pH effects on these AuNPs synthesis were studied. AuNPs originating from cell extract of *T. viride* were synthesized at a lower temperature (30°C) and higher pH (9) compared to that from cell extract of *H. lixii* (100°C and 7, respectively) (Mishra et al., 2013). Hemo-compatible and stable spherical-shaped 16- to 19-nm-sized AuNPs were synthesized from the cell surface proteins of *Rhizopus oryzae* sp. (Kitching et al., 2016); 10- to 30-nm-sized AuNPs were also synthesized from edible fungal species, i.e., *Pleurotus ostreatus*, where maximum yield was obtained at pH 3 (El Domany et al., 2018). Other fungal species, i.e., *Fusarium solani*, *Aspergillus* sp., *Cladosporium* sp. and *Colletotrichum* sp., have been exploited for green synthesis of AuNPs (Shankar et al., 2003; Clarance et al., 2020; Elegbede et al., 2020; Munawer et al., 2020). For the first time, Zhang et al. (2016) biosynthesized triangular, spherical and hexagonal AuNPs from *Magnusio mycesingens* (yeast). They found that these AuNPs had a great reducing potency against nitrophenols.

3.6.2 AgNP Synthesis from Fungi

The fungi-mediated biosynthesis of AgNPs may be extracellular or intracellular. The extracellular mechanism results in the production of free NPs in the dispersion, since metal precursor is added directly to the aqueous filtrate already containing the biomolecules of fungal species only (Sabri et al., 2016; Costa Silva et al.,

2017), whereas in terms of intracellular mechanism, the metal precursor is first added to the mycelial culture and then incorporated into the biomass. Accordingly, after the synthesis process, NPs' extraction is crucial for the latter's release by disrupting biomass (Rajput et al., 2016; Molnár et al., 2018). Since no extraction is required in the extracellular mechanism, this method is more viable and has been used widely (Azmath et al., 2016; Gudikandula et al., 2017). AgNPs have been synthesized from the fungal species *Fusarium oxysporum* and the effect of different pH medium on the shape of these AgNPs has been studied by Rajput et al. (2016). It was observed that spherical, rod-shaped, triangular and irregularly shaped AgNPs were obtained at pH 3, while spherical and monodisperse AgNPs were obtained at pH of 5 and 7, respectively. On the other hand, a combination of spherical and oblong AgNPs was obtained at pH 9. AgNPs have also been synthesized from other fungal species, i.e., *Penicillium polonicum*, *Trichoderma* sp., *Schizophyllum radiatum*, *Aspergillus* sp., marine yeast *Yarrowia lipolytica*, *Candida* sp. and genetically modified yeast *Pichia pastoris* (; Apte et al., 2013; Waghmare et al., 2015; Elahian et al., 2017; Neethu et al., 2018; Ramos et al., 2020). Among those fungal species exploited for AgNP biosynthesis, filamentous fungi attracted the researchers' attention owing to their great morphological characteristics and stability of synthesized AgNPs (Mukherjee et al., 2008). Many *Aspergillus* sp. have been exploited for AgNP biosynthesis. Wang et al. (2021) synthesized eco-friendly polydisperse spherical-shaped (1–24 nm) AgNPs originating from the extracellular supernatant of *Aspergillus sydowii*. They reported that these AgNPs had an antifungal activity against many pathogenic fungal species and an antiproliferative activity against MCF-7 and HeLa cells.

3.7   ALGAE-MEDIATED BIOSYNTHESIS OF NPs

Algal species are a great source of secondary metabolites; they can be easily cultivated, grow fast, can accrue heavy metal ions and can be modified into more pliable forms. The source of secondary metabolites obtained from algal species can potentially reduce and balance the metal precursors forming metallic, metal oxide and bimetallic NPs. The latter can be synthesized from algae intracellularly or extracellularly, based on the site of reducing agents and the location of synthesis. Among the various forms of algal species, the phaeophytes (brown), rhodophytes (red), bluegreen algae and micro- and macro-green algae are the most frequently studied for NP production.

3.7.1   AuNP Synthesis from Algae

Various algal species were reportedly used for AuNP synthesis. Among them *Turbinaria conoides* is considered the most prominent type of brown algae exploited for AuNP production. AuNPs of different shapes, i.e., spherical, triangular, rectangular and polydispersed, were produced by extracellular pathway from *T. conoides* using gold ion precursor and chloroauric acid (Khodashenas & Ghorbani, 2014). The brown algal species *Laminaria japonica* is another important species used in the biosynthesis of AuNPs. This species is an excellent fount of peptides, polyphenols, proteins, vitamins, fibers and carotenoids (Kushnerova et al., 2010). Functional groups combined with phytochemicals can act as reducing and capping agents. Spherical-shaped AuNPs (15–20 nm) were also synthesized extracellularly from *L. japonica* (Ghodake & Lee, 2011). Other brown algal species reported in the biosynthesis of AuNPs are *Fucus vesiculosus, Padina gymnospora, Dictyota bartayresiana, Sargassum myriocystum, Sargassum wightii, Ecklonia cava, Stoechospermum marginatum* and *Cystoseira baccata*. By using chloroauric acid as a precursor salt from red algae *Lemanea fluviatilis*, cubic close-packed and polydispersed crystalline AuNPs (5.9 nm) were synthesized (Singh et al., 2015; Murugesan et al., 2017). Another study reported the synthesis of spherical-shaped AuNPs extracellularly using reducing agents such as phenol, carbonyl and hydroxyl moiety from *Corallina officinalis* (El-Kassas & El-Sheekh, 2014). A simplistic biosynthesis of AuNPs has been reported from other red algal species, i.e., *Chondrus crispus, Kappaphycus alvarezii* and *Galaxaura elongata* (Castro et al., 2018). Numerous studies reported the involvement of protein as well as peptide as reducing agents in extracellular synthesis of cubic-, spherical- and octahedral-shaped AuNPs from a blue-green algae species *Spirulina platensis* (Kalabegishvili et al., 2012a, b; Iravani et al., 2018). Another type of AuNP (intracellular mono-dispersive triangular-shaped) was reportedly synthesized from blue-green algal species *Phormidium valderianum* at a wavelength of 530 nm using UV-visible spectroscopy (Mosulishvili et al., 2002). Some studies reported the use of cytoplasmic metabolites as reducing agents in the synthesis of 24-nm-sized hexagonal, spherical as well as cubic close-packed AuNPs produced extracellularly from *P. valderianum* (Parial et al., 2012; Khan et al., 2019). *Pithophora crispa* was widely used for AuNP synthesis using auric acid as a precursor besides extra- and intracellular peptides and proteins (Xie et al., 2007). *Prasiola crispa* and *Rhizoclonium fontinale*, members of green macro algae, were reportedly used for AuNP production (Parial & Pal, 2014). Due to their non-reproducibility at a perfect shape and size, the synthesis of AuNPs has been challenging. However, it is now possible to synthesize stable and reproducible AuNPs from green macro algae (Parial & Pal, 2014; Priyadharshini et al., 2014).

Besides monometallic NPs, bimetallic Ag-Au NPs have also been reportedly biosynthesized from red algae *Gracilaria edulis* (Al-Naamani et al., 2017) and from *S. platensis* using extracellular proteins (Mahdieh et al., 2012; González-Ballesteros et al., 2017).

3.7.2 AGNP BIOSYNTHESIS FROM ALGAE

Almost three-fourths of the data reported in the literature regarding biosynthesis of metallic NPs are those of algae-mediated biosynthesis origin. Due to their surpassing physicochemical characteristics over their bulk forms, AgNPs showed a significant value in different industries, i.e., textile, dental alloy, paints, jewelry and medical industries (Vaidyanathan et al., 2009). Various brown algal species, i.e., *Gelidiella acerosa, Desmarestia menziesii, Sargassum polycystum, Padina pavonica, Cystophora moniliformis* and *T. conoides*, were reported for AgNP biosynthesis (Ghodake & Lee, 2011; Azizi et al., 2013). The 96-nm-sized and spherical-shaped

AgNPs showed an extensive antibacterial action upon *E. coli*, *Candida albicans*, *Staphylococcus aureus* as well as *Staphylococcus epidermidis* synthesized extracellularly from *T. conoides*. The organic components, i.e., free hydroxyl, carbonyl, amines and polyamines from *Turbinaria* sp. (*T. ornata* and *T. conoides*) used in AgNP biosynthesis, served reportedly as reducing agents of silver's salt precursor (Khalil et al., 2014; Kumar-Krishnan et al., 2015).

The biosynthesis of AgNPs from various red algal species has also been reported in the literature. The size and shape of NPs help in their biomedical applications; 20- to 60-nm-sized and spherical-shaped AgNPs were successfully synthesized from various red algal species (Pugazhendhi et al., 2017). These extracellularly synthesized AgNPs play a significant role in the medical field, since they exhibit an antimicrofouling activity (Pugazhendhi et al., 2017; Vadlapudi & Amanchy, 2017).

Spherical-shaped (2- to 8-nm-sized) AgNPs have been reportedly synthesized from *S. platensis*, a cyanobacteria with multicellular trichomes. *S. platensis* is a great source of natural vitamins, iron, beta carotene, indispensable amino acids as well as fatty acids, which possibly help in the reduction as well as capping processes of NPs (Doshi et al., 2007). AgNPs of various shapes and sizes have been reportedly synthesized from blue-green algal species such as *Microchaete diplosiphon* (round-shaped NPs of 80 nm size), spherical-shaped and 10- to 25-nm-sized NPs from *Oscillatoria willei*, octahedral-shaped and 200-nm-sized NPs from *Plectonema boryanum* and pentagonal-shaped and 38- to 88-nm-sized NPs from *Cylindrospermum stagnale* (Husain et al., 2015). AgNPs have also been reportedly synthesized from green algae (both micro and macro), which when examined with various microscopic and spectroscopic techniques, i.e., SEM, DLS, EDX, XRD and FTIR, displayed variable species-dependent physicochemical characteristics (Lengke et al., 2006a; Mubarak-Ali et al., 2011; Jena et al., 2014).

3.8 VIRUS-MEDIATED FABRICATION OF NPs

Viruses are obligate intracellular parasites. They depend on their host for their growth as they don't possess enzymes. Nowadays, viruses are used as a template for manufacturing hybrid nanomaterials to nano smart devices (Steinmetz and Manchester, 2011; Khan et al., 2013). A broad diversity of animal and plant viral NPs have been studied and applied in medical technology (Lewis et al., 2006; Fischlechner & Donath, 2007; Young et al., 2008). Viruses have a protein covering outside their genetic material made up of capsomeres and called capsid. A capsid is used as biotemplate in metal fabrication (Merzlyak & Lee, 2006; Nam et al., 2006; Lee et al., 2009). It is clear that the formation of NPs from viruses is directly influenced by the size of capsomeres (Wang et al., 2002; Alexandridis, 2011). Various characteristics of plant viruses such as isotropy, gene diversity and susceptibility are used for the formation of NPs. Due to these characteristics, plant viruses are more prominently used for NP fabrication (Steinmetz & Manchester, 2011). Tobacco mosaic virus (TMV) particles were used for the fabrication of TMV/ zinc oxide field effect transistor (Sanctis et al., 2015). In another study, Au- and AgNPs were fabricated using squash leaf curl China virus as a template (Singh et al., 2008) (Table 3.1).

TABLE 3.1

Various NPs Produced by a Broad Range of Microorganisms

Microorganisms	Nanoparticle Produced	Shape, Size	Reference
Bacteria			
Bacillus subtilis	Titanium dioxide	10–30 nm	Dhandapani et al. (2012)
Bacillus mycoides	Titanium dioxide	Spherical, 40–60 nm	Órdenes-Aenishanslins et al. (2014)
Lactobacillus sp.	Titanium dioxide	50–100 nm	Ahmad et al. (2013; 2014)
Escherichia coli	Silver	5–50 nm	Saeed et al. (2020)
Thermophilic *Bacillus* sp.	Silver	9–32 nm	Deljou and Goudarzi (2016)
Gordonia amicalis	Silver	5–25 nm	Sowani et al. (2016)
Lactobacillus acidophilus	Silver	45–60 nm	Namasivayam et al. (2010)
Corynebacterium glutamicum	Silver	Irregular, 15 nm	Gowramma et al. (2015)
Bacillus cereus	Silver	Spherical, 4–5 nm	Kalimuthu et al. (2008)
Thermomonospora	Silver	8 nm	Ahmad et al. (2003)
Ureibacillus thermosphaericus	Silver	50.7 nm	Juibari et al. (2011)
Bacillus licheniformis	Silver	50 nm	Babu and Gunasekaran (2009)
Brevibacterium casei	Silver and gold	Spherical, 10–50 nm	Kalishwaralal et al. (2010)
Acinetobacter sp.	Gold	15–40 nm	Wadhwani et al. (2016)
Lactobacillus kimchicus	Gold	5–30 nm	Markus et al. (2016)
Paracoccus haeundaensis	Gold	20.93 nm	Patil et al. (2019)
Pseudomonas aeruginosa	Gold	15–30 nm	Husseiny et al. (2007)
Micrococcus yunnanensis	Gold	53.8 nm	Jafari et al. (2018)
Rhodococcus sp.	Gold	Spherical, 5–15 nm	Ahmad et al. (2003)
Rhodopseudomonas capsulata	Gold	Spherical, 10–20 nm	He et al. (2007)
Shewanella oneidensis	Gold	10–20 nm	Konishi et al. (2007)
Shewanella algae	Gold	Spherical, 12–17 nm	Suresh et al. (2011)
Mycobacterium sp.	Gold	5–55 nm	Camas et al. (2018)
Aeromonas hydrophila	Zinc oxide	57.7 nm	Jayaseelan et al. (2013)
Lactobacillus plantarum	Zinc oxide	7–19 nm	Selvarajan and Mohanasrinivasan (2013)
Lactobacillus sporogenes	Zinc oxide	145.70 nm	Mishra et al. (2013)
Bacillus licheniformis	Zinc oxide	250 nm	Tripathi et al. (2014)
Bacillus megaterium	Zinc oxide	Rod and cubic, 45–95 nm	Saravanan et al. (2018b)
Pseudomonas aeruginosa	Zinc oxide	Spherical, 35–80 nm	Singh et al. (2014)
Rhodococcus pyridinivorans	Zinc oxide	Roughly spherical, 100–120 nm	Kundu et al. (2014)
Staphylococcus aureus	Zinc oxide	Acicular, 10–50 nm	Rauf et al. (2017)

(Continued)

TABLE 3.1 (*Continued*)
Various NPs Produced by a Broad Range of Microorganisms

Microorganisms	Nanoparticle Produced	Shape, Size	Reference
Streptomyces sp.	Zinc oxide	Spherical, 20–50 nm	Balraj et al. (2017)
Sphingobacterium thalpophilum	Zinc oxide	Triangular, 40 nm	Rajabairavi et al. (2017)
Fungi			
Fusarium oxysporum	Silver	Spherical, 5–50 nm	Senapati et al. (2004)
Aspergillus fumigatus	Silver	Spherical, 5–25 nm	Bhainsa and D'Souza (2006)
Aspergillus flavus	Silver	Spherical, 8–9 nm	Vigneshwaran et al. (2007)
Trichoderma viride	Silver	Spherical, 5–40 nm	Fayaz et al. (2010)
Phanerochaete chrysosporium	Silver	Pyramidal, 50–200 nm	Vigneshwaran et al. (2006)
Verticillium sp.	Silver	Spherical, 25–32 nm	Senapati et al. (2004)
Candida utilis	Gold	NA	Gericke and Pinches (2006a)
Verticillium luteoalbum	Gold	NA	Gericke and Pinches (2006b)
Yarrowia lipolytica	Gold	Triangular, 15 nm	Agnihotri et al. (2009)
Alternaria alternata	Zinc oxide	Hexagonal, spherical and triangular, 45–150 nm	Sarkar et al. (2014)
Fusarium sp.	Zinc oxide	Triangular, >100 nm	Velmurugan et al. (2010)
Aspergillus aeneus	Zinc oxide	Spherical, 100–140 nm	Jain et al. (2013)
Aspergillus niger	Zinc oxide	Spherical, 61 nm	Kalpara et al. (2018)
Aspergillus fumigatus	Zinc oxide	Spherical and hexagonal, 1.2–6.8 nm	Raliya and Tarafdar (2013)
Aspergillus terreus	Zinc oxide	Spherical, 54.8–82.6 nm	Baskar et al. (2013)
Candida albicans	Zinc oxide	25 nm	Shamsuzzaman et al. (2017)
Algae			
Chlorella vulgaris	Silver	9–20 nm	Jianping et al. (2007)
Oscillatoria willei	Silver	Spherical, 100–200 nm	Mubarak-Ali et al. (2011)
Spirulina platensis	Silver	Spherical, 11.6 nm	Mahdieh et al. (2012)
Pterocladia capillacea	Silver	Spherical, 7 nm	El-Rafie and Zahran (2013)
Jania rubens	Silver	Spherical, 12 nm	El-Rafie and Zahran (2013)
Colpomenia sinuosa	Silver	Spherical, 20 nm	El-Rafie and Zahran (2013)
Ulva fasciata	Silver	Spherical, 7 nm	El-Rafie and Zahran (2013)

(Continued)

TABLE 3.1 (*Continued*)
Various NPs Produced by a Broad Range of Microorganisms

Microorganisms	Nanoparticle Produced	Shape, Size	Reference
Galaxaura elongata	Gold	Hexagonal and triangular, 3.8–77.5 nm	Abdel-Raouf et al. (2017)
Plectonema boryanum	Gold	Cubic, 10–25 nm	Lengke et al. (2006b)
Sargassum wightii	Gold	Planar, 8–12 nm	Singaravelu et al. (2007)
Sargssum myriocystum	Zinc oxide	36 nm	Nagarajan and Kuppusamy (2013)
Chlorella	Zinc oxide	Hexagonal, 19.44 nm	Khalafi et al. (2019)

3.9 FACTORS INFLUENCING THE MICROBIAL SYNTHESIS OF NPs

Various factors must be taken into consideration during synthesis and characterization of NPs. Such factors could be modified to produce shape- and size-desired NPs. The shape and size of the particle are the prime factors in determining NP characteristics. They can also affect the chemical properties of NPs (Baer et al., 2013). The melting point of NPs decreases with decreased particle size (Akbari et al., 2011). The temperature of the reaction is another significant factor that affects the synthesis of NPs and plays an important role in the determination of NP size, shape, stability and yield. The formed nature of NPs is determined by the temperature of the reaction medium. For biological synthesis of NPs, either ambient temperature or a temperature <100°C is required compared to the synthesis via chemical route (it requires a temperature <350°C) and physical route (it requires temperatures usually above 350°C). Since temperature also affects the shape and size of NPs, spherical-shaped ones are chiefly produced at low temperatures while rod-shaped ones are produced at a higher reaction temperature using fungal extract from *Neurospora crassa* (Gericke and Pinches, 2006a).

The pH of the reaction medium greatly affects the size, texture and shape of NPs. Soni and Prakash (2011) reportedly studied the shape and size of fungal species-mediated synthesizing AgNPs and proved that the latter were greatly affected by the pH of the solution medium. Controlled synthesis of NPs of desired size could be made by varying the pH of the solution medium already used for the synthesis process. Large-sized NPs are generally synthesized in acidic medium (Dubey et al., 2010). pH alters the electrical charges of biomolecules, which could in turn alter their capping and reducing properties, thereby affecting NPs synthesis (Khalil et al., 2014). Pressure is another important factor that affects NPs. The shape and size of NPs could be affected by the pressure applied to the reaction medium. It has been reported that the reduction of metal ions occurs at a much faster rate at ambient pressure (Tran et al., 2013).

The time span for which the reaction medium is incubated reportedly influences the quality, properties and type of NPs synthesized by microbe-mediated processes as in the case of green synthesis of AgNPs using glucose and collagen (Darroudi et al., 2011). The properties of NPs synthesized by the biological route are also altered with time and affected by light exposure and storage conditions (Kuchibhatla et al., 2012). The nature of NPs synthesized is greatly influenced by the surrounding environment. In microbe-based synthesis process, NPs form a layer that makes them larger and thicker (Lynch et al., 2007). Besides that, the surrounding environment also influences the structure and composition of synthesized NPs. Different intra- and extracellular proteins and enzymes produced in variable quantities by different microorganisms also affect the microbial synthesis of NPs (Haverkamp et al., 2007).

3.10   APPLICATION OF MICROBE-BASED SYNTHESIZED NPs

Synthesized NPs from microorganisms, i.e., bacteria, fungi, algae and viruses, have a wide range of applications in different fields such as electronics, catalysis, energy science, magnetics, biomedicine and so on. Because of their interesting properties, i.e., reliability, feasibility, environmental friendly and non-toxicity, biosynthesized NPs have used widely in recent years over those synthesized by different methods. Microbial cells are regarded as ideal for production of various nanostructures and instruments for nano-sciences, as they possess a small size, a physiological variation, a controlled culture and are manipulatable genetically (Villaverde, 2010).

AgNPs have been used widely in the biomedical industry for drug delivery, as dressings for wound healing, in contraceptive devices, disease diagnosis and treatment (Moore, 2006; Skirtach et al., 2006; Sibbald et al., 2007; Uchihara, 2007; Chen & Schluesener, 2008) and in water treatment, agriculture and antimicrobial gel formulation. Microbe-mediated synthesized AgNPs possess different properties, i.e., antimicrobial activity against various microorganisms inclusive of Gram-positive as well as Gram-negative bacteria, fungi and yeast (Abdeen et al., 2014), antifouling and antibiofilm activities. Because of their antimicrobial activity, green NPs are also mixed in various materials of daily use, i.e., toothpaste, cosmetics, deodorants, humidifiers and water purification systems (Baker et al., 2005). The antibacterial activities of microbe-based NPs were fascinating for their large-scale utilization in textile coatings, health industry, food storage industry and environment. Bacteria-mediated synthesized NPs possess a great larvicidal potential against *Aedes aegypti*, the dengue vector (Debabov et al., 2013).

NPs play a significant role as an agent in site-specific drug delivery therapy due to its potential to change the physical, chemical and biological attributes in relation to surface:volume of the target. Green NPs could also be applied in the treatment of various human disorders, i.e., cardiovascular, Parkinson's and Alzheimer's diseases, diabetes, neurological disorders, tuberculosis and osteoporosis (Gupta et al., 2017). AuNPs have a wide range of applications in the biomedical field, separation sciences, pharmaceuticals and disease diagnosis for therapeutic drug delivery, delivery of genetic material (Paciotti et al., 2004), hyperthermia therapy, and in the area of biosensor technology (Kreibig & Vollmer, 2013), due to their variable size, shape and surface properties. AuNPs synthesized from *Sesbania drummondii* possess a catalytic activity, which could help in the reduction of nitrogen-containing aromatic compounds during water purification process (Zhang et al., 2020).

3.11 CONCLUSION

The microbe-mediated biosynthesis of NPs has been widely used as an alternative and much convenient approach to physical and chemical ones. It is more feasible, cost-effective, non-toxic and environmental friendly. Various metallic and bimetallic NPs have been reportedly synthesized from bacteria, fungi, algae and viruses being mainly applied in biomedicine, electronics, energy science, magnetics and various industries. However, scant studies related to virus-mediated biosynthesis of NPs are available, which could be further explored with new advancing characterization techniques. The application of such NPs in various fields could be also refined.

REFERENCES

Abdeen, S., Geo, S., Sukanya Praseetha, P.K., & Dhanya, R.P. (2014). Biosynthesis of silver nanoparticles from actinomycetes for therapeutic applications. *Int J Nano Dimens.* 5, 155–162.

Abdel-Raouf, N., Al-Enazi, N.M., & Ibraheem, B.M. (2017).Green biosynthesis of gold nanoparticles using *Galaxaura elongata* and characterization of their antibacterial activity. *Arab J Chem.* 10, S3029–S3039.

Agnihotri, M., Joshi, S., Kumar, A.R., Zinjarde, S., & Kulkarni, S. (2009). Biosynthesis of gold nanoparticles by the tropical marine yeast *Yarrowia lipolytica* NCIM 3589. *Mater Lett.* 63, 1231–1234.

Agri, U., Chaudhary, P., & Sharma, A. (2021). In vitro compatibility evaluation of agriusable nanochitosan on beneficial plant growth-promoting rhizobacteria and maize plant. *Natl Acad Sci Lett.* 44, 555–559.

Agri, U., Chaudhary, P., Sharma, A., & Kukreti, B. (2022). Physiological response of maize plants and its rhizospheric microbiome under the influence of potential bioinoculants and nanochitosan. *Plant Soil.* 474, 451–468.

Ahmad, A., Senapati, S., Khan, M.I., Kumar, R., & Sastry, M. (2003). Extracellular biosynthesis of monodisperse gold nanoparticles by a novel extremophilic actinomycete, *Thermomonospora* sp. *Langmuir.* 19, 3550–3553.

Ahmad, R., Khatoon, N., & Sardar, M. (2013). Biosynthesis, characterization and application of TiO_2 nanoparticles in biocatalysis and protein folding. *J Proteins Prot.* 4, 115–121.

Ahmad, R., Khatoon, N., & Sardar, M. (2014). Antibacterial effect of green synthesized TiO_2 nanoparticles. *Adv Sci Lett.* 20, 1616–1620.

Akbari, B., Tavandashti, M.P., & Zandrahimi, M. (2011). Particle size characterization of nanoparticles—a practical approach. *Iran J Mater Sci Eng.* 8, 48–56.

Alexandridis, P. (2011). Gold nanoparticle synthesis, morphology control, and stabilization by functional polymers. *Chem Eng Technol.* 14, 15–38.

Alghuthaymi, M.A., Almoammar, H., Rai, M., Said-Galiev, E., & Abd-Elsalam, K.A. (2015). Myconanoparticles: synthesis and their role in phytopathogens management. *Biotechnol Equip.* 29, 221–236.

Al-Naamani, L., Dobretsov, S., Dutta, J., & Burgess, J.G. (2017). Chitosan-zinc oxidenano composite coatings for the prevention of marine biofouling. *Chemosphere.* 168, 408–417.

Apte, M., Sambre, D., Gaikawad, S., Joshi, S., Bankar, A., Kumar, A.R., & Zinjarde, S. (2013). Psychrotrophic yeast *Yarrowia lipolyticancyc* 789 mediates the synthesis of antimicrobial silver nanoparticles via cell-associated melanin. *AMB Exp.* 3, 32.

Arshad, H., Sadaf, S., & Hassan, U. (2022). De-novo fabrication of sunlight irradiated silver nanoparticles and their efficacy against *E. coli* and *S. epidermidis*. *Sci Rep*. 12, 676.

Azizi, S., Namvar, F., Mahdavi, M., Ahmad, M.B., & Mohamad, R. (2013). Biosynthesis of silver nanoparticles using brown marine macroalga, *Sargassum muticum* aqueous extract. *Materials*. 6, 5942–5950.

Azmath, P., Baker, S., Rakshith, D., & Satish, S. (2016). Mycosynthesis of silver nanoparticles bearing antibacterial activity. *Saudi Pharm J*. 24, 140–146.

Babu, M.M.G. & Gunasekaran, P. (2009). Production and structural characterization of crystalline silver nanoparticles from *Bacillus cereus* isolate. *Colloids Surf B Biointerfaces*. 74, 191–195.

Baer, D.R., Engelhard, M.H., Johnson, G.E., Laskin, J., Lai, J., Mueller, K., Manusamy, P., Thevuthesan, S., Wang, H., & Washton, N. (2013). Surface characterization of nanomaterials and nanoparticles: important needs and challenging opportunities. *J Vac Sci Technol A: Vac Surf*, 31, 050820.

Baker, C., Pradhan, A., Pakstis, L., Pochan, D., & Shah, S. (2005). Synthesis and antibacterial properties of silver nanoparticles. *J Nanosci Nanotechnol*. 5, 244–249.

Balraj, B., Senthilkumar, N., Siva, C., Krithikadevi, R., Julie, A., VethaPotheher, I., & Arulmozhi, M. (2017). Synthesis and characterization of zinc oxide nanoparticles using marine *Streptomyces* sp. with its investigations on anticancer and antibacterial activity. *Res Chem Intermed*. 43, 2367–2376.

Baskar, G., Chandhuru, J., Fahad, K.S., & Praveen, A.S. (2013). Mycological synthesis, characterization and antifungal activity of zinc oxide nanoparticles. *Asian J Pharm Technol*. 3, 142–146.

Bennur, T., Khan, Z., Kshirsagar, R., Javdekar, V., & Zinjarde, S. (2016). Biogenic gold nanoparticles from the Actinomycete *Gordonia amarae*: application in rapid sensing of copper ions. *Sensors Actuators B: Chem*. 233, 684–690.

Bhainsa, K.C., & D'Souza, S.F. (2006). Extracellular biosynthesis of silver nanoparticles using the fungus *Aspergillus fumigatus*. *Colloids Surf B Biointerfaces*. 47, 160–164.

Bhatt, P., Pandey, S.C., Joshi, S., Chaudhary, P., Pathak, V.M., Huang, Y., Wu, X., Zhou, Z., & Chen, S. (2022). Nanobioremediation: a sustainable approach for the removal of toxic pollutants from the environment. *J Hazard Mater*. 427, 128033.

Bukhari, I.S., Hamed, M.M., Al-Agamy, M.H., Gazwi, H.S.S., Radwan, H.H., & Youssif, A.M. (2021). Biosynthesis of copper oxide nanoparticles using *Streptomyces* MHM38 and its biological applications. *J Nanomater*. 2021, 6693302.

Camas, M., Camas, A.S., & Kyeremeh, K. (2018). Extracellular synthesis and characterization of gold nanoparticles using *Mycobacterium* sp. BRS2A-AR2 isolated from the aerial roots of the Ghanaian mangrove plant, *Rhizophora racemosa*. *Indian J Microbiol*. 58, 214–221.

Castro, L., Blázquez, M.L., Muñoz, J.A., González, F. & Ballester, A. (2018). Biological synthesis of metallic nanoparticles using algae. *IET Nanobiotechnol*. 7, 109–116.

Chaudhary, P., Chaudhary, A., Bhatt, P., Kumar, G., Khatoon, H., Rani, A., Kumar, S., & Sharma, A. (2022). Assessment of soil health indicators under the influence of nanocompounds and *Bacillus* spp. in field condition. *Front Environ Sci*. 9, 769871.

Chaudhary, P., Chaudhary, A., Parveen, H., Rani, A., Kumar, G., Kumar, A., & Sharma, A. (2021e). Impact of nanophos in agriculture to improve functional bacterial community and crop productivity. *BMC Plant Biol*. 21, 519.

Chaudhary, P., Khati, P., Chaudhary, A., Gangola, S., Kumar, R., & Sharma, A. (2021a). Bioinoculation using indigenous *Bacillus* spp. improves growth and yield of *Zea mays* under the influence of nanozeolite. *3 Biotech*. 11, 11.

Chaudhary, P., Khati, P., Chaudhary, A., Maithani, D., Kumar, G., & Sharma, A. (2021d). Cultivable and metagenomic approach to study the combined impact of nanogypsum and *Pseudomonas taiwanensis* on maize plant health and its rhizospheric microbiome. *PLoS One*. 16, e0250574.

Chaudhary, P., Khati, P., Gangola, S., Kumar, A., Kumar, R., & Sharma, A. (2021c). Impact of nanochitosan and *Bacillus* spp. on health, productivity and defence response in *Zea mays* under field condition. *3 Biotech.* 11, 237.

Chaudhary, P., & Sharma, A. (2019). Response of nanogypsum on the performance of plant growth promotory bacteria recovered from nanocompound infested agriculture field. *Environ Ecol.* 37, 363–372.

Chaudhary, P., Sharma, A., Chaudhary, A., Khati, P., Gangola, S., & Maithani, D. (2021b). Illumina based high throughput analysis of microbial diversity of rhizospheric soil of maize infested with nanocompounds and *Bacillus* sp. *Appl Soil Ecol.* 159, 103836.

Chen, X., & Schluesener, H.J. (2008). Nanosilver: a nanoproduct in medical application. *Toxicol Lett.* 176, 1–12.

Chen, X., & Wen, J. (2012). In situ wet-cell TEM observation of gold nanoparticle motion in an aqueous solution. *Nanoscale Res Lett.* 7, 1–6.

Chevali, V., & Kandare, E. (2016). Rigid biofoam composites as eco-efficient construction materials. In F. Pacheco-Torgal, V. Ivanov, N. Karak, H. Jonkers (Eds.) *Biopolymers and Biotech Admixtures for Eco-Efficient Construction Materials.* Woodhead Publishing, 275–304.

Chugh, D., Viswamalya, V.S., & Das, B. (2021). Green synthesis of silver nanoparticles with algae and the importance of capping agents in the process. *J Genet Eng Biotechnol.* 19, 126.

Clarance, P., Luvankar, B., Sales, J., Khusro, A., Agastian, P., Tack, J.C., Al Khulaifi, M.M., Al-Shwaiman, H.A., Elgorban, A.M., & Syed, A. (2020). Green synthesis and characterization of gold nanoparticles using endophytic fungi *Fusarium solani* and its in-vitro anticancer and biomedical applications. *Saudi J Biol Sci.* 27, 706–712.

Correa-Llantén, D.N., Muñoz-Ibacache, S.A., Castro, M.E., Munoz, P.A., & Blamey, J.A. (2013). Gold nanoparticles synthesized by *Geobacillus* sp. strain ID17 a thermophilic bacterium isolated from Deception Island, Antarctica. *Microb Cell Fact.* 12, 75.

Costa Silva, L.P., Oliveira, J.P., Keijok, W.J., Silva, A.R., Aguiar, A.R., Guimarães, M.C., Ferraz, C.M., Araujo, J.V., Tobias, F.L., & Braga, F.R. (2017). Extracellular biosynthesis of silver nanoparticles using the cell-free filtrate of *nematophagus* fungus *Duddingtonia flagrans*. *Int J Nanomed.* 12, 6373–6381.

Dahoumane, S.A., Mechouet, M., Wijesekera, K., Filipe, C.D.M., Sicard, C., Bazylinski, D.A., & Jeffryes, C. (2017). Algae-mediated biosynthesis of inorganic nanomaterials as a promising route in nanobiotechnology-a review. *Green Chem.* 19(3), 552–587.

Darroudi, M., Ahmad, M.B., Zamiri, R., Zak, A.K., Abdullah, A.H. & Ibrahim, N.A. (2011). Time-dependent effect in green synthesis of silver nanoparticles. *Int J Nanomed.* 6, 677–681.

Debabov, V.G., Voeikova, T.A., Shebanova, A.S., Shaitan, K.V., Emelyanova, L.K., Novikova, L.M., & Kirpichnikov, M.P. (2013). Bacterial synthesis of silver sulfide nanoparticles. *Nanotechnol Russ.* 8, 269–276.

Deepty, M., Srinivas, C., Kumar, E.R., Mohan, N.K., Prajapat, C.L., Rao, T.V., Meena, S.S., Verma, A., & Sastry, D. (2019). XRD, EDX, FTIR and ESR spectroscopic studies of co-precipitated Mn- substituted Zn– ferrite nanoparticles. *Ceram Int.* 45, 8037–8044.

Deljou, A., & Goudarzi, S. (2016). Green extracellular synthesis of the silver nanoparticles using thermophilic *Bacillus* Sp. AZ1 and its antimicrobial activity against several human pathogenetic bacteria. *Iran J Biotechnol.* 14, 25–32.

Dhandapani, P., Maruthamuthu, S., & Rajagopal, G. (2012). Bio-mediated synthesis of TiO_2 nanoparticles and its photocatalytic effect on aquatic biofilm. *J Photochem Photobiol B Biol.* 110, 43–49.

Doshi, H., Ray, A., & Kothari, I. (2007). Bioremediation potential of live and dead *Spirulina*: spectroscopic, kinetics and SEM studies. *Biotechnol Bioeng.* 96, 1051–1063.

Dubey, S., Lahtinen, M., & Sillanpää, M. (2010). Tansy fruit mediated greener synthesis of silver and gold nanoparticles. *Process Biochem.* 45, 1065–1071.

El Domany, E.B., Essam, T.M., Ahmed, A.E., & Farghali, A.A. (2018). Biosynthesis physico-chemical optimization of gold nanoparticles as anti-cancer and synergetic antimicrobial activity using *Pleurotusostreatus* fungus. *J Appl Pharm Sci.* 8, 119–128.

Elahian, F., Reiisi, S., Shahidi, A., & Mirzaei, S.A. (2017). High-throughput bioaccumulation, biotransformation, and production of silver and selenium nanoparticles using genetically engineered *Pichiapastoris. Nanomed Nanotechnol Biol Med.* 13, 853–861.

Elegbede, J.A., Lateef, A., Azeez, M.A., Asafa, T.B., Yakeen, T.A., Oladipo, I.C., Aina, D.A., Beukes, L. S., & Guegium-Kana, E.B. (2020). Biofabrication of gold nanoparticles using xylanases through valorization of corncob by *Aspergillus niger* and *Trichoderma longibrachiatum*: antimicrobial, antioxidant, anticoagulant and thrombolytic activities. *Waste Biomass Valoriz.* 11, 781–791.

El-Kassas, H.Y., & El-Sheekh, M.M. (2014). Cytotoxic activity of biosynthesized gold nanoparticles with an extract of the red seaweed *Corallina officinalis* on the MCF-7 human breast cancer cell line. *Asian Pac J Cancer Prev.* 15, 4311–4317.

El-Rafie, H.M., & Zahran, M.K. (2013). Green synthesis of silver nanoparticles using polysaccharides extracted from marine macroalgae. *Carb Polym.* 96, 403–410.

Fayaz, A.M., Balaji, K., Girilal, M., Yadav, R., Kalaichelvan, P.T., & Venketesan, R. (2010). Biogenic synthesis of silver nanoparticles and their synergistic effect with antibiotics: a study against gram-positive and gram-negative bacteria. *Nanomed Nanotechnol Biol Med.* 6, 103–109.

Fernández-Llamosas, H., Castro, L., Blázquez, M. L., Díaz, E., & Carmona, M. (2017). Speeding up bioproduction of selenium nanoparticles by using *vibrio natriegens* as microbial factory. *Sci Rep.* 7, 16046.

Fischlechner, M., & Donath, E. (2007). Viruses as building blocks for materials and devices. *Angew Chem.* 46, 3184–3193.

Gao, C., Wang, Y., Ye, Z., Lin, Z., Ma, X., & He, Q. (2020). Biomedical micro-/nanomotors: from overcoming biological barriers to *in vivo* imaging. *Adv Mater.* 33, 2000512.

Gericke, M., & Pinches, A. (2006a). Microbial production of gold nanoparticles. *Gold Bull.* 39, 22–28.

Gericke, M., & Pinches, A. (2006b). Biological synthesis of metal nanoparticles. *Hydrometall.* 83,132–140.

Ghodake, G., & Lee, D.S. (2011). Biological synthesis of gold nanoparticles using the aqueous extract of the brown algae *Laminaria japonica. J. Nanoelectron Opto Electro.* 6, 268–271.

Gholami-Shabani, M., Akbarzadeh, A., Norouzian, D., Amini, A., Gholami-Shabani, Z., & Imani, A. (2014). Antimicrobial activity and physical characterization of silver nanoparticles green synthesized using nitrate reductase from *Fusarium oxysporum. Appl Biochem Biotechnol.* 172, 4084–4098.

Gmoshinski, I.V., Khotimchenko, S.A., Popov, V.O., Dzantiev, B.B., Zherdev, A.V., Demin, V.F., & Buzulukov, Y.P. (2013). Nanomaterials and nanotechnologies: methods of analysis and control. *Russ Chem Rev.* 82, 48.

Gomaa, E.Z. (2022). Microbial mediated synthesis of zinc oxide nanoparticles, characterization and multifaceted applications. *J Inorg Organomet Polym Mater.* 32, 4114–4132.

González-Ballesteros, N., Prado-López, S., Rodríguez-González, J.B., Lastra, M., & Rodríguez-Argüelles, M.C. (2017). Green synthesis of gold nanoparticles using brown algae *Cystoseira baccata:* its activity in colon cancer cells. *Colloids Surf B Biointerfaces.* 153, 190–198.

Gowramma, B., Keerthi, U., Rafi, M., & Muralidhara, R.D. (2015). Biogenic silver nanoparticles production and characterization from native stain of *Corynebacterium* species and its antimicrobial activity. *3 Biotech.* 5, 195–201.

Gudikandula, K., Vadapally, P., & Charya, M.A.S. (2017). Biogenic synthesis of silver nanoparticles from white rot fungi: their characterization and antibacterial studies. *Open Nano.* 2, 64–78.

Gupta, A., Saleh, N.M., Das, R., Landis, R.F., Bigdeli, A., Motamedchaboki, K., Compos, A.R., Pomeroy, K., Mahmoud, M., & Rotello, V.M. (2017). Synergistic antimicrobial therapy using nanoparticles and antibiotics for the treatment of multidrug-resistant bacterial infection. *Nano Futures.* 1, 015004.

Hall, J.B., Dobrovolskaia, M.A., Patri, A.K., & McNeil, S.E. (2007). Characterization of nanoparticles for therapeutics. *Nanomedicine.* 2, 789–803.

Haverkamp, R.H., Marshall, A.T., & Van Agterveld, D. (2007). Pick your carats: nanoparticles of gold-silver-copper alloy produced in vivo. *J Nanoparticle Res.* 9, 697–700.

Hayazawa, N., Tarun, A., Taguchi, A., & Furusawa, K. (2012). Tip-enhanced Raman spectroscopy. In C.S.S.R. Kumar (Ed.) *Raman Spectroscopy for Nanomaterials Characterization* (pp. 445–476). Springer, Berlin, Heidelberg.

He, S., Guo, Z., Zhang, Y., Zhang, S., Wang, J., & Gu, N. (2007). Biosynthesis of gold nanoparticles using the bacteria *Rhodopseudomonas capsulate. Mater Lett.* 61, 3984–3987,

Hinterdorfer, P., Garcia-Parajo, M.F., & Dufrene, Y.F. (2012). Single-molecule imaging of cell surfaces using near-field nanoscopy. *Acc Chem Res.* 45, 327–36.

Horikoshi, S., & Serpone, N. (2017). Microwave-assisted synthesis of nanoparticles. In G. Cravotto & D. Carnaroglio (Eds.) *Microwave Chemistry* (pp. 248–269). De Gruyter, Berlin, Germany.

Husain, S., Sardar, M., & Fatma, T. (2015). Screening of cyanobacterial extracts for synthesis of silver nanoparticles. *World J Microbiol Biotechnol.* 31, 1279–1283.

Husseiny, M.I., El-Aziz, M.A., Badr, Y., & Mahmoud, M.A. (2007). Biosynthesis of gold nanoparticles using *Pseudomonas aeruginosa. Spectrochim Acta A Mol Biomol Spectrosc.* 67, 1003–1006.

Ianoul, A., & Johnston, L.J. (2007). Near-field scanning optical microscopy to identify membrane microdomains. *Methods Mol Biol.* 400, 469–480.

Ibrahim, S., Ahmad, Z., Manzoor, M. Z., Mujahid, M., Faheem, Z., & Adnan, A. (2021). Optimization for biogenic microbial synthesis of silver nanoparticles through response surface methodology, characterization, their antimicrobial, antioxidant, and catalytic potential. *Sci Rep.* 11, 1–18.

Iravani, S., Thota, S., & Crans, D. (2018). *Metal Nanoparticles: Synthesis and Applications in Pharmaceutical Sciences* (pp. 15–32). Wiley, New York.

Jafari, M., Rokhbakhsh-Zamin, F., Shakibaie, M., Moshafi, M.H., Ameri, A., Rahimi, H.R., & Forootanfar, H. (2018). Cytotoxic and antibacterial activities of biologically synthesized gold nanoparticles assisted by *Micrococcus yunnanensis* strain J2. *Biocatalysis Agric Biotechol.* 15, 245–253.

Jain, N., Bhargava, A., Tarafdar, J.C., Singh, S.K., & Panwar, J. (2013). A biomimetic approach towards synthesis of zinc oxide nanoparticles. *Appl Microbiol Biotechnol.* 97, 859–869.

Jayaseelan, C., Ramkumar, R., Rahuman, A.A., & Perumal, P. (2013). Green synthesis of gold nanoparticles using seed aqueous extract of *Abelmoschus esculentus* and its antifungal activity. *Ind Crops Prod.* 45, 423–429.

Jeevan, P., Ramya, K., & Rena, A.E. (2012). Extracellular biosynthesis of silver nanoparticles by culture supernatant of *Pseudomonas aeruginosa. Indian J Biotechnol.* 11, 72–76.

Jena, J., Pradhan, N., Nayak, R.R., Dash, B.P., Sukla, L.B., Panda, P.K., & Mishra, B.K. (2014). Microalga *Scenedesmus sp.*: a potential low-cost green machine for silver nanoparticle synthesis. *J Microbiol Biotechnol.* 24, 522–533.

Jianping, X., Jim, Y.L., Daniel, I.C.W., & Yen, P.T. (2007). Identification of active biomolecules in the high-yield synthesis of single-crystalline gold nanoplates in algal solutions. *Small.* 3, 668–672.

Jones, R.R., Hooper, D.C., Zhang, L., Wilverson, D., & Valev, V.K. (2019). Raman techniques: fundamentals and frontiers. *Nanoscale Res Lett.* 14, 231.

Juibari, M.M., Abbasalizadeh, S., Jouzani, G.S., & Noruzi, M. (2011). Intensified biosynthesis of silver nanoparticles using a native extremophilic *Ureibacillus thermosphaerius* strain. *Materials Lett.* 65, 1014–1017.

Kalabegishvili, T., Kirkesali, E. & Rcheulishvili, A. (2012a). *Synthesis of Gold Nanoparticles by Blue- Green Algae Spirulina Platensis* (No. JINR-E--14-2012-31). Frank Lab. of Neutron Physics.

Kalabegishvili, T.L., Kirkesali, E.I., Rcheulishvili, A.N., Ginturi, E.N., Murusidze, I.G., Pataraya, D.T., Gurielidze, M.A., Tsertsvadze, G.I., Gabunia, V.N., & Lomidze, L.G. (2012b). Synthesis of gold nanoparticles by some strains of arthrobacter genera. *Mater Sci Eng A.*, 2, 164–173.

Kalimuthu, K., Babu, S.R., & Venkataraman, D., Bilal, M., & Gurunathan, S. (2008). Biosynthesis of silver nanocrystals by *Bacillus licheniformis*. *Colloids Surf B Biointerfaces.* 65. 150–3.

Kalishwaralal, K., Venkataraman, D., Pandian, S.R.K., Muniasamy, K., BarathManiKanth, S., Karthikeyan, B., & Gurunathan, S. (2010). Biosynthesis of silver and gold nanoparticles using *Brevibacterium* casei. *Colloids Surf B Biointerfaces.* 77. 257–62.

Khalafi, T., Buazar, F., & Ghanemi, K. (2019). Phycosynthesis and enhanced photocatalytic activity of zinc oxide nanoparticles toward organosulfur pollutants. *Sci Rep.* 9, 6866.

Khalil, M.M., Ismail, E.H., El-Baghdady, K.Z., & Mohamed, D. (2014). Green synthesis of silver nanoparticles using olive leaf extract and its antibacterial activity. *Arab J Chem.* 7, 1131–1139.

Khan, A.A., Fox, E.K., Gorzny, M.L., Nikulina, E., Brougham, D.F., Wege, C., & Bittner, A.M. (2013). pH control of the electrostatic binding of gold and iron oxide nanoparticles to Tobacco mosaic virus. *Langmuir.* 29, 2094–2098.

Khan, A.U., Khan, M., Malik, N., Cho, M.H., & Khan, M.M. (2019). Recent progress of algae and blue–green algae-assisted synthesis of gold nanoparticles for various applications. *Bioprocess Biosyst Eng.* 42, 1–15.

Khan, A.U., Malik, N., Khan, M., Cho, M.H., & Khan, M. M. (2018). Fungi-assisted silver nanoparticle synthesis and their applications. *Bioprocess Biosyst Eng.* 41, 1–20

Khati, P., Bhatt, P., Kumar, R., & Sharma, A. (2018). Effect of nanozeolite and plant growth promoting rhizobacteria on maize. *3Biotech.* 8, 141.

Khati, P., Sharma, A., Chaudhary, P., Singh, A.K., Gangola, S., & Kumar, R. (2019). High-throughput sequencing approach to access the impact of nanozeolite treatment on species richness and evens of soil metagenome. *Biocatalysis Agric Biotechnol.* 20, 101249.

Khodashenas, B., & Ghorbani, H.R. (2014). Synthesis of silver nanoparticles with different shapes. *Arab J Chem.* 12, 1823–1838.

Kitching, M., Choudhary, P., Inguva, S., Guo, Y., Ramani, M., Das, S.K., & Marsili, E. (2016). Fungal surface protein mediated one-pot synthesis of stable and hemocompatible gold nanoparticles. *Enzym Microb Technol.* 95, 76–84.

Klaus, T., Joerger, R., Olsson, E., & Granqvist, C-G. (1999). Silver-based crystalline nanoparticles, microbially fabricated. *Proc Natl Acad Sci.* 96, 13611–13614.

Konishi, Y., Ohno, K., Saitoh, N., Nomura, T., Nagamine, S., Hishida, H. Takahashi, Y., & Uruga, T. (2007). Bioreductive deposition of platinum nanoparticles on the bacterium *Shewanella algae*. *J Biotechnol.* 128, 648–53.

Kreibig, U., & Vollmer, M. (2013). *Optical Properties of Metal Clusters. Springer Science & Business Media*, Berlin, Heidelberg.

Kuchibhatla, S.V.N.T., Karakoti, A.S., Baer, D.R., Samudrala, S., Engelhard, M.H., Amonette, J.E., Thevuthasan, S., & Seal, S. (2012). Influence of aging and environment on nanoparticle chemistry: implication to confinement effects in nanoceria. *J Phys Chem C.* 116, 14108–14114.

Kukreti, B., Sharma, A., Chaudhary, P., Agri, U., & Maithani, D. (2020). Influence of nanosilicon dioxide along with bioinoculants on *Zea mays* and its rhizospheric soil. *3 Biotech.* 10, 345.

Kumar, A., & Dixit, C.K. (2017). Methods for characterization of nanoparticles. In S. Nimesh, R. Chandra, & N. Gupta (Eds.) *Advances in Nanomedicine for the Delivery of Therapeutic Nucleic Acids* (pp. 43–58). Woodhead Publishing.

Kumar, P., Pahal, V., Gupta, A., Vadhan, R., Chandra, H., & Dubey, R.C. (2020). Effect of silver nanoparticles and *Bacillus cereus* LPR2 on the growth of *Zea mays*. *Sci Rep.* 10, 20409.

Kumar-Krishnan, S., Prokhorov, E., Hernández-Iturriaga, M., Mota-Morales, J.D., Vázquez-Lepe, M., Kovalenko, Y., Sanchez, I.C., & Luna-Bárcenas, G. (2015). Chitosan/silver nanocomposites: synergistic antibacterial action of silver nanoparticles and silver ions. *Eur Polym J.* 67, 242–251.

Kumari, S., Sharma, A., Chaudhary, P., & Khati, P. (2020). Management of plant vigor and soil health using two agriusable nanocompounds and plant growth promotory rhizobacteria in fenugreek. *3 Biotech.* 10, 461.

Kundu, D., Hazra, C., Chatterjee, A., Chaudhari, A., & Mishra, S. (2014). Extracellular biosynthesis of zinc oxide nanoparticles using *Rhodococcus pyridinivorans* NT2: multifunctional textile finishing, biosafety evaluation and in vitro drug delivery in colon carcinoma. *J Photochem Photobiol B Biol.* 140, 194–204.

Kushnerova, N., Fomenko, S., Sprygin, V., Kushnerova, T., Khotimchenko, Y.S., Kondrat'eva, E., & Drugova, L. (2010). An extract from the brown alga *Laminaria japonica*: a promising stress-protective preparation. *Russ J Mar Biol.* 36, 209–214.

Lee, Y.J., Yi, H., Kim, W.J., Kang, K., Yun, D.S., Strano, M.S., Ceder, G., & Belcher, A.M. (2009). Fabricating genetically engineered high-power lithium-ion batteries using multiple virus gene. *Science* 324, 1051–1055.

Lengke, M. F., Fleet, M.E., & Southam, G. (2006a). Morphology of gold nanoparticles synthesized by filamentous cyanobacteria from gold(I)-thiosulfate and gold(III)-chloride complexes. *Langmuir.* 22, 2780–2787.

Lengke, M.F., Ravel, B., Fleet, M.E., Wanger, G., Gordon, R.A., & Southam, G. (2006). Mechanisms of gold bioaccumulation by filamentous cyanobacteria from gold (III)–chloride complex. *Environ Sci Technol.* 40, 6304–6309.

Lewis, J.D., Destito, G., Zijlstra, A., Gonzalez, M.J., & Quigley, J.P. (2006). Viral nanoparticles as tools for intravital vascular imaging. *Nat Med.* 12, 354–360.

Li, X., Xu, H., Chen, Z.S., & Chen, G. (2011). Biosynthesis of nanoparticles by microorganisms and their applications. *J Nanomater.* 8, 1–16.

Lynch, T., Cedervall, M., Lundqvist, C., Cabaleiro-Lago, S., Linse, & Dawson, K.A. (2007). The nanoparticle-protein complex as a biological entity; a complex fluids and surface science challenge for the 21st century. *Adv Colloid Interface Sci.* 134–135, 167–174.

Ma, Y., Nolte, R.J.M., & Cornelissen, J.J.L.M. (2012) Virus-based nanocarriers for drug delivery. *Adv Drug Deliv Rev.* 64, 811.

Mahdieh, M., Zolanvari, A.A., & Azimee, A.S. (2012). Green biosynthesis of silver nanoparticles by *Spirulina platensis*. *Sci Iran.* 19, 926–929.

Manor, J., Feldblum, E.S., Zanni, M.T., & Arkin, I.T. (2012). Environment polarity in proteins mapped noninvasively by FTIR spectroscopy. *J Phys Chem Lett.* 3, 939–44.

Markus, J., Mathiyalagan, R., Kim, Y.J., Abbai, R., Singh, P., Ahn, S., Perez, Z., Hurh, J., & Yang, D.C. (2016). Intracellular synthesis of gold nanoparticles with antioxidant activity by probiotic *Lactobacillus kimchicus* DCY51[T] isolated from Korean kimchi. *Enzyme Microbial Technol.* 95, 85–93.

Marooufpour, N., Alizadeh, M., Hatami, M., & Lajayer, B.A. (2019). Biological synthesis of nanoparticles by different groups of bacteria. In R. Prasad (Ed.) *Microbial Nanobionics* (pp. 63–85). Springer, Cham.

McNaught, A.D., & Wilkinson, A. (1997). *Compendium of Chemical Terminology*. Blackwell Science, Oxford.

Menon, S., Rajeshkumar, S., & Venkat-Kumar, S. (2017). A review on biogenic synthesis of gold nanoparticles, characterization, and its applications. *Resource-Efficient Technol*. 3, 516–527.

Merzlyak, A., & Lee, S.W. (2006). Phage as template for hybrid materials and mediators for nanomaterials synthesis. *Curr Opin Chem Biol*. 10, 246–252.

Mishra, M., Paliwal, J.S., Singh, S.K., Selvarajan, E., Subathradevi, C., & Mohanasrinivasan, V. (2013). Studies on the inhibitory activity of biologically synthesized and characterized zinc oxide nanoparticles using *Lactobacillus sporogens* against *Staphylococcus aureus*. *J Pure Appl Microbiol*. 7, 1–6

Moghaddam, A. B., Namvar, F., Moniri, M., Tahir, P., Azizi, S., & Mohamad, F. (2015). Nanoparticles biosynthesized by fungi and yeast: a review of their preparation, properties, and medical applications. *Molecules*. 20, 16540–16565.

Molnár, Z., Bódai, V., Szakacs, G., Erdélyi, B., Fogarassy, Z., Sáfrán, G., Varg, T., Konya, Z., Toth-Szelea, E., Szucs, R., & Lagzi, I. (2018). Green synthesis of gold nanoparticles by thermophilic filamentous fungi. *Sci Rep*. 8, 3943.

Moore, K. (2006). A new silver dressing for wounds with delayed healing. *Wounds UK*. 2, 70–78.

Mosulishvili, L., Kirkesali, E., Belokobylsky, A., Khizanishvili, A., Frontasyeva, M., Pavlov, S., & Gundorina, S. (2002). Experimental substantiation of the possibility of developing selenium-and iodine-containing pharmaceuticals based on blue–green algae *Spirulina platensis*. *J Pharm Biomed*. 30, 87–97.

Mubarak-Ali, D., Sasikala, M., Gunasekaran, M., & Thajuddin, N. (2011). Biosynthesis and characterization of silver nanoparticles using marine cyanobacterium *Oscillatoria willei* NTDM 01. *Dig J Nanomater Biostruct*. 6, 385–390.

Mukherjee, P., Roy, M., Mandal, B.P., Dey, G.K., Mukherjee, P.K., Ghatak, J., Tyagi, A.K., & Kale, S.P. (2008). Green synthesis of highly stabilized nanocrystalline silver particles by a non-pathogenic and agriculturally important fungus *T. asperellum*. *Nanotechnology*. 19, 075103.

Munawar, U., Raghavendra, V.B., Ningaraju, S., Krishna, K.L., Ghosh, A.R., Melappa, G., & Pugazhendhi, A. (2020). Biofabrication of gold nanoparticles mediated by the endophytic *Cladosporium* species: photodegradation, in vitro anticancer activity and in vivo antitumor studies. *Int J Pharm*. 588, 119729.

Murugesan, S., Bhuvaneswari, S., & Sivamurugan, V. (2017). Green synthesis, characterization of silver nanoparticles of a marine red alga *Spyridia fusiformis* and their antibacterial activity. *Int J Pharm Sci*. 9, 192–197.

Nagarajan, S., & Kuppusamy, A.K. (2013). Extracellular synthesis of zinc oxide nanoparticle using seaweeds of Gulf of Mannar, India. *J Nanobiotechnol*. 11, 39.

Nam, K.T., Kim, D.W., Yoo, P.J., Chiang, C.Y., Meethong, N., Hammond, P.T., Chiang, Y.M., & Belcher, A.M. (2006). Virus-enabled synthesis and assembly of nanowires for lithium ion battery electrodes. *Science*. 312, 885–888.

Namasivayam, S.K.R., Gnanendra, E.K., & Reepika, R. (2010). Synthesis of silver nanoparticles by Lactobaciluus acidophilus 01 strain and evaluation of its in vitro genomic DNA toxicity. *Nano-Micro Lett*. 2, 160–163.

Nangia, Y., Wangoo, N., Goyal, N., Shekhawat, G., & Suri, C.R. (2009). A novel bacterial isolate *Stenotrophomonas maltophilia* as living factory for synthesis of gold nanoparticles. *Microb Cell Fact*. 8, 39.

Neethu, S., Midhun, S.J., Sunil, M., Soumya, S., Radhakrishnan, E., & Jyothis, M. (2018). Efficient visible light induced synthesis of silver nanoparticles by *Penicillium polonicumara 10* isolated from *Chetomorphaantennina* and its antibacterial efficacy against *Salmonella enteric* serovar typhimurium. *J Photochem Photobiol B Biol*. 180, 175–185.

Netala, V. R., Bethu, M. S., & Pushpalatah, B. (2016). Biogenesis of silver nanoparticles using endophytic fungus *Pestalotiopsis microspora* and evaluation of their antioxidant and anticancer activities. *Int J Nanomed.* 11, 5683–5696.

Órdenes-Aenishanslins, N.A., Saona, L.A., Durán-Toro, V.M., Monras, J.P., Bravo, D.M., & Perez- Donoso, J.M. (2014). Use of titanium dioxide nanoparticles biosynthesized by *Bacillus mycoides* in quantum dot sensitized solar cells. *Microb Cell Fact.* 13, 90.

Paciotti, G.F., Myer, L., Weinreich, D., Goia, D., Pavel, N., McLaughlin, R.E., & Tamarkin, L. (2004). Colloidal gold: a novel nanoparticle vector for tumor directed drug delivery. *Drug Del.* 11, 169–183.

Pal, T., Mohiyuddin, S., & Packirisamy, G. (2018). Facile and green synthesis of multicolor fluorescence carbon dots from curcum: *in vitro* and *in vivo* bioimaging and other application. *ACS Omega.* 3, 831–843.

Parial, D., & Pal, R. (2014). Green synthesis of gold nanoparticles using cyanobacteria and their characterization. *Indian J Appl Res.* 4, 69–72.

Parial, D., Patra, H.K., Roychoudhury, P., Dasgupta, A.K., & Pal, R. (2012). Gold nanorod production by cyanobacteria—a green chemistry approach. *J Appl Phycol.* 24, 55–60.

Parveen, K., Banse, V., & Ledwani, L. (2016). Green synthesis of nanoparticles: their advantages and disadvantages. In *AIP Conference Proceedings* (Vol. 1724, No. 1, p. 020048). AIP Publishing LLC.

Patri, A., Dobrovolskaia, M., Stern, S., McNeil, S., & Amiji, M. (2006). Preclinical characterization of engineered nanoparticles intended for cancer therapeutics. In M.M. Amiji (Ed.) *Nanotechnology for Cancer Therapy* (pp. 105–137). CRC Press, Boca Raton.

Patil, M.P., Bayaraa, E., Subedi, P. Piad, L.L.A., Tarte, N.H., & Kim, G. (2019). Biogenic synthesis, characterization of gold nanoparticles using *Lonicera japonica* and their anticancer activity on HeLa cells. *J Drug Deliv Sci Technol.* 51, 83–90.

Popovic, Z.V., Dohcevic-Mitrovic, Z., Scepanovic, M., Grujic-Brojcin, M., & Askrabic, S. (2011). Raman scattering on nanomaterials and nanostructures. *Ann Phys.* 523, 62–74.

Priyadharshini, R.I., Prasannaraj, G., Geetha, N., & Venkatachalam, P. (2014). Microwave-mediated extracellular synthesis of metallic silver and zinc oxide nanoparticles using macro-algae *(Gracilariaedulis)* extracts and its anticancer activity against human PC3 cell lines. *Appl Biochem Biotechnol.* 174, 2777–2790.

Pugazhendhi, A., Prabakar, D., Jacob, J.M., Karuppusamy, I., & Saratale, R.G. (2017). Synthesis and characterization of silver nanoparticles using *Gelidium amansii* and its antimicrobial property against various pathogenic bacteria. *Microb Pathog.* 114, 41–45.

Rajabairavi, N., Raju, C.S., Karthikeyan, C., Varutharaju, K., Nethaji, S., Hameed, A.S., & Shajahan, A. (2017). Biosynthesis of novel zinc oxide nanoparticles (ZnO NPs) using endophytic bacteria *Sphingobacterium thalpophilum*. In J. Ebeneza (Ed.) *Recent Trends in Materials Science and Applications* (pp. 245–254). Springer, Cham.

Rajput, S., Werezuk, R., Lange, R.M.S., & McDermott, M.T. (2016). Fungal isolate optimized for biogenesis of silver nanoparticles with enhanced colloidal stability. *Langmuir.* 32, 8688–8697.

Raliya, R., & Tarafdar, J.C. (2013). ZnO nanoparticle biosynthesis and its effect on phosphorous-mobilizing enzyme secretion and gum contents in cluster bean (*Cyamopsis tetragonoloba* L.). *Agric Res.* 22, 48–57.

Ramos, M.M., Morais, E.d.S., Sena, I.d.S., Lima, A.L., de Oliveira, F.R., de Freitas, C.M., Fernandes, C.P., de Carvalho, J.C.T., & Ferreira, I.M. (2020). Silver nanoparticle from whole cells of the fungi *Trichoderma spp.* Isolated from Brazilian amazon. *Biotechnol Lett.* 42, 833–843.

Ratner, B.D., Hoffman, A.S., Schoen, F.J., & Lemons, J.E. (2004). *Biomaterials Science: An Introduction to Materials in Medicine*. Elsevier. Academic Press.

Rauf, M.A., Owais, M., Rajpoot, R., Ahmad, F., Khan, N., & Zubair, S. (2017). Biomimetically synthesized ZnO nanoparticles attain potent antibacterial activity against less susceptible: *S. aureus* skin infection in experimental animals. *RSC Adv.* 7, 36361–36373.

Rodrigues, M.C., Rolim, W.R., Viana, M.M., Souza, T.R., Gonçalves, F., Tanaka, C.J., Bueno-Silva, B., & Seabra, A.B. (2020). Biogenic synthesis and antimicrobial activity of silica-coated silver nanoparticles for esthetic dental applications. *J Dentist.* 96, 103327.

Sabri, M.A., Umer, A., Awan, G.H., Hassan, M.F., & Hasnain, A. (2016). Selection of suitable biological method for the synthesis of silver nanoparticles. *Nanomater Nanotechnol.* 6, 1–20.

Sadowski, Z., Maliszewska, I., Polowczyk, I., Kozlecki, T., & Grochowalska, B. (2008). A Biosynthesis of colloidal-silver particles using microorganisms. *Polish J Chem.* 82, 377–382.

Saeed, S., Iqbal, A., & Ashraf, M.A. (2020). Bacterial-mediated synthesis of silver nanoparticles and their significant effect against pathogens. *Environ Sci Pollut Res.* 27, 37347–37356.

Salem, S.S., & Fouda, A. (2021). Green synthesis of metallic nanoparticles and their prospective biotechnological applications: an overview. *Biol Trace Elem Res.* 199, 344–370.

Sanctis, S., Hoffmann, R.C., Eiben, S., Schneider, J.J., & Beilstein, J. (2015). Microwave assisted synthesis and characterisation of a zinc oxide/tobacco mosaic virus hybrid material. An active hybrid semiconductor in a field-effect transistor device. *Nanotechnol.* 6, 785–791.

Sapsford, K.E., Tyner, K.M., Dair, B.J., Deschamps, J.R., & Medintz. I.L. (2011). Analyzing nanomaterial bioconjugates: a review of current and emerging purification and characterization techniques. *Anal Chem.* 83, 4453–88.

Saravanan, M., Barik, S.K., MubarakAli, D., Prakash, P., & Pugazhendhi, A. (2018a). Synthesis of silver nanoparticles from *Bacillus brevis* (ncim 2533) and their antibacterial activity against pathogenic bacteria. *Microb Pathog.* 116, 221–226.

Saravanan, M., Gopinath, V., Chaurasia, M.K., Syed, A., Ameen, F., & Purushothaman, N. (2018b). Green synthesis of anisotropic zinc oxide nanoparticles with antibacterial and cytofriendly properties. *Microb Pathog.* 115, 57–63.

Sarkar, J., Ghosh, M., Mukherjee, A., Chattopadhyay, D., & Acharya, K. (2014). Biosynthesis and safety evaluation of ZnO nanoparticles. *Bioprocess Biosyst Eng.* 37, 165–71.

Sayyid, N.H., & Zghair, Z.R. (2021). Biosynthesis of silver nanoparticles produced by *Klebsiella pneumoniae*. *Materials Today Proceed.* 42, 2045–2049.

Selvarajan, E., & Mohanasrinivasan, V. (2013). Biosynthesis and characterization of ZnO nanoparticles using *Lactobacillus plantarum* VITES07. *Mater Lett.* 112, 180–182.

Senapati, S., Mandal, D., Ahmad, A., Mandal, D., Senapati, S., Sainkar, S.R., Khan, M.I., Parishcha, R., Ajaykumar, P.V., Alam, M., Kumar, R., & Sastry, M. (2004). Fungus mediated synthesis of silver nanoparticles: a novel biological approach. *Indian J Phys A.* 78A, 101–105

Shamsuzzaman, M.A., Khanam, H., & Aljawfi, R.N. (2017). Biological synthesis of ZnO nanoparticles using *C. albicans* and studying their catalytic performance in the synthesis of steroidal pyrazolines. *Arab J Chem.* 10, S1530–S1536.

Shankar, S.S., Ahmad, A., Pasricha, R., & Sastry, M. (2003). Bioreduction of chloroaurate ions by geranium leaves and its endophytic fungus yields gold nanoparticles of different shapes. *J Mater Chem.* 13, 1822–1826.

Sharma, N., Pinnaka, A.K., Raje, M., Fnu, A., Bhattacharyya, M.S., & Choudhury, A.R. (2012). Exploitation of marine bacteria for production of gold nanoparticles. *Microbial Cell Fact.* 11, 86.

Shetty, P., Supraja, N., Garud, M., & Prasad, T.N.V.K.V. (2014). Synthesis, characterization and antimicrobial activity of *Alstonia scholaris* bark-extract-mediated silver nanoparticles. *J Nanostructure Chem.* 4, 161–170.

Shunmugam, R., Balusamy, S.R., Kumar, V., Menon, S., Lakshmi, T., & Perumalsamy, H. (2021). Biosynthesis of gold nanoparticles using marine microbe (Vibrio alginolyticus) and its anticancer and antioxidant analysis. *J King Saud Univ Sci.* 33, 101260.

Sibbald, R.G., Contreras-Ruiz, J., Coutts, P., Fierheller, M., Rothman, A., & Woo, K. (2007). Bacteriology, inflammation, and healing: a study of nanocrystalline silver dressings in chronic venous leg ulcers. *Adv Skin Wound Care.* 20, 549–558.

Sikora, A., Bartczak, D., Geissler, D., Kestens, V., Roebben, G., Ramaye, Y., Varga, Z., Palmai, M., Sgard, A.G., Goenaga-Infante, H., & Minelli, C. (2015). A systematic comparison of difference techniques to determine the zeta potential of silica nanoparticles in biological medium. *Anal Methods.* 7, 9835–9843.

Singaravelu, G., Arockiamary, J.S., Kumar, V.G., & Govindaraju, K. (2007). A novel extracellular synthesis of monodisperse gold nanoparticles using marine alga, *Sargassum wightii Greville. Colloids Surf B Biointerfaces.* 57, 97–101.

Singh, B.N., Rawat, A.K., Khan, W., Naqvi, A.H., & Singh, B.R. (2014). Biosynthesis of stable antioxidant ZnO nanoparticles by *Pseudomonas aeruginosa* rhamnolipids. *PloS One.* 9, e106937.

Singh, C.R., Kathiresan, K., & Anandhan, S. (2015). A review on marine based nanoparticles and their potential applications. *Afr J Biotechnol.* 14, 1525–1532.

Singh, H., Du, J., Singh, P., & Yi, T.H. (2018). Extracellular synthesis of silver nanoparticles by Pseudomonas sp. THG-LS1.4 and their antimicrobial application. *J Pharma Anal.* 8, 258–264.

Singh, R., Raj, S.K., & Prasad, V. (2008). Molecular characterization of a strain of squash leaf curl China virus from north India. *J Phytopathol.* 156, 222–228.

Singh, R., Wagh, P., Wadhwani, S., Gaidhani, S., Kumbhar, A., Bellare, J., & Chopade, B.A. (2013). Synthesis, optimization, and characterization of silver nanoparticles from *Acinetobacter calcoaceticus* and their enhanced antibacterial activity when combined with antibiotics. *Int J Nanomed.* 8, 4277.

Skirtach, A.G., Muñoz Javier, A., Kreft, O., Köhler, K., Piera Alberola, A., Möhwald, H., Parak, W. J., & Sukhorukov, G.B. (2006). Laser-induced release of encapsulated materials inside living cells. *Angew Chemie.* 45, 4612–4617.

Song, Y., Zhang, Z., Elsayed-Ali, H.E., Wang, H., Henry, L.L., Wang, Q., Zou, S., & Zhang, T. (2011). Identification of single nanoparticles. *Nanoscale.* 3, 31–44.

Soni, N., & Prakash, S. (2011). Factors affecting the geometry of silver nanoparticles synthesis in *Chrysosporium tropicum* and *Fusarium oxysporum. Am J Nanotechnol.* 2, 112–121.

Sowani, H., Mohite, P., Damale, S., Kulkarni, M., & Zinjarde, S. (2016). Carotenoid stabilized gold and silver nanoparticles derived from the Actinomycete *Gordonia amicalis* HS-11 as effective free radical scavengers. *Enzyme Microb Technol.* 95, 164–173.

Srinath, B.S., Namratha, K., & Byrappa, K. (2018). Eco-friendly synthesis of gold nanoparticles by *Bacillus subtilis* and their environmental applications. *Advanced Sci Lett.* 24, 5942–5946.

Steinmetz, N.F., & Manchester, M. (2011). *Viral Nanoparticles: Tools for Materials Science & Biomedicine* (1st edition). Jenny Stanford Publishing.

Stillman, N.R., Kovacevic, M., Igor, B., & Sabine, H. (2020). *In silico* modelling of cancer nanomedicine, across scales and transport barriers. *NPJ Comput Mater.* 6, 1–10.

Suba, S., Vijayakumar, S., Nilavukkarasi, M., Vidhya, E., & Punitha, V.N. (2022). Eco synthesized silver nanoparticles as a next generation of nanoproduct in multidisciplinary applications. *Environ Chem Ecotoxicol.* 4, 13–19.

Suresh, A.K., Pelletier, D.A., Wang, W., Broich, M.L., Moon, J.W., Gu, B., Allison, D.P., Joy, D.C., Phelps, T.J., & Doktycz, M.J. (2011). Biofabrication of discrete spherical gold nanoparticles using the metal-reducing bacterium *Shewanella oneidensis. Acta Biomater.* 7, 2148–2152.

Suzuki, E. (2002). High-resolution scanning electron microscopy of immunogold-labelled cells by the use of thin plasma coating of osmium. *J Microsc.* 208, 153–157.

Tiede, K., Boxall, A.B., Tear, S.P., Lewis, J., David, H., & Hassellov, M. (2008). Detection and characterization of engineered nanoparticles in food and the environment. *Food Addit Contam Part A Chem Anal Control Expo Risk Assess.* 25, 795–821.

Tran, Q.H., Nguyen, V.Q., & Le, A. (2013). Silver nanoparticles: synthesis, properties, toxicology, applications and perspectives. *Adv Nat Sci Nanosci Nanotechnol* 4, 033001.

Tripathi, R.M., Bhadwal, A.S., Gupta, R.K., Singh, P., Shrivastav, A., & Shrivastav, B.R. (2014). ZnO nanoflowers: novel biogenic synthesis and enhanced photocatalytic activity. *J Photochem Photobiol B Biol.* 141, 288–95.

Uchihara, T. (2007). Silver diagnosis in neuropathology: principles, practice and revised interpretation. *Acta Neuropathol.* 113, 483–499.

Vadlapudi, V., & Amanchy, R. (2017). Synthesis, characterization and antibacterial activity of silver nanoparticles from Red Algae, *Hypneamu sciformis*. *Adv Biol Res.* 11, 242–249.

Vahabi, K., Mansoori, G.A., & Karimi, S. (2011). Biosynthesis of silver nanoparticles by fungus *Trichoderma reesei*. *Insciences J.* 1, 65–79.

Vaidyanathan, R., Kalishwaralal, K., Gopalram, S., & Gurunathan, S. (2009). Nanosilver— the burgeoning therapeutic molecule and its green synthesis. *Biotechnol Adv.* 27, 924–937.

Velmurugan, P., Shim, J., You, Y., Choi, S., Kamala-Kannan, S., Lee, K.J., Kim, H.J., & Oh, B.T. (2010). Removal of zinc by live, dead, and dried biomass of *Fusarium* spp. isolated from the abandoned-metal mine in South Korea and its perspective of producing nanocrystals. *J Hazard Mat.* 182, 317–324.

Vigneshwaran, N., Ashtaputre, N., Varadarajan, P., Nachane, R., Paralikar, K., & Balasubramanya, R. (2007). Biological synthesis of silver nanoparticles using the fungus *Aspergillus flavus*. *Mater Lett.* 61, 1413–1418.

Vigneshwaran, N., Kathe, A.A., Varadarajan, P.V., Nachane, R.P., & Balasubramanya, R.H. (2006). Biomimetics of silver nanoparticles by white rot fungus, *Phaenerochaete chrysosporium*. *Colloids Surf B.* 53, 55–59

Villaverde, A. (2010). Nanotechnology, bionanotechnology and microbial cell factories. *Microb Cell Fact.* 9, 53.

Wadhwani, S.A., Shedbalkar, U.U., Singh, R., Vashisth, P., Pruthi, V., & Chopade, B.A. (2016). Kinetics of synthesis of gold nanoparticles by *Acinetobacter* sp. SW30 isolated from environment. *Indian J Microbiol.* 56, 439–444.

Waghmare, S.R., Mulla, M.N., Marathe, S.R., & Sonawane, K.D. (2015). Ecofriendly production of silver nanoparticles using *Candida utilis* and its mechanistic action against pathogenic microorganisms. *3 Biotech.* 3, 33–38.

Wang, D., Liu, L., Zhou, Y., Wang, L., Zhang, Y., & Xue, B. (2021). Fungus-mediated green synthesis of nano-silver using *Aspergillus sydowii* and its antifungal/antiproliferative activities. *Sci Rep.* 11, 10356.

Wang, Q., Lin, T.W., Tang, L., Johnson, J.E., & Finn, M.G. (2002). Icosahedral virus particles as addressable nanoscale building blocks. *Angew Chem Int Ed.* 41, 459–462.

Xie, J., Lee, J.Y., Wang, D.I., & Ting, Y.P. (2007). Identification of active biomolecules in the high-yield synthesis of single-crystalline gold nanoplates in algal solutions. *Small.* 3, 672–682.

Young, M., Willits, D., Uchida, M., & Douglas, T. (2008). Plant viruses as biotemplates for materials and their use in nanotechnology. *Annu Rev Phytopathol.* 46, 361–384.

Zhang, D., MaXin, L., Huang, H., & Zhang, G. (2020). Green synthesis of metallic nanoparticles and their potential application to treat cancer. *Front Chem.* 8, 799.

Zhang, X., Qu, Y., Shen, W., Wang, J., Li, H., Zhang, Z., Li, S., & Zhou, J. (2016). Biogenic synthesis of gold nanoparticles by yeast *Magnusio mycesingens* lh-f1 for catalytic reduction of nitrophenols. *Colloids Surf A Physicochem Eng Asp.* 497, 280–285.

4 Green Synthesis of Plant-Based Nanoparticles

Shalini Tiwari and Barkha Sharma
Govind Ballabh Pant University of Agriculture & Technology

CONTENTS

4.1 INTRODUCTION

"Nanotechnology is concerned with the segregation, consolidation, and deformation of materials by one atom or by one molecule." The scientific advancement of the twenty-first century was the development of nanotechnology (Adeyemi et al., 2022). The interdisciplinary scope of this topic includes the development, administration, and application of materials with a scale less than 100 nm (Anjum et al., 2019). An entirely new field of study, materials science and engineering, has been opened up by the remarkable daily increase in nanotechnology, which includes microbiology, nano-biotechnology, and surface-enhanced Raman scattering (SERS) (Chaudhary & Sharma, 2019; Letchumanan et al., 2021; Kumari et al., 2021). In different fields including agriculture and engineering (such as environmental, electronic, and mechanical), biomedical and therapeutic sciences, chemical manufacturing industries, space industries, biofuel and energy science, catalysis, photo-electrochemical applications and remediation purpose, nanotechnology is becoming more and more significant (Narayanan & Sakthivel, 2011; Agri et al., 2022; Chaudhary et al., 2022; Bhatt et al., 2022). Nanoparticles (NPs) are extremely miniscule in size and have a high surface-to-volume ratio. They are of great interest as they display different physical and chemical properties from bulk materials with the same chemical makeup,

DOI: 10.1201/9781003345565-4

including biological, electrical, and absorption (Clark & Macquarrie, 2008; Agri et al., 2021; Kumari et al., 2020; Kukreti et al., 2020). Applications for nanoparticles include wireless electronic logic, memory systems, optics, antimicrobial activities, computer transistors, biosensors, stress management, crop production, and soil health maintenance (Nath & Banerjee, 2013; Khati et al., 2017; 2018; 2019a, b; Chaudhary et al., 2021a, b, c, d, e). Nanoparticle synthesis can be done using organisms ranging from prokaryotes to eukaryotes. Actually, living organisms have the ability to synthesize metal nanoparticles, which has stepped up an advanced and exciting path for the development of these natural nano-factories (Mubarak Ali et al., 2011).

Green nanotechnology is a modern, rapidly growing, and effective nanotechnology field that, through research, development, and innovation, is assisting in the advancement of innovative, dependable, and sustainable solutions to global concerns (Jadoun et al., 2021). Indeed, green nanotechnology is predicated on reducing the risks associated with the production and application of nanomaterials (Paosen et al., 2017). The objective of green nanotechnology is to create unique and profitable nanomaterials with long-term benefits by using living organisms or their constituent components (such as cells, tissues, or organs) as a raw material (Iravani, 2011). A wide range of nanostructured materials have recently been synthesized naturally, without utilizing obnoxious chemicals or organic solvents and opting for an environment-friendly approach (Hussain et al., 2016). Due to their unique qualities that can only be accessed by these nanostructures, the recently manufactured metal nanoparticles (MNPs), one of the nano-objects that have been produced using green nanotechnology technologies and are possibly the most promising, are widely used as functional materials (Srikar et al., 2016). Recently, synthesis of metal nanoparticles using plants and their parts as raw extract is a topic of interest across the globe. Researchers from all over the world are interested in the biological synthesis, also known as the "green synthesis," of metal nanoparticles, particularly gold and silver nanoparticles, using plants (plant tissue, plant extracts, and whole living plants). Plants or their extracts can act as both reducing and capping agents in this process (Yadi et al., 2018). Plant or its extract-based metal NP synthesis is especially cost-effective, making it a practical and affordable choice for large-scale metal NP synthesis (Dikshit et al., 2021). Although it is environmentally favourable, the bioreduction of metal NPs by mixtures of metabolites found in plant extracts (such as proteins, enzymes, polysaccharides, and organic acids) is chemically complex (Roy and Das, 2015). This chapter has covered the green synthesis of all possible metal nanoparticles using plants because of the significant and important roles that plants play in bioinspired procedures for producing them.

4.2 GREEN SYNTHESIS OF PLANT-BASED NANOPARTICLES: GLOBAL PERSPECTIVE

Metal ions from salt solutions are chemically reduced to synthesize metal nanoparticles (MNPs), which are then stabilized by adding a stabilizing agent, also referred to as a capping agent or stabilizer. Strong bases, such as sodium borohydride ($NaBH_4$) or sodium hydroxide ($NaOH$), are frequently used in this process (Alabdallah & Hasan, 2021). However, the substances used as reducing agents and solvents to dissolve the stabilizers are frequently hazardous, and if any residues of them are found in the nanosystems, they may have a negative and destructive effect on both human health and the environment (Gour & Jain, 2019). Therefore, this may result in a number of issues regarding the safety of MNP applications as well as a quest for creative alternative methods aimed at their synthesis. One of these options is the green synthesis of MNPs using biological systems, which is based on the principle of green chemistry (Behzad et al., 2021).

Prokaryotic (microorganisms) or eukaryotic (including plants and animals) or parts of them can be utilized in the advanced green synthesis of MNPs, which may take place through intracellular or extracellular pathways (Peralta-Videa et al., 2016). The mass green synthesis of MNPs is facilitated by the reduction of a target metal ion, which is encouraged by primary and secondary metabolites from biological components. The reducing agents or other molecules present during the production of MNPs may also create a layer (coating) that stabilizes them, lowering or eliminating their propensity to aggregate or grow irregularly (Hano & Abbasi, 2021). Additionally, particular experimental parameters (such as temperature, pH, and reagent concentration) may have an impact on the creation and characteristics of MNPs produced using green synthesis techniques (Behzad et al., 2021).

The majority of MNPs produced using green synthesis techniques exhibit traits that are desirable from a sustainability perspective, including being environmentally friendly (using less toxic chemicals and solvents), quick and easy to make (fewer steps), biocompatible (capable of being used without any prior purification directly in target applications), biodegradable (capable to be hydrolysed by biological pathways), cheap cost, and high yield (Khan et al., 2019). Processes for producing nanoparticles using plant extracts are easily scalable and may be less expensive as compared to the very expensive technologies based on microbiological processes. Plant extracts could serve as both reducing and stabilizing agents for the synthesis of nanoparticles. It is recognized that the plant extract's source impacts the nanoparticles' properties (Wahab et al., 2021) (Figure 4.1).

4.3 CHARACTERIZATION OF PLANT-BASED NANOPARTICLES

Usually, the size, shape, surface area, and dispersity of nanoparticles are their distinguishing features. Numerous applications depend mainly on the uniformities of these properties (Patel et al., 2015). The following methods of identifying nanoparticles are widespread: UV-visible spectrophotometry, dynamic light scattering (DLS), scanning electron microscopy (SEM), transmission electron microscopy (TEM), X-ray diffraction (XRD), and Fourier transform infrared spectroscopy (FTIR) (Jiang et al., 2009).

One of the frequently employed techniques is UV-visible spectroscopy. Wavelengths of light between 300 and 800 nm are frequently used to characterize various metal nanoparticles in the size range of 2–100 nm (Brito et al., 2022). The silver and gold nanoparticles are each characterized using spectroscopic absorption measurements in the 400–450 nm and 500–550 nm wavelength ranges, respectively (Feldheim & Foss, 2002). The dynamic light scattering (DLS) technique is used to determine the surface charge and size distribution of the particles floating in a liquid

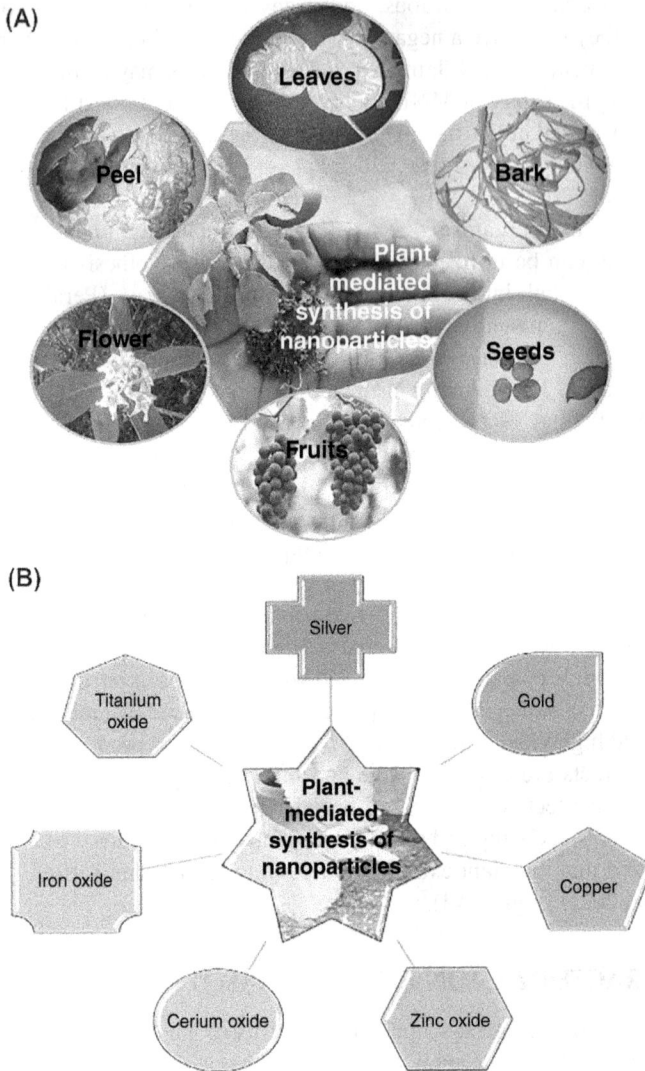

FIGURE 4.1 Different plant parts used (A) for the green synthesis of various nanoparticles (B). (Taken with permission from Kumar and Rajeshkumar (2018), Licence Number: 5394160401790.)

(Jiang et al., 2009). Another way of characterization that is frequently employed is electron microscopy. At the nanoscale to micrometre scale, morphological characterization is carried out using transmission electron microscopy (TEM) and scanning electron microscopy (SEM). TEM has a 1,000-fold higher resolution when compared to SEM, which makes it more preferential than SEM (Schaffer et al., 2009).

With the help of FTIR spectroscopy, the surface chemistry can be analysed. Organic functional groups that adhere to nanoparticle surfaces, such as carbonyls and hydroxyls, as well as other surface chemical residues, are detected using FTIR (Chithrani et al., 2006). XRD is used to characterize and identify the phase of the crystallization in the nanoparticles. When X-rays enter the nanomaterial, they produce a diffraction pattern that is weighed against guidelines for gathering structural information (Kumar et al., 2010).

4.4 GREEN SYNTHESIS OF NANOPARTICLES

The processes used for nanoparticle synthesis typically include either a "bottom-up" or a "top-down" strategy. The production of nanoparticles occurs in top-down synthesis, by sizing down from an appropriate starting material (Thakkar et al., 2010). Top-down approaches create goods with defective surface structures, which is a severe drawback given how important the surface chemistry and other physical characteristics of nanoparticles are (Mallikarjuna et al., 2011).

In the process of "bottom-up synthesis," nanoparticles are joined by smaller building blocks including atoms, molecules, and smaller particles. Bottom-up synthesis produces the final particle by first creating the nanostructured building components of the nanoparticles. Chemical and biological manufacturing techniques are mostly used in bottom-up synthesis (Dhillon et al., 2012). The bottom-up method is the most likely mechanism for synthesizing nanoparticles. Microorganism-based synthesis methods have received the most attention among biological synthesis methods. Naturally, microbial synthesis is easily scalable, safe for the environment, and consistent with the product's use in therapeutic applications; yet, microbes-based synthesis is considerably more expensive than that of plant extracts (Park et al., 2011) (Figure 4.2).

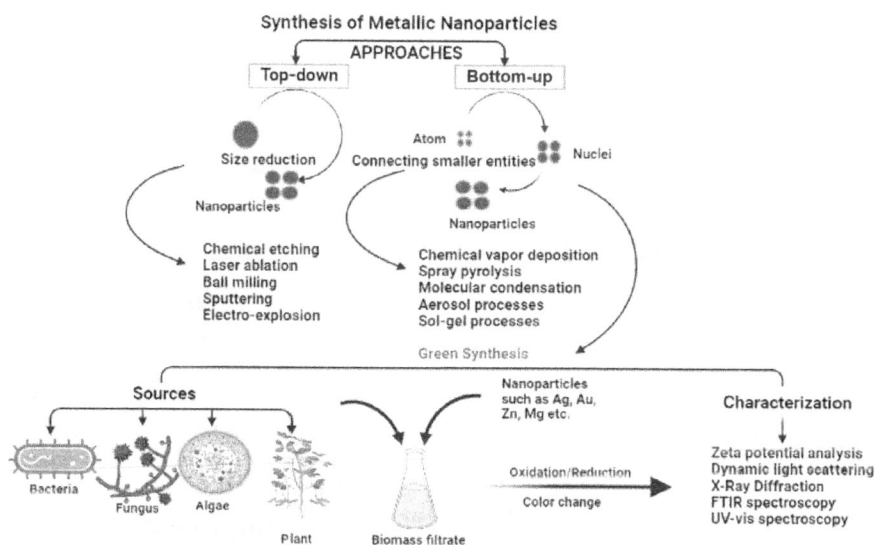

FIGURE 4.2 Schematic representation of green synthesis (BIOINSPIRED) of nanoparticles (FTIR, Fourier transform infrared spectroscopy.)

TABLE 4.1

Plant-Based Nanoparticles with Their Antimicrobial Properties Against Various Pathogens

Plant	Nanoparticles	Antimicrobial Properties Against Microorganisms	References
Curcuma longa	AgNPs	Escherichia coli BL-21 strain	Sathishkumar et al. (2010)
Dioscorea bulbifera	AgNPs	MDROs: Acinetobacter baumannii, Pseudomonas aeruginosa, E. coli	Ghosh et al. (2010)
Catharanthus roseus	AgNPs	Bacillus subtilis, Staphylococcus aureus, E. coli, Klebsiella pneumoniae, Candida albicans	Malabadi et al. (2012)
Brillantaisia owariensis, Crossopteryx febrifuga, Senna siamea	AgNPs	S. aureus, E. coli, P. aeruginosa	Kambale et al. (2020)
Fagonia indica	AgNPs	E. coli, Citrobacter amalonaticus, Shigella sonnei, Salmonella typhi	Adil et al. (2019)
Urtica dioic	AgNPs	Bacillus cereus, B. subtilis, S. aureus, Staphylococcus epidermidis, E. coli, K. pneumoniae, Serratia marcescens, Salmonella typhimurium	Jyoti et al. (2016)
Ocimum tenuiflorum	NiNPs	K. pneumoniae, S. typhi and E. coli, S. epidermidis, B. subtilis and Fungi: Candida albicans	Jeyaraj Pandian et al. (2016)
Piper guineense	AuNPs	S. aureus, Streptococcus pyogenes	Shittu et al. (2017)
Zea mays	AgNPs	B. cereus, Listeria monocytogenes, S. aureus, E. coli, S. typhimurium	Rajkumar et al. (2019)
Typha angustifolia	AgNPs	E. coli, K. pneumoniae	Gurunathan (2015)

4.4.1 GREEN SYNTHESIS USING PLANTS

In order to create nanoparticles, plant extracts are correctly combined with a metal salt solution at room temperature, and the rapid reaction time is seen. Silver, gold, and numerous other metal nanoparticles have all been synthesized using the above-mentioned procedure (Li et al., 2011). The creation of nanoparticles, as well as their quantity and other features, is influenced by a number of variables, including the kind of plant and plant part extract, the concentration of the metal salt, pH, temperature, and incubation period (Dwivedi & Gopal, 2010).

According to Prasad and Elumalai (2011), silver nanoparticles were synthesized utilizing a *Polyalthia longifolia* leaf extract. The presence of biochemical molecules such as phenolics, terpenoids, polysaccharides, and flavone chemicals in the extract was credited with the reduction and exhibits a maximum bactericidal activity at a concentration of 45 μg/mL. The leaf extract of *Pelargonium graveolens* was used in the synthesis of silver nanoparticles; the particles formed quickly and a steady size of 16–40 nm was attained. Silver nitrate and the natural monoterpene alcohol geraniol ($C_{10}H_{18}O$), which is present in some plants, are combined to create silver nanoparticles that are consistently formed and range in size from 1 to 10 nm (Huang et al., 2007). At a dosage of 5 μg/mL, it was discovered that these nanoparticles might limit the growth of the cancer cell line (fibrosarcoma WEHI 164) by up to 60%.

The presence of H^+ ions, NAD^+, and ascorbic acid in the leaf extract of *Desmodium triflorum* was attributed to the reduction of silver ions to nanoparticles. The usage of *Datura metel* leaf extract for the synthesis of very stable silver nanoparticles (16–40 nm) has also been reported. The leaf extract contains a high amount of various metabolites such as alkaloids, enzymes, amino acids, alcoholic compounds, and polysaccharides which are mainly responsible for the reduction of silver ions to nanoparticles (Table 4.1). Also, the quinol and chlorophyll pigments present in the leaf extract may have helped in the reduction and stabilization of the silver nanoparticles (Ahmad et al., 2011). Leaf extract of *Euphorbia hirta* was used to create spherical-shaped silver nanoparticles (40–50 nm) (Jha et al., 2009). Silver nanoparticles with strong antibacterial properties were made using *Citrus sinensis* peel extract. Average size of particles generated at 60°C was roughly 10 nm, but the size of particles increased abruptly to 35 nm when the reaction temperature was dropped to 25°C (Vilchis-Nestor et al., 2008).

According to Castro et al. (2011), excellent gold nanowire synthesis was seen when sugar beet pulp was used as a raw material which acts as a reductant at room temperature. Chains and nanowires were later created by joining the basic nanoparticles The parameters of the reaction, specifically the pH, had an impact on the development of the nanowires and nanorods. *Azadirachta indica* leaf extract has the potential to produce excellent bimetallic Ag-Au core-shell nanoparticles when Ag^+ and Au^+ ions are reduced in aqueous solution. The extract's reducing sugars may have helped the nanoparticles become more stable (Shankar et al., 2004). *Aloe barbadensis* and *Camellia sinensis* extracts have been used to create gold and silver nanoparticles. The initial amounts of the metal salts and *Clonorchis sinensis* extract had an effect on the optical properties of the nanoparticles. Using latex from the Euphorbiaceae family, silver and copper nanoparticles (5–10 nm) were synthesized. These nanoparticles had good bactericidal action against both Gram-negative and Gram-positive bacteria (Chandran et al., 2006; Vilchis-Nestor et al., 2008) (Table 4.2).

Bayberry tannin aqueous solution was utilized in a one-step, Au-Pd core-shell nanoparticle synthesis that occurred at room temperature. Tannin caused the reduction of Au^{3+} ions to Au nanoparticles when combined with an amalgam of Au^{3+} and Pd^{2+}. The gold nanoparticles then worked as seeds to help a Pd shell grow (Huang et al., 2011). Amarnath et al. (2011) revealed that the palladium nanoparticles have high antimicrobial efficacy, which is stabilized by chitosan and grape polyphenols. Coffee and tea extracts could be used to make palladium nanoparticles, which ranged in size from 20 to 60 nm (Nadagouda & Varma, 2008). An extract of *Annona squamosa* L. peel was used to create palladium nanoparticles (75–85 nm) (Roopan et al., 2012). Lee et al. (2011) used magnolia leaf extract to create copper nanoparticles

TABLE 4.2
Biosynthesis of Nanoparticles Using Various Parts of Plants and Its Extract

Plant	Plant Part	Nanoparticle (NP)	Size of NP (nm)	Shape of NP	References
Rheum turkestanicum	Leaves	CeO_2	30	Cubic	Sabouri et al. (2022)
Pistacia chinensis	Seed	Au	10–100	Spherical	Alhumaydhi (2022)
Carissa spinarum	Leaves	ZnO	45.76	Cubic	Saka et al. (2022)
Sanvitalia procumbens	Whole	ZnO	19	Spherical	Rashid et al. (2022)
Andrographis paniculata	Leaves	V_2O_5	58	Spherical	Selvakumar et al. (2022)
Moringa oleifera	Leaves	CuO	–	Microspheres	Kalaiyan et al. (2021)
Moringa oleifera	Leaves	MgO	20–50	Spherical	Fatiqin et al. (2021)
Rhazya stricta	Leaves	ZnO	70–90	Hexagonal	Najoom et al. (2021)
Lippia adoensis	Leaves	ZnO	19.78	Spherical/Hexagonal	Demissie et al. (2020)
Moringa oleifera	Leaves	FeO	15	Rod	Aisida et al. (2020)
Cassia fistula	Leaves	ZnO	2.72	Spherical	Naseer et al. (2020)
Melia azedarach	Leaves	ZnO	3	Spherical	Naseer et al. (2020)
Handelia trichophylla	Flowering parts	Ag	20–50	Spherical	Yazdi et al. (2019)
Allium ampeloprasum	Leaves	Ag	2–43	Spherical	Khoshnamvand et al. (2019)
Filicium decipiens	Leaves	Pd	2–22	Spherical	Sharmila et al. (2017)
Clerodendrum phlomidis	Leaves	Ag	23–42	Spherical	Sriranjani et al. (2016)
Ipomoea pes-caprae	Leaves	Ag	14	Spherical	Arulmoorthy et al. (2015)
Erythrina indica	Leaves	Ag	70–90	Spherical	Kalainila et al. (2014)
Ammannia baccifera	Aerial	Ag	10–30	Hexagonal	Suman et al. (2013)

(40–100 nm) which were toxic to adenocarcinomic human alveolar basal epithelial cells (A549 cells) and had good antimicrobial activity against *Escherichia coli.*

4.5 FACTORS AFFECTING PLANT-BASED GREEN SYNTHESIS

The two main challenges that are frequently encountered in the biogenesis of nanoparticles are controlling the size and shape of the particle and establishing monodispersity in the solution phase. There are a number of variables that could directly affect or indirectly impede the phytosynthesis of metal nanoparticles (Akhtar et al., 2013). To overcome these difficulties, efforts have been made to regulate growth factors such as pH, temperature, and incubation time.

4.5.1 pH

From the findings of prior studies, it is clear that pH may be a very important factor in the process of biosynthesis of nanoparticles. The colloidal gold biosynthesis from alfalfa biomass is influenced by pH, and researchers found that the pH shift affected the size of the nanoparticles (Gardea-Torresdey et al., 1999). A synthetic methodology must be optimized in order to efficiently synthesize nanoparticles because different plant extracts, and even extracts from different regions of the same plant, may have varied pH values. According to several researchers, lower pH (2–4) is preferable over higher pH for the formation of bigger nanoparticles. According to Armendariz et al. (2004) the size of gold nanoparticles with *Avena sativa* is highly dependent on pH and affects its synthesis. At pH 2, large-sized nanoparticles (25–85 nm) were produced, albeit in small amounts, while pH 3 and 4 produced considerable amounts of smaller-sized nanoparticles. They postulated that gold nanoparticles would choose to agglomerate to form larger nanoparticles at low pH (pH 2) instead of nucleation for the new nanoparticle synthesis. At pH 3 and 4, on the other hand, there are more functional groups (carbonyl and hydroxyl) that can bind gold (Au). As a result, newer Au (III) complexes would attach to the biomass at once, nucleating separately and generating nanoparticles that are relatively small in size.

At high pH, silver undergoes a rapid and inclusive reduction, and at various pH levels, silver nanoparticles that have been synthesized exhibit a negative zeta potential (Dwivedi & Gopal, 2010). They discovered that increasing the pH causes the absolute value of the negative zeta potential to increase, resulting in the formation of highly dispersed nanoparticles. The fact that there was only a very tiny difference in the zeta potential values between pH 2 and 10 in their investigation with *Chenopodium album* suggests that silver and gold nanoparticles are stable throughout a wider pH range.

4.5.2 TEMPERATURE

It is undeniable that the synthesis of silver nanoparticles increases with increasing temperature. At higher temperatures, gold nanoparticles formed at a faster rate. It was discovered that at higher temperatures, nanorod- and platelet-shaped gold nanoparticles form, whereas spherical nanoparticles form at lower temperatures (Kumar et al.,

2010). In a study utilizing Tansy (*Tanacetum vulgare*) fruit extract as a temperature probe from 25°C to 150°C, the absorption peak's sharpness increased for both silver and gold nanoparticles. This is because as the temperature increases the rate of reaction also increases, which improves the process of nanoparticle synthesis as the size of synthesized nanoparticles determines how sharp the absorbance peak is, as with higher temperature, the particle size decreases, which causes the plasmon resonance band of silver and gold nanoparticles to sharpen (Gericke & Pinches, 2006).

4.5.3 INCUBATION TIME

According to earlier studies, contact or incubation time may also have an impact on how nanoparticles are synthesized (Dwivedi & Gopal, 2010). This is the amount of time needed to finish the reaction's various steps. It was seen that the UV absorption spectra peaks of *Chenopodium* leaf extract were sharpened as contact time was increased. They reported that nanoparticles appeared within a short time interval of 15 minutes of the reaction and they increased a maximum of up to 2 hours, but after that only slight variations were seen. The nanoparticle instability was developed because it takes a certain amount of time for nanoparticles to fully nucleate and then become stable. Within 10 minutes of the reaction starting, silver and gold nanoparticle formation began in the Tansy fruit-mediated synthesis. Additionally, they discovered that a longer contact time is what causes the peaks in both silver and gold nanoparticles to become sharper. During their experiment, this research team found that 60 minutes was the ideal amount of time to allow the reaction for the synthesis of silver nanoparticles to be complete (Dubey et al., 2010).

4.6 MECHANISMS BEHIND PLANT-BASED GREEN SYNTHESIS

The metabolic pathways for synthesis of metal and bimetallic NPs must be explained in order to develop a reasonable strategy in this area. Therefore, it is crucial to conduct research and analysis on the underlying molecular mechanism in order to regulate the physical and chemical properties such as size, shape, and crystallinity of the metal NPs (Nadagouda & Varma, 2008). Several theories for the synthesis of nanoparticles have been put forth in recent years. More study is needed since the precise process leading the plant-based metal nanoparticles is unknown (Elavazhagan & Arunachalam, 2011). Numerous additional theories regarding the mechanism by which metal nanoparticles participate in phytosynthesis have been put forth since the initial data were published in the 1960s and the late 1990s.

The National Chemical Laboratory (NCL), Pune, India, research team documented their efforts over the past 10 years to explain the mechanism behind silver and gold NP synthesis from a variety of plants, including geranium, neem, aloe vera, lemon grass, tamarind, and chickpea (Rautaray et al., 2005). They made the assumption that terpenoids present in the geranium extract play a crucial role in the reduction of silver ions and their oxidation to carbonyl groups. To support the hypothesis, in FTIR analysis, they assigned a band of $1,748\,\mathrm{cm}^{-1}$ to the ester C=O group of chlorophyll (Shankar et al., 2003). Use of neem leaf broth for the synthesis of metallic and bimetallic nanoparticles, reducing sugars present in the broth, was responsible for the reduction of metal ions, leading to the synthesis of corresponding metal NPs. It was thought that the secondary metabolites such as flavonoids and terpenoids which are components of neem leaf broth act as the surface-active molecules stabilizing the NPs. Their research team discovered that the reducing sugar (aldoses) present in the extract of lemon grass (*Cymbopogon flexuosus*) was responsible for the reduction of aqueous $AuCl_4^-$ for the synthesis of gold nano triangles. After reduction, it was discovered that aldehydes/ketones bound to the developing nanoparticles, giving them a "liquidlike" appearance that allowed for room temperature sintering. The aldehydes and ketones found in the extracts were thought to be crucial in determining how well the shape of gold NPs was evaluated (Shankar et al., 2004).

Two techniques including inductive coupled plasmon (ICP) and XRD spectroscopy were used in the study with *Chilopsis linearis* for the analysis of phytoextraction by gold. According to the authors of this paper, salt was translocated inside the plant after being transported in its ionic form across the root membrane, where it underwent a reduction process to become metal elements (Gardea-Torresdey et al., 2005). The synthesis of silver nanoparticles using the extract of citrus lemon confirmed the ability of citric acid to function as both the chief reducing agent and possibly the stabilizing agent. Research on *Citrullus colocynthis* found that the chemicals firmly connected to the synthetically made silver nanocomposites may be polyphenols with bound amide regions and aromatic rings (Narayanan & Sakthivel, 2011). Gold nanoparticle reduction and stabilization are two tasks that fenugreek seed extract is capable of performing. They explained that seed extract rich in flavonoids can act as powerful reducing agents and might be the reason behind the reduction of chloroauric acid, also proteins having carboxylate groups can act as a surfactant which may work as an adherent to the top surface of gold (Au) nanoparticles (Narayanan & Sakthivel, 2008).

4.7 APPLICATIONS OF PLANT-BASED NANOPARTICLES

In addition to the biological sciences, where they are currently in high demand, NPs have a broad usage in electronics, biomedical, environment, energy, and other sectors of the economy. NPs have been intensively studied in every sector and are incredibly intriguing for biological applications, including the most well-known Ag and Au NPs. In comparison to NPs produced chemically, green NPs obtained from plants typically have a wider range of applications and less negative side effects.

- Protection of human health through nanomedicine (antimicrobial, antiparasitic, antiproliferative, anti-apoptotic, anti-oxidative, anti-inflammatory activities, etc.) (Al-Radadi, 2022).
- Agriculture (target-specific delivery of biomolecules, precision farming with controlled agrochemical release, more effective nutrient uptake, detection and management of plant diseases, etc.) (Alabdallah & Hasan, 2021) (Figure 4.3).
- Food science and engineering (processes for processing, storing, and packaging food) (Singh et al., 2017).

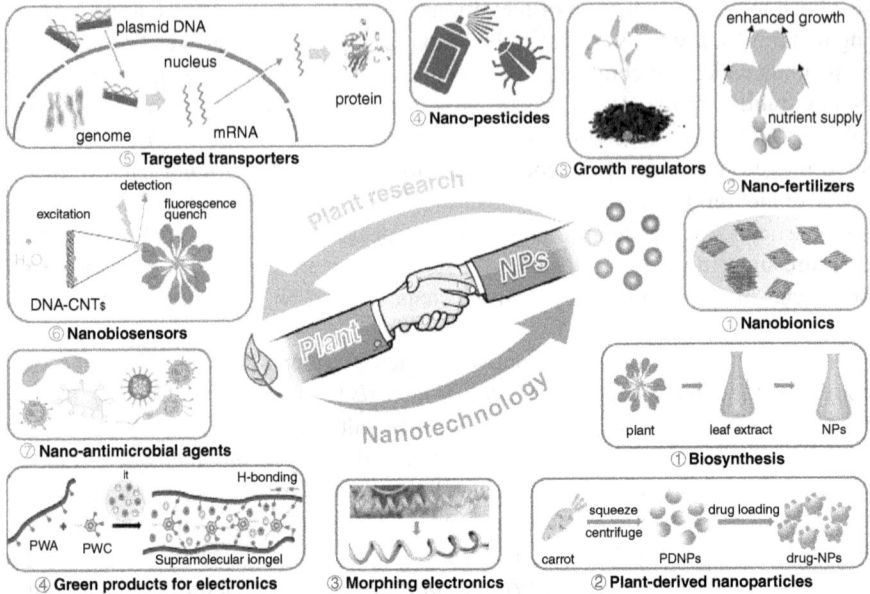

FIGURE 4.3 The various functions of NPs in plants are represented by the green boxes. These functions include nanobionics, nano-fertilizers, growth regulators, nano-pesticides, targeted transporters, nanobiosensors, and nano-antimicrobial agents. The various functions of plants in NP-based applications are represented by the red boxes. These functions include the extraction of NPs from plants, the biosynthesis of NPs with plant extract, the design of morphing electronics inspired by plants, and green electronics. (Taken with permission from Hu and Xianyu (2021), Quote Number: 501761692.)

- Environmental engineering (biocatalysts, photocatalysts, biosensors, etc.) (Mukherjee & Mishra, 2021).
- Cosmetics (sunscreen, lotions, anti-aging, bioactive compounds delivery, nanoemulsion, etc.) (Paiva-Santos et al., 2021).

4.8 CONCLUSION

In the recent literature, there has been a significant expansion of the database of numerous plants employed in the synthesis of various nanoparticles. This chapter describes the creation and characterization of phytogenic nanoparticles, as well as the physicochemical variables that affect nanoparticle synthesis, such as pH, temperature, and incubation time. When characterizing nanoparticles, techniques such as TEM, SEM, EDS, DLS, and FTIR were used. A fascinating and recently found branch of nanotechnology that significantly impacts the environment, promotes the long-term sustainability of nano-science, and uses plants to create green nanoparticles is the use of plants in the manufacture of these particles. These green plant-based NPs have the potential to be used in various biological fields such as cosmetics, agriculture, waste water treatment, dye degradation, catalysis, agriculture, food packaging, therapeutics, biotechnology, optics, and electronics. A detailed understanding of the biochemical principles behind the synthesis of plant-mediated nanoparticles is necessary to make this strategy economically competitive with the known techniques.

REFERENCES

Adeyemi, J.O., Oriola, A.O., Onwudiwe, D.C., & Oyedeji, A.O. (2022). Plant extracts mediated metal-based nanoparticles: synthesis and biological applications. *Biomolecules.* 12, 627.

Adil, M., Khan, T., Aasim, M., Khan, A.A., & Ashraf, M. (2019). Evaluation of the antibacterial potential of silver nanoparticles synthesized through the interaction of antibiotic and aqueous callus extract of *Fagonia indica*. *AMB Exp.* 9(1), 1–12.

Agri, U., Chaudhary, P., & Sharma, A. (2021). In vitro compatibility evaluation of agriusable nanochitosan on beneficial plant growth-promoting rhizobacteria and maize plant. *Natl Acad Sci Lett.* 44, 555–559.

Agri, U., Chaudhary, P., Sharma, A., & Kukreti, B. (2022). Physiological response of maize plants and its rhizospheric microbiome under the influence of potential bioinoculants and nanochitosan. *Plant Soil.* 474, 451–468.

Ahmad, N., Sharma, S., Singh, V.N., Shamsi, S.F., Fatma, A., & Mehta, B.R. (2011). Biosynthesis of silver nanoparticles from *Desmodium triflorum*: a novel approach towards weed utilization. *Biotechnol Res Int.* 2011, 454090.

Aisida, S.O., Madubuonu, N., Alnasir, M.H., Ahmad, I., Botha, S., & Maaza, M. (2020). Biogenic synthesis of iron oxide nanorods using *Moringa oleifera* leaf extract for antibacterial applications. *Appl Nanosci.* 10, 305–315.

Akhtar, M.S., Panwar, J., & Yun, Y.S. (2013). Biogenic synthesis of metallic nanoparticles by plant extracts. *ACS Sustain Chem Eng.* 1, 591–602.

Alabdallah, N.M., & Hasan, M.M. (2021). Plant-based green synthesis of silver nanoparticles and its effective role in abiotic stress tolerance in crop plants. *Saudi J Biol Sci.* 28, 5631–5639.

Alhumaydhi, F.A. (2022). Green synthesis of gold nanoparticles using extract of *Pistacia chinensis* and their *in vitro* and *in vivo* biological activities. *J Nanomater.* 2022, 5544475.

Al-Radadi, N.S. (2022). Biogenic proficient synthesis of (Au-NPs) via aqueous extract of Red Dragon Pulp and seed oil: characterization, antioxidant, cytotoxic properties, antidiabetic anti-inflammatory, anti-Alzheimer and their anti-proliferative potential against cancer cell lines. *Saudi J Biol Sci.* 29, 2836–2855.

Amarnath, K., Kumar, J., Reddy, T., Mahesh, V., Ayyappan, S. R., & Nellore, J. (2011). Synthesis and characterization of chitosan and grape polyphenols stabilized palladium nanoparticles and their antibacterial activity. *Colloids Surf B Biointerfaces.* 92, 254–261.

Anjum, S., Anjum, I., Hano, C., & Kousar, S. (2019). Advances in nanomaterials as novel elicitors of pharmacologically active plant specialized metabolites: current status and future outlooks. *RSC Adv.* 9, 40404–40423.

Armendariz, V., Herrera, I., Jose-Yacaman, M., Troiani, H., Santiago, P., & Gardea-Torresdey, J. L. (2004). Size controlled gold nanoparticle formation by *Avena sativa* biomass: use of plants in nanobiotechnology. *J Nanopart Res.* 6, 377–382.

Arulmoorthy, M.P., Vasudevan, S., Vignesh, R., Rathiesh, A.C., & Srinivasan, M. (2015). Green synthesis of silver nanoparticle using medicinal sand dune plant *Ipomoea pescaprae* leaf extract. *Int J Sci Invent Today.* 4, 384–92.

Behzad, F., Naghib, S.M., Tabatabaei, S.N., Zare, Y., & Rhee, K.Y. (2021). An overview of the plant-mediated green synthesis of noble metal nanoparticles for antibacterial applications. *J Ind Eng Chem.* 94, 92–104.

Bhatt, P., Pandey, S.C., Joshi, S., Chaudhary, P., Pathak, V.M., Huang, Y., Wu, X., Zhou, Z., & Chen, S. (2022). Nanobioremediation: a sustainable approach for the removal of toxic pollutants from the environment. *J Hazard Mater.* 427, 128033.

Brito, S.D.C., Malafatti, J.O.D., Arab, F.E., Bresolin, J.D., Paris, E.C., de Souza, C.W. O., & Ferreira, M.D. (2022). One-pot synthesis of CuO, ZnO, and Ag nanoparticles: structural, morphological, and bactericidal evaluation. *Inorg Nano-Metal Chem.* 1–11. https://doi. org/10.1080/24701556.2022.2078358

Castro, L., Blázquez, M.L., Muñoz, J.A., González. F., Garcia-Balboa, C., Ballester, A. (2011). Biosynthesis of gold nanowires using sugar beet pulp. *Process Biochem.* 46, 1076–82.

Chandran, S.P., Chaudhary, M., Pasricha, R., Ahmad, A., & Sastry, M. (2006). Synthesis of gold nanotriangles and silver nanoparticles using *Aloe vera* plant extract. *Biotech Progr.* 22, 577–583.

Chaudhary, P., Chaudhary, A., Bhatt, P., Kumar, G., Khatoon, H., Rani, A., Kumar, S., & Sharma, A. (2022). Assessment of soil health indicators under the influence of nano-compounds and *Bacillus* spp. in field condition. *Front Environ Sci.* 9, 769871.

Chaudhary, P., Chaudhary, A., Parveen, H., Rani, A., Kumar, G., Kumar, A., & Sharma, A. (2021e). Impact of nanophos in agriculture to improve functional bacterial community and crop productivity. *BMC Plant Biol.* 21, 519.

Chaudhary, P., Khati, P., Chaudhary, A., Gangola, S., Kumar, R., & Sharma, A. (2021a). Bioinoculation using indigenous *Bacillus* spp. improves growth and yield of *Zea mays* under the influence of nanozeolite. *3 Biotech.* 11, 11.

Chaudhary, P., Khati, P., Chaudhary, A., Maithani, D., Kumar, G., & Sharma, A. (2021d). Cultivable and metagenomic approach to study the combined impact of nanogypsum and *Pseudomonas taiwanensis* on maize plant health and its rhizospheric microbiome. *PLoS One.* 16, e0250574.

Chaudhary, P., Khati, P., Gangola, S., Kumar, A., Kumar, R., & Sharma, A. (2021c). Impact of nanochitosan and *Bacillus* spp. on health, productivity and defence response in *Zea mays* under field condition. *3 Biotech.* 11, 237.

Chaudhary, P., & Sharma, A. (2019). Response of nanogypsum on the performance of plant growth promotory bacteria recovered from nanocompound infested agriculture field. *Environ Ecol.* 37, 363–372.

Chaudhary, P., Sharma, A., Chaudhary, A., Khati, P., Gangola, S., & Maithani, D. (2021b). Illumina based high throughput analysis of microbial diversity of rhizospheric soil of maize infested with nanocompounds and *Bacillus* sp. *Appl Soil Ecol.* 159, 103836.

Chithrani, B.D., Ghazani, A.A., & Chan, W.C. (2006). Determining the size and shape dependence of gold nanoparticle uptake into mammalian cells. *Nano Lett.* 6, 662–668.

Clark, J.H., & Macquarrie, D.J. (Eds.). (2008). *Handbook of Green Chemistry and Technology.* John Wiley & Sons, Blackwell Science Ltd.

Demissie, M.G., Sabir, F.K., Edossa, G.D., & Gonfa, B.A. (2020). Synthesis of zinc oxide nanoparticles using leaf extract of *Lippia adoensis* (koseret) and evaluation of its anti-bacterial activity. *J Chem.* 2020, 7459042.

Dhillon, G.S., Brar, S.K., Kaur, S., & Verma, M. (2012). Green approach for nanoparticle biosynthesis by fungi: current trends and applications. *Crit Rev Biotechnol.* 32, 49–73.

Dikshit, P.K., Kumar, J., Das, A.K., Sadhu, S., Sharma, S., Singh, S., & Kim, B.S. (2021). Green synthesis of metallic nanoparticles: applications and limitations. *Catalysts.* 11, 902.

Dubey, S.P., Lahtinen, M., & Sillanpää, M. (2010). Tansy fruit mediated greener synthesis of silver and gold nanoparticles. *Process Biochem.* 45, 1065–1071.

Dwivedi, A.D., & Gopal, K. (2010). Biosynthesis of silver and gold nanoparticles using *Chenopodium album* leaf extract. *Colloids Surf A Physicochem Eng Asp.* 369, 27–33.

Elavazhagan, T., & Arunachalam, K.D. (2011). *Memecylon edule* leaf extract mediated green synthesis of silver and gold nanoparticles. *Int J Nanomed.* 6, 1265.

Fatiqin, A., Amrulloh, H., & Simanjuntak, W. (2021). Green synthesis of MgO nanoparticles using *Moringa oleifera* leaf aqueous extract for antibacterial activity. *Bull Chem Soc Ethiop.* 35, 161–170.

Feldheim, D.L., Foss, C.A. (2002). *Metal Nanoparticles: Synthesis, Characterization, and Applications*. CRC Press, Boca Raton, FL.

Gardea-Torresdey, J.L., Rodriguez, E., Parsons, J.G., Peralta-Videa, J.R., Meitzner, G., & Cruz-Jimenez, G. (2005). Use of ICP and XAS to determine the enhancement of gold phytoextraction by *Chilopsis linearis* using thiocyanate as a complexing agent. *Anal Bioanal Chem*. 382, 347–352.

Gardea-Torresdey, J.L., Tiemann, K.J., Gamez, G., Dokken, K., Tehuacanero, S., & Jose-Yacaman, M. (1999). Gold nanoparticles obtained by bio-precipitation from gold (III) solutions. *J Nanop Res*. 1, 397–404.

Gericke, M., & Pinches, A. (2006). Biological synthesis of metal nanoparticles. *Hydrometallurgy*. 83, 132–140.

Ghosh, M., Bandyopadhyay, M., & Mukherjee, A. (2010). Genotoxicity of titanium dioxide (TiO$_2$) nanoparticles at two trophic levels: plant and human lymphocytes. *Chemosphere*. 81(10), 1253–1262.

Gour, A., & Jain, N.K. (2019). Advances in green synthesis of nanoparticles. *Artif Cells Nanomed Biotechnol*. 47, 844–851.

Gurunathan, S. (2015). Biologically synthesized silver nanoparticles enhances antibiotic activity against Gram-negative bacteria. *J Ind Eng Chem*. 29, 217–226.

Hano, C., & Abbasi, B.H. (2021). Plant-based green synthesis of nanoparticles: Production, characterization and applications. *Biomolecules*. 12, 31.

Hu, J., & Xianyu, Y. (2021). When nano meets plants: a review on the interplay between nanoparticles and plants. *Nano Today*. 38, 101143.

Huang, J., Li, Q., Sun, D., Lu, Y., Su, Y., Yang, X., & Chen, C. (2007). Biosynthesis of silver and gold nanoparticles by novel sundried *Cinnamomum camphora* leaf. *Nanotechnology*. 18, 105104.

Huang, X., Wu, H., Pu, S., Zhang, W., Liao, X., Shi, B. (2011). One-step room-temperature synthesis of Au@Pd core–shell nanoparticles with tunable structure using plant tannin as reductant and stabilizer. *Green Chem*. 13, 950–957.

Hussain, I., Singh, N.B., Singh, A., Singh, H., & Singh, S.C. (2016). Green synthesis of nanoparticles and its potential application. *Biotechnol Lett*. 38, 545–560.

Iravani, S. (2011). Green synthesis of metal nanoparticles using plants. *Green Chem*. 13, 2638–2650.

Jadoun, S., Arif, R., Jangid, N.K., & Meena, R.K. (2021). Green synthesis of nanoparticles using plant extracts: a review. *Environ Chem Lett*. 19, 355–374.

Jeyaraj Pandian, C., Palanivel, R., & Dhanasekaran, S. (2016). Screening antimicrobial activity of nickel nanoparticles synthesized using *Ocimum sanctum* leaf extract. *J Nanopart*. 2016, 4694367.

Jha, A.K., Prasad, K., Kumar, V., & Prasad, K. (2009). Biosynthesis of silver nanoparticles using *Eclipta* leaf. *Biotech Prog*. 25, 1476–1479.

Jiang, J., Oberdörster, G., & Biswas, P. (2009). Characterization of size, surface charge, and agglomeration state of nanoparticle dispersions for toxicological studies. *J Nanopart Res*. 11, 77–89.

Jyoti, K., Baunthiyal, M., & Singh, A. (2016). Characterization of silver nanoparticles synthesized using *Urtica dioica* Linn. leaves and their synergistic effects with antibiotics. *J Radiat Res Appl Sci*. 9(3), 217–227.

Kalainila, P., Subha, V., Ernest Ravindran, R. S., & Sahadevan, R. (2014). Synthesis and characterization of silver nanoparticle from *Erythrina indica*. *Asian J Pharm Clin Res*. 7, 39–43.

Kalaiyan, G., Suresh, S., Prabu, K.M., Thambidurai, S., Kandasamy, M., Pugazhenthiran, N., & Muneeswaran, T. (2021). Bactericidal activity of *Moringa oleifera* leaf extract assisted green synthesis of hierarchical copper oxide microspheres against pathogenic bacterial strains. *J Environ Chem Engi*. 9, 104847.

Kambale, E.K., Nkanga, C.I., Mutonkole, B.P.I., Bapolisi, A.M., Tassa, D.O., Liesse, J. M. I., & Memvanga, P.B. (2020). Green synthesis of antimicrobial silver nanoparticles using aqueous leaf extracts from three Congolese plant species (Brillantaisia patula, *Crossopteryx febrifuga and Senna siamea*). *Heliyon*. 6(8), e04493.

Khan, T., Ullah, N., Khan, M.A., & Nadhman, A. (2019). Plant-based gold nanoparticles; a comprehensive review of the decade-long research on synthesis, mechanistic aspects and diverse applications. *Adv Colloid Interface Sci*. 272, 102017.

Khati, P., Bhatt, P., Kumar, R., & Sharma, A. (2018). Effect of nanozeolite and plant growth promoting rhizobacteria on maize. *3Biotech*. 8, 141.

Khati, P., Chaudhary, P., Gangola, S., & Sharma, P. (2019a). Influence of nanozeolite on plant growth promotory bacterial isolates recovered from nanocompound infested agriculture field. *Environ Ecol*. 37, 521—527.

Khati, P., Chaudhary, P., Gangola, S., Bhatt, P., & Sharma, A. (2017). Nanochitosan supports growth of *Zea mays* and also maintains soil health following growth. *3 Biotech*. 7, 81.

Khati, P., Sharma, A., Chaudhary, P., Singh, A.K., Gangola, S., & Kumar R. (2019b). High-throughput sequencing approach to access the impact of nanozeolite treatment on species richness and evens of soil metagenome. *Biocatal Agric Biotechnol*. 20, 101249.

Khoshnamvand, M., Huo, C., & Liu, J. (2019). Silver nanoparticles synthesized using *Allium ampeloprasum* L. leaf extract: characterization and performance in catalytic reduction of 4-nitrophenol and antioxidant activity. *J Mol Struct*. 1175, 90–96.

Kukreti, B., Sharma, A., Chaudhary, P., Agri, U., & Maithani, D. (2020). Influence of nanosilicon dioxide along with bioinoculants on *Zea mays* and its rhizospheric soil. *3 Biotech*. 10, 345.

Kumar, S., Sneha, M., & Yun, Y.S. (2010). Immobilization of silver nanoparticles synthesized using *Curcuma longa* tuber powder and extract on cotton cloth for bactericidal activity. *Biores Technol*. 101, 7958–7965.

Kumar, S.V., & Rajeshkumar, S. (2018). Plant-based synthesis of nanoparticles and their impact. In D.K. Tripathi, P. Ahmad, S. Sharma, D.K. Chauhan, & N.K. Dubey (Eds.) *Nanomaterials in Plants, Algae, and Microorganisms* (pp. 33–57). Academic Press.

Kumari, H., Khati, P., Gangola, S., Chaudhary, P., & Sharma, A. (2021). Performance of plant growth promotory rhizobacteria on maize and soil characteristics under the influence of TiO$_2$ nanoparticles. *Pantnagar J Res*. 19, 28–39.

Kumari, S., Sharma, A., Chaudhary, P., & Khati, P. (2020). Management of plant vigor and soil health using two agriusable nanocompounds and plant growth promotory rhizobacteria in Fenugreek. *3 Biotech*. 10, 461.

Lee, H. J., Lee, G., Jang, N. R., Yun, J. H., Song, J. Y., & Kim, B. S. (2011). Biological synthesis of copper nanoparticles using plant extract. *Nanotechnol*. 1, 371–374.

Letchumanan, D., Sok, S. P., Ibrahim, S., Nagoor, N.H., & Arshad, N.M. (2021). Plant-based biosynthesis of copper/copper oxide nanoparticles: an update on their applications in biomedicine, mechanisms, and toxicity. *Biomolecules*. 11, 564.

Li, X., Xu, H., Chen, Z.S., & Chen, G. (2011). Biosynthesis of nanoparticles by microorganisms and their applications. *J Nanomat*. 2011, 270974.

Malabadi, R.B., Naik, S.L., Meti, N.T., Mulgund, G.S., Nataraja, K., & Kumar, S.V. (2012). Silver nanoparticles synthesized by in vitro derived plants and callus cultures of *Clitoria ternatea*; Evaluation of antimicrobial activity. *Res Biotechnol*. 3(5), 26–38.

Mallikarjuna, K., Narasimha, G., Dillip, G.R., Praveen, B., Shreedhar, B., Lakshmi, C.S., & Raju, B.D.P. (2011). Green synthesis of silver nanoparticles using *Ocimum* leaf extract and their characterization. *Dig J Nanomater Biostructures*. 6, 181–186.

Mubarak Ali, D., Thajuddin, N., Jeganathan, K., & Gunasekaran, M. (2011). Plant extract mediated synthesis of silver and gold nanoparticles and its antibacterial activity against clinically isolated pathogens. *Colloids Surf B Biointerfaces*. 85, 360–365.

Mukherjee, S., & Mishra, M. (2021). Application of strontium-based nanoparticles in medicine and environmental sciences. *Nanotechnol Environ Eng*. 6, 1–15.

Nadagouda, M.N., & Varma, R.S. (2008). Green synthesis of silver and palladium nanoparticles at room temperature using coffee and tea extract. *Green Chem*. 10, 859–862.

Najoom, S., Fozia, F., Ahmad, I., Wahab, A., Ahmad, N., Ullah, R., & Khan, A. A. (2021). Effective antiplasmodial and cytotoxic activities of synthesized Zinc oxide nanoparticles using *Rhazya stricta* Leaf Extract. *Evid Based Complement Alternat Med*. Elsevier. 2021, 5586740.

Narayanan, K.B., & Sakthivel, N. (2008). Coriander leaf mediated biosynthesis of gold nanoparticles. *Mat Lett*. 62, 4588–4590.

Narayanan, K.B., & Sakthivel, N. (2011). Extracellular synthesis of silver nanoparticles using the leaf extract of *Coleus amboinicus* Lour. *Mater Res Bull*. 46, 1708–1713.

Naseer, M., Aslam, U., Khalid, B., & Chen, B. (2020). Green route to synthesize zinc oxide nanoparticles using leaf extracts of *Cassia fistula* and *Melia azadarach* and their antibacterial potential. *Sci Rep*. 10, 1–10.

Nath, D., & Banerjee, P. (2013). Green nanotechnology–a new hope for medical biology. *Environ Toxicol Pharmacol*. 36, 997–1014.

Paiva-Santos, A.C., Herdade, A.M., Guerra, C., Peixoto, D., Pereira-Silva, M., Zeinali, M., & Veiga, F. (2021). Plant-mediated green synthesis of metal-based nanoparticles for dermopharmaceutical and cosmetic applications. *Int J Pharma*. 597, 120311.

Paosen, S., Saising, J., Septama, A.W., & Voravuthikunchai, S.P. (2017). Green synthesis of silver nanoparticles using plants from *Myrtaceae* family and characterization of their antibacterial activity. *Mater Lett*. 209, 201–206.

Park, Y., Hong, Y.N., Weyers, A., Kim, Y.S., & Linhardt, R.J. (2011). Polysaccharides and phytochemicals: a natural reservoir for the green synthesis of gold and silver nanoparticles. *IET Nanobiotechnol*. 5, 69–78.

Patel, P., Agarwal, P., Kanawaria, S., Kachhwaha, S., & Kothari, S.L. (2015). Plant-based synthesis of silver nanoparticles and their characterization. In M. Siddiqui, M. Al-Whaibi, & F. Mohammad (Eds.) *Nanotechnology and Plant Sciences* (pp. 271–288). Springer, Cham.

Peralta-Videa, J.R., Huang, Y., Parsons, J.G., Zhao, L., Lopez-Moreno, L., Hernandez-Viezcas, J.A., & Gardea-Torresdey, J.L. (2016). Plant-based green synthesis of metallic nanoparticles: scientific curiosity or a realistic alternative to chemical synthesis? *Nanotechnol Environ Eng*. 1, 1–29.

Prasad, T.N.V.K.V., & Elumalai, E. (2011). Biofabrication of Ag nanoparticles using *Moringa oleifera* leaf extract and their antimicrobial activity. *Asian Pacific J Trop Biomed*. 1, 439–442.

Rajkumar, T., Sapi, A., Das, G., Debnath, T., Ansari, A., & Patra, J.K. (2019). Biosynthesis of silver nanoparticle using extract of *Zea mays* (corn flour) and investigation of its cytotoxicity effect and radical scavenging potential. *J Photochem Photobiol B Biol*. 193, 1–7.

Rashid, Y., Ahmad, I., Ahmad, N., Aslam, M., & Alotaibi, A. (2022). Affective antidepressant, cytotoxic activities, and characterization of phyto-assisted zinc oxide nanoparticles synthesized using *Sanvitalia procumbens* aqueous extract. *BioMed Res Internat*. 2022, 1621372.

Rautaray, D., Sanyal, A., Bharde, A., Ahmad, A., & Sastry, M. (2005). Biological synthesis of stable vaterite crystals by the reaction of calcium ions with germinating chickpea seeds. *Crystal Growth Des*. 5, 399–402.

Roopan, S.M., Bharathi, A., Kumar, R., Khanna, V.G., & Prabhakarn, A. (2012). Acaricidal, insecticidal, and larvicidal efficacy of aqueous extract of *Annona squamosa* L peel as biomaterial for the reduction of palladium salts into nanoparticles. *Colloids Surf B Biointerfaces*. 92, 209–212.

Roy, S., & Das, T.K. (2015). Plant mediated green synthesis of silver nanoparticles-A. *Int J Plant Biol Res.* 3, 1044–1055.

Sabouri, Z., Sabouri, M., Amiri, M.S., Khatami, M., & Darroudi, M. (2022). Plant-based synthesis of cerium oxide nanoparticles using *Rheum turkestanicum* extract and evaluation of their cytotoxicity and photocatalytic properties. *Mater Technol.* 37, 555–568.

Saka, A., Tesfaye, J.L., Gudata, L., Shanmugam, R., Dwarampudi, L.P., Nagaprasad, N., & Rajeshkumar, S. (2022). Synthesis, characterization, and antibacterial activity of ZnO nanoparticles from fresh leaf extracts of Apocynaceae, *Carissa spinarum* L. (Hagamsa). *J Nanomat.* 2022, 6230298.

Sathishkumar, M., Sneha, K., & Yun, Y.S. (2010). Immobilization of silver nanoparticles synthesized using Curcuma longa tuber powder and extract on cotton cloth for bactericidal activity. *Bioresour Technol.* 20, 7958–7965.

Schaffer, B., Hohenester, U., Trügler, A., & Hofer, F. (2009). High-resolution surface plasmon imaging of gold nanoparticles by energy-filtered transmission electron microscopy. *Phys Rev B.* 79, 041401.

Selvakumar, V., Muthusamy, K., Mukherjee, A., Chelliah, R., Barathikannan, K., Oh, D. H., & Karuppannan, S. (2022). Antibacterial efficacy of phytosynthesized multi-metal oxide nanoparticles against drug-Resistant foodborne pathogens. *J Nanomater.* 2022, 6506796.

Shankar, S.S., Ahmad, A., Pasricha, R., & Sastry, M. (2003). Bioreduction of chloroaurate ions by geranium leaves and its endophytic fungus yields gold nanoparticles of different shapes. *J Mater Chem.* 13, 1822–1826.

Shankar, S.S., Rai, A., Ahmad, A., & Sastry, M. (2004). Rapid synthesis of Au, Ag, and bimetallic Au core–Ag shell nanoparticles using Neem (*Azadirachta indica*) leaf broth. *J Colloid Interface Sci.* 275, 496–502.

Sharmila, G., Fathima, M.F., Haries, S., Geetha, S., Kumar, N.M., & Muthukumaran, C. (2017). Green synthesis, characterization and antibacterial efficacy of palladium nanoparticles synthesized using *Filicium decipiens* leaf extract. *J Mol Struct.* 1138, 35–40.

Shittu, K.O., Bankole, M.T., Abdulkareem, A.S., Abubakre, O.K., & Ubaka, A.U. (2017). Application of gold nanoparticles for improved drug efficiency. *Adv Natl Sci Nanosci Nanotechnol.* 8(3), 035014.

Singh, T., Shukla, S., Kumar, P., Wahla, V., Bajpai, V.K., & Rather, I.A. (2017). Application of nanotechnology in food science: perception and overview. *Front Microbiol.* 8, 1501.

Srikar, S.K., Giri, D.D., Pal, D.B., Mishra, P.K., & Upadhyay, S.N. (2016). Green synthesis of silver nanoparticles: a review. *Green Sustain Chem.* 6, 34–56.

Sriranjani, R., Srinithya, B., Vellingiri, V., Brindha, P., Anthony, S.P., Sivasubramanian, A., & Muthuraman, M.S. (2016). Silver nanoparticle synthesis using *Clerodendrum phlomidis* leaf extract and preliminary investigation of its antioxidant and anticancer activities. *J Mol Liq.* 220, 926–930.

Suman, T.Y., Elumalai, D., Kaleena, P.K., & Rajasree, S.R. (2013). GC–MS analysis of bioactive components and synthesis of silver nanoparticle using *Ammannia baccifera* aerial extract and its larvicidal activity against malaria and filariasis vectors. *Ind Crops Prod.* 47, 239–245.

Thakkar, K.N., Mhatre, S.S., & Parikh, R.Y. (2010). Biological synthesis of metallic nanoparticles. *Nanomed Nanotechnol Biol Med.* 6, 257–262.

Vilchis-Nestor, A.R., Sánchez-Mendieta, V., Camacho-López, M.A., Gómez-Espinosa, R.M., Camacho-López, M.A., & Arenas-Alatorre, J.A. (2008). Solventless synthesis and optical properties of Au and Ag nanoparticles using *Camellia sinensis* extract. *Mat Lett.* 62, 3103–3105.

Wahab, S., Khan, T., Adil, M., & Khan, A. (2021). Mechanistic aspects of plant-based silver nanoparticles against multi-drug resistant bacteria. *Heliyon.* 7, e07448.

Yadi, M., Mostafavi, E., Saleh, B., Davaran, S., Aliyeva, I., Khalilov, R., & Milani, M. (2018). Current developments in green synthesis of metallic nanoparticles using plant extracts: a review. *Artif Cells Nanomed Biotechnol.* 46, S336–S343.

Yazdi, M.E.T., Amiri, M.S., Hosseini, H.A., Oskuee, R.K., Mosawee, H., Pakravanan, K., & Darroudi, M. (2019). Plant-based synthesis of silver nanoparticles in *Handelia trichophylla* and their biological activities. *Bull Mater Sci.* 42, 1–8.

5 Concepts of Nano-Based Agriculture

Swati Lohani and Priyanka Khati
ICAR-Vivekananda Parvatiya Krishi Anusandhan Sansthan

Anuj Sharma
CSIR-Central Institute of Medicinal and Aromatic Plants

Jyotsana Maura
DRDO

CONTENTS

5.1 INTRODUCTION

Agricultural practices have been part of human life ever since human civilization started. Agriculture is the key source of food, which is the basic necessity of any living entity. At present, we are dependent on the agriculture sector not only for food availability but also for every human life aspects that are also dependent on agriculture and agricultural products via direct or indirect means. More than 45% of the global population is directly dependent on agricultural activities for their livelihood. In India which is an agriculture-based nation, a major section of population (around 70%) depends on agriculture. Besides, the agriculture sector is the provider of raw materials for many industries such as jute and tobacco, cotton fabric, sugar, edible/-non-edible oils, rice and fruit processing, etc. Furthermore, agriculture and agricultural products are major contributors to any country's socio-economic development (Dixon & Jorgenson, 2012).

Globally, at present, there are many challenges facing agriculturalists such as global population increment, climate change, urbanization, agrochemical waste caused by runoff, accumulation of pesticides and fertilizers and many environmental ecological hazards. In addition to these problems, the agriculture sector has limited resources such as production land, water and soil, etc. (Chen & Yada, 2011). Therefore, it is a matter of great concern to fulfil the global food demand with these limited resources without damaging the environment. In the current situation, farmers emphasize on more food production to fulfil the increasing food demand; therefore, they use highly toxic agrochemicals such as chemical fertilizers, herbicides, fungicides and insecticides for the large-scale production and protection of crops. It is a well-known fact that these chemical substances can have inadvertent and detrimental environmental, ecological and social impacts. These substances are used in huge amount than their actual need because, during the application, a major amount of pesticides and fertilizers is lost due to the process of volatilization, degradation and photolysis. In case of pesticides, less than 0.1% of the applied amount acts effectively against the target organisms and may also kill beneficial insects and can also cause resistance in target organisms (Roco, 1999). Furthermore, the extreme and improper use of these chemical pesticides and fertilizers has polluted groundwater and surface water by increasing levels of nutrients and toxins, resulting in poor health, eutrophication of aquatic habitats, decreasing fishery land causing death of other water bodies.

Additionally, agricultural practices such as irrigation, excessive tillage and fertilizer overdose have caused severe damage to the nutrient profile in soils (Knorr et al., 2005). The weathering of soil is a result of flood irrigation and poor drainage practices, declining the soil quality by changing soil physicochemical properties such as soil pH, bulk density, porosity, moisture content, etc. (Mukhopadhyay, 2005). There are many ecological threats related to agricultural sector such as soil nutrient depletion, nutrient deficiency in soil, etc. This situation gets aggravated as the soil temperature rises across ecosystems (Karn et al., 2009). Therefore, according to the current situation and demand, we need to explore and practice such methods and technologies which are eco-friendly, cost-effective, easy to use and which have the ability to enhance the crop production and yield that can fulfil the global food demand. Therefore, in order to manage these problems, we need to practice one of the frontier technologies such as nano-based agriculture and nanotechnology (Eapen & D'souza, 2005; Chaudhary et al., 2021a).

5.2 NANO-AGRICULTURE AND ITS IMPORTANCE OVER CONVENTIONAL FARMING

Nanoscience and nanotechnologies are making very significant impacts due to their innovative scientific advancements which is the major achievement of the twenty-first century (Narrod & Abbott, 2011; Scott & Chen, 2013). Nanoscience deals with the matter considered at the nanoscale (1–100 nm), their physical and chemical properties, their biological impact and their execution in favour of human welfare (Holdren,

2011). According to the US-EPA, ingredients that contain particles with dimensions measuring approximately 1–100 nm are nanoparticles. Nanotechnology refers to the process of building stuff from the bottom-up with atomic precision so as to gain benefits such as lighter weight, higher strength and greater chemical reactivity. Fundamental properties of matter at nanoscale may differ due to small size and high reactivity, from corresponding bulk material. For example, carbon in the form of graphite is soft and fragile, but in the form of carbon nanotube, it is 117 times stronger than steel (Cheng et al., 2006). Precise and novel use of such nano-based technology results in the progress of innovative and new properties that can be used to solve many technical, economic and social problems. Today this technology is being used on a larger scale in the fields of medicine, food packaging, formulation, remediation and electronics (Bhatt et al., 2022). Comparative to these sectors, use of nanotechnology in agriculture is in its natal state. There are many research aspects which should be studied. Research works on the delivery system of agricultural chemicals based on nanotechnology and their field applications have been rapidly conducted by many countries (Joseph & Morrison, 2006). Nowadays, nano materials are being used to increase fertilization (or fertilizer) use efficiency that led to enhanced crop yields, reducing pesticide quantity, smart agrochemical delivery, early rapid and smart monitoring via nanosensors for assessing plant health and soil quality, regulating agricultural food security by detecting toxic chemicals and pathogens in food (Moraru et al., 2003; Chau et al., 2007; Chaudhary & Sharma, 2019; Chaudhary et al., 2021b).

There are many challenges which are faced by farmers and also by agricultural scientists due to conventional farming. Conventional farming polluted and potentially damaged the environment and caused many health hazards to humans and other animals and elicited stagnation in crop yield, nutrient depletion and reduction in organic matter of soil. The agricultural sector is always benefited from advanced techniques of science, which help us to mitigate these problems and challenges more efficiently. Nano-based agriculture and agriculture techniques are one of the best ways which help us to manage these challenges (Fraceto et al., 2016; Chaudhary et al., 2022a) (Table 5.1 and Figure 5.1).

Nano particles have some unique properties such as good sorption capacity, higher surface-to-volume ratio and controlled release kinetics of supplies to targeted sites, which make them useful implements for growth of plants and yield enhancement (Chaudhary et al., 2021c, d). In conventional farming, agrochemicals are used directly and in larger input amount than the needed amount and most of these chemicals either run off or leach and contaminate the ground water, whereas in nanotechnology-based farming, agri-products and chemicals can be used as a smart delivery system. Nano-based agrochemicals ensure nutrient and water use efficiencies through controlled and slow-release pattern (Agri et al., 2021). This leads to minimal losses and leaching of inputs to ground water. They are designed in such a way that they release the required ingredients according to the biological demands and when encountered with environmental stress (Kalia & Kaur, 2019; Khati et al., 2019a, b). The nanotechnologic intervention in farming has visions for improving the efficiency of nutrient use through nano-formulations of fertilizers, nano-pesticide-mediated IPM practices, nano-biosensors for smart monitoring and application of inputs in real time for precision farming (Sangeetha et al., 2017).

TABLE 5.1

Comparison between Nano-Agriculture and Conventional Agriculture

Properties	Nano-Agriculture	Conventional Agriculture
Nutrient uptake efficiency	Improved efficiency due to nano dimensions	Decreased efficiency as bulk amount is not available to roots
Input dose amount	Very low	Very high
Effective time of treatment	Short	Long
Smart monitoring of stress and nutrient management	Yes	No
Controlled delivery system	Present	Absent
Soil microbial activity	High	Low
Toxicity	Evident in some crops	Mostly evident in many crops

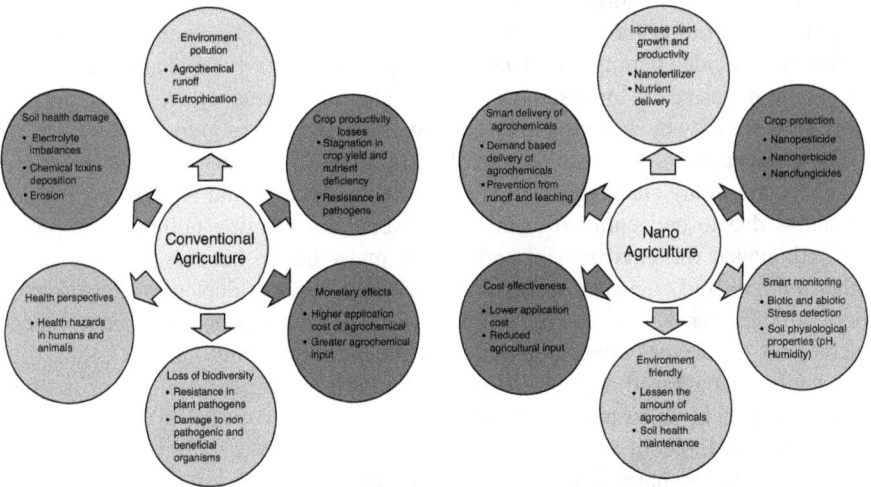

FIGURE 5.1 Comparison between conventional agriculture practices and nano-agriculture practices.

5.3 NANOTECHNOLOGY INTERVENTION IN AGRICULTURE

5.3.1 TYPES OF NANOPARTICLES USED IN AGRICULTURE

A wide variety of nanoparticles are used in the agricultural sector. Some of the most common nanoparticles used in agriculture are depicted in Figure 5.2.

5.3.2 NANOPARTICLES AND THEIR BENEFICIARY EFFECT IN AGRICULTURE

5.3.2.1 Crop Improvement and Yield Enhancement

Nanomaterial application in agriculture mainly aims to enhance crop yield through optimized nutrient management. The altered NP formulation not only improves

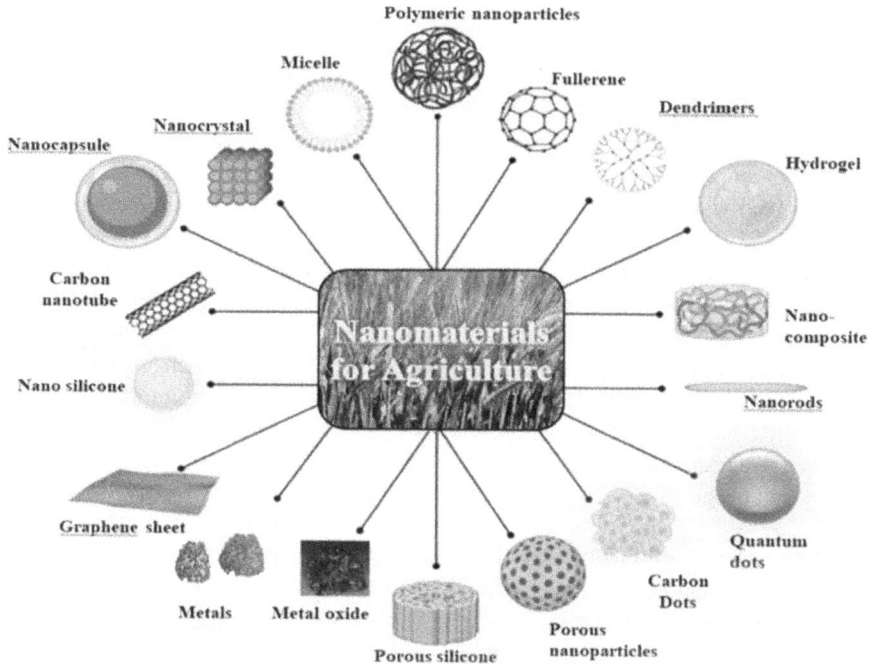

FIGURE 5.2 Nanoparticles used for delivery of agrochemicals.

nutrient use efficacy but the plant nutrients can also be retained. There are a variety of nanoparticles that improve crop production such as carbon nanomaterials, metal and metal oxide NPs and nanofertilizers (Chaudhary et al., 2021e; Agri et al., 2022). Plant nutrition has a huge impact on agricultural production and crop quality. Maintaining nutritional status by fertilizer application accounts for almost 40%–60% of total global food output (Roberts, 2009). Fertilizers serve an important role in agricultural production and development; nevertheless, the nutrient utilization efficiency of commonly used fertilizers is still poor, and a considerable amount is wasted due to runoff, leaching and microbial immobilization. Fertilizers lose a large amount of their nutrient to the environment which causes significant economic and resource losses (Wu & Liu, 2008). To address these issues, the use of nano-based fertilizers is a cutting-edge nanotechnology method that will change the fertilizer industry in the near future.

5.3.2.1.1 Carbon Nanomaterials (CNMs)

Carbon nanomaterials (CNMs) are among the significant group of nanomaterials because of their chemical, thermal, electrical and mechanical properties. CNMs can be used as a carrier for fertilizers owing to their stable molecular arrangement, dispersion uniformity and lesser toxicity during the application (Zhu et al., 2022). In a study on tomato plant, the effect of single-walled carbon nanotubes (SWCNTs), multi-walled carbon nanotubes (MWCNTs), graphene and bulk activated carbon (AC) was observed and it was discovered that single- and multi-walled CNTs facilitated improved seed germination, and greater plant biomass was also obtained due to better water uptake and retention efficiency. It was also perceived that MWCNTs act at the molecular and gene expression level such as the aquaporin genes which were significantly changed and were responsible for the working of water channel proteins (Khodakovskaya et al., 2011). It was inferred from the experiments that these nanotubes produced an aligned network of vascular tissue that consequently resulted in amplified water uptake efficiency (Tripathi et al., 2011). The use of CNTs for improved water transport in dry/arid zone was thought to be considerably more effective and likely for increasing crop biomass and yield of maize plants (Tiwari et al., 2014). Likewise, the study on wheat plants showed that at a concentration of 150 mg L^{-1} water-soluble carbon nano-dot improved root growth (Tiwari et al., 2014). Similarly, the study on bitter melon after using fullerenol-C$_{60}$ (OH)$_{20}$ results in 54%– 128% higher biomass and fruit yield, increased level of insulin which are antidiabetic compounds, and cucurbitacin-B and lycopene compounds which are anticancerous compounds (Kole et al., 2013). These studies suggest that carbon NPs may be considered as a remarkable achievement for improvement in plant growth, thus accelerating quality and progress of agro-nanotechnology.

5.3.2.1.2 Metal and Metal Oxide-Based Nanoparticles

Impact assessment of metal-based NPs such as silver (Ag), gold (Au), copper (Cu), titanium (Ti), iron (Fe) and molybdenum (Mo) on various crops was done and it was found that these particles can affect various plant growth parameters (Kumari et al., 2021). Recently the use of SNPs has gained interest of most of the researchers as it has been observed in different studies that they can alter plant growth at any level (Yan & Chen 2019). In a study in which soil of wheat plants was applied with SNPs (25–50 ppm), the outcome was improved plant height and fresh as well as dry weights with respect to control. The SNPs enhanced seminal roots, and even at lower concentration (at 25 parts per million), it increased the grain number/spike and thus improved the grain yield (Razzaq et al., 2016). In another study wherein SNPs not only improved root and shoot length but various other parameters such as fresh weight, chlorophyll contents and vigour index were significantly improved after applying nanoparticles on Indian mustard seedlings (Sharma et al., 2012). Besides SNPs, Au NPs are also considered as an auspicious tool as they have been used long ago for targeted delivery of DNA in plant cells. It has been reported that Au nanoparticles (10 μg/mL) in *Arabidopsis* improve seed production, vegetative growth and reactive oxygen species (ROX) activity (Kumar et al., 2013). Application of Au NPs (1,000 μM concentration) in flame lily (*Gloriosa superba*), which is an endangered medicinal plant, showed 39.67% increment in seed germination rate due to increased permeability of water, oxygen and uptake of gold ions. Gold ions stimulated gibberellic acid (GA3) activity after interacting with embryo cells and stimulate the release of α-amylase enzyme in aleurone layer that leads to conversion of starch into simple sugars. The other effects noted were root initiation, elongation of nodes, higher leaf numbers and increase in biomass (2.40-fold) and fresh weight (5.18-fold) (Gopinath et al., 2014). Some metal-based NPs such as iron (Fe), copper (Cu) and molybdenum (Mo) were recorded to help in improving crop production and yield. Iron (Fe) is known as a significant element for the development of plants and plays a key role in the photosynthetic reactions. Iron acts as a cofactor for various enzymes that improve the performance of photosystem and also contribute to RNA synthesis (Malakouti & Tehrani, 2005). Improvement in essential oils and better flower yield was reported after applying Fe NPs at a concentration of 1 g/L (Amuamuha et al., 2012) on marigold (*Calendula officinalis*). Similarly, different concentrations (10–50 ppm) of Cu nanoparticles were tested for plant growth enhancement in wheat plant, and it was observed that there was a major rise in leaf area, chlorophyll content, number of grains/spike, number of spikes/pot, weight of 100 grain and grain yield (Hafeez et al., 2015).

Metal oxide NPs such as titanium oxide (TiO_2), zinc oxide (ZnO), ferric oxide (Fe_2O_3), cerium oxide (CeO_2) and silicon oxide (SiO_2) have also been used for crop production enhancement (Kukreti et al., 2020; Kumari et al., 2021). They have significant potential to increase germination of seeds and enhance plant growth and yield (Razzaq et al., 2016). ZnO nanoparticles have been proved to enhance different plant growth parameters in various plants. In a study where 6-month-old onion bulbs were exposed to different concentrations (10–40 ppm), treatment was given three times with an interval of 15 days. It was observed that at the concentrations of 20 and 30 ppm plants showed significantly higher growth and flowering was achieved 12–14 days prior in comparison to control, and high-quality and healthy seeds were obtained (Laware & Raskar, 2014). Similarly, application of ZnO NPs (500–1,000 ppm) on spinach leaves reportedly increased the width, length, surface area and colour of leaves in comparison to control and also elevated protein levels and dietary fibre contents (Kisan et al., 2015). Morteza et al. (2013) reported higher yield and pigment production in *Zea mays*, after spraying nano-TiO_2 at the reproductive stage. Iron oxide (Fe_2O_3) nanoparticles have also been found to increase nutrient intake, enhanced photosynthesis and thus growth in peanut plant (Liu et al., 2005). Similarly, nano-Fe_2O_3 (0.5 ppm) exposure led to increased leaf and pod dry weight in soybean (Sheykhbaglou et al., 2010). The impacts of CeO_2 NPs on *Coriandrum sativum* were analysed for their growth and biochemical assays, and resulted in increased root and shoot length, improved biomass, higher ascorbate peroxidase activity in roots and higher catalase activity in shoots (Morales et al., 2013). Likewise, CeO_2 NPs (0–500 mg/kg) effected agronomic traits such as yield and nutrition positively in wheat, and an overall improvement of growth, biomass and yield production was observed. Besides this, nutrient composition such as amino acids was also affected which was evident from increase in linolenic acids (6.17%) over control (Rico et al., 2014). SiO_2 NPs in tomato plant significantly improved germination, vigour index and fresh weight of seedling (Siddiqui & Al-Whaibi, 2014).

5.3.2.1.3 Nanofertilizers

Currently, the estimated world population is seven billion and it may reach approximately around nine billion by 2050. For fulfilling the increasing food needs, farmers use intensive agricultural farming practices that eventually lead to reduction of soil fertility and reduction of crop yields (Huang et al., 2017). As a result of these practices about 40% of the world's agricultural land is suffering from severe loss in soil fertility and had an adverse effect on soil intrinsic nutrient equilibrium (Kale & Gawade, 2016). It is apparent that conventional fertilizers are manifested by sever losses via evaporation, leaching and self-degradation (50%–70%) which finally reduces their efficiency and increases the production cost, and a major portion of the toxic waste from conventional fertilizers remains in the soil and leach into water bodies, thereby contaminating drinking and ground water (VanDijk & Meijerink, 2014; Dubey & Mailapalli, 2016). It is a well-known fact that utilization of fertilizers is inevitable for soil fertility, crop growth and productivity. To attain higher yield in agriculture, it is prerequisite that soil fertilizers should be managed precisely (Keller et al., 2013; Davarpanah et al., 2016; Li et al., 2018b). Therefore, nanofertilizers prove to be an advanced practice in the agricultural sector. Nanofertilizers have immense potential for crop growth and nutrition as they are released in a controlled manner and minimize differential losses (Godfray et al., 2010; Kumari et al., 2020). It is evident that nanoscale nitrogenous fertilizers and some porous nanomaterials, such as clay, zeolites or chitosan, synchronize with crop requirements and release nitrogen in accordance and, thus, improve plant uptake mechanism (Millán et al., 2008; Khati et al., 2017a, b; 2018; Shukla et al., 2019). Nanofertilizers are available in different forms and according to their mode of action, they can be classified as nanocomposite fertilizers, slow-release fertilizers or magnetic fertilizers to deliver a large range of micro- and macro-nutrients with advantageous properties (Abdel-Aziz et al., 2016). Encapsulation is one of the methods for the construction of nanofertilizers in which targeted nutrients (core particles) are encapsulated within a polymeric membrane via physical or chemical methods (Couvreur et al., 1995). Nanofertilizers provide a balanced source of crop nutrition throughout different developmental stages of the plant that results in improved agricultural production. Nanofertilizers have wide applications for crop productivity. Carbon NPs along with fertilizers can improve wheat grain yields (28.81%), rice (10.29%), spring maize (10.93%) and soybean (16.74%) (Subramanian et al., 2015). The chitosan-NPK fertilizers effectively enhanced the mobilization index, harvest index and the crop index of wheat yield parameters. Nanotechnology also aids in the delivery of plant hormones, which are critical in agriculture for enhancing crop productivity and quality. The major obstacle of using plant growth-promoting rhizobacteria (PGPR) is their deterioration and decline in field condition due to environmental factors such as excess heat and light. Nanomaterials have the potential to solve such issues as they can entrap or encapsulate the bioagents by improving their stability and survival (Rossi et al., 2014; Campos et al., 2014; Grillo et al., 2016). Various nanocarrier systems based on nanochitosan and nanozeolite have been used for crop protection and gene delivery as well (Kashyap et al., 2015).

5.4 NANOPARTICLES IN PLANT PROTECTION

The application of nanotechnology in plant pathology offers new insights towards crop protection. There are various nanoparticles known for their toxicity towards pathogens and insects due to several mechanisms. Nanomaterials when applied with safety measures and following specific guidelines related to dosage and use can give best results in agriculture production system with negligible environmental effect and no toxicity to human health (Mousavi & Rezaei, 2011). Nanomaterials are applied efficiently for safe management of pesticides, herbicides and fungicides at lower doses to cover big plant surfaces and thousands of plants (Kuzma & VerHage,

2006). Nanoparticles can be used for protecting plants against various pests, insects, pathogens (fungi, bacteria, microbial pathogens) and can also be used for weed control and for increasing induced systemic resistance in plants. Nanoscale products are highly reactive and target specific as compared to the bulk counterparts; thus, very less amount of nanomaterials have the potential to show better crop protection (Debnath et al., 2011).

5.4.1 NANOPESTICIDES

Agricultural productivity is largely affected by destruction caused due to pests and pathogens (Tyagi et al., 2022). The conventional pesticides are majorly affected by premature degradation, runoff or leaching, therefore they are less efficient and always require higher amount than the actual need. Nanoparticles may have a key role in the control of host pathogens and insect pests (Khot et al., 2012). Recently developed nano-encapsulated pesticide formulation has many benefiting properties such as slow and controlled release, better solubility, more specificity, stability and permeability (Bhattacharyya et al., 2016). These nanopesticides are formulated in such a way that the encapsulated active ingredient is inside the nanocoated material. Thus, the active ingredient is protected from early death stage and thus improves the pest control efficacy for a longer period. Yang et al. (2009) tested the impact of polyethylene glycol-coated NPs which was loaded with garlic oil against adult *Tribolium castaneum* insect. About 80% control efficacy against adult *T. castaneum* apparently had been observed due to the controlled release of the active components. Nano-based pesticide formulation led to lesser dosage of pesticides and also reduced human beings' exposure to them (Nuruzzaman et al., 2016). Thus, nano-based pesticides can lead towards the approach of designing non-toxic pesticide delivery systems to increase food production globally and on the other hand drop the adverse environmental influences to the ecosystem (de Oliveira et al., 2014; Kah et al., 2013; Kah & Hofmann, 2014; Bhattacharyya et al., 2016; Grillo et al., 2016). Few chemical companies (such as Karate ZEON, Subdue MAXX, Ospray's Chyella and BASF) have openly promoted nanoscale pesticides for sale (Gouin, 2004).

5.4.2 NANOHERBICIDES

Crop weeds are considered as the biggest threat for damaging the crop because they utilize greater amount of nutrient for their growth that would otherwise be available to crop plants (Chaudhary et al., 2022b). Conventionally used techniques and methods such as eradicating weeds manually by hand consume a lot of time and require a lot of effort and manpower. Constant use of herbicides for more extensive time span may cause resistance in the weeds. Apart from these consequences, chemical herbicides, which are commercially available nowadays, can also cause damage to non-targeted plants and decline the soil fertility. Application of nanoherbicides can be an alternative, healthier and eco-friendly practice for weed control. The other benefit of using nano herbicide is that they don't contaminate the soil while the other herbicides leave toxic remains in the soil (Pérez-de-Luque, 2017). In a study of mustard plants, conventionally used atrazine herbicide and poly-epsilon caprolactone

(PCL)-encapsulated atrazine nanocapsules have been used. The atrazine nanocapsules were more prominent than the commercial atrazine formulation and illustrated more effective post-emergence herbicidal activity (Oliveira et al., 2015).

5.4.3 PROTECTION FROM PLANT DISEASES

Diseases in crop plants are one of the major factors in limiting crop productivity and yield generation. The causative agent of disease, either virus, bacteria, fungi or nematode, and their infection can spoil the crop to a larger extent and lead to reduced yield, poor yield quality and limited shelf life and thereafter economic loss. It is documented that most of the foremost crop damages incur by fungal infestation alone (Godfray et al., 2016). Though conventional fungicides work on pathogenic fungi and curb these kinds of losses, they can also simultaneously harm non-target living organisms and therefore create biodiversity imbalance (Nikunj et al., 2014). Ag NPs have antimicrobial/antifungal properties and thus can be used for disinfection purposes in infected crops (Baker et al., 2005). An experiment was conducted by Jo et al. (2009) to study the effectiveness of Ag NPs on two pathogenic fungi *Bipolaris sorokiniana* and *Magnaporthe grisea*, and they observed that after 3 hours of silver NP application the disease symptoms reduced significantly. It was also documented that both copper NPs and silver NPs can suppress the growth of fungal pathogens *Botrytis cinerea* and *Alternaria alternata* (Ouda, 2014). Studies showing antifungal activity of metal oxide nanoparticles such as ZnO and MgO NPs against *A. alternata*, *Fusarium oxysporum*, and *Mucor plumbeus* (Wani & Shah, 2012) were also evident. Similarly, Sarlak et al. (2014) designed an MWCNT-grafted poly citric acid hybrid material which encapsulated two conventional fungicide (zineb and mancozeb), and as a result, they observed it to be a more potent fungicide against *A. alternata* (Sarlak et al., 2014).

The ZnO and MgO NPs when applied at a concentration of 100 ppm showed a high inhibition rate in germination of fungal spores in *Rhizopus stolonifer*, *F. oxysporum*, *A. alternata*, and *M. plumbeus* (Wani & Shah, 2012). Similarly, the growth of pathogenic bacteria, *Pseudomonas aeruginosa*, and fungi, *Aspergillus flavus*, was significantly repressed when exposed to ZnO NPs (Jayaseelan et al., 2012).

5.5 NANOPARTICLES IN SMART MONITORING OF PLANTS

Smart monitoring system of plants proposes precise management of costly agrochemicals and detection of major plant stress events (biotic or abiotic) in a changing scenario. This monitoring helps in minimizing damages to ensure optimal crop productivity and food security (Table 5.1). The traditional techniques used for diagnosing and sensing were laborious and time consuming. The target specificity and good sensitivity of functionalized nanoparticles can be applied to design some stress detection devices for plants (Kashyap et al., 2017).

5.5.1 NANOBIOSENSOR

Nanomaterials due to their unique surface chemistry, targeted action, perfect electrical and thermal characteristics and higher sensitivities are suitable for good sensing systems for essential signalling molecules at the field and laboratory conditions (Yao et al., 2014). Nanosensors can be applied in numerous aspects such as crop cultivation, insect-pathogen detection and assessment of soil parameters. Adding to these advancements, NMs have the advantage of better sensitivity and faster detection. The binding of the target molecule with that of the biosensor can be measured via various direct and indirect means, such as recording colour changes and electrical potential. These smart nano materials can be used as nano-biosensors for the optimization of crop growth and exact detection of stress and limited resources and also communicating signals for real-time monitoring of plant health status (Giraldo et al., 2019). Another approach is array based, in which biomolecules are attached to a substrate and numerous analytes can be measured at the same time (Ghormade et al., 2011). The smart monitoring of crop plants is depicted in Figure 5.3. The current statuses of different nano-based products in the market are listed in Table 5.2.

Plants have an evolved system for stress sensing and signalling mechanism as they are sessile. There are many plant stress signalling molecules which are involved in plant stress response such as H_2O_2, Ca^{2+}, sugars, gas molecules (such as NO_2, CO, H_2S), plant hormones (such as abscisic acid, methyl salicylate, jasmonic acid, and ethylene) and volatile organic compounds such as isoprene. Plant nanobiotechnology is an effective approach in monitoring such signalling molecules in plant species. There are lots of nano-biosensors for different stress signalling molecules, for example, quantum dot is used as a sensor for detection of glucose (Li et al., 2018), hemin-complexed DNA aptamer-coated SWCNTs are used for H_2O_2 (Wu et al., 2020), carbon nanotubes coated with AT15 are used for NO (Zhang et al., 2011; Giraldo et al., 2014; 2015) and nanoneedle transistor-based sensors are used for Ca^{2+}

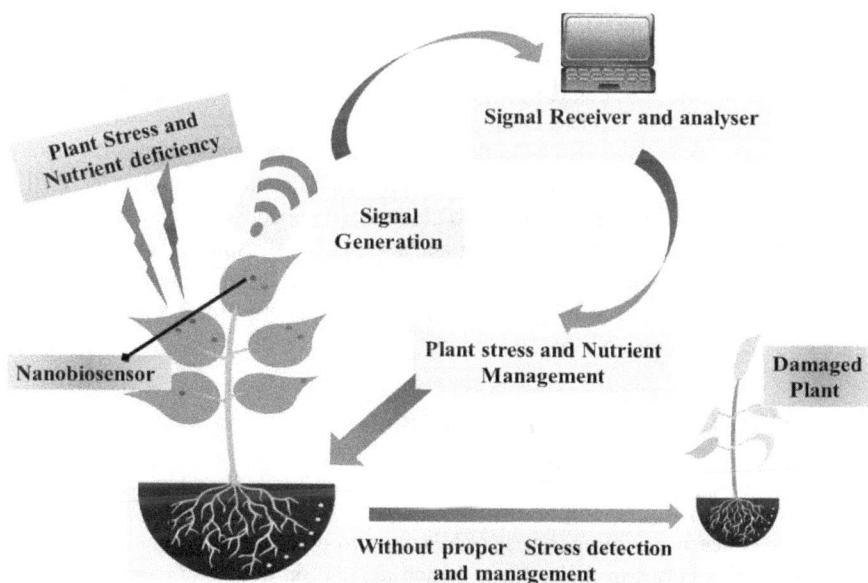

FIGURE 5.3 Smart monitoring of crop plants.

TABLE 5.2

Current Status of Nano-Agricultural Products

Application	Nano-Product	Content of Nanoparticles
Nano fertilizer	Nano-Gro™	Plant growth promotor
	TAG NANO (NPK, PhoS, Zinc, Cal, etc.)	Proteino-lacto-gluconate chelated with micronutrients, vitamins, probiotics, humic acid
	Nano Green	Extract of corn, Nano-green grain, soyabeans, potatoes, palm and coconut
	Nano Max NPK Fertilizer	Macro- and micronutrients, amino acids, organic carbon, trace elements, vitamins and probiotic chelated by organic acids
	Nano-Ag Answer®	Unknown
	Master Nano Chitosan Organic Fertilizer	Water-soluble liquid chitosan and phenolic compounds
Nanopesticides	NANOCU®	Copper NPs
	N/A	Fe, Zn, Mo, N NPs, silica NPs, polymeric NPs, ZnO NPs
Nanoherbicides	N/A	Copper-doped poly-epsilon caprolactone (PCL) nanocapsules
Nanosensors	N/A	Montmorillonite
	N/A	Graphene

(Son et al., 2011). Using TGA (thioglycolic acid)-QD and BA (boronic acid)-QD, the change of glucose level in the leaf of *Arabidopsis* plants was traced via fluorescence tracking quantum dots (Li et al., 2018). Nanosensors for monitoring temperature (Di Giacomo et al., 2015), humidity (Oren et al., 2017) and stomatal activities (Koman et al., 2017) have also been developed. Apart from stress signalling, nano-based biosensors are also reported for plant pathogen detection. Singh et al. (2010) used nanogold-based immunosensors which have the capacity to detect Karnal bunt disease in wheat (*Tilletia indica*) using surface plasmon resonance. *Xanthomonas axonopodi*, causative agent for bacterial spot disease in Solanaceae plants, can be detected by fluorescence Si NPs (Yao et al., 2009). The ethylene hormone for fruit ripening is detected by CNT-based sensors.

5.6 ETHICAL AND SAFETY CONCERNS RELATED TO NANOPARTICLES

Nanoparticles are of great importance despite ethical and safety issues always being a matter of concern regarding the use of NPs for agronomic applications. Application of nanoparticles varies in their effects which depend on their mode of application, concentration and the type of plant species as well. Therefore, there is a lack of transparent and precise framework for risk governance. It is a widely acknowledged fact that there is sparse availability of reliable information related to the safe use of NPs (Kah, 2015; Kah et al., 2019; Lauterwasser, 2005). There is an essentiality of thorough study on the impact of nano-based products and their appropriate dosage and effect on the environment (Hasler et al., 2015). It is clear that NPs such as CNTs improve seed germination by increasing water permeability and hence increase water uptake efficacy in tomato and rice (Wu et al., 2008; Nair et al., 2010), but hinder root elongation in tomato and lettuce (Canas et al., 2008). Similarly, TiO_2 NPs increase the photosynthesis rate and nitrogen metabolism and cause oxidative stress (Lei et al., 2008). Lettuce leaves internalize the Ag NPs after foliar spraying, hence the chances of transfer and accumulation in human body through the food chain are high, which can damage the cell membrane and DNA (Asha Rani et al., 2009; Larue et al., 2014; Gliga et al., 2014). As a result, toxicity and risk assessment studies are required to highlight the concerns connected with the spread and accumulation of NPs in the food chain, allowing everyone to design better policy decisions for their wider commercialization without hazardous impact (Tilman et al., 2002).

5.7 CONCLUSION

The global population is dependent on the agriculture sector for food, and overexploitation of the natural resources, and massive and uncontrolled use of agrochemicals created a major threat to food and energy supply. Nanotechnology is one of the emerging frontier technologies which have great and innovative potential for improving crop yield and maintaining environmental balance as well. Nanotechnology can resolve major agricultural problems such as crop production and yield enhancement by increasing nutrient use efficiency through nanofertilizers; nanoparticle element pest and pathogen management via nanopesticides, herbicides, fungicides; and monitoring of plant and soil health with the help of nano-biosensors, prior to any kind of disease threat and stress.

The field applications of nanotechnology are still in their early stage to attain sustainable agriculture. There are a wide range of aspects related to nanoparticle impact that should be studied and discovered by conducting research and experiments at greater extents. There is a need for thorough study and gathering the knowledge of utilizing NPs and NMs in the agriculture sector, their mode of action, acclimatization, toxicity, etc. Future researches must be emphasized towards risk assessment along with the benefits of nanotechnology to avoid the risk factors associated with it, validating the permissible level of nanoparticle dose within safety limits.

REFERENCES

Abdel-Aziz, H.M., Hasaneen, M.N., & Omer, A.M. (2016). Nano chitosan-NPK fertilizer enhances the growth and productivity of wheat plants grown in sandy soil. *Span J Agric Res*. 14(1), e0902–e0902.

Agri, U., Chaudhary, P., & Sharma, A. (2021). In vitro compatibility evaluation of agriusable nanochitosan on beneficial plant growth-promoting rhizobacteria and maize plant. *Natl Acad Sci Lett* 44, 555–559.

Agri, U., Chaudhary, P., Sharma, A., & Kukreti, B. (2022). Physiological response of maize plants and its rhizospheric microbiome under the influence of potential bioinoculants and nanochitosan. *Plant Soil*. 474, 451–468.

Amuamuha, L., Pirzad, A., & Hadi, H. (2012). Effect of varying concentrations and time of Nanoiron foliar application on the yield and essential oil of Pot marigold. *Int Res J App Basic Sci.* 3(10), 2085–2090.

Asha Rani, P.V., Low KahMun, G., Hande, M.P., & Valiyaveettil, S. (2009). Cytotoxicity and genotoxicity of silver nanoparticles in human cells. *ACS Nano.* 3(2), 279–290.

Baker, C., Pradhan, A., Paktis, L., Pochan, D.J., & Shah, S.I. (2005). Antimicrobial activity of silver nanoparticles. *J Nanosci Technol.* 5, 244–249.

Bhatt, P., Pandey, S.C., Joshi, S., Chaudhary, P., Pathak, V.M., Huang, Y., Wu, X., Zhou, Z., & Chen, S. (2022). Nanobioremediation: a sustainable approach for the removal of toxic pollutants from the environment. *J Hazard Mater.* 427, 128033.

Bhattacharyya, A., Duraisamy, P., Govindarajan, M., Buhroo, A.A., & Prasad, R. (2016). Nano-biofungicides: emerging trend in insect pest control. In Prasad, R. (Ed.) *Advances and Applications through Fungal Nanobiotechnology. Fungal Biology* (pp. 307–319). Springer, Cham.

Campos, E.V.R., de Oliveira, J.L., & Fraceto, L.F. (2014). Applications of controlled release systems for fungicides, herbicides, acaricides, nutrients, and plant growth hormones: a review. *Adv Sci Eng Med.* 6(4), 373–387.

Canas, J.E., Long, M., Nations, S., Vadan, R., Dai, L., Luo, M., Ambikapathi, R., Lee, E.H., & Olszyk, D. (2008). Effects of functionalized and nonfunctionalized single-walled carbon nanotubes on root elongation of selected crop species. *Nanomat Environ.* 2, 1922–1931.

Chau, C.F., Wu, S.H., & Yen, G.C. (2007). The development of regulations for food nanotechnology. *Trends Food Sci Technol.* 18(5), 269–280.

Chaudhary, P., Chaudhary, A., Bhatt, P., Kumar, G., Khatoon, H., Rani, A., Kumar, S., & Sharma, A. (2022a). Assessment of soil health indicators under the influence of nano-compounds and *Bacillus* spp. in field condition. *Front Environ Sci.* 9, 769871.

Chaudhary, P., Chaudhary, A., Parveen, H., Rani, A., Kumar, G., Kumar, A., & Sharma, A. (2021e). Impact of nanophos in agriculture to improve functional bacterial community and crop productivity. *BMC Plant Biol.* 21, 519.

Chaudhary, P., Khati, P., Chaudhary, A., Gangola, S., Kumar, R., & Sharma, A. (2021a). Bioinoculation using indigenous *Bacillus* spp. improves growth and yield of *Zea mays* under the influence of nanozeolite. *3 Biotech.* 11, 11.

Chaudhary, P., Khati, P., Chaudhary, A., Maithani, D., Kumar, G., & Sharma, A. (2021d). Cultivable and metagenomic approach to study the combined impact of nanogypsum and *Pseudomonas taiwanensis* on maize plant health and its rhizospheric microbiome. *PLoS One.* 16, e0250574.

Chaudhary, P., Khati, P., Gangola, S., Kumar, A., Kumar, R., & Sharma, A. (2021c). Impact of nanochitosan and *Bacillus* spp. on health, productivity and defence response in *Zea mays* under field condition. *3 Biotech.* 11, 237.

Chaudhary, P., & Sharma, A. (2019). Response of nanogypsum on the performance of plant growth promotory bacteria recovered from nanocompound infested agriculture field. *Environ Ecol.* 37, 363–372.

Chaudhary, P., Sharma, A., Chaudhary, A., Khati, P., Gangola, S., & Maithani, D. (2021b). Illumina based high throughput analysis of microbial diversity of rhizospheric soil of maize infested with nanocompounds and *Bacillus* sp. *Appl Soil Ecol.* 159, 103836.

Chaudhary, P., Singh, S., Chaudhary, A., Sharma, A., & Kumar, G. (2022b). Overview of bio-fertilizers in crop production and stress management for sustainable agriculture. *Front Plant Sci.* 13, 930340.

Chen, H., & Yada, R. (2011). Nanotechnologies in agriculture: new tools for sustainable development. *Trends Food Sci Technol.* 22, 585–594.

Cheng, M.M.C., Cuda, G., Bunimovich, Y.L., Gaspari, M., Heath, J.R., Hill, H.D., & Ferrari, M. (2006). Nanotechnologies for biomolecular detection and medical diagnostics. *Cur Opin Chem Biol.* 10(1), 11–19.

Couvreur, P., Dubernet, C., & Puisieux, F. (1995). Controlled drug delivery with nanoparticles: current possibilities and future trends. *Eur J Pharma Biophar.* 41, 2–13.

Davarpanah, S., Tehranifar, A., Davarynejad, G., Abadía, J., & Khorasani, R. (2016). Effects of foliar applications of zinc and boron nano-fertilizers on pomegranate (*Punicagranatum cv. Ardestani*) fruit yield and quality. *Sci Hortic.* 210, 57–64.

de Oliveira, J.L., Campos, E.V.R., Bakshi, M., Abhilash, P.C., & Fraceto, L.F. (2014). Application of nanotechnology for the encapsulation of botanical insecticides for sustainable agriculture: prospects and promises. *Biotechnol Adv.* 32(8), 1550–1561.

Debnath, N., Das, S., Seth, D., Chandra, R., Bhattacharya, S.C., & Goswami, A. (2011). Entomotoxic effect of silica nanoparticles against *Sitophilus oryzae* (L.). *J Pest Sci.* 84(1), 99–105.

Di Giacomo, R., Daraio, C., & Maresca, B. (2015). Plant nanobionic materials with a giant temperature response mediated by pectin-Ca^{2+}. *Proc Nat Acad Sci.* 112(15), 4541–4545.

Dixon, P.B., & Jorgenson, D. (Eds.). (2012). *Handbook of Computable General Equilibrium Modeling* (Vol. 1). Newnes, North Holland.

Dubey, A., & Mailapalli, D.R. (2016). Nanofertilisers, nanopesticides, nanosensors of pest and nanotoxicity in agriculture. In *Sustainable Agriculture Reviews* (pp. 307–330). Springer, Cham.

Eapen, S., & D'souza, S.F. (2005). Prospects of genetic engineering of plants for phytoremediation of toxic metals. *Biotechnol Adv.* 23(2), 97–114.

Fraceto, L.F., Grillo, R., de Medeiros, G.A., Scognamiglio, V., Rea, G., & Bartolucci, C. (2016). Nanotechnology in agriculture: which innovation potential does it have? *Front Environ Sci.* 4, 20.

Ghormade, V., Deshpande, M.V., & Paknikar, K.M. (2011). Perspectives for nano-biotechnology enabled protection and nutrition of plants. *Biotechnol Adv.* 29(6), 792–803.

Giraldo, J.P., Landry, M.P., Faltermeier, S.M., McNicholas, T.P., Iverson, N.M., Boghossian, A.A., & Strano, M.S. (2014). Plant nanobionics approach to augment photosynthesis and biochemical sensing. *Nat Mater.* 13(4), 400–408.

Giraldo, J.P., Landry, M.P., Kwak, S.Y., Jain, R.M., Wong, M.H., Iverson, N.M., & Strano, M.S. (2015). A ratiometric sensor using single chirality near infrared fluorescent carbon nanotubes: application to in vivo monitoring. *Small.* 11(32), 3973–3984.

Giraldo, J.P., Wu, H., Newkirk, G.M., & Kruss, S. (2019). Nanobiotechnology approaches for engineering smart plant sensors. *Nat Nanotechnol.* 14(6), 541–553.

Gliga, A.R., Skoglund, S., Odnevall-Wallinder, I., Fadeel, B., & Karlsson, H.L. (2014). Size-dependent cytotoxicity of silver nanoparticles in human lung cells: the role of cellular uptake, agglomeration and Ag release. *Particle Fibre Toxicol.* 11(1), 1–17.

Godfray, H.C.J., Beddington, J.R., Crute, I.R., Haddad, L., Lawrence, D., Muir, J.F., & Toulmin, C. (2010). Food security: the challenge of feeding 9 billion people. *Science.* 327(5967), 812–818.

Godfray, H.C.J., Mason-D'Croz, D., & Robinson, S. (2016). Food system consequences of a fungal disease epidemic in a major crop. *Philos Trans R Soc B Biol Sci.* 371(1709), 20150467.

Gopinath, K., Gowri, S., Karthika, V., & Arumugam, A. (2014). Green synthesis of gold nanoparticles from fruit extract of Terminalia arjuna, for the enhanced seed germination activity of Gloriosa superba. *J Nanostruc Chem.* 4(3), 1–11.

Gouin, S. (2004). Microencapsulation: industrial appraisal of existing technologies and trends. *Trends Food Sci Technol.* 15(7–8), 330–347.

Grillo, R., Abhilash, P.C., & Fraceto, L.F. (2016). Nanotechnology applied to bio-encapsulation of pesticides. *J Nanosci Nanotechnol.* 16(1), 1231–1234.

Hafeez, A., Razzaq, A., Mahmood, T., & Jhanzab, H.M. (2015). Potential of copper nanoparticles to increase growth and yield of wheat. *J Nanosci Adv Technol.* 1(1), 6–11.

Hasler, K., Bröring, S., Omta, S.W.F., & Olfs, H.W. (2015). Life cycle assessment (LCA) of different fertilizer product types. *Eur J Agron.* 69, 41–51.

Holdren, J.P. (2011). *The National Nanotechnology Initiative Strategic Plan Report at Subcommittee on Nanoscale Science, Engineering and Technology of Committee on Technology.* National Science Technology Council (NSTC), Arlington.

Huang, M., Wang, Z., Luo, L., Wang, S., Hui, X., He, G., & Malhi, S.S. (2017). Soil testing at harvest to enhance productivity and reduce nitrate residues in dryland wheat production. *Field Crops Res.* 212, 153–164.

Jayaseelan, C., Rahuman, A.A., Kirthi, A.V., Marimuthu, S., Santhoshkumar, T., Bagavan, A., & Rao, K.B. (2012). Novel microbial route to synthesize ZnO nanoparticles using *Aeromonas hydrophila* and their activity against pathogenic bacteria and fungi. *Spectrochim Acta Part A Mol Biomol Spectrosc.* 90, 78–84.

Jo, Y.K., Kim, B.H., & Jung, G. (2009). Antifungal activity of silver ions and nanoparticles on phytopathogenic fungi. *Plant Dis.* 93(10), 1037–1043.

Joseph, T., & Morrison, M. (2006). Nanotechnology in Agriculture and Food. A Nanoforum report, Institute of Nanotechnology May 2006, www.nanoforum.org

Kah, M. (2015). Nanopesticides and nanofertilizers: emerging contaminants or opportunities for risk mitigation? *Front Chem.* 3, 64.

Kah, M., Beulke, S., Tiede, K., & Hofmann, T. (2013). Nanopesticides: state of knowledge, environmental fate, and exposure modeling. *Crit Rev Environ Sci Technol.* 43, 1823–1867.

Kah, M., & Hofmann, T. (2014). Nanopesticide research: current trends and future priorities. *Environ Int.* 63, 224–235.

Kah, M., Tufenkji, N., & White, J.C. (2019). Nano-enabled strategies to enhance crop nutrition and protection. *Nat Nanotechnol.* 14(6), 532–540.

Kale, A.P., & Gawade, S.N. (2016). Studies on nanoparticle induced nutrient use efficiency of fertilizer and crop productivity. *Green Chem Technol Lett.* 2, 88–92.

Kalia, A., & Kaur, H. (2019). Nano-biofertilizers: harnessing dual benefits of nano-nutrient and bio-fertilizers for enhanced nutrient use efficiency and sustainable productivity. In S. Ranjan, N. Dasgupta, & E. Lichtfouse (Eds.) *Nanoscience for Sustainable Agriculture* (pp. 51–73). Springer, Cham.

Karn, B., Kuiken, T., & Otto, M. (2009). Nanotechnology and in situ remediation: a review of the benefits and potential risks. *Environ Health Perspect.* 117(12), 1813–1831.

Kashyap, P.L., Kumar, S., & Srivastava, A.K. (2017). Nanodiagnostics for plant pathogens. *Environ Chem Lett.* 15(1), 7–13.

Kashyap, P.L., Xiang, X., & Heiden, P. (2015). Chitosan nanoparticle based delivery systems for sustainable agriculture. *Int J Biol Macromol.* 77, 36–51.

Keller, A.A., McFerran, S., Lazareva, A., & Suh, S. (2013). Global life cycle releases of engineered nanomaterials. *J Nanop Res.* 15(6), 1–17.

Khati, P., Bhatt, P., Kumar, R., & Sharma, A. (2018). Effect of nanozeolite and plant growth promoting rhizobacteria on maize. *3Biotech.* 8(141), 1–12.

Khati, P., Chaudhary, P., Gangola, S., & Sharma A. (2019a). Influence of nanozeolite on plant growth promotory bacterial isolates recovered from nanocompound infested agriculture field. *Environ Ecol.* 37 (2), 521–527.

Khati, P., Gangola, S., Bhatt, P., & Sharma A. (2017a). Nanochitosan induced growth of *Zea Mays* with soil health maintenance. *3 Biotech.* 7(81), 1–9.

Khati, P., Sharma A., Chaudhary, P., Singh, A.K., Gangola, S., & Kumar, R. (2019b). High-throughput sequencing approach to access the impact of nanozeolite treatment on species richness and evenness of soil metagenome. *Biocatal Agric Biotechnol.* 20, 101249.

Khati, P., Sharma A., Gangola, S., Kumar, R., Bhatt, P., & Kumar, G. (2017b). Impact of some agriusable nanocompounds on soil microbial activity: an indicator of soil health. *Clean Soil Air Water.* 45 (5), 1–7.

Khodakovskaya, M.V., de Silva, K., Nedosekin, D.A., Dervishi, E., Biris, A.S., Shashkov, E.V., & Zharov, V.P. (2011). Complex genetic, photothermal, and photoacoustic analysis of nanoparticle-plant interactions. *Proc Nat Acad Sci.* 108(3), 1028–1033.

Khot, L.R., Sankaran, S., Maja, J.M., Ehsani, R., & Schuster, E.W. (2012). Applications of nano-materials in agricultural production and crop protection: a review. *Crop Prot.* 35, 64–70.

Kisan, B., Shruthi, H., Sharanagouda, H., Revanappa, S.B., & Pramod, N.K. (2015). Effect of nano-zinc oxide on the leaf physical and nutritional quality of spinach. *Agrotechnology.* 5(1), 135.

Knorr, W., Prentice, I.C., House, J.I., & Holland, E.A. (2005). Long-term sensitivity of soil carbon turnover to warming. *Nature.* 433(7023), 298–301.

Kole, C., Kole, P., Randunu, K.M., Choudhary, P., Podila, R., Ke, P.C., & Marcus, R.K. (2013). Nanobiotechnology can boost crop production and quality: first evidence from increased plant biomass, fruit yield and phytomedicine content in bitter melon (*Momordica charantia*). *BMC Biotechnol.* 13(1), 1–10.

Koman, V.B., Lew, T.T., Wong, M.H., Kwak, S.Y., Giraldo, J.P., & Strano, M.S. (2017). Persistent drought monitoring using a microfluidic-printed electro-mechanical sensor of stomata in planta. *Lab Chip.* 17(23), 4015–4024.

Kukreti, B., Sharma, A., Chaudhary, P., Agri, U., & Maithani, D. (2020). Influence of nanosilicon dioxide along with bioinoculants on *Zea mays* and its rhizospheric soil. *3 Biotech.* 10, 345.

Kumar, V., Guleria, P., Kumar, V., & Yadav, S.K. (2013). Gold nanoparticle exposure induces growth and yield enhancement in *Arabidopsis thaliana*. *Sci Total Environ.* 461, 462–468.

Kumari, H., Khati, P., Gangola, S., Chaudhary, P., & Sharma A. (2021). Performance of plant growth promotory rhizobacteria on maize and soil characteristics under the influence of TiO_2 nanoparticles. *Pantnagar Res J.* 19, 28–39.

Kumari, S., Sharma A., Chaudhary, P., & Khati, P. (2020). Management of plant vigour and soil health using two agriusable nanocompounds and plant growth promontory rhizobacteria in Fenugreek. *3 Biotech.* 10(461), 1–11.

Kuzma, J., & VerHage, P. (2006). *Nanotechnology in Agriculture and Food Production: Anticipated Applications.* Project on Emerging Nanotechnologies, Washington, DC.

Larue, C., Castillo-Michel, H., Sobanska, S., Trcera, N., Sorieul, S., Cécillon, L., & Sarret, G. (2014). Fate of pristine TiO_2 nanoparticles and aged paint-containing TiO_2 nanoparticles in lettuce crop after foliar exposure. *J Hazard Mater.* 273, 17–26.

Lauterwasser, C. (2005). Small Sizes that Matter: Opportunities and Risks of Nanotechnologies. Report in Co-Operation with the OECD International Futures Programme. http://www.oecd.org/dataoecd/32/1/44108334.pdf.

Laware, S.L., & Raskar, S. (2014). Influence of zinc oxide nanoparticles on growth, flowering and seed productivity in onion. *Int J Cur Microbiol Sci.* 3(7), 874–881.

Lei, Z., Mingyu, S., Xiao, W., Chao, L., Chunxiang, Q., Liang, C., & Fashui, H. (2008). Antioxidant stress is promoted by nano-anatase in spinach chloroplasts under UV-B radiation. *Bio Trace Elem Res.* 121(1), 69–79.

Li, C., Li, Y., Li, Y., & Fu, G. (2018a). Cultivation techniques and nutrient management strategies to improve productivity of rain-fed maize in semi-arid regions. *Agric Water Manag.* 210, 149–157.

Li, J., Wu, H., Santana, I., Fahlgren, M., & Giraldo, J.P. (2018b). Standoff optical glucose sensing in photosynthetic organisms by a quantum dot fluorescent probe. *ACS Appl Mater Interfaces.* 10(34), 28279–28289.

Liu, X.M., Zhang, F.D., Zhang, S.Q., He, X.S., Fang, R., Feng, Z., & Wang, Y.J. (2005). Effects of nano-ferric oxide on the growth and nutrients absorption of peanut. *Plant Nutr Fert Sci.* 11, 14–18.

Malakouti, M., & Tehrani, M. (2005). Micronutrient role in increasing yield and improving the quality of agricultural products. *Tarbiat Modarres Press.* 46, 1200–1250.

Millán, G., Agosto, F., Vázquez, M., Botto, L., Lombardi, L., & Juan, L. (2008). Use of clino-ptilolite as a carrier for nitrogen fertilizers in soils of the Pampean regions of Argentina. *Cienc Investig Agrar.* 35(3), 293–302.

Morales, M.I., Rico, C.M., Hernandez-Viezcas, J.A., Nunez, J.E., Barrios, A.C., Tafoya, A., & Gardea-Torresdey, J.L. (2013). Toxicity assessment of cerium oxide nanoparticles in cilantro (*Coriandrum sativum* L.) plants grown in organic soil. *J Agric Food Chem.* 61(26), 6224–6230.

Moraru, C.I., Panchapakesan, C.P., Huang, Q., Takhistov, P., Liu, S., & Kokini, J.L. (2003). Nanotechnology: a new frontier in food science understanding the special properties of materials of nanometer size will allow food scientists to design new, healthier, tastier, and safer foods. *Nanotech.* 57(12), 24–29.

Morteza, E., Moaveni, P., Farahani, H.A., & Kiyani, M. (2013). Study of photosynthetic pigments changes of maize (*Zea mays* L.) under nano TiO$_2$ spraying at various growth stages. *Springer Plus.* 2(1), 1–5.

Mousavi, S.R., & Rezaei, M. (2011). Nanotechnology in agriculture and food production. *J Appl Environ Biol Sci.* 1(10), 414–419.

Mukhopadhyay, S.S. (2005). Weathering of soil minerals and distribution of elements: pedochemical aspects. *Clay Res.* 24(2), 183–199.

Nair, R., Varghese, S.H., Nair, B.G., Maekawa, T., Yoshida, Y., & Kumar, D.S. (2010). Nanoparticulate material delivery to plants. *Plant Sci.* 179(3), 154–163.

Narrod, C.A., & Abbott, L. (2011). Agriculture, food, and water nanotechnologies for the poor: Opportunities and Const. (No. 19).

Nikunj, P., Purvi, D., Niti, P., Anamika, J. & Gautam, H.K. (2014). Agronanotechnology for plant fungal disease management: a review. *Int J Curr Microbiol App Sci.* 3(10), 71–84.

Nuruzzaman, M.D., Rahman, M.M., Liu, Y., & Naidu, R. (2016). Nanoencapsulation, nano-guard for pesticides: a new window for safe application. *J Agric Food Chem.* 64(7), 1447–1483.

Oliveira, H.C., Stolf-Moreira, R., Martinez, C.B.R., Grillo, R., de Jesus, M.B., & Fraceto, L.F. (2015). Nanoencapsulation enhances the post-emergence herbicidal activity of atrazine against mustard plants. *PLoS One.* 10(7), e0132971.

Oren, S., Ceylan, H., Schnable, P.S., & Dong, L. (2017). High resolution patterning and transferring of graphene-based nanomaterials onto tape toward roll to roll production of tape based wearable sensors. *Adv Mat Technol.* 2(12), 1700223.

Ouda, S.M. (2014). Antifungal activity of silver and copper nanoparticles on two plant pathogens, *Alternaria alternata* and *Botrytis cinerea. Res J Microbiol.* 9(1), 34.

Pérez-de-Luque, A. (2017). Interaction of nanomaterials with plants: what do we need for real applications in agriculture? *Front Environ Sci.* 5, 12.

Razzaq, A., Ammara, R., Jhanzab, H. M., Mahmood, T., Hafeez, A., & Hussain, S. (2016). A novel nanomaterial to enhance growth and yield of wheat. *J Nanosci Technol.* 2(1), 55–58.

Rico, C. M., Lee, S. C., Rubenecia, R., Mukherjee, A., Hong, J., Peralta-Videa, J.R., & Gardea-Torresdey, J. L. (2014). Cerium oxide nanoparticles impact yield and modify nutritional parameters in wheat (*Triticum aestivum* L.). *J Agric Food Chem.* 62(40), 9669–9675.

Roberts, T.L. (2009). The role of fertilizer in growing the world's food. *Better Crops.* 93, 2.

Roco, M.C. (1999). Towards a US national nanotechnology initiative. *J Nano Res.* 1(4), 435.

Rossi, M., Cubadda, F., Dini, L., Terranova, M. L., Aureli, F., Sorbo, A., & Passeri, D. (2014). Scientific basis of nanotechnology, implications for the food sector and future trends. *Trends Food Sci Technol.* 40(2), 127–148.

Sangeetha, J., Thangadurai, D., Hospet, R., Purushotham, P., Karekalammanavar, G., Mundaragi, A. C., & Harish, E.R. (2017). Agricultural nanotechnology: concepts, benefits, and risks. In R. Prasan, M. Kumar, & V. Kumar (Eds.) *Nanotechnology* (pp. 1–17). Springer, Singapore.

Sarlak, N., Taherifar, A., & Salehi, F. (2014). Synthesis of nanopesticides by encapsulating pesticide nanoparticles using functionalized carbon nanotubes and application of new nanocomposite for plant disease treatment. *J Agric Food Chem.* 62(21), 4833–4838.

Scott, N., & Chen, H. (2013). Nanoscale science and engineering for agriculture and food systems. *Indus Biotechnol.* 9(1), 17–18.

Sharma, P., Bhatt, D., Zaidi, M.G.H., Saradhi, P.P., Khanna, P.K., & Arora, S. (2012). Silver nanoparticle-mediated enhancement in growth and antioxidant status of *Brassica juncea*. *Appl Biochem Biotechnol.* 167(8), 2225–2233.

Sheykhbaglou, R., Sedghi, M., Shishevan, M.T., & Sharifi, R.S. (2010). Effects of nano-iron oxide particles on agronomic traits of soybean. *Not Sci Biol.* 2(2), 112–113.

Shukla, P., Chaurasia, P., Younis, K., Qadri, O.S., Faridi, S.A., & Srivastava, G. (2019). Nanotechnology in sustainable agriculture: studies from seed priming to post-harvest management. *Nanotechnol Environ Eng.* 4(1), 1–15.

Siddiqui, M.H., & Al-Whaibi, M.H. (2014). Role of nano-SiO_2 in germination of tomato (*Lycopersicum esculentum* seeds Mill.). *Sau J Biol Sci.* 21(1), 13–17.

Singh, S., Singh, M., Agrawal, V.V., & Kumar, A. (2010). An attempt to develop surface plasmon resonance based immunosensor for Karnal bunt (*Tilletia indica*) diagnosis based on the experience of nano-gold based lateral flow immuno-dipstick test. *Thin Solid Films.* 519(3), 1156–1159.

Son, D., Park, S.Y., Kim, B., Koh, J.T., Kim, T.H., An, S., & Hong, S. (2011). Nanoneedle transistor-based sensors for the selective detection of intracellular calcium ions. *ACS Nano.* 5(5), 3888–3895.

Subramanian, K. S., Manikandan, A., Thirunavukkarasu, M., & Rahale, C. S. (2015). Nano-fertilizers for balanced crop nutrition. In M. Rai, C. Ribeiro, L. Mattoso, & N. Duran (Eds.) *Nanotechnologies in Food and Agriculture* (pp. 69–80). Springer, Cham.

Tilman, D., Cassman, K. G., Matson, P. A., Naylor, R. & Polasky, S. (2002). Agricultural sustainability and intensive production practices. *Nature.* 418(6898), 671–677.

Tiwari, D.K., Dasgupta-Schubert, N., Villaseñor Cendejas, L.M., Villegas, J., Carreto Montoya, L., & BorjasGarcía, S.E. (2014). Interfacing carbon nanotubes (CNT) with plants: enhancement of growth, water and ionic nutrient uptake in maize (*Zea mays*) and implications for nanoagriculture. *Appl Nanosci.* 4(5), 577–591.

Tripathi, S., Sonkar, S. K., & Sarkar, S. (2011). Growth stimulation of gram (*Cicer arietinum*) plant by water soluble carbon nanotubes. *Nanoscale.* 3(3), 1176–1181.

Tyagi, J., Chaudhary, P., Mishra, A., Khatwani, M., Dey, S., & Varma, A. (2022). Role of endophytes in abiotic stress tolerance: with special emphasis on *Serendipita indica*. *Int J Environ Res.* 16, 62.

VanDijk, M., & Meijerink, G.W. (2014). A review of food security scenario studies: gaps and ways forward. In T.J. Achterbosch, M. van Dorp, W.F. van Driel, J.J. Groot, J. van der Lee, A. Verhagen, & I. Bezlepkina (Eds.) *The Food Puzzle: Pathways to Securing Food for All* (pp. 30–32). Wageningen UR. https://edepot.wur.nl/349629.

Wani, A.H. & Shah, M.A. (2012). A unique and profound effect of MgO and ZnO nanoparticles on some plant pathogenic fungi. *J Appl Pharma Sci.* 2, 40–44.

Wu, L. & Liu, M. (2008). Preparation and properties of chitosan-coated NPK compound fertilizer with controlled-release and water-retention. *Carbohyd polym.* 72, 240–247. 10.1016/j.carbpol.2007.08.020.

Wu, P., Chen, X., Hu, N., Tam, U. C., Blixt, O., Zettl, A. & Bertozzi, C.R. (2008). Biocompatible carbon nanotubes generated by functionalization with glycodendrimers. *Angew Chem Int Edit*, 47, 5022–5025.

Wu, H., Nißler, R., Morris, V., Herrmann, N., Hu, P., Jeon, S.J. & Giraldo, J.P. (2020). Monitoring plant health with near-infrared fluorescent H_2O_2 nanosensors. *Nano Lett.* 20(4), 2432–2442.

Yan, A., & Chen, Z. (2019). Impacts of silver nanoparticles on plants: a focus on the phytotoxicity and underlying mechanism. *Int J Mol Sci.* 20(5), 1003.

Yang, F.L., Li, X.G., Zhu, F. & Lei, C.L. (2009). Structural characterization of nanoparticles loaded with garlic essential oil and their insecticidal activity against *Tribolium castaneum* (Herbst) (Coleoptera: Tenebrionidae). *J. Agric Food Chem* 57:10156–10162.

Yao, J., Yang, M., & Duan, Y. (2014). Chemistry, biology, and medicine of fluorescent nanomaterials and related systems: new insights into biosensing, bioimaging, genomics, diagnostics, and therapy. *Chem Rev.* 114(12), 6130–6178.

Yao, K.S., Li, S.J., Tzeng, K.C., Cheng, T.C., Chang, C.Y., Chiu, C.Y., & Lin, Z.P. (2009). Fluorescence silica nanoprobe as a biomarker for rapid detection of plant pathogens. In *Advanced Materials Research*, 79, 513–516.

Zhang, J., Boghossian, A.A., Barone, P.W., Rwei, A., Kim, J.H., Lin, D., & Strano, M.S. (2011). Single molecule detection of nitric oxide enabled by d (AT) 15 DNA adsorbed to near infrared fluorescent single-walled carbon nanotubes. *J Am Chem Soc.* 133(3), 567–581.

Zhu, L., Chen, L., Gu, J., Ma, H., & Wu, H. (2022). Carbon-based nanomaterials for sustainable agriculture: their application as light converters, nanosensors, and delivery tools. *Plants.* 11(4), 511.

6 Nanoparticles and Their Effect on Plant Health and Productivity

Pragati Srivastava
Govind Ballabh Pant University of Agriculture & Technology

Parul Chaudhary and Heena Parveen
Graphic Era Hill University and National
Dairy Research Institute

CONTENTS

6.1   INTRODUCTION

Nanoparticles (NPs) have a minuscule structure ranging in size from 1 to 100 nm, its classification based on its physicochemical properties, shapes or sizes. Fullerenes, metal NPs, ceramic NPs and polymeric NPs are certain exclusive examples of different groups of NPs (Khan et al., 2019). NPs' small size and high surface-to-volume ratio make it unique in its physical and chemical properties such as biochemical reactivity and toughness. Nanotechnology is a potential progressing field with huge

DOI: 10.1201/9781003345565-6

application in basic and applied sciences (Ali et al., 2014). It is widely accepted in the agriculture sector for its significant contribution in disease treatment in crop plants, new tools for pathogen detection, food security and site-directed delivery of agro-chemicals in nutrient-deficient crops, hence enhancing the plant growth and making it resistant to environmental stresses (Echiegu, 2017; Kumari et al., 2020). NPs are extensively used in agriculture, electronics industry, energy conservation, remedia-tion, medicine, cosmetic industry and also environmental engineering (Bhatt et al., 2022; Chaudhary et al., 2022a). Its remarkable application in sustainable agricul-ture as nanocapsules, nanoparticles, nanoemulsion and viral capsid for improving site-directed delivery of active ingredients for providing immunity against pest and pathogens is significant. It also enhances the bioavailability of nutrients and inhibits nutrient loss that occurred in conventional fertilization. The water holding capacity is corrected via zeolite and nanoclay NPs for their slow release into plants for proper water retention (Chaudhary & Sharma, 2019). Various electrochemical biosensors such as carbon nanotubes and nanofibres have been designed for sensitive detection of environmental factors: temperature, pH and moisture to improvise plant growth and immunity (Mali et al., 2020).

NPs that make up nanofertilizers have the capability to raise plant growth by stimu-lating seed germinating rate, increasing plant sprout, root and shoot time. NPs have the tendency to interact at the cellular level and even at their organelles to increase the activ-ity of enzymes engaged in biochemical and physiological plant processes (Jaskulski et al., 2022). Many studies reported observable increment in plant yield and biomass production. The gene involved in the physiological processes of respiration, transpira-tion, photosynthesis and chlorophyll content is also upregulated (Sanzari et al., 2019; Hu et al., 2020). NPs have antioxidant properties to overcome oxidative stress in plants caused due to various environmental problems such as heavy metals, salinity, high temperature and ultraviolet (UV) radiation (Kumar et al., 2020). NPs have the compe-tence to harvest electrons for enhancing photosynthetic machinery in plants; they also have the property of converting NIR and UV to visible light to dilate light spectrum for enhanced synthesis of nutrients from carbon dioxide and water in light limiting condi-tions (Wu & Li, 2021). To feed the growing population the agriculture sector needs to be rationalized in terms of productivity and higher yielding crops, and nanotechnology can serve as a potential tool for transforming agriculture by maximizing its output and minimizing the input of labour, hazardous chemical fertilizers and having a detailed study of monitoring environmental conditions (Kamle et al., 2020). Today the chief applications of nanotechnology such as nanofertilizers and nanopesticides as amend-ment are highlighted for boosting plant growth (Yadu et al., 2021).

6.2   WHY NANOFERTILIZERS OVER CHEMICAL AND BIOFERTILIZERS (PROS AND CONS)?

Prolonged use of chemical fertilizers may disturb the microbiology of the soil, making it acidic in nature, hence soil fertility is lost and the growing crop is deficient in essen-tial micronutrients, thus posing a health issue for humankind. Biofertilizer was con-sidered a good replacement for chemical fertilizer as it is eco-friendly, cost-effective, easily accessible to plants, has enhanced activity of plant growth promotion and does not affect the microorganism in the soil (Chaudhary et al., 2022b). Microbial amendments for improving soil fertility and enhancing plant growth is the basic approach of biofertilizers. They are isolated from different agro-ecological zones and commercialized (Mahapatra et al., 2022). It may have free-living nitrogen-fixing bacteria like *Azotobacter, Rhodospirillum, Anabaena, Nostoc, Azospirillum,* Cyanobacteria, Symbiotic nitrogen-fixing bacteria like *Rhizobium, Frankia, Pseudomonas, Bacillus* etc. (Din et al., 2019; Zarezadeh et al., 2020; Bandopadhyay, 2020). It's a sustainable agricultural approach over hazardous use of chemical fertilizers. Beside its so many advantages there are many hurdles in their maintenance and storage. Maintaining the inoculums' potential, strain specificity, quality checks, labour-intensive methods for large-scale production and contamination are always a concern which are certain stumbling blocks in their smooth production. Its efficiency keeps decreasing with respect to time and their storage needs intensive care and for achieving good production rate should be used before its expiry date (Chittora et al., 2020). Nanofertilizer is an advanced product over bio- and chemical fertilizers, to overcome these hurdles. Its unique physical properties such as increased surface area for reactivity, sustained and slow release of fertilizer leading to minimum release of toxic substance into the environment make them an ideal strategy for enhancing the agricultural sector (Shang et al., 2019; Khati et al., 2019a).

6.3 NANOPARTICLE-MEDIATED PLANT STRESS TOLERANCE

At the present time, the transient weather conditions such as frequent droughts, salinity-affected areas, uneven distribution of rainfall and temperature, and climate change have led to the deterioration of yield globally (Tyagi et al., 2022). Use of chemical fertilizers to cope up with the nutrient deficiency in soil has imparted many negative repercussions in soil. NPs especially Ag, Ti, Mn, Ce have come up as a green approach for enhancing both abiotic and biotic stress resistance capacity in plants. Their external application has proven to improve plant tolerance to drought, salinity, and low- and high-temperature stress (Alabdallah & Hasan, 2021). Conventional fertilizers used earlier were less bioavailable to plants because of their large size and lesser soluble phase, whereas NP fertilizers have enhanced bioavailability and dispersion capacity of minerals and nutrients in the soil. The pattern of nutrient release is precisely controlled via encapsulation which inhibits nutrient loss via leaching (Nagula & Usha, 2016). A recent study depicts the ZnO and TiO NP efficiency against heat stress-mediated antioxidant defence system in wheat (Thakur et al., 2021).

A series of events such as plant photosynthetic activity, activation of defensive enzyme and increment of anti-stress compounds take place when silica nanoparticles are used as nanofertilizer, hence strengthening the physical barrier (Wang et al., 2022). CeO_2 NPs when compared to the bulk counterpart is more efficient in scavenging reactive oxygen species (ROS). CeO_2 NPs have two oxidation states (Ce^{3+} and Ce^{4+}) where the number of surface oxygen vacancies keeps changing and is quickly regenerated via redox cycling reaction due to nanoscale lattice of CeO_2 NPs for conferring potent ROS scavenging ability (Walkey et al., 2015). Mn_3O_4 NPs, a new advanced nanoparticle with two oxidation states, have high ROS scavenging capacity tested *in vivo* compared to bulk Mn_3O_4 (Yao et al., 2018). The application of scavenging ROS via cerium NPs is given a term nanoceria, which is exceptionally catalytically active from that of bulk material. The buffering capacity of Ce due to its two oxidation states makes it appropriate for both oxidation and reduction reactions. Foliar application of CeO coated with polyacrylic acid increased the salinity stress in *Arabidopsis* and cotton (Wu et al., 2018; Liu et al., 2021). Also, its soil application increased salinity stress in *Brassica napus* (Rossi et al., 2016). It modulated the action of ion channels and transporters for better retention of mesophyll K content, therefore leading to the suppression of KOR (K^+ outward rectifying channels for K^+ loss) and also inciting the expression of HKT (high-affinity K^+ transporters for proton expulsion). Foliar application of Mn_3O_4 NPs in cucumber and sorghum increased tolerance to salt stress and drought and its leaf infiltration in *Arabidopsis thaliana* reduces UV, high light and temperature stress.

Carbon nanodots leaf infiltration in peanuts plants (50 mg/L, size 3.9 nm, zeta volt −16.9 mV) enhanced drought stress (Su et al., 2018). Hydroponic application of multi-walled carbon nanotubes in Murashige and Skoog medium improved salinity stress in *B. napus* (20 mg/L, 6–12 nm in size) and promoted the signalling of gaseous NO and transduction of aquaporins in broccoli (10 mg/L, 6–9 nm) (Martínez-Ballesta et al., 2016; Zhao et al., 2019). Soil application of silicon NPs (125 mg/L, 20–30 nm) upregulated the chlorophyll content and shoot biomass and increased recovery from drought stress (Ghorbanpour et al., 2020). Apart from abiotic stress, biotic stress unlike disease-causing pathogens, insects, microorganisms, nematodes, etc. pose a threat to agricultural crops, leading to dropdown in yield and nutrient-limiting crop. To overcome this issue excessive use of chemical fertilizers has worsened the soil fertility, directly affecting human health and the environment. Use of nanotechnology in the form of green pesticides, herbicides and insecticides can be an ecologically sound practice for improving crop yield and productivity. Among so many NPs reported silver and copper have proven to be significant suppressors for disease-causing pathogens and also reduce drought and salinity stress.

6.4   NANOPARTICLES AS ANTIMICROBIAL AGENT

Crop production and yield are affected by many disease-causing microorganisms, which ultimately leads to huge economic loss. Billions of dollars are wasted due to the loss incurred by plant parasitic nematodes worldwide (Khan et al., 2022). Root knot disease starts initiating in plants infected by the root knot nematode *Meloidogyne incognita*, mainly in crop plants like tomatoes, lettuces, peppers, etc. *Aspergillus flavus*, a filamentous, pathogenic and saprophytic fungi, produces aflatoxin which cause diseases in crops such as maize, cottonseed and peanuts, hence limiting the crop yield (Shakeel et al., 2018). It ultimately leads to human intoxication due to ingestion of nuts, grains and the products derived from the abovementioned crops infected with this fungal strain (Amaike & Keller, 2011). Agricultural production can be enhanced by the maintenance of global food security and controlling yield loss due to pathogenic microorganism. The use of nanoparticles as antimicrobial is a novel and most efficient approach as a control measure for the suppression of disease-causing pathogens in crop plants. They are known to have effective bactericidal, fungicidal and nematicidal activity (Ali et al., 2020). In bacteria, nanoparticles are reported to disrupt DNA replication, RNA synthesis and bacterial cell membranes; restrict the synthesis of peptidoglycan; reduce the cell viability hence distorting its shape; disturb the ROS metabolism leading to cellular damage; and disturb the mitochondrial functioning, and hence ATP generating machinery is hampered, apoplastic trafficking is blocked and depolarization of cell membrane takes place and inhibits toxin production (Castillo-Henríquez et al., 2020). NPs in fungi inhibit the growth of hyphae and sporangia; suppress spore germination, hyphal wall membrane breakage and thinning; disrupt cell permeability; block water channel in sporangial area; and denature cell wall lysing enzyme, and thus coating of the entire fungal pathogen takes place (Banik & Pérez-de-Luque, 2017). The viral machinery is disturbed via reducing % lesion growth, hence limiting the rate of pathogenesis, reducing polymerase activity, injuring viral protein shell stability, and inactivating glycoprotein envelope (Vargas-Hernandez et al., 2020). Metals such as silver, copper, titanium, zinc and gold are known to synthesize NPs having antimicrobial properties. NPs' ultrasmall size which is 250 times smaller than the bacterial cell makes it a remarkable antimicrobial agent which interacts with the negative charge on the cell wall or microorganism cell surface, leading to the disruption of cell integrity, permeability, osmoregulation, electron transport and respiration which ultimately leads to lyses of the cell and release of the cell component (Islam et al., 2019; Rashmi et al., 2020). NP's antimicrobial property is size and dose dependent which states that higher the concentration higher will be the ability to interact with cell organelles and bacterial genetic material (Manivasagan et al., 2022). NPs also have the ability of generating reactive oxygen species by interacting with NADH dehydrogenase once it enters the bacterial cell, leading to depletion of ATP currencies and hence disturbing the respiratory chain. Its interaction with proteins, sulphur, phosphorus-containing cell constituents and DNA disrupts the cell (Cui et al., 2012; Ahmad et al., 2016).

6.5 NANOPARTICLES EMPLOYED IN PLANT DISEASE MANAGEMENT

6.5.1 Silver NPs

Silver NPs' (Ag NPs) shape, size and flexible electrical, optical and antimicrobial properties make it the most promising NPs. Ag NPs have been reported to inhibit the growth of enormous yeast and mould species such as *Aspergillus brasiliensis, Candida glabrata, C. tropicalis, C. krusei, C. albicans* and *Penicillium oxalicum*. Its mechanism of action includes alteration of membrane structure and disruption of the cell wall leading to cell lysis and release of the cell components, mitochondrial dysfunction, protein destabilization and denaturation inside the cell, ROS toxicity and alteration of the phosphotyrosine profile and stimulation of cell signals by modulation (Sharma et al., 2019). Ag NPs (100 ppm) subsidize the growth of *Pectobacterium carotovorum*, causal agent of soft rot in *Beta vulgaris* (Ghazy et al., 2021). A study reported the antibacterial property of Ag NPs (5 mm conc) against *Xanthomonas oryzae* causing bacterial leaf streak in *Oryza sativa. Sclerotium rolfsii* activity was abolished via Ag NPs causing collar rot in

Cicer arietinum (Desai et al., 2021). The cytotoxic behaviour of Ag NPs is due to liberation of ROS leading to the increment of ROS and glutathione scale, which mediates detrimental action on pathogenic fungi and its conidial growth (Siddiqi et al., 2018). Ag NPs are not only effective against bacteria but have a broad range of antimicrobial activity showing inhibitory effects against viruses also; 10–20 nm size Ag NP was effective against suppression of *Sunn hemp rosette virus* (SHRV) in *Cyamopsis tetragonoloba*. Also, yellow mosaic virus activity was suppressed via diminishing the severity of disease concentration of pathogen, leave lesions and infection percentage. It also reduces the activity of Tobacco Mosaic Virus in *Nicotiana tabacum* (Vargas-Hernandez et al., 2020). BBTV (banana bunchy top virus) is the most dangerous viral disease of banana. Recent research demonstrated that after commencement of the infection, foliar application of Ag NPs at 40–60 ppm concentration resulted in a low incidence of disease severity, % pathogenicity and banana bunchy top virus concentration was also found to be decreasing (Mahfouz et al., 2020). Spherical 20–100 nm Ag NPs synthesized from *Gossypium hirsutum* showed inhibitory effect against *Xanthomonas campestris* and *Xanthomonas axonopodis*. Green synthesis of Ag NPs via seed exudates of *G. hirsutum* suppressed growth of *Rhizoctonia solani* and *Neofusicoccum parvum* (Vanti et al., 2020). Green synthesized silver NPs from two algae (*Turbinaria turbinata* and *Ulva lactusa*) were effective in suppressing activity of *Meloidogyne javanica*, a root knot nematode in brinjal. The perfect optimized dose is 12.75 mg/100 mL used for inhibiting the effect of the nematode (Abdellatif et al., 2016). Ag NPs in urea solution when applied to the field showed significant anti-pesticidal effects in crops and vegetables (Bapat et al., 2022).

6.5.2 GOLD NPS

Use of gold NPs (AuNPs) in plant disease management is well acquainted in achieving productive agriculture (Figure 6.1). AuNPs' usage as biosensors for plant disease identification is appreciable. Its morphological traits, distinctive shape, size, control geometry and stable nature make it a novel NP with broad-spectrum antimicrobial activity. Tip dieback disease identification using electrochemistry technique (with AuNP single-stranded DNA) was used; 214 ppm limit range of pathogen was detected via DNA hybridization assay (Darr et al., 2017). A current study demonstrated electrochemical impedance spectroscopy-based detection via anti-Au PVV IgG polyclonal antibodies of plum pox virus within detectable limit of 10 pg/mL (Dyussembayev et al., 2021). A recent study reported calorimetric surface functionalization assay via AuNP–ssDNA for the detection of bacterial wilt within detectable limit of 7.5 ng (Farber et al., 2019). A study demonstrated AuNP extraction from *Mentha piperita*, an essential oil having antifungal properties, with 60 nm size exhibiting a spherical structure against *Aspergillus niger* (Thanighaiarassu et al., 2014). Antifungal activities of AuNPs synthesized from *Abelmoschus esculentus* were demonstrated against phytopathogens *A. niger* and *A. flavus* (Hernández-Díaz et al., 2021). Its mode of action involves restriction of peptidoglycan synthesizing machinery, hence loss of cell integrity leading to cell porosity and leakage of the cell component; it also disturbs the uncoiling and transcription of DNA. Green synthesis of AuNPs (20–30 nm)

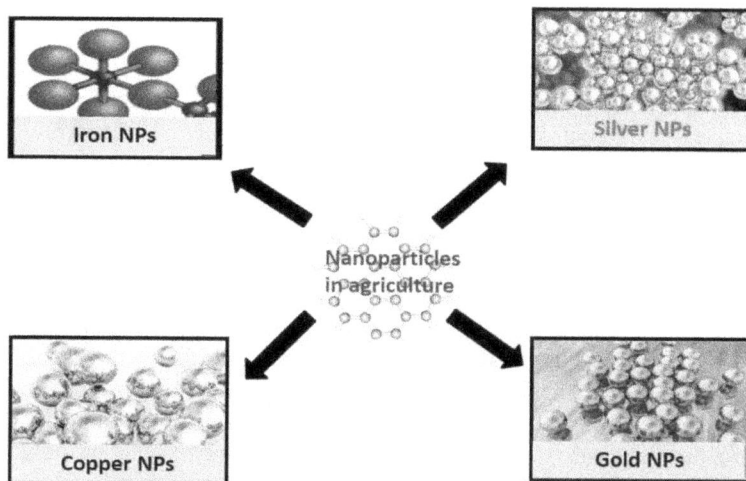

FIGURE 6.1 Different nanoparticles used in agriculture.

from *Citrus sinensis* peel, *Azadirachta indica* and *Mentha spicata* leaves, flower and leaf extract showed antimicrobial activity against *Pseudomonas aeruginosa* (Rao et al., 2017).

6.5.3 Copper NPs

Copper is an integral part of many enzymes, cofactors and proteins. Being an essential micronutrient, it has remarkable application in agricultural practices. It can be synthesized naturally or chemically. Copper NPs (CuNPs) have been attributed for their strong catalytic activity and porosity, which leads to higher reaction rate in lesser time. It closely interacts with pathogenic bacterial cell membrane lipids via oxidation for its degeneration, hence conferring antimicrobial property. The reliability and strength of CuNPs increase when it is used in combination with nanomaterial such as chitosan (i.e. Cu-CuNPs) which is proven to be an effective antifungal agent. It was effective in suppressing phytopathogenic growth of *Pythium aphanidermatum, Trichoderma viride, A. flavus, R. solani, Fusarium moniliforme, Fusarium oxysporum, Botrytis cinerea, Curvularia lunata* and *Alternaria alternata* (Vanti et al., 2020). CuONP (250 μL/L) was more effective than its CuNP counterpart in inhibiting the growth of *Pythium ultimum* causing pink rot in *Solanum tuberosum* (Hassan et al., 2019). Also, CuONPs at 30 mg/L concentration controlled the growth of *R. solani* (root rot) in *Solanum lycopersicum* (Shen et al., 2020). Shende et al. (2015) proved that CuNP and CuONP extracted from *Stachys lavandulifolia* and *Citrus medica* exhibited strong antimicrobial activity against *P. aeruginosa* and fungus *Fusarium* (*F. graminearum, F. culmorum,* and *F. oxysporium*), respectively. Additionally, a case study reported CuONP was effective in suppression of *Escherichia coli* showing inhibition zone of 18 nm and methicillin-resistant *Staphylococcus aureus* SRSA of 22 nm (Agarwal et al., 2017).

6.5.4 ZINC NPS

Zinc oxide NPs (ZnONPs) have a dual role in plant as essential micronutrient and suppressor for disease-causing pathogen in plants, hence its application in plants is an effective step in controlling plant resistance to biotic and abiotic stress. Zn as a cofactor, metal component and regulatory factor for many enzymes was engaged in physiological reaction for plant growth and development. Jamdagni et al. (2018) reported ZnONPs as a potent antimicrobial agent; its activity is directly proportional to its concentration and inversely to its particle size. Elmer and White (2018) reported the antibacterial and antifungal properties against *Sclerotinia sclerotiorum*, *Penicillium expansum*, *Rhizopus stolonifer*, *R. solani*, *Mucor plumbeus*, *A. alternata*, *F. oxysporum*, and *B. cinerea*. It disrupts the pathogen morphology by clumping and thinning of fungal hyphae at effective dosage of 9 mmol^{-1} which was confirmed via Zn acetate with TEM and HROM (high-resolution optical microscopy). It induces ultra-structural changes within the pathogen (Arciniegas-Grijalba et al., 2017). A recent study elucidated the inhibitory effect of ZnONPs at 1,500 ppm on *F. oxysporum* causing vascular wilt in *S. lycopersicum* (González-Merino et al., 2021).

6.5.5 IRON NPS

Fe is an essential vital micronutrient required for chlorophyll synthesis and also in the maintenance of chloroplast structure. It takes part in the plant's basic metabolic process of respiration and photosynthesis. Iron nanoparticles (FeNPs) are basically reported as a tool for bioremediation in metal-contaminated sites, due to its high magnetism, catalytic activity and low toxicity. FeNPs are used as bio-fortification tool for enhancing Fe bioavailability in plants and at present as antimicrobial agent. Inhibitory effect of green synthesis of Fe_2O_3 NPs isolated from leaf extract of *A. indica* was observed against *Diplodia seriata*, *Botryosphaeria dothidea*, and *Alternaria mali* from apple orchids (Ahmed et al., 2017; Devatha et al., 2018). Antifungal activity against *A. niger*, *A. alternata*, *Penicillium chrysogenum*, *Cladosporium herbarum* and *Trichothecium roseum* was observed by Fe_2O_3 NPs, which basically interrupted the spore germination and initiated coating of the pathogen due to increase of volume and surface area ratio and decreased mycelium proliferation (Parveen et al., 2018).

6.6 NP-MEDIATED NOVEL STRATEGIES FOR PLANT DISEASE MANAGEMENT

NP-mediated plant stress tolerance capability is an astounding tool or process for plant disease management, but there is a certain bio-safety concern regarding this procedure (Figure 6.2). Recently use of Ce NPs for enhancing plant stress tolerance raised bio-safety issues due to aggregation of its particles on plants, leading to impairment of plant response to stress and leading to toxicity (Tan et al., 2017; Spicer et al., 2018). One appropriate way to overcome this problem is to conjugate the NPs with suitable material like chitosan to improve dispersibility and stability for nano-enabled agriculture. Use of Cu-Chitosan NP (374 nm, 23 mV, 0.12%) improved its dispersibility demonstrated via FTIR (Saharan et al., 2015). Its foliar application enhanced

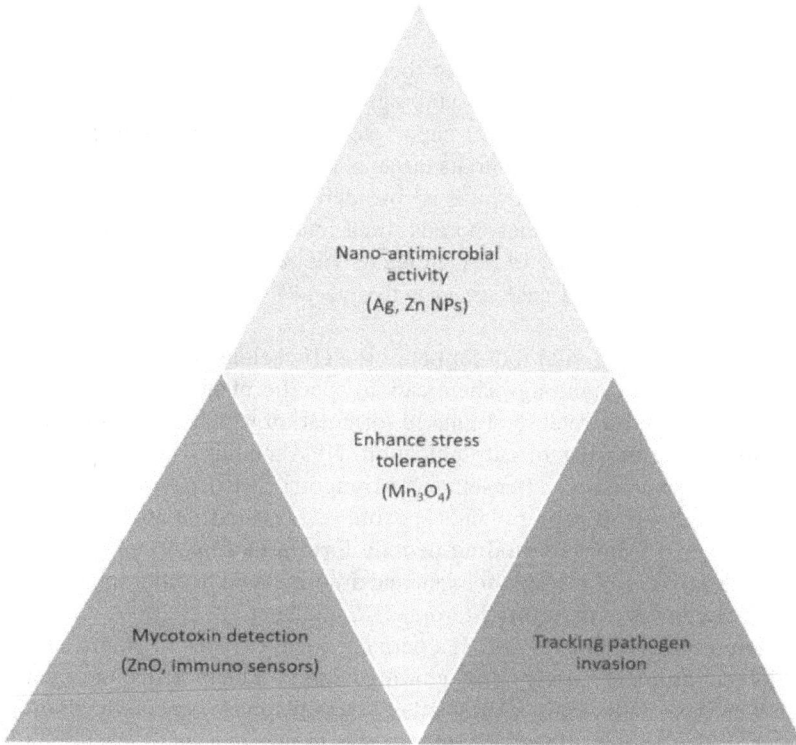

FIGURE 6.2 Plant disease management using nanotechnology.

the seedling length and fresh weight in tomato and shows antifungal effect against *A. solani* and *F. oxysporum*. An ecologically sound approach for sustainable agriculture is to design nanoparticles excluding heavy metals with increased dispersibility, for example, new Mn_3O_4 NPs tested in plants in Mn-deficient condition have proven high reactive oxygen species scavenging capacity, and it also enhances plant tolerance to salinity (Lu et al., 2020).

6.7 SURFACE MODIFICATION OF NANOFERTILIZER FOR PRECISE FARMING

Since decades commercial fertilizers have been used for enhancing crop productivity and disease management in plants. But the question is, how efficiently and precisely it reached its target plant. A recent study demonstrated that 30%–50% of chemical fertilizers increase crop productivity and the rest 40%–90% is lost in between application and plant target due to leaching and run-off to the environment, which raise a bio-safety concern for the contaminating environment due to its toxic effect (Stewart et al., 2005; Hu et al., 2020). Nanopesticides and nanofertilizers are also a pesticide formulation of essential micronutrients (N, P, K) and nano-engineered particles providing nutrients with maximum efficacy, providing antimicrobial capacity and improved capacity to stress tolerance to the target plant (Adisa et al., 2019). Nano agrichemical efficacy provides a median gain of 20%–30% when compared to their bulk counterpart (Kah et al., 2018). Targeted and controlled release of nano-agrochemicals is attained by its surface alteration in a stimulus-responsive nanomaterial which is solely responsible for its slow and controlled release into its target plant. Stimulus such as pH, temperature, light, enzymes and redox status is considered (Camara et al., 2019). A study demonstrated use of hollow mesoporous silica complexed with GA_3 (GA_3HMSN) Fe_3O_4 system for the release of GA_3 in plants via pH stimulus greater than 5 but less than 4 and promoted cabbage growth and 44% enhanced stem growth (Li et al., 2019a).

NP can be used as a vital tool for obtaining efficacious result for targeted and controlled release of nano-agrochemicals to specific plant organelles, organ or tissue to minimize the release of unused formulation which may lead to toxicity and highlight the process of eutrophication. NPs' mediated delivery in animal cell is already demonstrated (Anselmo & Mitragotri, 2016), paving a way to carry out experiments for targeting plant's specific cell compartments, organelles. B-cyclodextrin and truncated guiding peptide form a biofilm on QD for targeted and precise delivery of methyl viologen and ascorbic acid to chloroplast to assess the redox chemistry within the chloroplast (Santana et al., 2020). A recent study documented targeted delivery of NPs with chemical cargoes in plants, mediated by a biorecognition motif known as guiding peptide. This practice of organelles-targeted delivery will improve organelles' stress response especially chloroplast for tuning plant photosynthesis. Plants are able to utilize a small fraction of light from the visible range for carrying out photosynthesis. UV light causes detrimental effect on plants, and IR and NIR are merely absorbed by plants (Antonaru et al., 2020). During extensive cropping period, the leaves from the bottom region are not able to acquire sunlight in appropriate proportion. In order to improve the plant's efficiency for visible light uptake in low sunlight, the intensities (cloudy or rainy day) can be corrected by increasing visible light spectrum for absorption to enhance photosynthetic activity. Use of nanoparticles in conversion of UV and NIR light to visible range can lead to sudden drift in photosynthetic activity in light-compromised condition (Loo et al., 2019). An emerging biophotonics and nanomedicine field makes use of UCNPs (up-conversion nanoparticles) and DCNPs (down-conversion nanoparticles) for the conversion of UV and NIR to visible light (Wiesholler et al., 2019). Yb, Nd, Er-doped UCNPs are used for the conversion of IR light (excited under 808 and 980 nm) to visible light (both 510–570 and 640–700 nm). PEI-capped b-NaYF4:Gd^{3+}, Tb^{3+}DCNP, is used for converting UV light at 273 nm to visible light (at about 480–630 nm) (Malik et al., 2019). A recent study reported use of carbon dot nanobiosensor for enhancing light capturing capacity and its quantum yields. Carbon dots CD1:0.2 comprised citric acid and ethanolamine in molar ratios 1:0.2. Its foliar application was used to convert UV radiation (300–370 nm) to PAR (Photosynthetically Active Radiation) (up to 550 nm). An increment of 19% in shoot length, and 64% and 62% in dry weights of shoot and root were observed in rice plants with respect to control (Li et al., 2021).

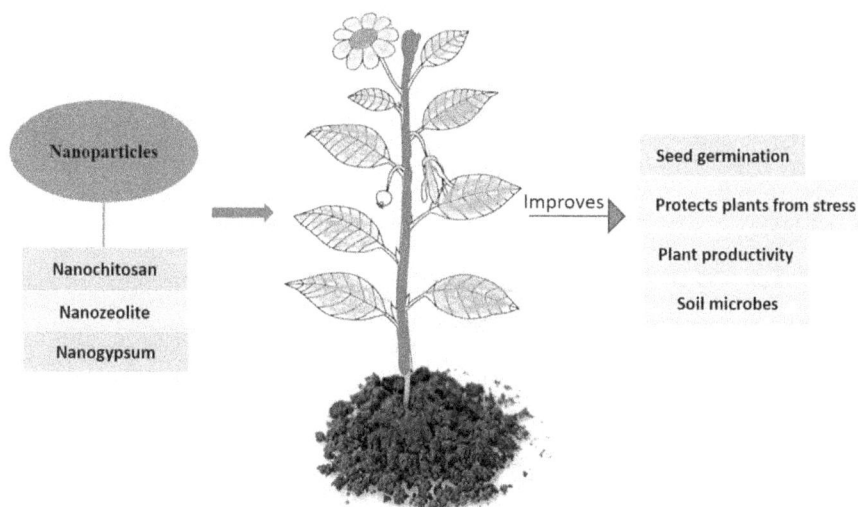

FIGURE 6.3   Nanoparticles for seed germination and maintenance of plant productivity.

6.8 NANOPARTICLES FOR SEED GERMINATION

The seedling phase for many perennial crops is prone to environmental stresses, mainly drought, salinity and heat (Finch-Savage & Bassel, 2016). Seed vigour helps in the establishment of the crop, improves its germination rate and increases stress tolerance capacity against environmental stress (Figure 6.3). Multiple strategies and techniques have been considered and brought into action for increasing the germination rate. Techniques such as halo priming, osmopriming, biopriming, solid matrix priming, chemopriming and thermopriming are used for increasing the seed germination rate (Paparella et al., 2015). The use of seed nanopriming is an emerging approach because it enables enhancement in crop production due to seed's increased germination capacity, its enhanced adaptability to abiotic stresses such as drought, salinity and heat (Table 6.1). An et al. (2020) reported a 41% increment in seedling fresh weight when seeds were nanoprimed with CeO_2 nanoparticles (21 nm, 500 mg/L) with respect to water primed control under osmotic stress. Nanoparticle-mediated seed priming especially with CeO_2 increases seed resistance capacity by modulation of signal transduction pathway involved in ROS and iron homeostasis. Its mechanism of action involves reduction in electrolyte leakage via ZnONPs (1) reduced peroxidation of lipids; (2) enhanced plant water content and photosynthetic activity via Fe_2O_3; and (3) reduced abnormalities in micronuclei and chromosomes via Ag NPs. Increase in SOD and POD activity is via ZnONPs. Priming seeds with Ag NPs increases amylase activity, resulting in increased soluble sugar content and upregulation of genes involved in water uptake via aquaporins (Mahakham et al., 2017). Fe_2O_3 NPs help in breaking the seed dormancy via reducing oxophytodienoic acid level (Kasote et al., 2019). Nanochitosan and nanozeolite (50 ppm) improved the maize seed germination as reported by Khati et al. (2017, 2018). Nanochitosan and nanozeolite also

TABLE 6.1

Effect of Nanoparticles on Seed Germination

Nanoparticles	Role of Nanoparticles	References
Silver NPs	Improved seed germination of *Festuca ovina*	Abbasi Khalaki et al. (2019)
Zinc sulphide NPs	Improve seed germination and plant growth of mung bean	Thapa et al. (2019)
Zinc oxide NPs	Improve seed germination and plant performance of wheat plants	Rai-Kalal and Jajoo (2021)
Multi-walled carbon nanotubes	Increased seed germination of *Alnus subcordata*	Rahimi et al. (2016)
Silver NPs	Improved seed germination of *Triticum aestivum*	Abou-Zeid and Ismail (2018)
Silica NPs	Improved seed germination and plant growth of wheat plants under salinity stress	Mushtaq et al. (2017)
Zinc oxide NPs	Improved seed germination in rice seeds	Sharma et al. (2021)
SiO_2	Improved maize seed germination	Kukreti et al. (2020)

improved maize seed enzyme activity and microbial population of soil which help plant development and production (Khati et al., 2019b; Agri et al., 2021, 2022). Silver NPs improved seed germination, watermelon yield and quality of fruits as reported by Acharya et al. (2020). Chitosan NPs at low concentration improved seed germination of wheat due to the better adsorption on seed surface (Li et al., 2019b). Nanosilica improves germination rate and plant height of wheat grass (Azimi et al., 2014). A lower amount of CuONPs also improved seedling development and metabolism in *Vigna radiata* (Singh et al., 2017). Silver NPs promote seed germination and shoot length in *Thymus vulgaris* crop (Ghavam, 2019). The positive impact of copper, zinc and silver nitrate NPs improved seed germination and root length of wheat plants (Bayat et al., 2022). Application of metal oxide NPs (100 mg/L) improved germination and growth of Solanaceae crops as reported by Younes et al. (2020).

6.9 NANOPARTICLES FOR PLANT PRODUCTIVITY, BIOCHEMICAL AND PHYSIOLOGICAL CHANGES IN PLANTS

Plant's physiological, morphological and biochemical traits tend to improve via usage of NPs in small concentrations. Stimulation of plant germination, photosynthesis and increased crop resistance to salinity, drought or toxic metals is reported via zinc oxide NPs (Kandhol et al., 2022). Del Buono et al. (2021) reported the enhancement of physiological and biochemical traits in maize by the use of biogenic ZnONPs extracted from a novel plant aquatic species, *Lemna minor* (duckweed). Lettuce plant's physiological parameters such as plant transpiration (E), intrinsic efficiency of water use (EUAi), carboxylation efficiency (k) and leaf CO_2 assimilation increased via foliar application of CuONPs (Pelegrino et al., 2021). Ullah et al. (2020) demonstrated that TiO$_2$-NPs at different concentrations (0, 30, 50 and 100 mg/kg) had a positive effect on wheat crop. Physiological parameters such as: antioxidant enzyme activities (superoxide dismutase, peroxidase) and contents of crude protein, H$_2$O$_2$, malondialdehyde content and metals/nutrients (Al, Ca, Mg, Fe, Zn and Cu) were increased drastically at 50 mg/kg. TiO$_2$-NPs without P fertilizer significantly enhanced the root and shoot length by 63% and 26%, respectively. The bioavailability of micronutrient heavy metals (Zn, Cu and Fe) and Al was also set off. Chaudhary et al. (2022a) assessed the indicators for soil health such as physicochemical analysis; total bacterial counts including N, P and K solubilizing bacteria; and also depicted the soil enzyme activities like fluorescein diacetate hydrolysis, alkaline phosphatase, β-glucosidase, dehydrogenase, amylase and aryl esterase under the influence of nanocompounds and *Bacillus* spp. in field condition. Iron oxide NPs are also used as nanofertilizers to improve iron availability to plants, antioxidant enzymes and increase peanut plant biomass (Rui et al., 2016). Titanium NPs improve the growth of root system, essential oil content and yield in *Mentha piperita* and *Thymus vulgaris* as reported by Fazeli-Nasab et al. (2018). Engineered iron oxide NPs such as hematite improved maize plant growth and chlorophyll content (Pariona et al., 2017). Multi-walled carbon nanotubes improve plant growth by increasing the production of photosynthesis rate and proteins in *Arabidopsis* plants (Fan et al., 2018). Application of zinc oxide NPs improved chlorophyll content (a and b) and increases the production of wheat grains (Adil et al., 2022). Zinc NPs (6–18 mg/L) increased the shoot and root dry weight, chlorophyll, carotenoids and total phenols of olive tree plants (Regni et al., 2022). Zinc oxide NPs have positive influence on growth parameters of pomegranate and improved plant biomass and chlorophyll content as compared to control (El-Mahdy & Elazab, 2020). ZnONPs (20 mg/L) boosted tomato growth by raising the chlorophyll content and genes involved in carbon metabolism and secondary metabolite synthesis (Sun et al., 2020). Nano zinc oxide (80 ppm) had the best outcomes and caused the greatest increases in plant height, yield and biomass accumulation (Sheoran et al., 2021). Table 6.2 summarizes the application of NPs in crop plant's health and productivity.

6.10   CONCLUSION

Many applications of nanoparticles have been reported and have practical implications in agriculture for enhanced plant protection and crop productivity: nanodiagnostics, nanoherbicides, nanopesticides, pesticide remediation, induced resistance in plants, nanoantimicrobial (bacteria, fungi, insects) are among some salient approaches used for attaining sustainable agriculture. Nanotechnology confers prospects in enhancing stress tolerating capacity of plants, seed nanopriming, use of encapsulated nanofertilizers for carrying precision farming to avoid environmental toxicity and wastage of left-over formulation. Keeping the utility of nanoparticles in mind there is a foremost requirement of exhaustively studying its dosage requirement in agriculture, as it easily enters the food chain and may directly affect the human population. Creation of safety measure policies and guidelines for its usage should be maintained. Its detailed comprehensive health effects, its impact on the environment

TABLE 6.2
Application of Nanoparticles in Maintenance of Plant Health and Productivity

Nanoparticles	Size/Concentration	Plant Species	Application Mode	Results	References
Ag NPs	30–100 nm/60 mg/L	*Helianthus annuus*	Foliar spray	Promote secondary metabolite content, plant yield and seed quality	Batool et al. (2021)
Ag NPs	55 nm/0.75 mg/mL	*Abelmoschus esculentus*	Foliar spray	Increase percent germination, number of leaves per plant, shoot and root length	Azeez et al. (2022)
TiO_2 NPs	20 nm/20 mg/L	*Oryza sativa*	Foliar spray	Increase the rice biomass, improved superoxide dismutase, peroxidase, catalase and ascorbate activity	Rizwan et al. (2019)
TiO_2 NPs	10–100 nm/40 mg/L	*Triticum aestivum*	Foliar spray	Enhanced leaf and root surface area, plant fresh and dry weight	Satti et al. (2021)
SiO_2 NPs	20 nm/400 mg/L	*Zea mays*	Seed inhibition	Enhanced seed germination, uptake of nitrogen, phosphorous and potassium	Rezaei et al. (2020)
ZnONPs	200 nm/100 ppm	*Solanum melongena*	Foliar spray	Improve membrane stability index, relative water content, photosynthetic efficiency and plant yield	Semida et al. (2021)
CeO_2 NPs	4 nm/500 mg/kg	*Zea mays*	Hydroponic	Promote photosynthesis and gas exchange, nutrient uptake and upregulated hydrogen peroxidase enzyme	Fox et al. (2020)
CeO_2 NPs	7 nm/25 mL	*Spinacia oleracea*	Irrigated water	Enhanced percent germination and significantly increased catalase activity	Singh et al. (2021)
TiO_2	50 nm/10ppm	*Zea mays*	Seed inoculation	Improved plant and soil health parameters	Kumari et al. (2021)
Nanozeolite and nanochitosan	80 nm/50ppm	*Zea mays*	Seed inoculation	Improved plant productivity, soil enzyme activities and microbial population	Chaudhary et al. (2021a, b, c)
Nanogypsum	10 nm/50ppm	*Zea mays*	Seed inoculation	Improved seed germination and plant health parameters	Chaudhary et al. (2021d)
Nanophos	Liquid formulation	*Zea mays*	Seed inoculation	Improved seed germination and plant health parameters	Chaudhary et al. (2021e)
Silicon dioxide	15 nm/10 ppm/pot	*Zea mays*	Seed inoculation	Improved seed germination, plant health, soil physicochemical properties and soil enzymes	Kukreti et al. (2020)
TiO_2	20–30 nm/100 mg/L/pot	*Dracocephalum moldavica L.*		Protects plants from salinity stress and improved their growth	Gohari et al. (2020)

and its safety clearance from the environment are required. A detailed study needs to be done on its toxicity level to overcome risk assessments and establish a reference guide for its toxicokinetics with proper concern to biological species before its utility on field.

REFERENCES

Abbasi Khalaki, M., Ghorbani, A., & Dadjou, F. (2019). Influence of nano-priming on *Festuca ovina* seed germination and early seedling traits under drought stress, in laboratory condition. *Ecopersia*. 7, 133–139.

Abdellatif, K.F., Abdelfattah, R.H., & El-Ansary, M.S.M. (2016). Green nanoparticles engineering on root-knot nematode infecting eggplants and their effect on plant DNA modification. *Iran J Biotechnol*. 14, 250.

Abou-Zeid, H., & Ismail, G. (2018). The role of priming with biosynthesized silver nanoparticles in the response of *Triticum aestivum* L. to salt stress. *Egypt J Bot*. 58(1), 73–85.

Acharya, P., Jayaprakasha, G.K., Crosby, K.M., Jifon, J.L., & Patil, B.S. (2020). Nanoparticle-mediated seed priming improves germination, growth, yield, and quality of watermelons (*Citrullus lanatus*) at multi-locations in Texas. *Sci Rep*. 10, 1–16.

Adil, M., Bashir, S., Bashir, S., Aslam, Z., Ahmad, N., Younas, T., Asghar, R.M.A., Alkahtani, J., Dwiningsih, Y., & Elshikh, M.S. (2022). Zinc oxide nanoparticles improved chlorophyll contents, physical parameters, and wheat yield under salt stress. *Front Plant Sci*. 13, 932861.

Adisa, I.O., Pullagurala, V.L.R., Peralta-Videa, J.R., Dimkpa, C.O., Elmer, W.H., Gardea-Torresdey, J.L., & White, J.C. (2019). Recent advances in nano-enabled fertilizers and pesticides: a critical review of mechanisms of action. *Env Sci Nano*. 6, 2002–2030.

Agarwal, H., Kumar, S.V., & Kumar, R.S. (2017). A review on green synthesis of zinc oxide nanoparticles–an eco-friendly approach. *Resource-Efficient Technol*. 3, 406–413.

Agri, U., Chaudhary, P., & Sharma, A. (2021). In vitro compatibility evaluation of agriusable nanochitosan on beneficial plant growth-promoting rhizobacteria and maize plant. *Natl Acad Sci Lett*. 44, 555–559.

Agri, U., Chaudhary, P., Sharma, A., & Kukreti, B. (2022). Physiological response of maize plants and its rhizospheric microbiome under the influence of potential bioinoculants and nanochitosan. *Plant Soil*. 474, 451–468.

Ahmad, A., Syed, F., Imran, M., Khan, A.U., Tahir, K., Khan, Z.U.H., & Yuan, Q. (2016). Phytosynthesis and antileishmanial activity of gold nanoparticles by *Maytenus royleanus*. *J Food Biochem*. 40, 420–427.

Ahmed, S., Chaudhry, S.A., & Ikram, S. (2017). A review on biogenic synthesis of ZnO nanoparticles using plant extracts and microbes: a prospect towards green chemistry. *J Photochem Photobiol B Biol*. 166, 272–284.

Alabdallah, N.M., & Hasan, M.M. (2021). Plant-based green synthesis of silver nanoparticles and its effective role in abiotic stress tolerance in crop plants. *Saudi J Biol Sci*. 28, 5631–5639.

Ali, M., Ahmed, T., Wu, W., Hossain, A., Hafeez, R., Islam Masum, M., & Li, B. (2020). Advancements in plant and microbe-based synthesis of metallic nanoparticles and their antimicrobial activity against plant pathogens. *Nanomaterial*. 10, 1146.

Ali, M.A., Rehman, I., Iqbal, A., Din, S., Rao, A. Q., Latif, A., & Husnain, T. (2014). Nanotechnology, a new frontier in agriculture. *Adv life Sci*. 1, 129–138.

Amaike, S., & Keller, N.P. (2011). *Aspergillus flavus*. *Annu Rev Phytopathol*. 49, 107–133.

An, J., Hu, P., Li, F., Wu, H., Shen, Y., White, J.C., & Giraldo, J.P. (2020). Emerging investigator series: molecular mechanisms of plant salinity stress tolerance improvement by seed priming with cerium oxide nanoparticles. *Environ Sci Nano*. 7, 2214–2228.

Anselmo, A.C., & Mitragotri, S. (2016). Nanoparticles in the clinic. *Bioengin Transl Med.* 1, 10–29.

Antonaru, L.A., Cardona, T., Larkum, A.W., & Nürnberg, D.J. (2020). Global distribution of a chlorophyll *f* cyanobacterial marker. *ISME J.* 14, 2275–2287.

Arciniegas-Grijalba, P.A., Patiño-Portela, M.C., Mosquera-Sánchez, L.P., Guerrero-Vargas, J.A., & Rodríguez-Páez, J.E. (2017). ZnO nanoparticles (ZnO-NPs) and their antifungal activity against coffee fungus *Erythricium salmonicolor*. *Appl Nanosci.* 7, 225–241.

Azeez, L., Aremu, H.K., & Olabode, O.A. (2022). Bioaccumulation of silver and impairment of vital organs in *Clarias gariepinus* from co-exposure to silver nanoparticles and cow dung contamination. *Bull Environ Contam Toxicol.* 108, 694–701.

Azimi, R., Borzelabad M.J., Feizi, H., & Azimi, A. (2014). Interaction of SiO_2 nanoparticles with seed prechilling on germination and early seedling growth of tall wheatgrass (*Agropyron elongatum* L.). *Polish J Chem Technol.* 16, 25–29.

Bandopadhyay, S. (2020). Application of plant growth promoting *Bacillus thuringiensis* as biofertilizer on *Abelmoschus esculentus* plants under field condition. *J Pure Appl Microbiol.* 14, 1287–1294.

Banik, S., & Pérez-de-Luque, A. (2017). In vitro effects of copper nanoparticles on plant pathogens, beneficial microbes and crop plants. *Span J Agric Res.* 15, e1005.

Bapat, M.S., Singh, H., Shukla, S.K., Singh, P.P., Vo, D.V.N., Yadav, A., & Kumar, D. (2022). Evaluating green silver nanoparticles as prospective biopesticides: an environmental standpoint. *Chemosphere.* 286, 131761.

Batool, S.U., Javed, B., Zehra, S.S., Mashwani, Z.U.R., Raja, N.I., Khan, T., & Alamri, S. (2021). Exogenous applications of bio-fabricated silver nanoparticles to improve bio-chemical, antioxidant, fatty acid and secondary metabolite contents of sunflower. *Nanomaterials.* 11, 1750.

Bayat, M., Zargar, M., Murtazova, K.M.-S.; Nakhaev, M.R., & Shkurkin, S.I. (2022). Ameliorating seed germination and seedling growth of nano-primed wheat and flax seeds using seven biogenic metal-based nanoparticles. *Agronomy.* 12, 811.

Bhatt, P., Pandey, S.C., Joshi, S., Chaudhary, P., Pathak, V.M., Huang, Y., Wu, X., Zhou, Z., & Chen, S. (2022). Nanobioremediation: a sustainable approach for the removal of toxic pollutants from the environment. *J Hazard Mater.* 427, 128033.

Camara, M.C., Campos, E.V.R., Monteiro, R.A., do Espirito Santo Pereira, A., de Freitas Proença, P.L., & Fraceto, L.F. (2019). Development of stimuli-responsive nano-based pesticides: emerging opportunities for agriculture. *J Nanobiotechnol.* 17, 1–19.

Castillo-Henríquez, L., Alfaro-Aguilar, K., Ugalde-Álvarez, J., Vega-Fernández, L., Montes de Oca-Vásquez, G., & Vega-Baudrit, J.R. (2020). Green synthesis of gold and silver nanoparticles from plant extracts and their possible applications as antimicrobial agents in the agricultural area. *Nanomaterials.* 10, 1763.

Chaudhary, P., Chaudhary, A., Bhatt, P., Kumar, G., Khatoon, H., Rani, A., Kumar, S., & Sharma, A. (2022a). Assessment of soil health indicators under the influence of nano-compounds and *Bacillus* spp. in field condition. *Front Environ Sci.* 9, 769871.

Chaudhary, P., Chaudhary, A., Parveen, H., Rani, A., Kumar, G., Kumar, A., & Sharma, A. (2021e). Impact of nanophos in agriculture to improve functional bacterial community and crop productivity. *BMC Plant Biol.* 21, 519.

Chaudhary, P., Khati, P., Chaudhary, A., Gangola, S., Kumar, R., & Sharma, A. (2021a). Bioinoculation using indigenous *Bacillus* spp. improves growth and yield of *Zea mays* under the influence of nanozeolite. *3 Biotech.* 11, 11.

Chaudhary, P., Khati, P., Chaudhary, A., Maithani, D., Kumar, G., & Sharma, A. (2021d). Cultivable and metagenomic approach to study the combined impact of nanogypsum and *Pseudomonas taiwanensis* on maize plant health and its rhizospheric microbiome. *PLoS One.* 16, e0250574.

Chaudhary, P., Khati, P., Gangola, S., Kumar, A., Kumar, R., & Sharma, A. (2021c). Impact of nanochitosan and *Bacillus* spp. on health, productivity and defence response in *Zea mays* under field condition. *3 Biotech.* 11, 237.

Chaudhary, P., & Sharma, A. (2019). Response of nanogypsum on the performance of plant growth promotory bacteria recovered from nanocompound infested agriculture field. *Environ Ecol.* 37, 363–372.

Chaudhary, P., Sharma, A., Chaudhary, A., Khati, P., Gangola, S., & Maithani, D. (2021b). Illumina based high throughput analysis of microbial diversity of rhizospheric soil of maize infested with nanocompounds and *Bacillus* sp. *Appl Soil Ecol.* 159, 103836.

Chaudhary, P., Singh, S., Chaudhary, A., Sharma, A., & Kumar, G. (2022b). Overview of bio-fertilizers in crop production and stress management for sustainable agriculture. *Front Plant Sci.* 13, 930340.

Chittora, D., Meena, M., Barupal, T., Swapnil, P., & Sharma, K. (2020). Cyanobacteria as a source of biofertilizers for sustainable agriculture. *Biochem Biophys Rep.* 22, 100737.

Cui, Y., Zhao, Y., Tian, Y., Zhang, W., Lü, X., & Jiang, X. (2012). The molecular mechanism of action of bactericidal gold nanoparticles on *Escherichia coli*. *Biomaterial.* 33, 2327–2333.

Darr, J.A., Zhang, J., Makwana, N.M., & Weng, X. (2017). Continuous hydrothermal synthesis of inorganic nanoparticles: applications and future directions. *Chem Rev.* 117, 11125–11238.

Del Buono, D., Di Michele, A., Costantino, F., Trevisan, M., & Lucini, L. (2021). Biogenic ZnO nanoparticles synthesized using a novel plant extract: application to enhance physiological and biochemical traits in maize. *Nanomaterials.* 11, 1270.

Desai, P., Jha, A., Markande, A., & Patel, J. (2021). Silver nanoparticles as a fungicide against soil-borne sclerotium rolfsii: a case study for wheat plants. In H. Sarma, S.J. Joshi, R. Prasad, & J. Jampilek (Eds.) *Biobased Nanotechnology for Green Applications* (pp. 513–542). Springer, Cham.

Devatha, C.P., Jagadeesh, K., & Patil, M. (2018). Effect of Green synthesized iron nanoparticles by *Azardirachta Indica* in different proportions on antibacterial activity. *Environ Nanotechnol Monit Manag.* 9, 85–94.

Din, M., Nelofer, R., Salman, M., Khan, F. H., Khan, A., Ahmad, M., & Khan, M. (2019). Production of nitrogen fixing *Azotobacter* (SR-4) and phosphorus solubilizing *Aspergillus niger* and their evaluation on *Lagenaria siceraria* and *Abelmoschus esculentus*. *Biotechnol Rep.* 22, e00323.

Dyussembayev, K., Sambasivam, P., Bar, I., Brownlie, J.C., Shiddiky, M.J., & Ford, R. (2021). Biosensor technologies for early detection and quantification of plant pathogens. *Front Chem.* 9, 144.

Echiegu, E.A. (2017). Nanotechnology applications in the food industry. In R. Prasad, V. Kumar, & M. Kumar (Eds.) *Nanotechnology* (pp. 153–171). Springer, Singapore.

El-Mahdy, M.T., & Elazab, D.S. (2020). Impact of Zinc oxide nanoparticles on pomegranate growth under in vitro conditions. *Russ J Plant Physiol.* 67, 162–167

Elmer, W., & White, J.C. (2018). The future of nanotechnology in plant pathology. *Annu Rev Phytopathol.* 56, 111–133.

Fan, X., Xu, J., Lavoie, M., Peijnenburg, W., Zhu, Y., Lu, T., Fu, Z., Zhu, T., & Qian H. (2018). Multiwall carbon nanotubes modulate paraquat toxicity in *Arabidopsis thaliana*. *Environ Pollut.* 233, 633–641

Farber, C., Mahnke, M., Sanchez, L., & Kurouski, D. (2019). Advanced spectroscopic techniques for plant disease diagnostics. A review. *TrAC Trends Anal Chem.* 118, 43–49.

Fazeli-Nasab, B., Sirousmehr, A.R., & Azad, H. (2018). Effect of titanium dioxide nanoparticles on essential oil quantity and quality in *Thymus vulgaris* under water deficit. *J Med Plants Res.* 2, 125–133.

Finch-Savage, W.E., & Bassel, G.W. (2016). Seed vigour and crop establishment: extending performance beyond adaptation. *J Exp Bot*. 67, 567–591.

Fox, J.P., Capen, J.D., Zhang, W., Ma, X., & Rossi, L. (2020). Effects of cerium oxide nanoparticles and cadmium on corn (*Zea mays* L.) seedlings physiology and root anatomy. *Nano Impact*. 20, 100264.

Ghavam, M. (2019). Effect of silver nanoparticles on tolerance to drought stress in *Thymus daenensis* Celak and *Thymus vulgaris* L. in germination and early growth stages. *Environ Stresses Crop Sci*. 12, 555–566.

Ghazy, N.A., El-Hafez, A., Omnia, A., El-Bakery, A.M., & El-Geddawy, D.I. (2021). Impact of silver nanoparticles and two biological treatments to control soft rot disease in sugar beet (*Beta vulgaris* L). *Egypt J Biol Pest Control*. 31, 1–12.

Ghorbanpour, M., Mohammadi, H., & Kariman, K. (2020). Nanosilicon-based recovery of barley (*Hordeum vulgare*) plants subjected to drought stress. *Environ Sci Nano*. 7, 443–461.

Gohari, G., Mohammadi, A., Akbari, A., Panahirad S., Dodpour, M.R., Fotopoulos, V., & Kimura, S. (2020). Titanium dioxide nanoparticles (TiO$_2$ NPs) promote growth and ameliorate salinity stress effects on essential oil profile and biochemical attributes of *Dracocephalum moldavica*. *Sci Rep*. 10, 912.

González-Merino, A.M., Hernández-Juárez, A., Betancourt-Galindo, R., Ochoa-Fuentes, Y.M., Valdez-Aguilar, L.A., & Limón-Corona, M.L. (2021). Antifungal activity of zinc oxide nanoparticles in *Fusarium oxysporum-Solanum lycopersicum* pathosystem under controlled conditions. *J Phytopathol*. 169(9), 533–544.

Hassan, S.E.D., Fouda, A., Radwan, A.A., Salem, S.S., Barghoth, M.G., Awad, M.A., El-Gamal, M.S. (2019). Endophytic actinomycetes *Streptomyces* spp. mediated biosynthesis of copper oxide nanoparticles as a promising tool for biotechnological applications. *J Biol Inorg Chem*. 24, 377–393.

Hernández-Díaz, J.A., Garza-García, J.J., Zamudio-Ojeda, A., León-Morales, J.M., López-Velázquez, J.C., & García-Morales, S. (2021). Plant-mediated synthesis of nanoparticles and their antimicrobial activity against phytopathogens. *J Sci Food Agric*. 101, 1270–1287.

Hu, P., An, J., Faulkner, M.M., Wu, H., Li, Z., Tian, X., & Giraldo, J.P. (2020). Nanoparticle charge and size control foliar delivery efficiency to plant cells and organelles. *ACS Nano*. 14, 7970–7986.

Islam, N.U., Jalil, K., Shahid, M., Rauf, A., Muhammad, N., Khan, A., & Khan, M.A. (2019). Green synthesis and biological activities of gold nanoparticles functionalized with *Salix alba*. *Arab J Chem*. 12, 2914–2925.

Jamdagni, P., Khatri, P., & Rana, J.S. (2018). Green synthesis of zinc oxide nanoparticles using flower extract of Nyctanthes arbor-tristis and their antifungal activity. *J King Saud Univ-Sci*. 30, 168–175.

Jaskulski, D., Jaskulska, I., Majewska, J., Radziemska, M., Bilgin, A., & Brtnicky, M. (2022). Silver nanoparticles (AgNPs) in urea solution in laboratory tests and field experiments with crops and vegetables. *Materials*. 15, 870.

Kah, M., Kookana, R.S., Gogos, A., & Bucheli, T.D. (2018). A critical evaluation of nanopesticides and nanofertilizers against their conventional analogues. *Nat Nanotechnol*. 13, 677–684.

Kamle, M., Mahato, D.K., Devi, S., Soni, R., Tripathi, V., Mishra, A.K., & Kumar, P. (2020). Nanotechnological interventions for plant health improvement and sustainable agriculture. *3 Biotech*. 10, 1–11.

Kandhol, N., Jain, M., & Tripathi, D.K. (2022). Nanoparticles as potential hallmarks of drought stress tolerance in plants. *Physiol Plant*. 174, e13665.

Kasote, D.M., Lee, J.H., Jayaprakasha, G.K., & Patil, B.S. (2019). Seed priming with iron oxide nanoparticles modulate antioxidant potential and defense-linked hormones in watermelon seedlings. *ACS Sustain Chem Eng*. 7, 5142–5151.

Khan, I., Saeed, K., & Khan, I. (2019). Nanoparticles: properties, applications and toxicities. *Arab J Chem*. 12, 908–931.

Khan, M., Khan, A. U., Rafatullah, M., Alam, M., Bogdanchikova, N., & Garibo, D. (2022). Search for effective approaches to fight microorganisms causing high losses in agriculture: application of *P. lilacinum* metabolites and myco synthesised silver nanoparticles. *Biomolecules*. 12, 174.

Khati, P., Bhatt, P., Kumar, R., & Sharma, A. (2018). Effect of nanozeolite and plant growth promoting rhizobacteria on maize. *3Biotech*. 8, 141.

Khati, P., Chaudhary, P., Gangola, S., & Sharma, P. (2019a). Influence of nanozeolite on plant growth promotory bacterial isolates recovered from nanocompound infested agriculture field. *Environ Ecol*. 37, 521–527.

Khati, P., Chaudhary, P., Gangola, S., Bhatt, P., & Sharma, A. (2017). Nanochitosan supports growth of *Zea mays* and also maintains soil health following growth. *3 Biotech*. 7, 81.

Khati, P., Sharma, A., Chaudhary, P., Singh, A.K., Gangola, S., & Kumar R. (2019b). High-throughput sequencing approach to access the impact of nanozeolite treatment on species richness and evens of soil metagenome. *Biocatal Agric Biotechnol*. 20, 101249.

Kukreti, B., Sharma, A., Chaudhary, P., Agri, U., & Maithani, D. (2020). Influence of nanosilicon dioxide along with bioinoculants on *Zea mays* and its rhizospheric soil. *3 Biotech*. 10, 345.

Kumar, H., Bhardwaj, K., Nepovimova, E., Kuča, K., Singh Dhanjal, D., Bhardwaj, S., & Kumar, D. (2020). Antioxidant functionalized nanoparticles: a combat against oxidative stress. *Nanomaterials*. 10, 1334.

Kumari, H., Khati, P., Gangola, S., Chaudhary, P., & Sharma, A. (2021). Performance of plant growth promotory rhizobacteria on Maize and soil characteristics under the influence of TiO$_2$ nanoparticles. *Pantnagar J Res*. 19, 28–39.

Kumari, S., Sharma, A., Chaudhary, P., & Khati, P. (2020). Management of plant vigor and soil health using two agriusable nanocompounds and plant growth promotory rhizobacteria in Fenugreek. *3 Biotech*. 10, 461.

Li, R., He, J., Xie, H., Wang, W., Bose, S.K., Sun, Y., Hu, J., & Yin, H. (2019a). Effects of chitosan nanoparticles on seed germination and seedling growth of wheat (*Triticum aestivum* L.). *Int J Biol Macromol*. 126, 91–100.

Li, X., Han, J., Wang, X., Zhang, Y., Jia, C., Qin, J., & Yang, Y. W. (2019b). A triple-stimuli responsive hormone delivery system equipped with pillararene magnetic nanovalves. *Mater Chem Front*. 3, 103–110.

Li, Y., Pan, X., Xu, X., Wu, Y., Zhuang, J., Zhang, X., & Liu, Y. (2021). Carbon dots as light converter for plant photosynthesis: augmenting light coverage and quantum yield effect. *J Hazard Mat*. 410, 124534.

Liu, J., Li, G., Chen, L., Gu, J., Wu, H., & Li, Z. (2021). Cerium oxide nanoparticles improve cotton salt tolerance by enabling better ability to maintain cytosolic K+/Na+ ratio. *J Nanobiotechnol*. 19, 1–16.

Loo, J.F.C., Chien, Y.H., Yin, F., Kong, S.K., Ho, H.P., & Yong, K.T. (2019). Upconversion and downconversion nanoparticles for biophotonics and nanomedicine. *Coord Chem Rev*. 400, 213042.

Lu, L., Huang, M., Huang, Y., Corvini, P.F.X., Ji, R., & Zhao, L. (2020). Mn$_3$O$_4$ nanozymes boost endogenous antioxidant metabolites in cucumber (*Cucumis sativus*) plant and enhance resistance to salinity stress. *Environ Sci Nano*. 7, 1692–1703.

Mahakham, W., Sarmah, A. K., Maensiri, S., & Theerakulpisut, P. (2017). Nanopriming technology for enhancing germination and starch metabolism of aged rice seeds using phytosynthesized silver nanoparticles. *Sci Rep*. 7, 1–21.

Mahapatra, D.M., Satapathy, K.C., & Panda, B. (2022). Biofertilizers and nanofertilizers for sustainable agriculture: phycoprospects and challenges. *Sci Total Environ*. 803, 149990.

Mahfouz, A.Y., Daigham, G.E., Radwan, A.M., & Mohamed, A.A. (2020). Eco-friendly and superficial approach for synthesis of silver nanoparticles using aqueous extract of Nigella sativa and *Piper nigrum* L seeds for evaluation of their antibacterial, antiviral, and anticancer activities a focus study on its impact on seed germination and seedling growth of *Vicia faba* and *Zea mays*. *Egypt Pharm J*. 19, 401.

Mali, S.C., Raj, S., & Trivedi, R. (2020). Nanotechnology a novel approach to enhance crop productivity. *Biochem Biophys Rep*. 24, 100821.

Malik, M., Padhye, P., & Poddar, P. (2019). Down conversion luminescence-based nanosensor for label-free detection of explosives. *ACS Omega*. 4, 4259–4268.

Manivasagan, P., Venkatesan, J., Senthilkumar, K., Sivakumar, K., Kim, S.K. (2022). Biosynthesis, Antimicrobial and Cytotoxic Effect of Silver Nanoparticles Using a Novel Nocardiopsis sp. MBRC-1. Available online: https://www.hindawi.com/journals/bmri/2013/287638/

Martínez-Ballesta, M., Zapata, L., Chalbi, N., & Carvajal, M. (2016). Multiwalled carbon nanotubes enter broccoli cells enhancing growth and water uptake of plants exposed to salinity. *J Nanobiotechnol*. 14, 1–14.

Mushtaq, N., Jamil, M., Riaz, G.L., Hornyak, N., Ahmed, S.S., Ahmed, M.N., Shahwani, M., & Malghani, N.K. (2017). Synthesis of silica nanoparticles and their effect on priming of wheat (*Triticum aestivum* L.) under salinity stress. *Biol Forum*. 9(1):150–157.

Nagula, S.A.I.N.A.T.H., & Usha, P.B. (2016). Application of nanotechnology in soil and plant system with special reference to nanofertilizers. *Adv Life Sci*. 1, 5544–5548.

Paparella, S., Araújo, S.S., Rossi, G., Wijayasinghe, M., Carbonera, D., & Balestrazzi, A. (2015). Seed priming: state of the art and new perspectives. *Plant Cell Rep*. 34, 1281–1293.

Pariona N., Martinez A.I., Hdz-García H.M., Cruz L.A., & Hernandez-Valdes, A. (2017). Effects of hematite and ferrihydrite nanoparticles on germination and growth of maize seedlings. *Saudi J Biol Sci*. 24, 1547–1554.

Parveen, S., Wani, A.H., Shah, M.A., Devi, H.S., Bhat, M.Y., & Koka, J.A. (2018). Preparation, characterization and antifungal activity of iron oxide nanoparticles. *Microb Pathogen*. 115, 287–292.

Pelegrino, M.T., Pieretti, J.C., Lange, C.N., Kohatsu, M.Y., Freire, B.M., Batista, B. L., & Seabra, A.B. (2021). Foliar spray application of CuO nanoparticles (NPs) and S-nitroso-glutathione enhances productivity, physiological and biochemical parameters of lettuce plants. *J Chem Technol Biotechnol*. 96, 2185–2196.

Rahimi, D., Kartoolinejad, D., Nourmohammadi, K., & Naghdi, R. (2016). Increasing drought resistance of *Alnus subcordata* C.A. Mey. seeds using a nano priming technique with multi-walled carbon nanotubes. *J Forest Sci*. 62(6), 269–278.

Rai-Kalal, P., & Jajoo, A. (2021). Priming with zinc oxide nanoparticles improve germination and photosynthetic performance in wheat. *Plant Physiol Biochem*. 160, 341–351.

Rao, Y., Inwati, G.K., & Singh, M. (2017). Green synthesis of capped gold nanoparticles and their effect on Gram-positive and Gram-negative bacteria. *Future Sci OA*. 3, FSO239.

Rashmi, B.N., Harlapur, S.F., Avinash, B., Ravikumar, C.R., Nagaswarupa, H.P., Kumar, M.A., & Santosh, M.S. (2020). Facile green synthesis of silver oxide nanoparticles and their electrochemical, photocatalytic and biological studies. *Inorg Chem Commun*. 111, 107580.

Regni, L., Del Buono, D., Micheli, M., Facchin, S.L., Tolisano, C., & Proietti, P. (2022). Effects of biogenic ZnO nanoparticles on growth, physiological, biochemical traits and antioxidants on olive tree in vitro. *Horticultura*. 8, 161.

Rezaei, S, N., Ariaii, P., & Charmchian Langerodi, M. (2020). The effect of encapsulated plant extract of hyssop (*Hyssopus officinalis* L.) in biopolymer nanoemulsions of *Lepidium perfoliatum* and *Orchis mascula* on controlling oxidative stability of soybean oil. *Food Sci Nutr*. 8, 1264–1271.

Rizwan, M., Ali, S., Malik, S., Adrees, M., Qayyum, M.F., Alamri, S.A., & Ahmad, P. (2019). Effect of foliar applications of silicon and titanium dioxide nanoparticles on growth, oxidative stress, and cadmium accumulation by rice (*Oryza sativa*). *Acta Physiol Plant.* 41, 1–12.

Rossi, L., Zhang, W., Lombardini, L., & Ma, X. (2016). The impact of cerium oxide nanoparticles on the salt stress responses of *Brassica napus* L. *Environ Poll.* 219, 28–36.

Rui, M., Ma, C., Hao, Y., Guo, J., Rui, Y., Tang, X., Zhao, Q., Fan, X., Zhang, Z., Hou, T., & Zhu, S. (2016). Iron oxide nanoparticles as a potential iron fertilizer for peanut (*Arachis hypogaea*). *Front Plant Sci.* 7, 815.

Saharan, V., Sharma, G., Yadav, M., Choudhary, M.K., Sharma, S.S., Pal, A., & Biswas, P. (2015). Synthesis and in vitro antifungal efficacy of Cu–chitosan nanoparticles against pathogenic fungi of tomato. *Int J Biolog Macromol.* 75, 346–353.

Santana, I., Wu, H., Hu, P., & Giraldo, J.P. (2020). Targeted delivery of nanomaterials with chemical cargoes in plants enabled by a biorecognition motif. *Nat Commun.* 11, 1–12.

Sanzari, I., Leone, A., & Ambrosone, A. (2019). Nanotechnology in plant science: to make a long story short. *Front Bioeng Biotechnol.* 7, 120.

Satti, S.H., Raja, N.I., Javed, B., Akram, A., Mashwani, Z.U.R., Ahmad, M.S., & Ikram, M. (2021). Titanium dioxide nanoparticles elicited agro-morphological and physicochemical modifications in wheat plants to control *Bipolaris sorokiniana*. *PloS One.* 16, e0246880.

Semida, W.M., Abdelkhalik, A., Mohamed, G.F., Abd El-Mageed, T.A., Abd El-Mageed, S.A., Rady, M.M., & Ali, E.F. (2021). Foliar application of zinc oxide nanoparticles promotes drought stress tolerance in eggplant (*Solanum melongena* L.). *Plants.* 10, 421.

Shakeel, Q., Lyu, A., Zhang, J., Wu, M., Li, G., Hsiang, T., & Yang, L. (2018). Biocontrol of *Aspergillus flavus* on peanut kernels using *Streptomyces yanglinensis* 3–10. *Front Microbiol.* 9, 1049.

Shang, Y., Hasan, M., Ahammed, G. J., Li, M., Yin, H., & Zhou, J. (2019). Applications of nanotechnology in plant growth and crop protection: a review. *Molecules.* 24, 2558.

Sharma, D., Afzal, S., & Singh, N.K. (2021). Nanopriming with phytosynthesized zinc oxide nanoparticles for promoting germination and starch metabolism in rice seeds. *J Biotechnol.* 336, 64–75.

Sharma, V.K., Sayes, C.M., Guo, B., Pillai, S., Parsons, J.G., Wang, C., & Ma, X. (2019). Interactions between silver nanoparticles and other metal nanoparticles under environmentally relevant conditions: a review. *Sci Total Environ.* 653, 1042–1051.

Shen, Y., Borgatta, J., Ma, C., Elmer, W., Hamers, R.J., & White, J.C. (2020). Copper nanomaterial morphology and composition control foliar transfer through the cuticle and mediate resistance to root fungal disease in tomato (*Solanum lycopersicum*). *J Agric Food Chem.* 68, 11327–11338.

Shende, S., Ingle, A. P., Gade, A., & Rai, M. (2015). Green synthesis of copper nanoparticles by *Citrus medica Linn. (Idilimbu)* juice and its antimicrobial activity. *World J Microbiol Biotechnol.* 31, 865–873.

Sheoran, P., Grewal, S., Kumari, S., & Goel, S. (2021). Enhancement of growth and yield, leaching reduction in *Triticum Aestivum* using biogenic synthesized zinc oxide nanofertilizer. *Biocatal Agric Biotechnol.* 32, 101938.

Siddiqi, K. S., Husen, A., & Rao, R. A. (2018). A review on biosynthesis of silver nanoparticles and their biocidal properties. *J Nanobiotechnol.* 16, 1–28.

Singh, A., Singh, N.B., Hussain, I., Singh, H., & Yadav, V. (2017). Synthesis and characterization of copper oxide nanoparticles and its impact on germination of *Vigna radiata* (L.) R. Wilczek. *Tropical Plant Res.* 4, 246–253.

Singh, R.P., Handa, R., & Manchanda, G. (2021). Nanoparticles in sustainable agriculture: an emerging opportunity. *J Control Rel.* 329, 1234–1248.

Spicer, C.D., Jumeaux, C., Gupta, B., & Stevens, M.M. (2018). Peptide and protein nanoparticle conjugates: versatile platforms for biomedical applications. *Chem Soc Rev.* 47, 3574–3620.

Stewart, W.M., Dibb, D.W., Johnston, A.E., & Smyth, T.J. (2005). The contribution of commercial fertilizer nutrients to food production. *Agronomy J.* 97, 1–6.

Su, L.X., Ma, X.L., Zhao, K.K., Shen, C.L., Lou, Q., Yin, D.M., & Shan, C.X. (2018).
Carbon nanodots for enhancing the stress resistance of peanut plants. *ACS Omega.* 3,
17770–17777.

Sun, L., Wang, Y., Wang, R., Wang, R., Zhang, P., Ju, Q., & Xu, J. (2020). Physiological, transcriptomic, and metabolomic analyses reveal zinc oxide nanoparticles modulate plant
growth in tomato. *Environ Sci Nano.* 7, 3587–3604.

Tan, W., Du, W., Barrios, A.C., Armendariz Jr, R., Zuverza-Mena, N., Ji, Z., & Gardea-
Torresdey, J.L. (2017). Surface coating changes the physiological and biochemical
impacts of nano-TiO$_2$ in basil (*Ocimum basilicum*) plants. *Environ Poll.* 222, 64–72.

Thakur, S., Asthir, B., Kaur, G., Kalia, A., & Sharma, A. (2021). Zinc oxide and titanium dioxide nanoparticles influence heat stress tolerance mediated by antioxidant defense system
in wheat. *Cereal Res Commun.* 50(3), 385–396.

Thanighaiarassu, R.R., Sivamai, P., Devika, R., & Nambikkairaj, B. (2014). Green synthesis
of gold nanoparticles characterization by using plant essential oil *Menthapiperita* and
their antifungal activity against human pathogenic fungi. *J Nanomed Nanotechnol.* 5, 1.

Thapa, M., Singh, M., Ghosh, C.K., Biswas, P.K., & Mukherjee, A. (2019). Zinc sulphide
nanoparticle (nZnS): a novel nano-modulator for plant growth. *Plant Physiol Biochem.*
142, 73–83.

Tyagi, J., Chaudhary, P., Mishra, A., Khatwani, M., Dey, S., & Varma, A. (2022). Role of
endophytes in abiotic stress tolerance: with special emphasis on *Serendipita indica*. *Int
J Environ Res.* 16, 62.

Ullah, S., Adeel, M., Zain, M., Rizwan, M., Irshad, M.K., Jilani, G., & Rui, Y. (2020).
Physiological and biochemical response of wheat (*Triticum aestivum*) to TiO$_2$ nanoparticles in phosphorous amended soil: a full life cycle study. *J Environ Manag.* 263, 110365.

Vanti, G.L., Masaphy, S., Kurjogi, M., Chakrasali, S., & Nargund, V.B. (2020). Synthesis and
application of chitosan-copper nanoparticles on damping off causing plant pathogenic
fungi. *Int J Biol Macromol.* 156, 1387–1395.

Vargas-Hernandez, M., Macias-Bobadilla, I., Guevara-Gonzalez, R. G., Rico-Garcia, E.,
Ocampo-Velazquez, R. V., Avila-Juarez, L., & Torres-Pacheco, I. (2020). Nanoparticles
as potential antivirals in agriculture. *Agriculture.* 10, 444.

Walkey, C., Das, S., Seal, S., Erlichman, J., Heckman, K., Ghibelli, L., & Self, W.T. (2015).
Catalytic properties and biomedical applications of cerium oxide nanoparticles. *Environ
Sci Nano.* 2, 33–53.

Wang, L., Ning, C., Pan, T., & Cai, K. (2022). Role of silica nanoparticles in abiotic and biotic
stress tolerance in plants: a review. *Int J Mol Sci.* 23, 1947.

Wiesholler, L.M., Frenzel, F., Grauel, B., Würth, C., Resch-Genger, U., & Hirsch, T. (2019).
Yb, Nd, Er-doped upconversion nanoparticles: 980 nm versus 808 nm excitation.
Nanoscale. 11, 13440–13449.

Wu, H., & Li, Z. (2021). Recent advances in nano-enabled agriculture for improving plant
performance. *Crop J.* 10(1), 1–12.

Wu, H., Shabala, L., Shabala, S., & Giraldo, J.P. (2018). Hydroxyl radical scavenging by
cerium oxide nanoparticles improves *Arabidopsis* salinity tolerance by enhancing leaf
mesophyll potassium retention. *Environ Sci Nano.* 5, 1567–1583.

Yadu, B., Xalxo, R., Chandra, J., Kumar, M., Chandrakar, V., & Keshavkant, S. (2021).
Applications of nanomaterials to enhance plant health and agricultural production. In V.P.
Singh, S. Singh, D.K. Tripathi, S.M. Prasad, & D.K. Chauhan (Eds.) *Plant Responses to
Nanomaterials* (pp. 1–19). Springer, Cham.

Yao, J., Cheng, Y., Zhou, M., Zhao, S., Lin, S., Wang, X., & Wei, H. (2018). ROS scavenging
Mn$_3$O$_4$ nanozymes for in vivo anti-inflammation. *Chem Sci.* 9, 2927–2933.

Younes, N.A., Hassan, H.S., Elkady, M.F., Hamed, A.M., & Dawood, M.F. (2020). Impact of synthesized metal oxide nanomaterials on seedlings production of three *Solanaceae* crops. *Heliyon.* 6, 3188–3198.

Zarezadeh, S., Riahi, H., Shariatmadari, Z., & Sonboli, A. (2020). Effects of cyanobacterial suspensions as bio-fertilizers on growth factors and the essential oil composition of chamomile, *Matricaria chamomilla* L. *J Appl Phycol.* 32, 1231–1241.

Zhao, G., Zhao, Y., Lou, W., Su, J., Wei, S., Yang, X., & Shen, W. (2019). Nitrate reductase-dependent nitric oxide is crucial for multi-walled carbon nanotube-induced plant tolerance against salinity. *Nanoscale.* 11, 10511–10523.

7 Nanotechnology-Mediated Agronomic Biofortification
A Way Forward

*Sanjay Kumar Sanadya, Rahul Sharma,
Sandeep Manuja, and Akashdeep Singh*
Chaudhary Sarwan Kumar Himachal
Pradesh Krishi Vishvavidyalaya

Mahendar Singh Bhinda
ICAR-Vivekananda Parvatiya Krishi Anusandhan Sansthan

Asha Kumari
ICAR-Indian Agricultural Research Institute

CONTENTS

DOI: 10.1201/9781003345565-7

7.1 INTRODUCTION

Globally, the majority of the population is dependent on cereal crops such as maize, rice and wheat for human consumption; however, lack of nutrition and costly vegetables or fruits cannot sustain human health. Besides, more than 0.82 billion of the population is still starving and some of the 2.5 billion people are malnourished and suffer from nutritional deficiency or toxicity, also known as 'hidden hunger'. In developing countries, malnutrition due to low nutritional foods leads to physical or mental weakness, stunting the growth of children and even resulting in death. As per statistical analysis for malnutrition globally, more than 809.9 million people are undernourished. From 2016 to 2018 alone, 0.194 billion undernourished population (24%) were in India, comprising a population of 46 million (30.9%) of the world's stunted children (below 5 years old) and 25.2 million (50.9%) of the world's wasted children (Food and Agriculture Organization of the United Nations, International Fund for Agricultural Development, United Nations Children's Fund, World Food Programme and World Health Organization, 2019). The energy density of the foods is the measure of the proportion of macronutrients supplied from the given food source. With the advent of industrialization, there is more demand for high-energy-dense foods which are rich in fatty acids, leading to an unhealthy lifestyle causing issues such as obesity, overweight and cardiac-related diseases. Consequently, there is an urgent need to incorporate foods rich in nutrition into the diets which can be achieved by fortifying our staple grains with the essential nutrients, vitamins and oils either through genetic approaches or by agronomic methods; this process is referred to as biofortification. It is the process of enrichment of the micronutrients, vitamins, desirable minerals or phytochemical components using breeding and agronomic practices in crop plants. The accessibility of the fortified grains was limited to rural areas with limited access to low-energy-dense foods. By employing breeding techniques along with agronomic interventions, it is possible to reduce the anti-nutritional level, thereby enhancing the bioavailability of minerals, vitamins, etc. The policies so far helped achieve food security but the focus is shifting to achieving nutritional security amidst self-sufficiency.

Nanotechnology alters crystalline particles into nano-sized particles that increase surface area (León-Silva et al., 2018; Kopittke et al., 2019). However, it is mostly exploited in electronics sectors, optical devices, remediation, water purification and health care sectors (Bhatt et al., 2022). The prospects of nanotechnology in the field of agriculture are vast but the research is limited. Due to its unique configuration, nanoscience would help evolve agricultural science and promote sustainability (Qureshi et al., 2018; Chhipa, 2019). The past few decades have revealed the potential of nanotechnology in improving the management practices and this led to change in the agricultural systems (Subramanian & Tarafdar 2011; Wang et al., 2012; Tarafdar, et al., 2013; Rico et al., 2014; Siddiqui et al., 2014; Zhao et al., 2014, 2020; Rico et al., 2015; Manjunatha et al., 2016; Saxena et al., 2016; Karimi & Fard 2017; López-Vargas et al., 2018; Hoang et al., 2019; Mittal et al., 2020; Younis et al., 2020; Weber et al., 2021; Verma et al., 2022). It enables the employment of a sophisticated agrochemical delivery system that is safe, precise and simple to use. Fertilizer application at critical stages of crop growth is an important management practice that influences the yield of any crop. The bulk of fertilizer is lost due to the environment, thereby threatening our dwindling resources. Subsequently, by employing nanotechnology it is possible to create nanofertilizers that are quickly absorbed by the soil and improve its quality, helping the plant to grow faster (Chaudhary et al., 2021a, b; 2022). Nanofertilizers are more effective than most traditional fertilizers due to their greater surface area in comparison to their volume which allows for gradual release and improves the nutrient-use efficiency (Chaudhary & Sharma, 2019; Agri et al., 2022). Encapsulated nanoparticles, nanogypsum, nano clays and zeolites improve fertilizer responsiveness, plant health, soil fertility restitution, and decrease environmental degradation and agroecology deprivation (Manjunatha et al., 2016; Kumari et al., 2020; Chaudhary et al., 2021c). Silica, ZnO, TiO_2, Al_2O_3, CeO_2 and FeO_2 are some of the commercialized nanoparticles in the market for improvement in productivity and micronutrient enrichment (Prasad et al., 2017; Kumari et al., 2021). This chapter outlines the roles of nanofertilizers in fortifying crop plants, utilizing conventional breeding and mineral fertilization but predominately focuses on the recent advancements and improvements in nutritional content in crop plants through fertilization with nanoparticles/nanofertilizers.

7.2 AGRONOMIC *VERSUS* GENETIC BIOFORTIFICATION

Chemical fertilizer is made up of a total of 18 components that are required for optimum plant growth and may be easily absorbed by plants (Iqbal & Umar, 2019). Biofortification is one of several techniques and procedures that alleviates micronutrient deficiencies. Biofortification is a novel crop-based technique for enhancing bioavailable micronutrients, vitamins and other elements in food items by combining agronomic and genetic approaches to living a healthy lifestyle (Figure 7.1). Bouis (2003) suggested that the biofortified varieties must be bred with the concept of enhancing the levels of micronutrients in the crop plant while reducing the levels of toxins, besides improving the yield. Biofortification exploiting conventional plant breeding approaches depends upon genetic variation (either naturally existing or to create new genetic variations) present in the gene pool; however, variations are considered as limiting factors focusing on yield rather than nutritional traits. Sometimes, this can be overcome by the pre-breeding method; however, it takes more time for the cultivars. For example, in wheat crops, selenium element enhancement needs to transfer traits from wild relatives of *Aegilops* species through the wide crossing. From a commercial viewpoint, a trait of interest can be introduced by mutation breeding but it will require a large number of mutant populations that enhances cost. Similarly, breeding approaches are being utilized in staple or non-staple crops, for example, QPM maize, Fe- and Zn-enriched rice cultivars and pulses or oilseed varieties such as *NuLin®*, a linseed cultivar with high alpha-linolenic acid (Omega-3 acid) (Morris, 2007; Flax Council of Canada, 2015). The different methods of agronomic biofortifications are listed in Table 7.1.

Genetic biofortification acts as a lasting and sustainable alternative for enhancing the bioavailability of nutrients. Mitchell-Olds (2010) reported that Genome-Wide Association Study (GWAS) is the most advanced tool able to assess the effect of many different single-nucleotide polymorphisms (SNPs) in an unrelated population.

FIGURE 7.1 Agronomic biofortification.

TABLE 7.1
Methods of Agronomic Biofortification

S. No.	Method	Principle
1.	Seed Priming	The seeds are treated with bioagents/nanocompounds/bioformulations/ nanobioformulation before the sowing
2.	Foliar Spray	The foliar parts are treated through spray of bioagent/nanocompounds in vegetative stage of plants which can penetrate the cell membranes
3.	Soil fertilization	Soil before the sowing of seeds is treated with the biofortification agent (bioagent/nanocompounds)

Neelam et al. (2011) successfully introgressed the QTLs for iron content into modern wheat from wild ancestor *Aegilops* species. Despite advantages, there are several limitations, the first of which is limited genetic diversity (Garg et al., 2018; Jha & Warkentin 2020). Another limitation of plant breeding is the time required for releasing a stable new variety (at least 6–7 generations of selection) (Glenn et al., 2017). Additionally, the results of domestication and breeding for economic traits might result in the loss of some beneficial traits including nutritional content (Brozynska et al., 2016). However, ancillary gene pools are the best alternative to recover what has been lost. Mabesa et al. (2013) found that through genotype-by-environment interaction (G×E interaction), a biofortified crop over time may lose its enhanced nutritional trait. Sivasakthi et al. (2019) stated stay-green phenotype in chickpeas is related to superior levels of lutein and beta-carotene which were lost during the domestication process. Wild relatives of crops are underrepresented and inadequate in their gene banks (Castañeda-Álvarez et al., 2016) and linkage drags (Li et al., 2008), and pre- and post-zygotic barriers often lead to poor yield or nutritional quality (Greene et al., 2018). However, several strategies have been introduced such as phenotyping platforms, speed breeding, molecular markers and genomic-assisted breeding approaches, which can accelerate the release of a cultivar (Ahmar et al., 2020). However, these methods have a higher input cost than conventional breeding and therefore are not widely employed (Ahmar et al., 2020). Despite several limitations (Figure 7.2), genetic approach-based biofortification research is being popularized among scientific communities and researchers are working on several crops to develop biofortified cultivars. One of the recent examples of genetic biofortification is in India in the year 2020–2021, when 17 cultivars were released in eight different crops and are being popularized among farmers. Besides conventional breeding, possible agronomic measures are being studied and employed for enhancing zinc, iron, selenium, copper and iodine contents in the major cereal crops. Agronomic biofortification of crops with iron, zinc and selenium using fertilizers would help the fight against hunger. Moreover, many reports evidence that biofortification not only enriches micronutrients in plants but also plays a significant role in synthesizing secondary compounds (Skrypnik et al., 2019; Newman et al., 2021; Puccinelli et al., 2021). Despite increasing crop yield and nutritional content, agronomic biofortification also deals with improving soil fertility. The method employs the use of mineral fertilizers to increase micronutrient content in crop plants. Therefore, in countries with developing economies, agronomic biofortification could be more beneficial than genetic biofortification (Yang et al., 2002). It is to apply fertilizer nutrients to the crop plants to enhance the levels of macro- and micronutrients in the economic part for a brief period resulting in improved nutritional status in humans (Cakmak & Kutman, 2017). Daniels (1996) earlier reported that the organic ones are more available for a human than the inorganic form of minerals, as they can be absorbed more easily, and are less excreted and less toxic (Garg et al., 2018). Macronutrients such as NPK make an important contribution to higher crop yields (Erisman et al., 2008). Apart from the advantages of the agronomic approach to biofortified crops, there are several limitations as well (Figure 7.2) such as loss of soil fertility, water and soil pollution, algal blooms and a decline in biodiversity (Jewell et al., 2020). Overuse of inorganic fertilizers would add financial problems to the small and marginal farmers due to

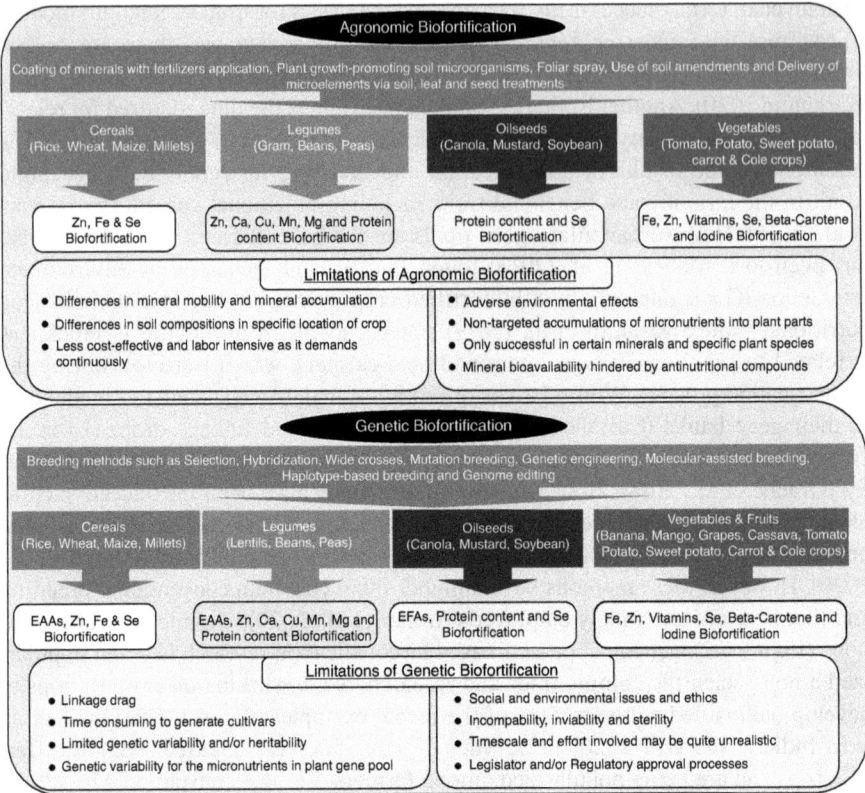

FIGURE 7.2 Agronomic and genetic biofortification.

their expensive nature (FAOSTAT, 2021). Stage-specific application of fertilizer is also a limiting factor to achieve maximum nutrient content. Previously researchers reported that fertilizer application is stage-specific. For example, Phattarakul et al. (2012) and Kyriacou et al. (2016) found that foliar application increased zinc content over other application methods in rice and was also growth stage-specific because the early dough stage is suitable for zinc nutrient fertilizer application than other crucial growth stages.

Similarly, Rodrigo et al. (2014) found that a higher amount of selenium content in wheat grain can be possible when applied between booting and heading growth stages with the foliar method. Furthermore, Rouphael and Kyriacou (2018) reported that for effective biofortification several factors are responsible such as soil cultivation practices, time of application, chemical structure and amount of fertilizers. Signore et al. (2018) compared hydroponic foliar fertilizer application to field foliar fertilizer application in carrots and found comparable results of toxic levels of iodine in both application methods. Germ et al. (2019) conducted research to compare selenium content in buckwheat application of fertilizer of selenium alone and combined with iodine and confirmed that selenium content was higher when fertilizer was applied in combination with selenium and iodine. The efficiency of the fertilizers to fortify crops with nutrients is influenced by numerous factors such as crop cultivar, environment and edaphic factors (Clark, 1983; White & Broadley, 2009). For example, the concentration of zinc varies when applied with zinc fertilizers in corn (Fahad et al., 2015) and rice genotypes (Saha et al., 2017). Mabesa et al. (2013) in their work observed that in soil with severe zinc deficiency, a rice genotype was able to achieve appreciable zinc content while agronomic manipulations were done using zinc-containing fertilizers. His study led us to the conclusion that it is possible to biofortify crops with the simplest of agronomic manipulations and one does not have to go through the expensive and time-consuming process of genetic biofortification, especially in developing countries with tied economies.

7.3 FERTILIZERS *VERSUS* NANOFERTILIZERS

The agricultural industry is being confronted with several global difficulties, including climate urbanization, resource sustainability, and pesticide and fertilizer build-up. Conventional fertilizers are more prone to losses due to leaching by water, volatilization due to high temperatures and fixation in the soil lattice compromising the crop growth and yield (Manjunatha et al., 2019). Fertilizers in enormous quantities are being applied by farmers to boost yield and productivity but even with best practice, it is difficult to get a higher use efficiency which seldom crosses 50%–70% while the rest is lost to the environment (Marchiol, 2019). As a result, fertilizers used to increase agricultural yield harm the environment and are inefficient economically (Ha et al., 2019). Large volumes of fertilizers, on the other hand, constitute a source of increased water contamination. Agricultural practices in Western Europe have polluted the waterways with NO_3^-, resulting from runoff from fields applied with nitrogenous fertilizers (Isermann, 1990), with the livestock industry contributing 73% of water pollution (Leip et al., 2015). Nitrogenous fertilizers are generally supplied in granular forms which are later integrated into the soil-plant system. It is taken up by the plants, fixed by the soil components or lost from the system (Jadon et al., 2018). Volatilization, denitrification and the production of N_2O are the main causes of nitrogen fertilizer loss (IPCC, 2015). Fertilizers account for 35%–40% of agricultural productivity; however, certain fertilizers have direct effects on plant development. The growing importance of nanofertilizers lies in the fact that they can enhance the use efficiency of various fertilizers besides conserving the resources and avoiding the degradation of the environment (Dimkpa et al., 2020; Mejias et al., 2021).

To put it another way, chemical fertilizers have negative environmental repercussions such as global warming by the release of greenhouse gases (GHGs) which require immediate attention; therefore, the hunt is for alternatives such as nanofertilizers. Nanotechnology has the potential to reduce greenhouse gases emissions considerably and thereby mitigate climate change (Kumari et al., 2022). Each fertilizer being inefficient in meeting the nutrient requirements for plant growth, development and nutritional composition, it becomes a fascinating effort to engineer materials to create nanofertilizers that deal with malnutrition and environmental-related issues. To address all of these issues more intelligently, nano-based fertilizer might be one of the solutions. Smaller particles provide for higher surface coverage. Plants and animals can even transmit nanoparticles through their cell walls. This procedure is used by nanotechnologists to provide fertilizers at the cellular level, which is more effective than standard fertilizer delivery methods. The nanofertilizers are synthesized by augmenting nutrients individually or in combination on nano-sized adsorbents. The nutrition can be contained inside nanomaterials such as nanotubes or nanoporous materials, covered with a thin protective polymer coating, and administered as nanoscale particles or emulsions. Because the nutrients are nanoscale, fortifying the plant with these nanonutrients appears to be a viable approach. Plants during their growth period absorb the nutrients, bridging the nutritional deficit gap (Agri et al., 2021). Nanofertilizers outperform traditional fertilizers in terms of nutritional performance, obviating the need for chemical fertilizers, improving drought and disease resistance, and posing fewer environmental risks. Because of their large surface area-to-volume ratio, they are easily absorbed by plants. Sizes and morphologies of nanoparticles, on the other hand, are critical factors in determining how much bio-accessibility plants have to soil (Chaudhary et al., 2021d). Nanoparticles cannot be triggered immediately before being taken up by plants; in its place, a sequence of processes ranging from oxidation to recombination may take place to deliver the micronutrients required by the plants. Because the nutrients are nanoscale, they appear to be a viable choice for supplementing the plant (Elemike et al., 2019). Fertilizer particles can be coated with nano-membranes to allow for a gradual and consistent release of nutrients. This procedure reduces the nutrient loss while increasing crop fertilizer efficiency (Subramanian & Tarafdar, 2009). Nanofertilizers comprising zinc, magnesium and iron are particularly useful in this area, as they provide critical micro- and macronutrients to plants. However, the size, concentration, composition and chemical properties of nanomaterials have an impact on their efficiency (Thakur et al., 2018).

Prasad et al. (2012) planned an experiment during 2008–2009 and 2009–2010 to compare yield performance using the application of fertilizers and nanofertilizers and found that a higher pod yield (30.5% and 38.8%) was obtained with the use of nanoscale zinc oxide along with NPK in comparison to application of NPK alone, respectively, while 29.5% and 26.3% higher pod yield was obtained with the application of chelated zinc at +NPK and NPK alone, respectively. Liang et al. (2013) reported that the carbon NP treatments at 25, 75 and 125 mg per pot enlarged the area of the leaf by 6.64%, 19.51% and 21.58% respectively as compared with conventional fertilizer. Raliya et al. (2014) conducted studies on root length, root area and chlorophyll content when applied with conventional fertilizers and nanofertilizers. They found a significant increase in the parameters under study with nano MgO (15 ppm) application in cluster bean over traditional fertilizers. Raliya and Tarafdar (2013) reported significantly higher rhizospheric microbial population (11.8%–13.8%) by application of zinc oxide nanoparticles during critical growth stage of cluster bean. Li et al. (2016) reported that soil with high adsorption capacity for metals and anionic nanoparticles increases the chance of their being available to the crop. With the advent of nanotechnology and its introduction to the agricultural systems, there is scope for next revolution in terms of nutritional security while maintaining the food security in order to feed the burgeoning population while also tackling hunger and environment sustainability. Similarly, there are several other studies that have reported the application of nanocompounds to improve plant and soil health under controlled and field conditions (Khati et al., 2017a, b; 2018; Chaudhary et al., 2021e).

7.4 NPs AND APPLICATIONS IN FIELD CROPS W.R.T. BIOFORTIFICATION

In order to improve the yield of crop plants, it is necessary to improve various physiological processes that are directly related to factors promoting the metabolism of the nutrients, growth of the seedling, and synthesis of carbohydrates and proteins which can be achieved through nanomaterials such as nanofertilizers (Zulfiqar et al., 2019). Furthermore, because of its target-bound delayed delivery, a lower volume of nanofertilizer application can efficiently boost nutrient content with minimal impact to the surrounding environment (Al-Mamun et al., 2021). Nanoparticles such as zeolites and nanoclays, which can improve fertilizer efficacy and restore soil fertility, can be a good source of biofortification. This is because zinc is one of the eight essential microelements for normal growth and development, enzymatic activity, biomass production, chlorophyll molecule stability, cell division, and functions as a regulatory co-factor in protein synthesis. Magnesium also helps with carbon dioxide fixation, chlorophyll and protein synthesis and other processes that improve plant output and development. Iron, on the other hand, plays an important function in increasing plant production, biomass, nitrogen fixation and development by delivering vital nutrients (Konate et al., 2017).

For enhanced development and plant output, many ways of applying nanofertilizers have been used, including soil mixing, foliar spray and seed priming. Similarly, NPs such as Zn/ZnO, FeO, TiO_2, CeO_2, Ag, Au, platinum, Se, Co, carbon nanotubes, etc. tested and employed for enriching nutrient levels in crops. Several nanoparticles are being used in crop plants and are summarized in Figure 7.3.

7.4.1 ZEOLITE-BASED NPs

Zeolite-based mineral components are safe for the soil and work in an effective manner to maintain available soil nutritional contents which allow efficient crop production and wide reduction in agricultural-related environmental concerns (Morales-Díaz et al., 2017). Zeolite also facilitates the slow release of nutrients into the soil and uptake by the plants, preventing nutrient loss that occurs in traditional systems (Guo, 2004; Khati et al., 2019a, b). Natural zeolite, which is made up of more than 50 mineral varieties that are primarily alkali and alkaline earth aluminosilicates, has recently been transformed into a nanofertilizer. This is mostly because of its widespread availability and low cost, and it has been used in the production of maize (Suppan, 2017; Eroglu et al., 2017). It is a clever transporter and regulator of main mineral fertilizers, in addition to delivering certain minerals for the plants. Because zeolite works as a transporter for nitrogen and potassium fertilizers, application volumes are lowered while productivity is increased (Polat et al., 2004). The zeolite component was discovered to combine with humus elements and boost agricultural production. The zeolite nanocomposites of NPK and other nutrients have influenced crop development (Harper, 2015). The inability of zeolite to absorb and

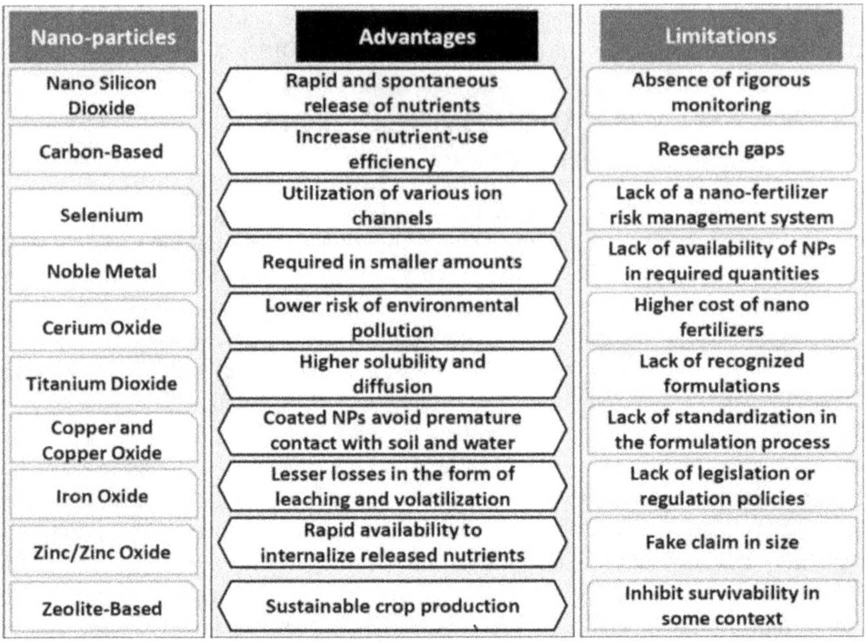

Nano-particles	Advantages	Limitations
Nano Silicon Dioxide	Rapid and spontaneous release of nutrients	Absence of rigorous monitoring
Carbon-Based	Increase nutrient-use efficiency	Research gaps
Selenium	Utilization of various ion channels	Lack of a nano-fertilizer risk management system
Noble Metal	Required in smaller amounts	Lack of availability of NPs in required quantities
Cerium Oxide	Lower risk of environmental pollution	Higher cost of nano fertilizers
Titanium Dioxide	Higher solubility and diffusion	Lack of recognized formulations
Copper and Copper Oxide	Coated NPs avoid premature contact with soil and water	Lack of standardization in the formulation process
Iron Oxide	Lesser losses in the form of leaching and volatilization	Lack of legislation or regulation policies
Zinc/Zinc Oxide	Rapid availability to internalize released nutrients	Fake claim in size
Zeolite-Based	Sustainable crop production	Inhibit survivability in some context

FIGURE 7.3 Important nanoparticles (NPs) and their advantages and limitations in ecosystem.

adsorb anions is one of its drawbacks; hence, it is supplemented with biopolymers in this regard (Morales-Díaz et al., 2017). 3D structure of zeolite facilitates strong cationic exchange, and improves porosity and surface area, resulting in more storage of positive and negative nutritional ions for a long time. Because of the efficiency of the hydrated negative inorganic ions present, these exposed surfaces have great interactions with cations and polar molecules. All of these attributes have made zeolite one of the most favoured fertilizer materials. All of these attributes have made zeolite one of the most favoured fertilizer materials (Table 7.2).

7.4.2 ZINC/ZINC OXIDE NPS

Zinc oxides (ZnO) and hydrated zinc sulphates ($ZnSO_4.H_2O$ or $ZnSO_4.7H_2O$) are the most common sources of zinc micronutrients for fertilizer fortification. Since zinc is found in most cellular enzymes and required for hormone synthesis, photosynthetic pigment chlorophyll formation, as well as glucose metabolism, therefore, the importance of zinc cannot be overlooked. Zinc oxides may be ingested, digested and stored in plant systems in nanoparticle form. Higher concentrations of zinc oxide NPs can damage plant development, resulting in intrinsic aberrations in seed germination, root growth and seedling biomass (Singh et al., 2013). Zinc is a micronutrient; thus, it is only required in modest amounts. Nonetheless, a deficit in this nutrient might stymie plant growth and development. The availability of micronutrients in the soil is the function of pH of the soil. Micronutrient availability diminishes when soil

TABLE 7.2

Application of Different Nanomaterials in Agriculture

Nanomaterial	Application	References
Zeolite-Based NPs	Facilitate slow release of inputs	Morales-Díaz et al. (2017); Suppan (2017); Eroglu et al. (2017); Morales-Díaz et al. (2017)
Zinc/Zinc Oxide NPs	Fertilizer fortification	Singh et al. (2013); Sadeghzadeh (2013)
Iron Oxide NPs	Increase agricultural productivity	Corredor et al. (2019); Zia et al. (2016); Hu et al. (2017)
Copper and Copper Oxide NPs	Fertilizers, plant growth regulators (PGRs), insecticides, herbicides and soil remediation additives in agriculture	Pandya-Lorch (2012); Xiong et al. (2017b); Xiong et al. (2017a); Pestovsky and Martínez-Antonio (2017)
Titanium Dioxide NPs	Improve nitrogen fixation and photosynthesis, hence enhancing the plants' overall growth efficiency	Asli and Neumann (2009); Zheng et al. (2005)
Cerium Oxide NPs	Crop nutritional quality enhancement and nutritional effects in agriculture	Gui et al. (2015); Cao et al. (2017)
Noble Metal NPs	Eliminate undesired microbes from soils, plants and hydroponic systems	Saurabh et al. (2015); Duhan et al. (2017); Kumar et al. (2013); Manikandan and Subramanian (2016); Parveen et al. (2016); Pestovsky and Martínez-Antonio (2017)
Selenium NPs	They've been widely used as supplements in plants and food	Skalickova et al. (2017); Uttam and Abioye (2017); Meetu and Shikha (2017)
Carbon-Based NPs	Improve water absorption and thus germination of seeds	Lahiani et al. (2015)
Nano-Silicon Dioxide NPs	Improve resistance to environmental stresses including biotic and abiotic stresses	Siddiqui and Al-Whaibi (2014); Avestan et al. (2016)

pH rises, with the exception of molybdenum. In the case of zinc, the process is that when soil pH rises, Zn is adsorbed on the soil surface along with other clay and inorganic elements, where it can be precipitated as $Zn(OH)_2$, $ZnCO_3$ and Zn_2SiO_4 (Sadeghzadeh, 2013). As a result of this behaviour, the entire system's solubility is weakened, resulting in inefficient mineral desorption. Nano zinc oxide serves various purposes: as a fertilizer in crop production but also as a mechanistic blocker of excess UV radiation during plant development and aiding in the recovery of lost nutrients in the soil.

7.4.3 IRON OXIDE NPS

Iron is one of the most prevalent elements in the Earth's crust, and it is found mostly in the oxide's forms such as magnetite, maghemite and hematite. It is employed in a variety of applications due to its low cost, although it has yet to be fully exploited in agriculture. Iron has recently been positioned as one of the transition metal nanoparticles of current attention for increased agricultural productivity due to its low toxicity, availability and superparamagnetic characteristics. Iron's pyrophoric and very reactive properties, on the other hand, limit its applicability (Zia et al., 2016). The iron deficiency in most meals is caused by low soil iron concentration, which is unavailable for plants to absorb. Iron fertilizers applied to calcareous soils are altered and may not be accessible for plant uptake. Iron-chelated fertilizers are utilized to remedy the issue. Iron fertilization in the form of hematite nanoparticles is gaining popularity in crops because of eradicating iron shortage in soils and crops in a rapid way. These nanoparticles play an important role in crop growth such as enhancing seed germination, root growth and chlorophyll content. When hematite NPs are put into the soil, iron is liberated from them, and they travel from the roots to other regions of the tissues to give them nutritional contents. When *Citrus maxima* shoots were subjected to both hematite NPs and Fe^{3+} treatment, their iron content increased relative to controls and Fe(II)-EDTA-treated plants (Hu et al., 2017). Carbon-coated FeNPs sprayed on leaves moved from the leaves to other regions of the plant, according to Corredor et al. (2019) and Hu et al. (2017).

7.4.4 COPPER AND COPPER OXIDE NPS

Due to its extensive use, copper is considered the third most important metal after iron and zinc and it is essential for the survival of most living organisms (Liu et al., 2018). Copper is an essential micronutrient that plants require in very small quantities. Bulk copper oxide has a long-lasting history in the agriculture sector, particularly as a fungicide and one of the most polluted minerals to contaminate the environment because of its insolubility, which allows it to quickly erode into water bodies. Copper has basically three oxidation states 0, +1 and +2, which exhibit various physicochemical properties. However, the advent of nanotechnology gives bright hope for copper utilization because it diminishes environmental pollution (Pandya-Lorch, 2012). Copper oxide NPs are used as fertilizers, plant growth regulators (PGRs), insecticides, herbicides and soil remediation additives in agriculture (Xiong et al., 2017b). Copper oxide NPs accumulate on lettuce and cabbage at concentrations up to 250 mg/L, resulting in decreased water content and vegetable growth (Xiong et al., 2017a). Because they are leafy vegetables, they are susceptible to air pollution in addition to absorbing copper oxide NPs from the soil, as the nanoparticles are deposited on their leaves, which is the component consumed by people. Because of the absorption from both the leaves and the soils, the dose on the veggies tends to grow. As it has been observed that 0.3 mg/L Cu^{2+} released from 1,000 mg/L copper nanoparticles stimulates plant development and is not hazardous to the plant (Pestovsky & Martínez-Antonio, 2017), the absorbed quantities are also dependent on the kind of plant, the soil and environmental conditions. Because of their oxidative nature, copper oxide NPs are more hazardous than copper nanoparticles, even at low concentrations, yet they have a favourable influence on the plants' photosynthetic activity. Copper oxide NPs have also been found to have antibacterial properties, which might impact microbial interactions in the soil since certain bacteria are resistant due to the release of Cu^{2+} ions, which bind to bacterial cellular membranes and cause harm (Xiong et al., 2017b).

7.4.5 Titanium Dioxide NPs

Zheng et al. (2005) carried out research to estimate the role of titanium dioxide nanoparticles in spinach and showed that TiO_2 improves nitrogen fixation and photosynthesis, hence enhancing the plants' overall growth efficiency. Asli and Neumann (2009) in their research applied titanium dioxide NPs (around 30 nm) in maize (*Zea mays*) and reported that these NPs failed to translocate because the nanoparticle sizes were not small enough to pass through the root cell pore having a width of 6.6 nm. However, in another investigation, Du et al. (2011) found that titanium dioxide NPs can penetrate wheat and some nanoparticles can enter through the root cells whereas others did not. On this basis, it was determined that the nanoparticles were polydispersed, with smaller particles less than 20 nm able to fill the root cells, but larger particles formed agglomerates in the soil medium and were unable to access the root cells. The presence of nanoparticles in the soil can have influence on the soil enzyme concentration which leads to inhibition of their activities and toxicity. As they are not translocated, their existence in the soil would have an impact on the ecosystem and soil environment resulting in environmental pollution.

7.4.6 Cerium Oxide NPs

Cerium oxide nanoparticles are among the most investigated nanomaterials for crop nutritional quality enhancement and nutritional effects in agriculture (Cao et al., 2018). The concentration used, the soil quality and the plant type all influence crops. When nanoparticles are sprayed in small amounts into the soil, they improve crop growth and nutritional value. They do, however, have certain negative consequences in larger quantities, depending on the plant's nature. Several studies support the effect of increasing cerium oxide NP concentrations on plants. One example is the accelerated growth of lettuce when treated with 100 mg/kg cerium oxide NPs, but the plant's development was hampered when the concentration was increased to 1,000 mg/kg cerium oxide NPs (Gui et al., 2015). A similar condition in a soybean plant found that cerium oxide NPs significantly boosted soybean photosynthetic rate when the soil moisture level was high, but not when the soil water content was low (Cao et al., 2017). The simple reason is that when the soil moisture content is low, the plant stomata shut down due to drought, preventing transpiration and CO_2 absorption by the plant (Cao et al., 2018). Positively charged cerium oxide NPs are better absorbed by plant roots due to their easy dissolution nature than negatively charged cerium oxide NPs, according to other systems that depict cerium oxide NP absorption by plants. The positive charge is easily dissolved and drawn to the negative charge in the plant rhizosphere which improves cerium transfer to other plant components.

7.4.7 NOBLE METAL NPS

Silver is the most extensively researched of the noble metal nanoparticles that have been employed in agriculture. Ag has a long history of antibacterial activity and is used to eliminate undesired microbes from soils, plants and hydroponic systems, with nanosilver providing an increased impact (Saurabh et al., 2015). Silver has been used to combat fungal rot and numerous plant diseases in addition to stimulating plant growth and seed germination (Duhan et al., 2017). However, there may be some undesirable consequences. In the soil system, oxidative dissolution of silver and polyvinylpyrrolidone-coated silver NPs had a detrimental impact on nitrification processes (Masrahi et al., 2014). Other noble metals, such as gold nanoparticles, have not been widely used in plant biotechnology or genetic engineering since they are not micronutrients. Gold NPs at low concentrations in soil have been shown to improve the shoot-to-length ratio in lettuce (*Lactuca sativa*) seedlings without changing microbial concentrations or producing toxicity (Pestovsky & Martinez-Antonio, 2017). Early flowering, improved pod length, increased chlorophyll, sugar content, act as free radical scavenger due to higher flavonoids content, antioxidant properties, improved seed germination, plant growth rate, and crop yield (Kumar et al., 2013; Manikandan & Subramanian, 2016; Parveen et al., 2016; Pestovsky & Martinez-Antonio, 2017). Mahakham et al. (2016) found that exposing mature seeds to gold nanoparticles (5–15 mg/L photosynthesized) increased germination and physiology without causing toxicity.

7.4.8 SELENIUM NPS

Selenium elements are found in many forms as halides, oxides, acids, selenoenzymes and selenium nucleic acids (Skalickova et al., 2017) that belong to sulphur family and can have oxidation states of -2, 0, $+2$, $+4$ and $+6$. The positively charged $(+2)$ oxidation state is the most stable oxidation state of selenides and it is in this state that they react with highly electropositive elements; the positive charged $(+6)$ oxidation state is an unstable condition and weak shielding effect of the d-orbitals introduced in selenium, they vary in several reactions with sulphur. Selenium deficiency affects over one billion people globally, making its supplementation or fertilization in plants and animals essential for human health (WHO, 2009). It was formerly thought to be hazardous; however, current study has revealed that toxicity depends upon the concentration of selenium (Uttam & Abioye, 2017; Meetu & Shikha, 2017). Selenium nanoparticles have intriguing physicochemical features and are highly bioavailable, with outstanding antibacterial, anticancer, antioxidant and other physiological actions. They've been frequently employed in nanomedicine because of their low toxicity profiles (Hosnedlova et al., 2018). They have been widely used as supplements in plants and food. When they come into touch with heavy metals such as mercury, cadmium and lead, they serve as detoxifying agents, and this ability has allowed them to protect live creatures from a variety of ailments. Using ligands to functionalize selenium compounds is simple. Selenocysteine interacting with glutathione peroxidase, for example, can neutralize free radicals in cells, thereby removing the radical's detrimental effects. SeNPs, on the other hand, do not interact with most molecules and are thus slowly released into biological systems (Skalickova et al., 2017).

7.4.9 CARBON-BASED NPS

Fullerenes, fullerenols, single-walled carbon nanotubes (SWCNTs) and multi-walled carbon nanotubes (MWCNTs) are organic or carbon-based nanomaterials that have been employed in crop development and act differently. Carbon nanotubes are aquaphobic in nature and cannot dissolve in aqueous conditions, instead forming aggregates due to strong van der Waals forces, whereas single-walled carbon nanotubes are hydrophilic in nature and due to their huge size can rarely permeate plant cell walls. Fullerenol is hydrophilic in nature and smaller in size and may be carried into plant tissues by apoplastic mechanisms (Lahiani et al., 2015). On the other hand, fullerenes and multi-walled carbon nanotubes (MWCNTs) can interrelate with the hydrophobic components of natural organic matter and make their way into plant tissues. Long-term exposure of crops to multi-walled carbon nanotubes had no harmful consequences, according to Lahiani et al. (2015) but had a favourable influence on the overall development of the plants (Lahiani et al., 2017). Single-walled carbon nanohorns have been tested in rice (*Oryza sativa*), tomato (*Solanum lycopersicum*), barley (*Hordium vulgare*), soybean (*Glycine max*), maize (*Zea mays*) and switchgrass to determine their germination and found that all but tomato germinated at greater concentrations of the single-walled carbon nanohorns (100 g/mL) (Lahiani et al., 2015). The tomato crop, on the other hand, showed the maximum rate of germination at lower nanomaterial concentrations (25 g/mL), whereas the soybean crop showed no early germination. There was no evidence of harm, and the findings revealed that various plants react differently to different quantities of nanomaterials.

7.4.10 NANO-SILICON DIOXIDE NPS

In comparison to other micronutrients, silicon has not been widely recognized as a necessary component for plant development. It does, however, provide plants with several benefits, extending from improved plant development and production to resistance to environmental stresses including biotic and abiotic stresses (Siddiqui & Al-Whaibi, 2014). It functions as a physico-mechanical barrier and controls plant physiological processes. Those crop plants have no silicon mineral considered a weak structure and are underdeveloped. Currently, silicon dioxide nanoparticles are being explored as one of the most fascinating new inorganic materials due to their exceptional chemical purity, ultrafine particle size, improved surface adsorption, energy, great dispersion and strong heat tolerance. Siddiqui and Al-Whaibi (2014) found that applying 8 g/L of 12 nm nano-silicon oxide increased seed germination in tomatoes (*S. lycopersicum*) (Siddiqui & Al-Whaibi, 2014). Nanostructured silicon dioxide has also been shown to aid plants in lowering transpiration rates and promoting green colouring and shoot growth (Avestan et al., 2016). Rice has a lower salt intake and transpiration bypass flow (Yeo et al., 1999). Nano-silicon dioxide is believed to have the ability to improve the growth rates of many plants, vegetables and improve soil microbial activities (Kukreti et al., 2020). It is a novel substance that has the potential to provide great food safety as well as a unique farming experience a silicon treatment.

7.5 ACHIEVEMENTS OF NANOTECHNOLOGIES W.R.T. BIOFORTIFICATION

Agricultural production surged in many nations throughout the world in the late 1960s as a result of the use of NPK-containing fertilizers, resulting in the Green Revolution and saving people from hunger. These fertilizers are crucial and necessary in the current context to boost agricultural productivity and preserve the human population from the next famine, as traditional practices will not be sufficient to sustain the world's current population of seven billion people (Graham et al., 2007). The application of micronutrient fertilizers to improve soil micronutrient status can help to reduce micronutrient insufficiency in people (Cakmak, 2008). The European Union's Health Grain Project, Harvest Plus initiative and the Bio Cassava Plus programme are using traditional plant breeding methods to produce biofortified cultivars (Garg et al., 2018). Biofortified orange-fleshed sweet potato cultivars, for example, were created to have a higher beta-carotene content to assist address vitamin A deficiency in underdeveloped nations (Low et al., 2017; International Potato Center, 2019). According to a nutritional analysis, Huang et al. (1999) developed 21 times increased beta-carotene content in orange-fleshed sweet potato type than other sweet potato varieties, which relieved vitamin A insufficiency in children in African countries, according to research (Van Jaarsveld et al., 2005; Hotz et al., 2012a, b). Another notable example is the iron-biofortified rice variety IR68144-2B-2–2-3 produced in collaboration with the HarvestPlus initiative of the International Rice Research Institute (IRRI), Los Banos (Gregorio et al., 2000). After processing and boiling, this rice variety has four to five times more iron content than other commercial rice varieties (Gregorio et al., 2000; Haas et al., 2005). This kind boosted ferritin and whole-body iron by 20% in Filipino women in a controlled diet trial (Haas et al., 2005).

Researchers from Iran were able to successfully develop the world's first nano-organic Fe-chelated fertilizer (The Free Library, 2009). Dr. Jagdish Chandra Tarafdar of the ICAR-Central Arid Zone Research Institute (CAZRI) in Jodhpur, Rajasthan, created the novel fertilizer variation. The fertilizer was created by devising a mechanism for breaking down the salts into nano-form using microbial enzymes. 'Nano fertilizer is eco-friendly and promotes soil aggregation, moisture retention, and carbon build-up since it is completely bio-sourced', Dr. Tarafdar added. It poses no health risk and is suitable for a wide range of varieties, including food grains, vegetables and horticulture. The ICAR-Indian Agricultural Research Institute (IARI) in New Delhi has signed a Memorandum of Understanding (MoU) with Prathista Industries Limited in Secunderabad to commercialize nanonutrients. In addition, some researchers have recently created and patented a nanofertilizer known as Nano-Leucite Fertilizer, which is eco-friendly and has the potential to minimize nutrient loss in food while increasing crop and food output (Weekly Technology Times, 2016). Mineral fertilization of crops in Finland (Aro et al., 1995), zinc fertilization in Turkey (Cakmak et al., 1999) and iodine fertilization in irrigation water in China demonstrate that it is possible in the developed world (Jiang et al., 1997).

7.6 CONCLUSIONS

The poor productivity of the soils due to less availability of nutrients greatly hamper food production, resulting in a financial burden on the farmers. Food biofortification using traditional breeding and agronomic practices have proven successful for a variety of crops and is an alternate way for poor and small marginal farmers to attain sustainability. As a result, biofortified foods have increased the nutrient content of diets and helped malnourished population. However, in some crops, there are substantial restrictions to conventional breeding and agronomic procedures that reduce the efficacy and worldwide applicability of biofortification technologies. In the agricultural industry, nanofertilizers have a considerable influence on crop plant productivity, resilience to abiotic stressors and nutritional status. As a result, prospective nanofertilizer applications in the agrifood biotechnology and horticulture industries must not be disregarded. Furthermore, in the present climate change scenario, the latent profits of nanofertilizers have flickered a lot of interest in increasing the production capacity of agricultural crops. Reduction in losses due to leaching and volatilization with the use of nanofertilizers are the primary economic benefits. Simultaneously, the well-known favourable influence on output and product quality has the potential to significantly boost producers' profit margins when this technology is used. Before commercializing nanofertilizers, the uncertainties surrounding nanomaterials' interactions with the environment and their possible impacts on human health must be thoroughly investigated. In order to present this unique frontier in sustainable agriculture, future studies must focus on producing complete information in these underexplored regions. Biosynthesized nanoparticle-based fertilizers and nanobiofertilizers should be investigated further as a viable method for increasing yields while being environmentally friendly. As a result, smart research and sound policy can pave the way for biofortification to be a huge success in the near future.

REFERENCES

Agri, U., Chaudhary, P., & Sharma, A. (2021). In vitro compatibility evaluation of agriusable nanochitosan on beneficial plant growth-promoting rhizobacteria and maize plant. *Natl Acad Sci Lett.* 44, 555–559.

Agri, U., Chaudhary, P., Sharma, A., & Kukreti, B. (2022). Physiological response of maize plants and its rhizospheric microbiome under the influence of potential bioinoculants and nanochitosan. *Plant Soil.* 474, 451–468.

Ahmar, S., Gill, R.A., Jung, K.H., Faheem, A., Qasim, M.U., Mubeen, M., & Zhou, W. (2020). Conventional and molecular techniques from simple breeding to speed breeding in crop plants: Recent advances and future outlook. *Int J Mol Sci.* 21, 2590.

Al-Mamun, M.R., Hasan, M.R., Ahommed, M.S., Bacchu, M.S., Ali, M.R., & Khan, M.Z.H. (2021). Nanofertilizers towards sustainable agriculture and environment. *Environ Technol Innov.* 23, 101658.

Aro, A., Alfthan, G., & Varo P. (1995). Effects of supplementation of fertilizers on human selenium status in Finland. *Analyst.* 120, 841–843.

Asli, S., & Neumann, P.M. (2009). Colloidal suspensions of clay or titanium dioxide nanoparticles can inhibit leaf growth and transpiration via physical effects on root water transport. *Plant Cell Environ.* 32, 577–584.

Avestan S., Naseri, L.A., Hassanzade, A., Sokri, S.M., & Barker, A.V. (2016). Effects of nano-silicon dioxide application on in vitro proliferation of apple rootstock. *J Plant Nutr.* 39, 850–855.

Bhatt, P., Pandey, S.C., Joshi, S., Chaudhary, P., Pathak, V.M., Huang, Y., Wu, X., Zhou, Z., & Chen, S. (2022). Nanobioremediation: A sustainable approach for the removal of toxic pollutants from the environment. *J Hazard Mater.* 427, 128033.

Bouis, H.W. (2003). Micronutrient fortification of plants through plant breeding: Can it improve nutrition in man at low cost? *Proc Nutr Soc.* 62(2), 403–411.

Brozynska, M., Furtado, A., & Henry, R.J. (2016). Genomics of crop wild relatives: Expanding the gene pool for crop improvement. *Plant Biotech J.* 14, 1070–1085.

Cakmak, I. (2008). Enrichment of cereal grains with zinc: Agronomic or genetic biofortifica-tion. *Plant Soil*, 302, 1–17.

Cakmak, I., Kalaycı, M., Ekiz, H., Braun, H.J., Kılınç, Y., & Yılmaz, A. (1999). Zinc deficiency as a practical problem in plant and human nutrition in Turkey: A NATO- science for stability project. *Field Crops Res.* 60, 175–188.

Cakmak, I., & Kutman, U.B. (2017). Agronomic biofortification of cereals with zinc: A review. *Eur J Soil Sci.* 69, 172–180.

Cao, Z., Rossi, L., Stowers, C., Zhang, W., Lombardini, L., & Ma, X. (2018). The impact of cerium oxide nanoparticles on the physiology of soybean (*Glycine max* (L.) under dif-ferent soil moisture conditions. *Environ Sci Pol Res.* 25, 930–939.

Cao, Z., Stowers, C., Rossi, L., Zhang, W., Lombardini, L., & Ma, X. (2017). Physiological effects of cerium oxide nanoparticles on the photosynthesis and water use efficiency of soybean (*Glycine max* (L.) Merr.). *Environ Sci Nano.* 4, 1086–1094.

Castañeda-Álvarez, N.P., Khoury, C.K., Achicanoy, H.A., Bernau, V., Dempewolf, H., Eastwood, R.J., Guarino, L., Harker, R.H., Jarvis, A., Maxted, N., Müller, J.V. Ramirez-Villegas, J., Sosa, C.C., Struik, P.C., Vincent, H., & Toll, J. (2016). Global conservation priorities for crop wild relatives. *Nat Plants.* 2, 1–6.

Chaudhary, P., Chaudhary, A., Bhatt, P., Kumar, G., Khatoon, H., Rani, A., Kumar, S., & Sharma, A. (2022). Assessment of soil health indicators under the influence of nano-compounds and *Bacillus* spp. in field condition. *Front Environ Sci.* 9, 769871.

Chaudhary, P., Chaudhary, A., Parveen, H., Rani, A., Kumar, G., Kumar, A., & Sharma, A. (2021e). Impact of nanophos in agriculture to improve functional bacterial community and crop productivity. *BMC Plant Biol.* 21, 519.

Chaudhary, P., Khati, P., Chaudhary, A., Gangola, S., Kumar, R., & Sharma, A. (2021a). Bioinoculation using indigenous *Bacillus* spp. improves growth and yield of *Zea mays* under the influence of nanozeolite. *3 Biotech.* 11, 11.

Chaudhary, P., Khati, P., Chaudhary, A., Maithani, D., Kumar, G., & Sharma, A. (2021d). Cultivable and metagenomic approach to study the combined impact of nanogypsum and *Pseudomonas taiwanensis* on maize plant health and its rhizospheric microbiome. *PLoS One.* 16, e0250574.

Chaudhary, P., Khati, P., Gangola, S., Kumar, A., Kumar, R., & Sharma, A. (2021c). Impact of nanochitosan and *Bacillus* spp. on health, productivity and defence response in *Zea mays* under field condition. *3 Biotech.* 11, 237.

Chaudhary, P., & Sharma, A. (2019). Response of nanogypsum on the performance of plant growth promotory bacteria recovered from nanocompound infested agriculture field. *Environ Ecol.* 37, 363–372.

Chaudhary, P., Sharma, A., Chaudhary, A., Khati, P., Gangola, S., & Maithani, D. (2021b). Illumina based high throughput analysis of microbial diversity of rhizospheric soil of maize infested with nanocompounds and *Bacillus* sp. *Appl Soil Ecol.* 159, 103836.

Chhipa, H. (2019). Mycosynthesis of nanoparticles for smart agricultural practice: A green and eco-friendly approach. In: Shukla, A., & Iravani, S. (eds.) *Green Synthesis,*

Characterization and Applications of Nanoparticles (1st edition). Elsevier Inc., pp. 87–109.

Clark, R.B. (1983). Plant genotype differences in the uptake, translocation, accumulation, and use of mineral elements required for plant growth. *Plant Soil.* 72, 175–196.

Corredor, E., Testillano, P.S., Coronado, M.J., González-Melendi, P., Fernández-Pacheco, R., Marquina, C., Ibarra, M.R., De La Fuente, J.M., Rubiales, D., Pérez-De-Luque, A., & Risueño, M.C. (2019). Nanoparticle penetration and transport in living pumpkin plants: In situ sub-cellular identification. *BMC Plant Biol.* 9, 1–11.

Daniels, LA. (1996). Selenium metabolism and bioavailability. *Biol Trace Elem Res.* 54(3), 185–199.

Dimkpa, C., Andrews, J., Fugice, U., Singh, P.S., Bindraban, W.H., Elmer, J.L., & Gardea-Torresdey, J.C. (2020). White Facile coating of urea with low-dose ZnO nanoparticles promotes wheat performance and enhances Zn uptake under drought stress. *Front Plant Sci.* 11, 1–12.

Du, W., Sun, Y., Ji, R., Zhu, J., Wu, J., & Guo, H. (2011). TiO_2 and ZnO nanoparticles negatively affect wheat growth and soil enzyme activities in agricultural soil. *J Environ Mont.* 13, 822.

Duhan, J.S., Kumar, R., Kumar, N., Kaur, P., Nehra, K., & Duhan, S. (2017). Nanotechnology: The new perspective in precision agriculture. *Biotech Rep.* 15, 11–23.

Elemike, E.E., Uzoh, I.M., Onwudiwe, D.C., & Babalola, O.O. (2019). The role of nanotechnology in the fortification of plant nutrients and improvement of crop production. *Appl Sci.* 9, 499.

Erisman, J.W., Sutton, M.A., Galloway, J.N., Klimont, Z., & Winiwarter W. (2008). How a century of ammonia synthesis changed the world? *Nat Geosc.* 1, 636–9.

Eroglu, N., Emekci, M., & Athanassiou, C.G. (2017). Applications of natural zeolites on agriculture and food production. *J Sci Food Agric.* 97, 3487–3499.

Fahad, S., Hussain, S., Saud, S., Hassan, S., Shan, D., Chen, Y., Deng, N., Khan, F., Wu, C., Wu, W., Shah, F., Ullah, B., Yousaf M., Ali, S., & Huang, J. (2015). Grain Cadmium and Zinc concentrations in maize Influenced by genotypic variations and Zinc fertilization. *Clean Soil Air Water.* 43, 1433–1440.

FAOSTAT. (2021). Food and Agricultural Commodities Production. Available online: http://www.faostat.fao.org

Flax Council of Canada. (2015). Varieties. In D. Morris (Ed.) *Growing Flax Production, Management and Diagnostic Guide.* Flax Council of Canada: Winnipeg, MB, Canada, pp. 49–53.

Food and Agriculture Organization of the United Nations, International Fund for Agricultural Development, United Nations Children's Fund, World Food Programme and World Health Organization (2019). *The State of Food Insecurity in the World 2019. Strengthening the Enabling Environment for Food Security and Nutrition.* FAO: Rome.

Garg, M., Sharma, N., Sharma, S., Kapoor, P., Kumar, A., Chunduri, V., & Arora, P. (2018). Biofortified crops generated by breeding, agronomy, and transgenic approaches are improving lives of millions of people around the world. *Front Nutr.* 5, 12.

Germ, M., Stibilj, V., Šircelj, H., & Jerše, A. (2019). Biofortification of common buckwheat microgreens and seeds with different forms of selenium and iodine. *J Sci Food Agric.* 99, 4353–4362.

Glenn, K.C., Alsop, B., Bell, E., Goley, M., Jenkinson, J., Liu, B., Martin, C., Parrott, W., Souder, C., Sparks, O., Urquhart, W., Ward, J.M., & Vicini J.L. (2017). Bringing new plant varieties to market: Plant breeding and selection practices advance beneficial characteristics while minimizing unintended changes. *Crop Sci.* 57, 2906–2921.

Graham, R.D., Welch, R.M., Saunders, D.A., Ortiz-Monasterio, I., Bouis, H.E., Bonierbale, M., de Haan, S., Burgos, G., Thiele, G., Liria, R., Meisner, C.A., Beebe, S.E., Potts,

M.J., Kadian, M., Hobbs, P.R., Gupta, R.K., & Twomlow, S. (2007). Nutritious subsistence food systems. *Adv Agron*. 92, 1–74.

Greene, S.L., Williams, K.A., Khoury, C.K., Kantar, M.B., & Marek, L.F. (2018). *North American Crop Wild Relatives: Conservation Strategies Volume 1*. Springer: Cham, p. 346.

Gregorio, G.B., Senadhira, D., Htut, H., & Graham, R.D. (2000). Breeding for trace minerals in rice. *Food Nut Bull*. 21, 382–396.

Gui, X., Zhang, Z., Liu, S., Ma, Y., Zhang, P., He, X., Li, Y., Zhang, J., Li, H., Rui, Y., Liu, L., & Cao, W. (2015). Fate and phytotoxicity of CeO_2 nanoparticles on lettuce cultured in the potting soil environment. *PLoS One*. 10, 1–10.

Guo, J. (2004). Synchrotron radiation, soft-X-ray spectroscopy and nanomaterials. *Inter J Nanotech*. 1, 193–225.

Ha, N.M.C., Nguyen, T.H., Wang S.L., & Nguyen, A.D. (2019). Preparation of NPK nanofertilizer based on chitosan nanoparticles and its effect on biophysical characteristics and growth of coffee in green house. *Res Chem Inter*. 45,151–63

Haas, J.D., Beard, J.L., Murray-Kolb, L.E., Del Mundo, A.M., Felix, A., & Gregorio, G.B (2005). Iron-biofortified rice improves the iron stores of non-anemic Filipino women. *J Nutr*. 135, 2823–2830.

Harper, T. (2015). *The Year of the Trillion Dollar Nanotechnology Market?* AZoNetwork UK Ltd: Manchester, UK.

Hoang, S.A., Nguyen, L.Q., Nguyen, N.H., Tran, C.Q., Nguyen, D.V., Le, N.T., Ha, C.V., Vu, Q.N., & Phan, C.M. (2019). Metal nanoparticles as effective promotors for maize production. *Sci Rep*. 9, 13925.

Hosnedlova, B., Kepinska, M., Skalickova, S., Fernandez, C., Ruttkay-Nedecky, B., Peng, Q., Baron, M., Melcova, M., Opatrilova, R., & Zidkova, J. (2018). Nano-selenium and its nanomedicine applications: A critical review. *Inter J Nanomed*. 13, 2107–2128.

Hotz, C., Loechl, C., De Brauw, A., Eozenou, P., Gilligan, D., Moursi, M., Munhaua, B., Van Jaarsveld, P., Carriquiry, A., & Meenakshi, J.V. (2012b). A large-scale intervention to introduce orange sweet potato in rural Mozambique increases vitamin A intakes among children and women. *Br J Nutr*. 108, 163–176.

Hotz, C., Loechl, C., Lubowa, A., Tumwine, J.K., Masawi, G.N., Baingana, R., Carriquiry, A., de BrauwMeenakshi, A., & Gilligan, D.O. (2012a). Introduction of beta-carotene-rich orange sweet potato in rural Uganda resulted in increased vitamin a intakes among children and women and improved vitamin a status among children. *J Nutri*. 142, 1871–1880.

Hu, J., Guo, H., Li, J., Wang, Y., Xiao, L., & Xing, B. (2017). Interaction of Fe_2O_3 nanoparticles with *Citrus maxima* leaves and the corresponding physiological effects via foliar application. *J Nanobiotechnol*. 15, 1–12.

Huang, A.S., Tanudjaja, L., & Lum, D. (1999). Content of alpha-, beta-carotene, and dietary fiber in 18 sweet potato varieties grown in Hawaii. *J Food Comp Anal*. 12, 147–151.

International Potato Center. (2019). *Sweetpotato Agri-Food Systems Program*. International Potato Center: Lima, Peru.

IPCC (2015). *Climate Change 2014: Mitigation of Climate Change, Summary for Policymakers Technical Summary, Part of the Working Group III Contribution to the Fifth Assessment Report of the Intergovernmental Panel on Climate Change*. Edenhofer, O., Pichs-Madruga, R., Sokona, Y., Minx, J. C., Farahani, E., Kadner, S. (eds.). Intergovernmental Panel on Climate Change: Geneva, Switzerland.

Iqbal, M., & Umar, S. (2019). Nano-fertilization to enhance nutrient use efficiency and productivity of crop plants. In: Husen, A. & Iqbal, M. (eds.) *Nanomaterials and Plant Potential*. Springer: Cham, pp. 473–505.

Isermann, K. (1990). Share of agriculture in nitrogen and phosphorus emissions into the surface waters of Western Europe against the background of their eutrophication. *Fertil Res*. 26, 253–269.

Jadon, P., Selladurai, R., Yadav, S.S., Coumar, M.V., Dotaniya, M.L., & Singh, A.K. (2018). Volatilization and leaching losses of nitrogen from different coated urea fertilizers. *J Soil Sci Plant Nutr.* 18, 1036–1047.

Jewell, C.P., Zhang, S.V., Gibson, M.J.S., Tovar-Méndez, A., Mcclure, B., & Moyle, L.C. (2020). Intraspecific genetic variation underlying postmating reproductive barriers between species in the wild tomato clade (*Solanum Lycopersicon*). *J Hered.* 111, 216–226.

Jha, A.B., & Warkentin, T.D. (2020). Biofortification of pulse crops: Status and future perspectives. *Plants.* 9, 73.

Jiang X.M., Cao X.Y., Jiang J.Y., Ma, T., James, D.W., & Rakeman, M.A. (1997). Dynamics of environmental supplementation of iodine: Four years' experience in iodination of irrigation water in Hotien, Xinjiang, China. *Arch Environ Occup Health.* 52(6), 399–408.

Karimi, E. & Fard, E.M. (2017). Nanomaterial effects on soil microorganisms. In: Ghorbanpour, M., Manika, K., & Varma, A. (eds) *Nanoscience and Plant–Soil Systems.* Springer: Cham, Switzerland, pp. 137–200.

Khati, P., Bhatt, P., Kumar, R., & Sharma, A. (2018). Effect of nanozeolite and plant growth promoting rhizobacteria on maize. *3Biotech.* 8(141), 1–12.

Khati, P., Chaudhary, P., Gangola, S., Bhatt, P., & Sharma, A. (2017a). Nanochitosan induced growth of *Zea Mays* with soil health maintenance. *3 Biotech.* 7(81), 1–9.

Khati, P., Chaudhary, P., Gangola, S., & Sharma A. (2019a). Influence of nanozeolite on plant growth promotory bacterial isolates recovered from nanocompound infested agriculture field. *Environ Ecol.* 37 (2), 521–527.

Khati, P., Sharma, A., Chaudhary, P., Singh, A K., Gangola, S., & Kumar, R. (2019b). High-throughput sequencing approach to access the impact of nanozeolite treatment on species richness and evenness of soil metagenome. *Biocatal Agric Biotechnol.* 20, 101249.

Khati, P., Sharma, A., Gangola, S., Kumar, R., Bhatt, P., & Kumar, G. (2017b). Impact of some agriusable nanocompounds on soil microbial activity: An indicator of soil health. *Clean Soil Air Water.* 45, 1–7.

Konate, A., He, X., Zhang, Z., Ma, Y., Zhang, P., Alugongo, G.M., & Rui, Y. (2017). Magnetic (Fe_3O_4) nanoparticles reduce heavy metals uptake and mitigate their toxicity in wheat seedling. *Sustainability.* 9, 5, 790.

Kopittke, P.M., Lombi, E., Wang, P., Schjoerring, J.K., & Husted S. (2019). Nanomaterials as fertilizers for improving plant mineral nutrition and environmental outcomes. *Environ Sci Nano.* 6, 3513–3524.

Kukreti, B., Sharma, A., Chaudhary, P., Agri, U., & Maithani, D. (2020). Influence of nanosilicon dioxide along with bioinoculants on *Zea mays* and its rhizospheric soil. *3 Biotech.* 10, 345.

Kumar, V., Guleria, P., Kumar, V., & Yadav, S.K. (2013). Gold nanoparticle exposure induces growth and yield enhancement in *Arabidopsis thaliana.* *Sci Total Environ.* 461, 462–468.

Kumari, A., Bhinda, M.S., Sharma, B., & Pariha, M. (2022). Climate change mitigation and nanotechnology: An overview. In: Faizan, M., Hayat, S. & Yu, F, (eds.) *Nanoparticles: A New Tool to Enhance Stress Tolerance.* Springer: Cham, pp. 33–60.

Kumari, H., Khati, P., Gangola, S., Chaudhary, P., & Sharma A. (2021). Performance of plant growth promotory rhizobacteria on maize and soil characteristics under the influence of TiO_2 nanoparticles. *Pant Res J.* 19(1), 28–39.

Kumari, S., Sharma A., Chaudhary, P., & Khati, P. (2020). Management of plant vigour and soil health using two agriusable nanocompounds and plant growth promontory rhizobacteria in Fenugreek. *3 Biotech.* 10(461), 1–11.

Kyriacou, M.C., Rouphael, Y., Di, F., Kyratzis, A., Scrio, F., Renna, M., De Pascale, S., & Santamaria, P. (2016). Micro-scale vegetable production and the rise of microgreens. *Trends Food Sci Technol.* 57, 103–115.

Lahiani, M.H., Chen, J., Irin, F., Puretzky, A.A., Green, M.J., & Khodakovskaya, M.V. (2015). Interaction of carbon nanohorns with plants: Uptake and biological effects. *Carbon.* 81, 607–619.

Lahiani, M.H., Nima, Z.A., Villagarcia, H., Biris, A.S., & Khodakovskaya, M.V. (2017). Assessment of effects of the long-term exposure of agricultural crops to carbon nanotubes. *J Agric Food Chem*. 66, 6654–6662.

Leip, A., Billen, G., Garnier, J., Grizzetti, B., Lassaletta, L., Reis, S., Simpson, D., Sutton, M.A., Wim de Vries, W., Franz Weiss, F., & Westhoek, H. (2015). Impacts of European livestock production: Nitrogen, sulphur, phosphorus and greenhouse gas emissions, land-use, water eutrophication and biodiversity. *Environ Res Lett*. 10, 115004.

León-Silva, S., Arrieta-Cortes, R. Fernández-Luqueño F., & López-Valdez, F. (2018). Design and production of nanofertilizers. In: Fernández-Luqueño, F. & López-Valdez, F. (eds.) *Agricultural Nanobiotechnology*. Springer: Cham, pp. 17–31.

Li, D., Pfeiffer, T.W., & Cornelius, P.L. (2008). Soybean QTL for yield and yield components associated with *Glycine soja* alleles. *Crop Sci*. 48, 571–581.

Li, H., Shan, C., Zhang, Y., Cai, J., Zhang, W., & Pan, B. (2016). Arsenate adsorption by hydrous ferric oxide nanoparticles embedded in cross-linked anion exchanger: Effect of the host pore structure. *ACS App Mater Inter*. 8, 3012–3020.

Liang, T., Yin, Q., Zhang, Y.L., Wang, B., Guo, W.M., Wang, J.W., & Xie, J.P. (2013). Effects of carbon nanoparticles application on the growth, physiological characteristics and nutrient accumulation in tobacco plants. *J Food Agric Environ*. 11, 954–958.

Liu, J., Dhungana, B., & Cobb, G.P. (2018). Environmental behavior, potential phytotoxicity, and accumulation of copper oxide nanoparticles and arsenic in rice plants. *Environ Toxicol Chem*. 37, 11–20.

López-Vargas, E.R., Ortega-Ortíz, H., Cadenas-Pliego, G., Romenus, K.D.A., de la Fuente, M.C., Benavides-Mendoza, A., & Juárez-Maldonado, A. (2018). Foliar Application of copper nanoparticles increases the fruit quality and the content of bioactive compounds in tomatoes. *Appl Sci*. 8, 1020.

Low, J.W., Mwanga, R.O.M., Andrade, M., Carey, E., & Ball, A.M. (2017). Tackling vitamin A deficiency with biofortified sweet potato in sub-Saharan Africa. *Global Food Sec*. 14, 23–30.

Mabesa, R.L., Impa, S.M., Grewal, D., & Johnson-Beebout, S.E. (2013). Contrasting grain-Zn response of biofortification rice (*Oryza sativa* L.) breeding lines to foliar Zn application. *Field Crops Res*. 149, 223–233.

Mahakham, W., Theerakulpisut, P., Maensiri, S., Phumying, S., & Sarmah, A.K. (2016). Environmentally benign synthesis of phytochemicals-capped gold nanoparticles as nano-priming agent for promoting maize seed germination. *Sci Total Environ*. 573, 1089–1102.

Manikandan, A., & Subramanian, K. (2016). Evaluation of zeolite based nitrogen nanofertilizers on maize growth, yield and quality on inceptisols and alfisols. *Int J Plant Soil Sci*. 9, 1–9.

Manjunatha, R., Naik, D., & Usharani K. (2019). Nanotechnology application in agriculture: A review. *J Pharmacogn Phytochem*. 8(3), 1073–1083.

Manjunatha, S.B., Biradar, D.P., & Aladakatti, Y.R. (2016). Nanotechnology and its applications in agriculture: A review. *J Farm Sci*. 29, 1–3.

Marchiol, L. (2019). Nanofertilizers: An outlook of crop nutrition in the fourth agricultural revolution. *Ital J Agron*. 14(3), 1367.

Masrahi, A., VandeVoort, A.R., & Arai, Y. (2014). Effects of silver nanoparticle on soil-nitrification processes. *Arch Environ Cont Toxicol*. 66, 504–513.

Meetu, G., & Shikha, G. (2017). An overview of selenium uptake, metabolism, and toxicity in plants. *Front Plant Sci*. 7, 2074.

Mejias, J.H., Salazar, F., Pérez Amaro, L., Hube, S., Rodriguez, M., & Alfaro, M. (2021). Nanofertilizers: A cutting-edge approach to increase nitrogen use efficiency in grasslands. *Front Environ Sci*. 9, 635114.

Mitchell-Olds, T. (2010). Complex-trait analysis in plants. *Genome Biol*. 11, 113.

Mittal, D., Kaur, G., Singh, P., Yadav, K., & Ali, S.A. (2020). Nanoparticle-based sustainable agriculture and food science: Recent advances and future outlook. *Front Nanotechnol.* 2, 10.

Morales-Díaz, A.B., Ortega-Ortíz, H., Juárez-Maldonado, A., Cadenas-Pliego, G., González-Morales, S., & Benavides-Mendoza, A. (2017). Application of nanoelements in plant nutrition and its impact in ecosystems. *Adv Nat Sci Nanosci Nanotechnol.* 8, 013001.

Morris, D. (2007). Description and composition of flax. In B. Tanwar, A. Goyal, Eds., *Flax-A Health and Nutrition Primer*, Flax Council of Canada: Winnipeg, MB, Canada, pp. 9–21.

Neelam, K., Rawat, N., Tiwari, V.K., Kumar, S., Chhuneja, P., Singh, K., Randhawa, S., & Dhaliwal, H.S. (2011). Introgression of group 4 and 7 chromosomes of *Ae. peregrina* in wheat enhances grain iron and zinc density. *Mol Breed.* 28, 623–634.

Newman, R.G., Moon, Y., Sams, C.E., Tou, J.C., & Waterland, N.L. (2021). Biofortifcation of sodium selenate improves dietary mineral contents and antioxidant capacity of culinary herb microgreens. *Front Plant Sci.* 12, 716437.

Pandya-Lorch, S.F.R. (2012). *About IFPRI and the 2020: Vision Initiative.* International Food Policy Research Institute: Washington, DC.

Parveen, K., Banse, V., & Ledwani, L. (2016). Green synthesis of nanoparticles: Their advantages and disadvantages. *AIP Conf Proc.* 1724, 020048.

Pestovsky, Y.S., & Martínez-Antonio, A. (2017). The Use of nanoparticles and nanoformulations in agriculture. *J Nanosci Nanotechnol.* 17, 8699–8730.

Phattarakul, N., Rerkasem, B., & Li, L.J. (2012). Biofortification of rice grain with zinc through zinc fertilization in different countries. *Plant Soil.* 361, 131–141.

Polat, E., Karaca, M., Demir, H., & Onus, N. (2004). Use of natural zeolite (clinoptilolite) in agriculture. *J Fruit Ornamental Plant Res.* 12, 183–189

Prasad, R., Bhattacharyya, A., Nguyen, Q.D. (2017). Nanotechnology in sustainable agriculture: Recent developments, challenges, and perspectives. *Front Microbiol.* 8, 1–13.

Prasad, T.N.V.K.V., Sudhakar, P., Sreenivasulu, Y., Latha, P, Munaswamy, V., Rajareddy, K., Sreeprasad, T.S., Sajanlal, P.R., & Pradeep, T. (2012). Effect of nanoscale zinc oxide particles on the germination, growth and yield of groundnut. *J Plant Nutr.* 35, 905–927.

Puccinelli, M., Pezzarossa, B., Pintimalli, L., & Malorgio, F. (2021). Selenium biofortifcation of three wild species, *Rumexcetosa* L., *Plantagocoronopus* L., and *Portulaca oleracea* L., grown as microgreens. *Agronomy.* 11, 1155.

Qureshi, A., Singh, D., & Dwivedi, S. (2018). Nano-fertilizers: A novel way for enhancing nutrient use efficiency and crop productivity. *Int J Curr Microbiol Appl Sci.* 7, 3325–3335.

Raliya, R., & Tarafdar, J.C (2013). ZnO nanoparticle biosynthesis and its effect on phosphorous mobilizing enzyme secretion and gum contents in Clusterbean (*Cyamopsis tetragonoloba*). *J Agric Res.* 2, 48–57.

Raliya, R., Tarafdar, J.C., Singh, S.K., Gautam, R., Choudhary, K., Maurino, V.G., & Saharan, V. (2014). MgO nanoparticles biosynthesis and its effect on chlorophyll contents in the leaves of clusterbean (*Cyamopsis tetragonoloba* L.). *Adv Sci Eng Med.* 6, 538–545.

Rico, C., Barrios, A.C., Tan, W., Rubenecia, R., Lee, S.C., Varela-Ramirez, A., Peralta-Videa, J.R., & Gardea-Torresdey, J.L. (2015). Physiological and biochemical response of soil-grown barley (*Hordeum vulgare* L.) to cerium oxide nanoparticles. *Environ Sci Poll Res.* 22, 10551–10558.

Rico, C., Lee, S.C., Rubenecia, R., Mukherjee, A., Hong, J., Peralta-Videa, J.R., & Gardea-Torresdey, J.L. (2014). Cerium oxide nanoparticles impact yield and modify nutritional parameters in wheat (*Triticum aestivum* L.). *J Agric Food Chem.* 62, 9669–9675.

Rodrigo, S., Santamaria, O., & Poblaciones, M.J. (2014). Selenium application timing: Influence in wheat grain and flour selenium accumulation under Mediterranean conditions. *J Agric Sci.* 6, 6.

Rouphael, Y., & Kyriacou, M.C. (2018). Enhancing quality of fresh vegetables through salinity eustress and biofortification applications facilitated by soilless cultivation. *Front Plant Sci.* 9, 1–6.

Sadeghzadeh, B. (2013). A review of zinc nutrition and plant breeding. *J Soil Sci Plant Nutr.* 13, 905–927.

Saha, S., Chakraborty, M., Padhan, D., Saha, B., Murmu, S., Batabyal, K., Seth, A., Hazra, G.C., Mandal, B., & Bell, R.W. (2017). Agronomic biofortification of zinc in rice: Influence of cultivars and zinc application methods on grain yield and zinc bioavailability. *Field Crops Res.* 210, 52–60.

Saurabh, S., Singh, B.K., Yadav, S.M., & Gupta, A.K. (2015). Applications of nanotechnology in agricultural and their role in disease management. *Res J Nanosci Nanotechnol.* 5, 15.

Saxena, R., Tomar, R.S., & Kumar, M. (2016). Exploring nano-biotechnology to mitigate abiotic stress in crop plants. *J Pharma Sci Res.* 8, 974.

Siddiqui, M.H., & Al-Whaibi, M.H. (2014). Role of nano-SiO_2 in germination of tomato (*Lycopersicum esculentum* seeds Mill.). *Saudi J Biol Sci.* 21, 13–17.

Siddiqui, M.H., Al-Whaibi, M.H., Faisal, M., & Al-Sahli, A.A. (2014). Nano-silicon dioxide mitigates the adverse effects of salt stress on *Cucurbita pepo* L. *Environ Toxicol Chem.* 33, 2429–2437.

Signore, A., Renna, M., D'Imperio, M., Serio, F. & Santamaria, P. (2018). Preliminary evidences of biofortification with iodine of "carota di polignano", an Italian carrot landrace. *Front Plant Sci.* 9, 1–8.

Singh, N.B., Amist, N., Yadav, K., Singh, D., Pandey, J.K., & Singh, S.C. (2013). Zinc oxide nanoparticles as fertilizer for the germination, growth and metabolism of vegetable crops. *J Nanoeng Nanomanuf.* 3, 353–364.

Sivasakthi, K., Marques, E., Kalungwana, N., Cordeiro, M., Sani, S.G.A.S., Udupa, S.M., Rather, I.A., Mir, R.R., Vadez, V., Vandemark, G.J., Gaur, P.M., Cook, D.R., Boesch, C., von Wettberg, E.J.B., Kholova, J., & Penmetsa, R.V. (2019). Functional dissection of the chickpea (*Cicer arietinum* L.) stay-green phenotype associated with molecular variation at an ortholog of Mendel's I gene for cotyledon color: Implications for crop production and carotenoid biofortification. *Int J Mol Sci.* 20, 55–62.

Skalickova, S., Milosavljevic, V., Cihalova, K., Horky, P., Richtera, L., & Adam, V. (2017). Selenium nanoparticles as a nutritional supplement. *Nutrition.* 33, 83–90.

Skrypnik, L., Novikova, A., & Tokupova, E. (2019). Improvement of phenolic compounds, essential oil content and antioxidant properties of sweet basil (*Ocimum basilicum* L.) depending on type and concentration of selenium application. *Plants.* 8, 458.

Subramanian, K.S., & Tarafdar, J.C. (2009). Nanotechnology in soil science. In: *Proceeding of the Indian Society of Soil Science Platinum Jubilee Celebration. December 22–25th,* ICAR-IARI, New Delhi, p. 199.

Subramanian, K.S., & Tarafdar, J.C. (2011). Prospects of nanotechnology in Indian farming. *Int J Agric Sci.* 81, 887–893.

Suppan, S. (2017). Applying nanotechnology to fertilizer the institute for agriculture and trade policy works locally and globally at the intersection of policy and practice to ensure fair and sustainable food, farm and trade systems. Available online: www.iatp.org

Tarafdar, A., Raliya, R., Wang, W.-N., Biswas, P., & Tarafdar, J.C. (2013). Green synthesis of TiO_2 nanoparticle using *Aspergillus tubingensis*. *Adv Sci Eng Med.* 5, 943–949.

Thakur, S., Thakur, T., & Kumar, R. (2018). Bio-nanotechnology and its role in agriculture and food industry. *J Mol Gen Med.* 12, 1–5.

The Free Library. (2009). First nano-organic iron chelated fertilizer invented in Iran. Retrieved May 03 2022 from https://www.thefreelibrary.com/First+Nano-Organic+Iron+Chelated+Fertilizer+Invented+in+Iran.-a0211779457

Uttam, S., & Abioye, F.L.S. (2017). Selenium in the soil-plant environment: A review. *Int J App Agric Sci.* 3, 1–18.

Van Jaarsveld, P.J., Faber, M., Tanumihardjo, S.A., Nestel, P., Lombard, C.J., & Benadé, A.J.S. (2005). Beta-carotene-rich orange-fleshed sweet potato improves the vitamin A status of primary school children assessed with the modified-relative-dose-response test. *Am J Clin Nutr.* 81, 1080–1087.

Verma, K.K., Song, X.-P., Joshi, A., Tian, D.-D., Rajput, V.D., Singh, M., Arora, J., Minkina, T., & Li, Y.R. (2022). Recent trends in nano-fertilizers for sustainable agriculture under climate change for global food security. *Nanomaterials.* 12, 173.

Wang, Q., Ma, X., Zhang, W., Pei, H., & Chen, Y. (2012). The impact of cerium oxide nanoparticles on tomato (*Solanum lycopersicum* L.) and its implications for food safety. *Metallomics.* 4, 1105–1112.

Weber, N.C., Koron, D., Jakopicˇ, J., Vebericˇ, R., Hudina, M., & Cesnik, H.B. (2021). Influence of nitrogen, calcium and nano-fertilizer on strawberry (*Fragaria ananassa* Duch.) Fruit inner and outer quality. *Agronomy.* 11, 997.

Weekly Technology Times. (2016). Nano-zinc oxide as a future fertilizer. Available online: https://technologytimes.pk/2016/04/27/nano-zinc-oxide-as-a-future-fertilizer/

White, P.J., & Broadley, M.R. (2009). Biofortification of crops with seven mineral elements often lacking in human diets-Iron, zinc, copper, calcium, magnesium, selenium and iodine. *New Phytologist.* 182, 49–84.

WHO. (2009). *Global Health Risks: Mortality and Burden of Disease Attributable to Selected Major Risks.* WHO: Geneva, Switzerland.

Xiong, T., Dumat, C., Dappe, V., Vezin, H., Schreck, E., Shahid, M., Pierart, A., & Sobanska, S. (2017b). Copper oxide nanoparticle foliar uptake, phytotoxicity, and consequences for sustainable urban agriculture. *Environ Sci Technol.* 51, 5242–5251.

Xiong, T.T., Dumat, C., Dappe, V., Vezin, H., Schreck, E., Sahid, M., Pierart, A., & Sobanksa, S. (2017a). Potential contamination of copper oxide nanoparticles and possible consequences on urban agriculture. *Environ Sci Technol.* 78, 5774–5782.

Yang, S.H., Moran, D.L., Jia, H.W., Bicar, E.H., Lee, M., & Scott, MP. (2002). Expression of a synthetic porcine alpha-lactalbumin gene in the kernels of transgenic maize. *Transgenic Res.* 11, 11–20.

Yco, A.R., Flowers, S.A., Rao, G., Welfare, K., Senanayake, N., & Flowers, T.J. (1999). Silicon reduces sodium uptake in rice (*Oryza sativa* L.) in saline conditions and this is accounted for by a reduction in the transpirational bypass flow. *Plant Cell Environ.* 22, 559–565.

Younis, A., Khattab, H., & Emam, M. (2020). Impacts of silicon and silicon nanoparticles on leaf ultrastructure and TaPIP1 and TaNIP2 gene expressions in heat stressed wheat seedlings. *Biol Plant.* 64, 343–352.

Zhao, L., Peralta-Videa, J.R., Rico, C.M., Hernandez-Viezcas, J.A., Sun, Y., Niu, G., Servin, A., Nunez, J.E., Duarte-Gardea, M., & Gardea-Torresdey, J.L. (2014). CeO_2 and ZnO nanoparticles change the nutritional qualities of cucumber (*Cucumis sativus*). *J Agric Food Chem.* 62, 2752–2759

Zhao, L., Wang, L.L.A., Zhang, H., Huang, M., Wu, H., Xing, B., Wang, Z., & Ji, R. (2020). Nano-biotechnology in agriculture: Use of nanomaterials to promote plant growth and stress tolerance. *J Agric Food Chem.* 68, 1935–1947.

Zheng, L., Hong, F., Lu, S., & Liu, C. (2005). Effect of Nano-TiO_2 on strength of naturally aged seeds and growth of spinach. *Biol Trace Elem Res.* 104, 83–91.

Zia, M., Phull, A.R., & Ali, J.S. (2016). Synthesis, characterization, applications, and challenges of iron oxide nanoparticles. *Nanotechnol Sci App.* 9, 49.

Zulfiqar, F., Navarro, M., Ashraf, M., Akram, N.A., & Munne-Bosch, S. (2019). Nanofertilizer use for sustainable agriculture: Advantages and limitations. *Plant Sci.* 289, 110270.

8 Nanoparticles and Their Application as Nano-bioformulations in Agriculture

Priyanka Khati and Swati Lohani
ICAR-Vivekananda Parvatiya Krishi Anusandhan Sansthan

Pratima Raypa
Govind Ballabh Pant University of Agriculture & Technology

Anuj Sharma
CSIR-Central Institute of Medicinal and Aromatic Plants

Asha Kumari
ICAR-Indian Agricultural Research Institute

Jyotsana Maura
DRDO

Vijaya Rani
ICAR-Indian Institute of Vegetable Research

Damini Maithani
IFTM University

CONTENTS

DOI: 10.1201/9781003345565-8

8.1 INTRODUCTION

Urbanization, energy and resource shortages, sustainable resource use, run-off, and the buildup of pesticides and fertilizers are just a few of the agricultural and environmental issues that nanotechnology has the ability to address (Chen & Yada, 2011; Ditta, 2012; Parisi et al., 2015). The usage of nanoscale materials in agricultural science is expanding as a result of the population's ongoing demand for higher agricultural yields and more efficient methods to optimize agricultural practices (Gogos et al., 2012; Chaudhary et al., 2021a; Chaudhary et al., 2022). In reality, the growth of precision farming and sustainable agriculture may be significantly impacted by nanotechnology. Ultimately, this seeks to increase agricultural output while reducing input (such as fertilizers, pesticides, and herbicides) by keeping an eye on environmental factors and taking focused action (Servin et al., 2015; Fraceto et al., 2016).

Due to their potential to enhance seed germination, growth, and plant protection through the controlled release of agrochemicals, nano-enhanced solutions have drawn increasing attention in this field (Agri et al., 2021; 2022). This has led to a decrease in the amount of chemical products used and a reduction in nutrient losses during fertilization. Additionally, nanotechnology in agriculture may offer creative ways to safeguard and improve soils and water, increasing food output and quality globally while also being environmentally benign (Biswal et al., 2012; Sonkaria et al., 2012; Ditta, 2012; Khot et al., 2012; Prasad et al., 2014; Sekhon, 2014; Chaudhary et al., 2021b, c).

It is believed that nanomaterials (NMs) are the best platform to drive the agri-nanotechnology revolution. These materials' incredibly small size (less than 100 nm) enables them to penetrate biological barriers and diffuse into plant tissue through root or foliar treatment, opening up new effective pathways for the delivery of nutrients and insecticides. Additionally, it is possible to perform surface engineering on NMs to give them suitable qualities and functions. Specifically, this will point them in the right direction within the plant or soil and provide clever release and distribution tactics. To increase crop productivity and growth, nanotechnology has produced nano-based formulations including nanofertilizers, nanopesticides, and nanoemulsions (Dasgupta et al., 2015; Poddar, 2018). The growth in productivity through effective nutrient uptake and delivery, water and soil remediation, genetic engineering of protein-encoding genes, improvement in soil characteristics, prevention against diseases, and enhanced crop productivity and growth are some of the most productive applications of nanotechnology in agricultural systems (Scrinis & Lyons 2007; Sekhon, 2014; Kukreti et al., 2020; Bhatt et al., 2022).

While nanoherbicides and nanopesticides can be used to efficiently manage weeds, pests, and herbs, nanofertilizers improve soil fertility by effectively supplying nutrients (Scrinis & Lyons, 2007; Dubey & Mailapalli, 2016; Chhipa, 2017). Nanosensors can be utilized for the detection of moisture level, nutrient concentration, soil water levels, and assessment of illnesses (Omanovic-Miklicanina & Maksimovic, 2016). The creation of insect-resistant, disease-resistant, and stress-tolerant crops is currently being focused on using nanoparticle (NP)-mediated gene delivery in plants for increased life (Wang et al., 2019; Jat et al., 2020). Numerous NMs, including carbon nanotubes, metal oxide nanoparticles, polymeric nanoparticles, and nanoemulsions with active substances, have demonstrated success in sustainable agriculture and food production (Mukherjee et al., 2016; Rastogi et al., 2017; Kaphle et al., 2018; Kaur et al., 2020).

8.2 NANOPARTICLES

The particle in nanoscale size is termed as nanoparticles. On the basis of various characteristics such as size and shape, origin, morphology, and physicochemical properties, nanoparticles are grouped into different types. For instance, metal semiconductor, carbon-based, polymeric, lipid-based and ceramic nanoparticles are some of them (Jeevanandam et al., 2018).

8.2.1 CLASSIFICATION BASED ON ORIGIN

According to their origin NPs can be classified as natural, synthetic (engineered), and incidental. Naturally produced nanoparticles exist in bodies of insects, plants, animals, and even in humans. Humans create engineered or synthetic nanoparticles with specific properties for specific applications. Incidental nanoparticles, on the other hand, are created as a by-product of industrial methods such as welding fumes, combustion processes, vehicle engine exhaust, and, by chance, forest fires.

8.2.1.1 Natural Nanoparticles

They are synthesized in the environment itself through either anthropogenic activities or biological species. The troposphere of the earth's atmosphere contains naturally occurring nanoparticles, as do the hydrosphere, comprising lakes, oceans, groundwater, and hydro channels; the lithosphere, comprising soils, rocks, and magma, in various stages of evolution process; and the biosphere, including higher organisms and microorganisms, including humans (Sharma et al., 2015). In addition to biological entities such as viruses, nanoparticles are also detected in fine sand, dust, volcanic ash, and ocean spray. Storms of dusts are the most frequent nanoparticle sources in barren deserts and terrestrial environments (Hochella et al., 2015). Up to 30,106 tonnes of NPs as volcanic ash can be released into the atmosphere during a single eruption, where they disperse globally and settle in the stratosphere and troposphere (Taylor, 2002). Many forest fires transport micron- and nano-sized particles through smokes and ashes, causing a variety of respiratory issues in humans and animals (Buseck & Adachi, 2008).

8.2.1.2 Engineered Nanoparticles

Synthetic (engineered) nanoparticles are produced through a range of physical, chemical, biological, and combination processes, vehicle exhaust and smoke, hydraulic grinding, and other ways. A large number of methodologies for risk assessment have recently been established to have an idea about the fate and behaviour of manufactured NPs in an array of environmental contexts. The key difficulty with synthetic NPs is figuring out whether the current information is enough to predict their performance or if they differ from natural NPs in any way that is environment related (Wagner et al., 2014). Quantum dots, nanoparticles of metals or metal oxides (e.g. Au and TiO_2), carbon nanotubes, and fullerenes of carbon are a few instances of engineered NPs.

Titanium dioxide NPs, carbon NPs, and hydroxyapatites are found in sports equipment, sunscreen, commercial cosmetics, and toothpastes (Weir et al., 2012; De Volder et al., 2013). Along with cosmetics, paints also most often contain NPs. Sunscreens and cosmetic creams often include a white pigment made of titanium oxide nanoparticles (NPs) larger than 100nm (Donaldson et al., 2004). Ag NPs are present in wet wipes, air disinfectant sprays, food storage containers, shampoos, and toothpaste. Investigations are being done to establish more databases about the potential dangers of NPs in various available products. Engineered nanoparticles pose a potential risk to human health as well as the environment.

8.2.1.3 Incidental Nanoparticles

Some of the usual human activities that cause NPs to form are simple combustion processes used in cooking, cars, fuel oil, coal for power generation, welding, ore-refining, aircraft engines, chemical manufacture, and smelting (Rogers et al.,

2005). Diesel engine exhaust and automobile exhaust emit particles between 20 and 130 nm, while gas engine particles range from 20 to 60 nm (Sioutas et al., 2005). In the atmosphere, diesel-generated particles contribute to more than 90% of the carbon comprising NPs (Kittelson, 2001); the prime source of nanoparticle entry in urban environments is vehicle exhausts. Anthropogenic activities that increase NPs in the atmosphere include cigarette smoking and building destruction. Smoke from cigarettes contains 100,000 chemical components distributed as nanoparticles (10–700 nm) (Ning et al., 2006).

Similar to this, when larger structures are demolished, nano- and micro-particulate matter <10 μm are liberated into the sky (Stefani et al., 2005). Around the area of a building demolition, NPs are released that include construction debris, respirable asbestos, fibres, lead, glass, and other harmful household goods.

8.2.2 Based on Chemical Nature

Based on the chemical nature of NPs they are differentiated into three categories: inorganic-based, organic-based, and composite-based nanoparticles.

8.2.2.1 Inorganic-Based Nanoparticles

Different metals and their corresponding metal oxide nanomaterials are included in inorganic-based nanoparticles. Examples of inorganic nanomaterials based on metal oxides include zinc oxide (ZnO), magnesium aluminium oxide ($MgAl_2O_4$), copper oxide (CuO), cerium oxide (CeO_2), iron oxide (Fe_2O_3), silica (SiO_2), and titanium dioxide (TiO_2). Examples of organic nanomaterials based on metal oxides include gold (Au), silver (Ag), and aluminium (Majhi & Yadav, 2021). They have a high surface energy and the capacity to adsorb tiny molecules. They are used in research fields, bioanalytical applications, and environmental applications as well as in the detection and imaging of biomolecules.

8.2.2.2 Organic-Based Nanoparticles

Both natural as well as artificial organic compound derivatives act as the backbone for organic nanoparticles (NPs). For example, cyclodextrins, protein aggregates, dendrimers, lipid bodies, milk emulsions, micelles and liposomes come under organic NPs. Numerous industrial goods also contain organic NPs, primarily in the food and cosmetics industries. Nano emulsions are used in the manufacture of several culinary products, including milk creams, pastries, and chocolates. Pharmaceutical formulations such as liposome vectors, polymer-protein hybrids, polymersomes, or polymer-drug conjugates, among others, also use organic NPs. Another crucial aspect of organic NPs is that they provide comparatively straightforward methods for encapsulating items (De Matteis, 2017). Organic NPs are the most appealing method for drug administration and biological applications since the molecules employed to generate them are biodegradable in nature.

8.2.2.3 Composite-Based Nanoparticles

Any combination of organic, metal, metal oxide, and/or carbon-based nanoparticles makes up composite nanoparticles, which also contain complex structures such as metal organic framework. They are employed in optical sensors, metal semiconductor junctions, modifiers of polymeric films for packaging, and other applications as highly active and specialized catalysts.

8.2.3 BASED ON NANOSCALE DIMENSIONS

On the basis of dimensions of nanoscale, nanomaterials are categorized into zero-, one-, two-, and three-dimensional, respectively (Ross et al., 2016).

8.2.3.1 Zero-Dimensional

No dimension in zero dimensions is larger than 100 nm because all three dimensions (x, y, and z) are nanoscale. It consists of molecules, clusters, fullerenes, rings, grains, powders, metal carbides, nanospheres, and nanoclusters, among other things.

8.2.3.2 One-Dimensional

Nanomaterials in the shape of needles can be created when two dimensions (x, y) are at the nanoscale and one (z) is outside the nanoscale. It includes nanofibers, nanotubes, nanorods, nanowires, etc., as well as spirals, belts, filaments, springs, columns, and needles.

8.2.3.3 Two-Dimensional

Two-dimensional objects have one dimension (x) that is nanoscale and two other dimensions (y, z) that are not. These 2D nanomaterials have a plate-like form and contain nanolayers, nanofilms, and nanocoatings, among other things, that are only a few nanometres thick.

8.2.3.4 Three-Dimensional

These nanoparticles are not contained in any dimension up to the nanoscale. Above 100 nm, they have three arbitrary dimensions. Nanoparticle dispersions, nanotubes, and nanowire bundles, as well as multinano layers (polycrystals) in which zero-, one-, and two-dimensional structural elements remain in close contact with one another and configure interfaces, are all components of the bulk three-dimensional nanoparticles.

8.3 SYNTHESIS AND CHARACTERISTICS OF NANOPARTICLES

Physical, chemical, and biological processes are used in three different ways to create nanoparticles. By using an atmospheric pressure tube furnace and the evaporation-condensation procedure, nanoparticles are created physically (Kruis et al., 2000). For the synthesis of NPs, common physical techniques include pyrolysis and spark discharge. Physical methods have the benefits of using radiation and speed as reducing agents and avoiding dangerous chemicals, but they also have drawbacks such as low yield, solvent contamination, high energy use, and uneven distribution (Tsuji et al., 2005). Some methods of synthesis of nanoparticles are listed in Table 8.1.

To produce nanoparticles for the chemical process, water or organic solvents are utilized. Reducing agents, metal precursors, and stabilizing/capping agents

TABLE 8.1
Methods for Synthesis of Nanoparticles

Methods	Example	Nanoparticles	Advantages
Biological	Plant extracts Microorganisms, i.e. Fungi Algae Bacteria Yeast	• *Pseudomonas stutzeri* was used to produce Ag NPs • Curcumin CuNP • Ag NP using *Phyllanthus emblica* fruit extract	• Role in bioreduction of metal ions and their stability • Fungicide and nanofertilizer • Antimicrobial activity against *Acidovorax oryzae* strain (Masum et al., 2019)
Chemical	Thermal decomposition Hydrothermal Microemulsion Coprecipitation	• Gold nanoparticle (GNPs) by chemical reduction method	• Small size and biocompatible GNPs (Hussain et al., 2020) • Minimizing surface oxidation and agglomeration
Physical	Gas phase deposition Powder ball milling Aerosol Electron beam lithography Laser ablation	• Silver nanoparticles	• No use of toxic chemicals • Pure and uniform shape and size

are the three key ingredients required for this procedure. Approaches such as top to down and bottom to up are generally used to manufacture nanoparticles. Nucleation and thereafter growth comprise two steps of the chemical process for reducing salts of nanoparticles (Deepak et al., 2011). The top to down approach involves mechanically grinding bulk metals, followed by stabilization using colloidal stabilizers. Chemical reductions, sono-decompositions, and electrochemical techniques are all examples of the bottom-up approach. In comparison to physical methods, this method's main benefit is its high yield. The ingredients utilized to create NPs, such as thio-glycerol, 2-mercaptoethanol, citrate, and borohydride, are toxic and dangerous, and the chemical processes are very expensive (Mallick et al., 2004). Another drawback is that the created particles do not meet expectations for purity since too many poisonous and dangerous by-products from the synthesis process must be removed, and it is also exceedingly challenging to prepare NPs with precise dimensions. Chemical processes consist of techniques such as electrochemical reduction, lithography, laser ablation, chemical reduction, laser irradiation, and thermal and sono-decomposition (Zhang et al., 2011). Low cost, high yield, and simplicity of production are benefits of chemical synthesis of nanoparticles. In order to get beyond the constraints of conventional procedures, it is now possible to synthesize NPs using green chemistry technologies. Recently, biologically mediated nanoparticle syntheses have gained popularity due to their reliability, dependability, affordability, and environmental friendliness. It primarily concentrated on the high-yield creation of nanoparticles (NPs) of specified sizes employing a variety of biological entities (Gurunathan et al., 2014). Also, nanoparticle synthesis is aided by bacteria's bio-sorption of metals including both gram-positive and -negative ones.

Brevibacterium casei, Lactobacillus strains, *Escherichia coli, Bacillus lichenifor-mis, Pseudomonas stutzeri* AG259, *Ganoderma neo-japonicum* Imazeki, *Fusarium oxysporum*, and plant extracts such as *Artemisia princeps, Allophylus cobbe*, and *Typha angustifolia* were all used in this green chemistry approach (Mullen et al., 1989; Nair & Pradeep, 2002; Kalimuthu et al., 2008; Kalishwaralal et al., 2010; Leung et al., 2010). A number of biomolecules are also utilized, including starch, biopolymers, fibrinolytic enzyme, and amino acids (Shankar & Rhim, 2015). For the synthesis of NPs in a biological manner, a reducing agent, a non-toxic substance, and essentially a solvent are prerequisites. The biological approaches offer advantages of amino acids, proteins, and secondary metabolite availability produced as by-products during the synthesis process. Nanomaterial size and forms depend upon the biological techniques, making them crucial for many biomedical applications. The abundance of biological resources, high density, stability, shorter turnaround times, and water solubility of produced nanoparticles are all benefits of biological processes (Thakkar et al., 2010). The morphology and structure of NPs, which are governed by the size aspect and form of the particles, govern the biological activity of NPs (Pal et al., 2007). Truncated-triangular nanoparticles are more efficient and have better characteristics. By optimizing the synthesis processes, which include the quantity of temperature, pH, precursors, and reducing and stabilizing agents, biological approaches are easier to manage than chemical ones in terms of shaping, sizing, and dispersion of the nanoparticles (Khodashenas & Ghorbani, 2019).

8.4 CHARACTERIZATION

Nanoparticles' behaviour, effectiveness, biodistribution, and safety are determined by their physicochemical nature. In order to assess the functional properties of produced particles, nanoparticle characterizations are crucial. Atomic force microscopy (AFM), UV-vis spectroscopy, transmission and scanning electron microscopy (TEM and SEM), dynamic light scattering (DLS), X-ray photoelectron spectroscopy (XPS), Fourier transform infrared spectroscopy (FTIR), and X-ray diffractometry (XRD) are some of the analytical methods used to characterize materials (Zhang et al., 2016).

8.4.1 UV-VISIBLE SPECTROSCOPY

This is a simple, rapid, highly sensitive, and selective method applicable to different classes of nanoparticles; in addition, it also requires very little time for dimensional measurement (Huang et al., 2007). Conduction and valence bands are found near to one another in nanoparticles, and electrons can travel freely there. The surface plasmon resonance (SPR) absorption band is formed due to the simultaneous oscillation of free electrons within the nanoparticles together with light wave resonance (Das et al., 2009). Factors such as dielectric medium, chemical environment, and size influence absorption of nanoparticles. For more than a year, nanoparticles were found to be stable (He et al., 2002).

8.4.2 X-Ray Diffraction (XRD)

Molecular and crystal structures can be analysed using X-ray diffraction (XRD), as can chemical species quantitatively resolved, diverse compounds qualitatively identified, isomorphous substitutions, particle sizes, and crystallinity measured, among other things (Waseda et al., 2011). Numerous diffraction patterns that reflect the physicochemical properties of crystal structures are created when X-rays reflect on crystal. The structural characteristics of materials including inorganic catalysts, biomolecules, glasses, polymers, superconductors, etc. are also analysed by XRD. Each of these materials has a distinct diffraction beam that defines and identifies it. To do this, diffracted beams are compared to reference databases in JCPDS (Joint Committee on Powder Diffraction Standards Library). Both nano and bulk are defined and identified by XRD.

8.4.3 Dynamic Light Scattering

It is a technique for measuring small particle distributions (size: 2–500 nm) that rely on light with particles' interaction (Tomaszewska et al., 2013). For characterization of nanoparticles, DLS is the most reliable method (Jans et al., 2009). It functions according to Rayleigh scattering from suspended particles of nanosize and takes into account the light scattered from laser through a suspension. Using its non-destructive approach, the average diameter of nanoparticles which are dispersed in solution can be estimated. Numerous particles can be tested at once, although a variety of sample-specific drawbacks are also found (Fissan et al., 2014).

8.4.4 Fourier Transform Infrared (FTIR) Spectroscopy

Reproducibility, quick data collection, strong signal, accuracy, reduced sample heating, and ratio of signal-to-noise are offered by FTIR (Kumar & Barth, 2010). This non-intrusive technique is frequently employed in both research and industrial area so as to know if biomolecules play a role in the nanoparticles' synthesis. It has also been utilized in testing materials with nanoscale dimension, *viz.* interactions in substrate-enzyme during catalytic reactions, thus ensuring the covalent attachment of functional groups to metals, nanotubes of carbon, and Au nanoparticles (Baudot et al., 2010).

8.4.5 Photoelectron Spectroscopy (XPS)

This is sometimes mentioned as electron spectroscopy and is a quantitative surface spectroscopic chemical method employed to calculate empirical formulas (Demathieu et al., 1999). High vacuum is used during the procedure. When nanoparticles are exposed to X-rays, they give out electrons, and the quantity of escaping electrons from the surface are counted to produce an XPS spectrum (Manna et al., 2001).

8.4.6 Scanning Electron Microscopy (SEM)

SEM is a surface imaging technique and is efficient in discerning the various particle size distributions, surface morphology of artificial particles at the nanoscale, and forms of nanoparticles (Johal & Johnson, 2011). By manually quantifying and counting the nanoparticles or by utilizing specialized software, SEM is used to determine the shape of nanoparticles and produce histograms from images. SEM's limitation is that it cannot resolve the internal structure; nonetheless, it can offer crucial details about the purity and level of aggregate formation of nanoparticles. Modern high-resolution SEM can be used to understand the morphology of NPs which are less than 10 nm in dimension (Wang, 2000).

8.4.7 Transmission Electron Microscopy (TEM)

It is a considered as the most precise and efficient technique to quantify particle size and form distribution in nanoparticles (Lin et al., 2014). Regarding the mechanism, its magnification is attained by dividing the length between the lens and specimen by the focal and image plane distance. Better spatial resolution and the capacity to perform extra analytical measures are two advantages that TEM has over SEM (Hall et al., 2007). The drawbacks are that they need a narrow sample section, a high vacuum, and took more time.

8.4.8 Atomic Force Microscopy (AFM)

This technique examines the dispersion and aggregation of nanoparticles along with their size, form, structure, and sorption. Contact, non-contact, and intermittent sample contact are its three scanning modes (Koh et al., 2008). With modern electron microscopy (EM) techniques, it is possible to study the real-time association of nanoparticles with lipid in membranes. Also, it can measure even sub-nanometre scale in fluids without requiring oxide-free conductive surfaces and without harming native surface textures (Gmoshinski et al., 2013), although its primary flaw is the overestimation of sample lateral size which may be because of the size of the cantilever.

8.4.9 Localized Surface Plasmon Resonance (LSPR)

Nanoparticle with metallic origin is directly activated by near-visible light LSPR, including coherent spatial oscillation of electrons. It is estimated through temperature, dielectric environment, and electronic characteristics of the nanoparticles. LSPR spectral peak frequency is extremely sensitive to the nanostructure medium because of its refractive index. Therefore, LSPR frequency changes are utilized as a technique for detecting molecular interactions close to the nanoparticle surfaces (Sannomiya et al., 2008). It is employed in tools such as bioimaging and biomolecular detection with higher sensitivity to examine the fundamental characteristics of nanoparticles. The link between the spectra and local structure is well understood via single-nanoparticle LSPR spectroscopy. Compared to nanoparticle arrays, single nanoparticles have better refractive index sensitivity (Lis & Cecchet, 2014).

8.5 NANOMATERIAL AS AGROCHEMICALS

8.5.1 NANOFERTILIZERS

The most common engineered materials used in nano-enabled fertilizers for the targeted release and delivery of nutrients are hydroxyapatite, mesoporous silica, nanogypsum, chitosan, calcite, carbon nanotubes, metal oxide NPs (zinc oxide NPs, copper oxide NPs, titanium dioxide NPs), magnetite, zeolites, and nanoclays for foliar and soil application (Chaudhary et al., 2021d). These materials have shown great potential in the slow release of nutrients and can replace conventional chemical fertilizers (Figure 8.1).

8.5.2 NANOPESTICIDES: PROTECTION AGAINST PESTS, PATHOGENS, AND WEEDS

Nanotechnology-based biopesticides have provide to be an effective and environmentally friendly pest control. Nano-biopesticides are best because of their small size, durability, large surface area, increased effectiveness, versatility, high solubility, and low toxicity. Nano-biopesticides in addition to target-specific pest control have lower toxicity than the chemical pesticides. The different nanoparticles and compounds used for the purpose of nanopesticide formulations are chitosan, copper, nanosilica, silver, aluminium, titanium, etc., which are known for their pesticidal effects (Figure 8.1).

8.5.3 NANO-BIOSENSORS

The concept of smart agriculture is mainly based on sensor systems that can sense the need of water, fertilizers, and other input and the danger of biotic and abiotic stress in real time. The smart sensor system based on nanocomposites through real-time sensory system not only helps in finding the problem in a targeted manner but also helps in efficient resource utilization and thus minimize the losses. The nano-compounds which are known for their good biocompatibility and their electronic and optical properties are mainly used for designing the nano-biosensors, for example, gold, silver, graphene oxide, quantum dots, carbon nanotubes, etc. (Figure 8.1).

8.6 NANO-BIOFORMULATIONS AND THEIR ADVANTAGES OVER CONVENTIONAL BIOFORMULATIONS

The term *bioformulations* means a biological agent (mostly microorganisms) with good plant growth promontory activities entrapped in a carrier material, which provide for its nutrition and water supply during field application (Figure 8.2). The nano-compounds as per some studies are reported to improve the shelf life of microbial bioagents through extension of stationary phase (Khati et al., 2019a).

According to Khati et al. (2017a, b; 2018; 2019b), the nanocompounds are also beneficial for plant growth improvement which can be an additional aspect in any biofor-mulation. As an advancement over convention bioformulations, nano-bioformulations are those formulations where either the carrier material is nanoparticle/composite/compound or the formulation is encapsulated into a nanosheet. These encapsulations ensure the slow release of nutrient and water to ensure longer accessibility of these supplies to the biological agent, hence increasing its shelf life. Longer shelf life in

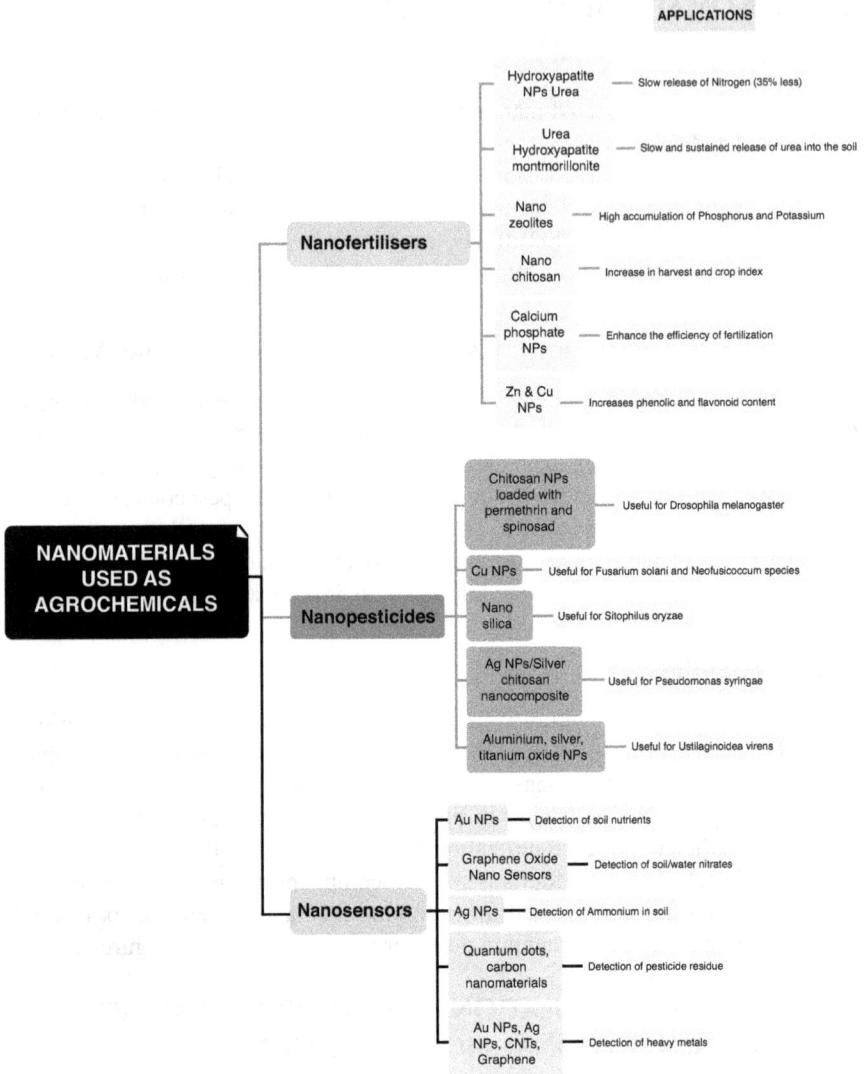

FIGURE 8.1 Some nanomaterial used as agrochemicals.

field application helps in better performance of the biological agent in field conditions. A synergistic effect of nanocompounds along with plant growth promoting bacteria on plant and soil health was evaluated by Chaudhary et al. (2021a, b, c, d) and Kumari et al. (2020). Similar effect of TiO_2 on plant heath of maize was evaluated by Kumari et al. (2021) who found that when the nanoparticles were applied at concentration of 50 ppm better plant health parameters were observed in comparison to control.

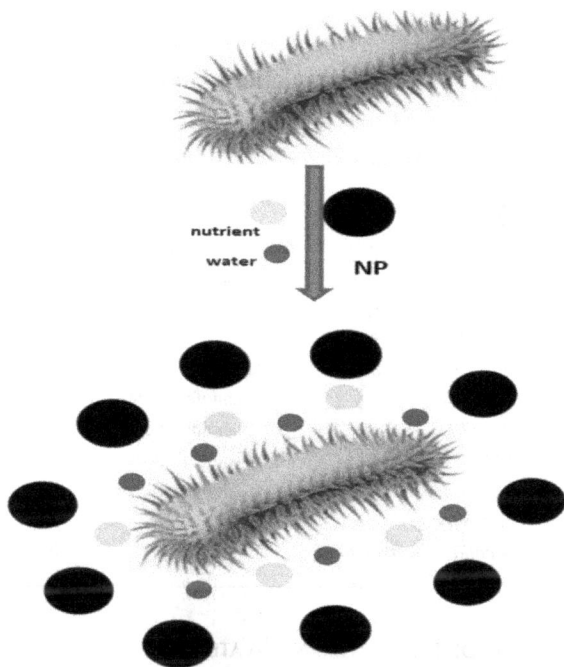

FIGURE 8.2   Synthesis of nano-bioformulation.

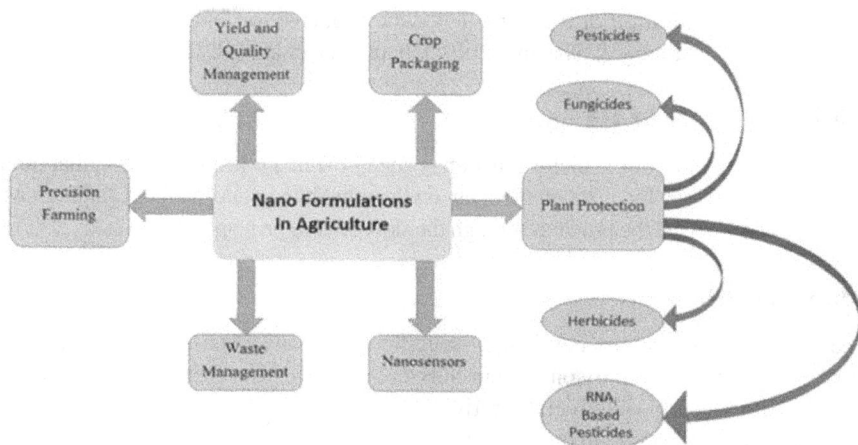

FIGURE 8.3   Application of nano-formulation/nano-bioformulation in agriculture.

The above-mentioned studies show the beneficial effects of nanocompounds/particles in bioformulation (Figure 8.3). The advantages of nano-bioformulations over conventional bioformulations are listed below. Some reports regarding the application of nanoparticles in crop protection are listed in Table 8.2.

TABLE 8.2

Recent Reports in Various Nanoformulations Employed for Crop Protection

Nanoformulations	Target Pest/Organisms	Action/Application	References
Ag/AgCl	*Macrophomina phaseolina*	Antifungal	Spagnoletti et al. (2019)
Zinc NPs	*Tetranychus urticae*	Antirepellant	de Oliveira et al. (2019)
Chitosan CuO	*Fusarium oxysporum*	Antifungal	Kaur et al. (2018)
Chitosan NPs	*Pyricularia grisea, Xanthomonas oryzae*	Antimicrobial	Sathiyabama and Muthukumar (2020)
Curcumin CuNP	*Fusarium oxysporum*	Fungicide and nanofertilizer	Sathiyabama et al. (2020)
ZnO and SiO$_2$ NPs	*Sitophilus oryzae*	Nanoinsecticide	Haroun et al. (2020)
Chitosan NPs	*Bidens pilosa*	Nanoherbicide	Sherkhane et al. (2018)
Ag NPs	*Xanthomonas axonopodis*	Nanobactericide	Maruyama et al. (2016)
Titanium NPs	*Nicotiana benthamiana*	Viricide	Shah et al. (2019)

8.6.1   SLOW RELEASE OF NUTRIENTS AND WATER

In comparison to conventional bioformulation where the nutrient and water are directly provided to the bioagent, the supplies are entrapped and released slowly in nano-bioformulations. This slow release helps the bioagent to use the nutrient and water for longer duration of time and hence they survive for longer duration (Khati et al., 2019a; Chaudhary & Sharma, 2019; Chaudhary et al., 2021e).

8.6.2   LONGER SHELF LIFE

The shelf life of most of the bioformulation is one of the major limiting factors that limit the application of any bioagent on filed, but on the other hand nano-bioformulations have specifically addressed the problem through the slow-release principle (Khati et al., 2019a).

8.6.3   COST-EFFECTIVE

The nano-bioformulations have target-specific action and thus they are applied in lesser quantities in comparison to their bulk counterparts. In addition to this the longer shelf life also adds into their effectiveness which make the nano-bioformulations much more cost-effective in comparison to other bioformulations.

8.7   CONCLUSION

Nanotechnology can be used for the reduction of agricultural problems including environmental effects and food security. These nano-systems have shown a high level of controlled-release behaviour, resulting in their effectiveness in long-term use. But the most significant stumbling block to nanotechnological application in agricultural progress lies within its ethical acceptance. Farmers in India do not know about nanotechnological applications and uses in agricultural facilities. Toxicity concerns of nanoparticles are yet another major issue. In addition, the NP-based agricultural research is of interest due to the delayed exposure of NPs in the natural system. So, it becomes the responsibility of researchers and scientists to develop a better understanding of nanotechnology along with their side effects and proper dosage for best results. As a result, different aspects of nano-biopesticides, including their present status, constraints, prospects, and regulatory network, must be checked regularly to ensure their successful use for the benefit of humanity. Nanotechnology in the field of agriculture has huge benefits if applied with precautions. Nano-bioformulations have the potential to address various issues in agriculture such as low productivity, biotic and abiotic stresses but on the other hand their use must be done with precaution and a complete knowledge of dosage and guidelines must be provided along with the product for the use of farmers.

REFERENCES

Agri, U., Chaudhary, P., & Sharma, A. (2021). In vitro compatibility evaluation of agriusable nanochitosan on beneficial plant growth-promoting rhizobacteria and maize plant. *Natl Acad Sci Lett*. 44, 555–559.

Agri, U., Chaudhary, P., Sharma, A., & Kukreti, B. (2022). Physiological response of maize plants and its rhizospheric microbiome under the influence of potential bioinoculants and nanochitosan. *Plant Soil*. 474, 451–468.

Baudot, C., Tan C.M., & Kong, J.C. (2010). FTIR spectroscopy as a tool for nano-material characterization. *Infared Phys Technol*. 53(6), 434–438.

Bhatt, P., Pandey, S.C., Joshi, S., Chaudhary, P., Pathak, V.M., Huang, Y., Wu, X., Zhou, Z., & Chen, S. (2022). Nanobioremediation: a sustainable approach for the removal of toxic pollutants from the environment. *J Hazard Mater*. 427, 128033.

Biswal, S.K., Nayak, A.K., Parida, U.K., & Nayak, P.L. (2012). Applications of nanotechnology in agriculture and food sciences. *Int J Sci Innov Discov*. 2, 21–36.

Buseck, P.R., & Adachi, K. (2008). Nanoparticles in the atmosphere. *Elements*. 4(6), 389–394.

Chaudhary, P., Chaudhary, A., Bhatt, P., Kumar, G., Khatoon, H., Rani, A., Kumar, S., & Sharma, A. (2022). Assessment of soil health indicators under the influence of nano-compounds and *Bacillus* spp. in field condition. *Front Environ Sci*. 9, 769871.

Chaudhary, P., Chaudhary, A., Parveen, H., Rani, A., Kumar, G., Kumar, A., & Sharma, A. (2021e). Impact of nanophos in agriculture to improve functional bacterial community and crop productivity. *BMC Plant Biol*. 21, 519.

Chaudhary, P., Khati, P., Chaudhary, A., Gangola, S., Kumar, R., & Sharma, A. (2021a). Bioinoculation using indigenous *Bacillus* spp. improves growth and yield of *Zea mays* under the influence of nanozeolite. *3 Biotech*. 11, 11.

Chaudhary, P., Khati, P., Chaudhary, A., Maithani, D., Kumar, G., & Sharma, A. (2021d). Cultivable and metagenomic approach to study the combined impact of nanogypsum and *Pseudomonas taiwanensis* on maize plant health and its rhizospheric microbiome. *PLoS One*. 16, e0250574.

Chaudhary, P., Khati, P., Gangola, S., Kumar, A., Kumar, R., & Sharma, A. (2021c). Impact of nanochitosan and *Bacillus* spp. on health, productivity and defence response in *Zea mays* under field condition. *3 Biotech*. 11, 237.

Chaudhary, P., & Sharma, A. (2019). Response of nanogypsum on the performance of plant growth promotory bacteria recovered from nanocompound infested agriculture field. *Environ Ecol*. 37, 363–372.

Chaudhary, P., Sharma, A., Chaudhary, A., Khati, P., Gangola, S., & Maithani, D. (2021b). Illumina based high throughput analysis of microbial diversity of rhizospheric soil of maize infested with nanocompounds and *Bacillus* sp. *Appl Soil Ecol.* 159, 103836.

Chen, H., & Yada, R. (2011). Nanotechnologies in agriculture: new tools for sustainable development. *Trends Food Sci Technol.* 22, 585–594.

Chhipa, H. (2017). Nanofertilizers and nanopesticides for agriculture, *Environ Chem Lett.* 15, 15–22.

Das, R., Nath, S.S., Chakdar, D., Gope, G., & Bhattacharjee, R.J. (2009). Preparation of silver nanoparticles and their characterization. *J Nanotechnol.* 5, 1–6.

Dasgupta, N., Ranjan, S., Mundekkad, D., Ramalingam, C., Shanker, R., & Kumar, A. (2015). Nanotechnology in agro-food: from field to plate. *Food Res Int.* 69, 381–400.

De Matteis, V. (2017). Exposure to inorganic nanoparticles: routes of entry, immune response, biodistribution and in vitro/in vivo toxicity evaluation. *Toxics.* 5(4), 29.

de Oliveira, J.L., Campos, E.V., Camara, M.C., Della Vechia, J.F., de Matos, S.T.S., de Andrade, D.J., Goncalves, K.C., Nascimento, J.D., Polanczyk, R.A., de Araujo, D.R. & Fraceto, L.F. (2019). Hydrogels containing botanical repellents encapsulated in zein nanoparticles for crop protection. *ACS Appl Nano Mater.* 3(1), 207–217.

De Volder, M.F., Tawfick, S.H., Baughman, R.H., & Hart, A.J. (2013). Carbon nanotubes: present and future commercial applications. *Science.* 339(6119), 535–539.

Deepak, V., Umamaheshwaran, P.S., Guhan, K., Nanthini, R.A., Krithiga, B., Jaithoon, N.M., & Gurunathan, S. (2011). Synthesis of gold and silver nanoparticles using purified URAK. *Colloids Surfaces B Biointer.* 86(2):353–358.

Demathieu, C., Chehimi, M.M., Lipskier, J.F., Caminade, A.M., & Majoral, JP. (1999). Characterization of dendrimers by X-ray photoelectron spectroscopy. *Appl Spectro.* 53(10), 1277–1281.

Ditta, A. (2012). How helpful is nanotechnology in agriculture? *Adv Nat Sci Nanosci Nanotechnol.* 3, 033002.

Donaldson, K., Stone, V., Tran, C.L., Kreyling, W., & Borm, P.J. (2004). *Nanotoxicol Occup Environ Med.* 61(9), 727–728.

Dubey, A., & Mailapalli, D. R. (2016). Nanofertilizers, nanopesticides, nanosensors of pests and nanotoxicity in agriculture. In E. Lichtfouse (ed.), *Book Sustainable Agriculture Reviews*, Springer International Publishing, Switzerland, pp. 307–330.

Fissan, H., Ristig, S., Kaminski, H., Asbach, C., & Epple, M. (2014). Comparison of different characterization methods for nanoparticle dispersions before and after aerosolization. *Anal Methods.* 6(18), 7324–7334.

Fraceto, L.F., Grillo, R., deMedeiros, G.A., Scognamiglio, V., Rea, G., & Bartolucci, C. (2016.) Nanotechnology in agriculture: which innovation potential does it have? *Front Environ Sci.* 4, 20.

Gmoshinski, I.V., Khotimchenko, S.A., Popov, V.O., Dzantiev, B.B., Zherdev, A.V., Demin, V.F., & Buzulukov, Y.P. (2013). Nanomaterials and nanotechnologies: methods of analysis and control. *Russian Chem Rev.* 82(1), 48.

Gogos, A., Knauer, K., & Bucheli, T.D. (2012). Nanomaterials in plant protection and fertilization: current state, foreseen applications, and research priorities. *J Agric Food Chem.* 60, 9781–9792.

Gurunathan, S., Han, J., Park, J.H., & Kim, J.H. (2014). A green chemistry approach for synthesizing biocompatible gold nanoparticles. *Nanoscale Res Lett.* 9(1), 1–11.

Hall, J.B., Dobrovolskaia, M.A., Patri, A.K. & McNeil, S.E. (2007). Characterization of nanoparticles for therapeutics. *Nanomedicine.* 2(6), 789–803.

Haroun, S.A., Elnaggar, M.E., Zein, D.M., & Gad, R.I. (2020). Insecticidal efficiency and safety of zinc oxide and hydrophilic silica nanoparticles against some stored seed insects. *J Plant Prot Res.* 60, 77–85.

He, R., Qian, X., Yin, J., & Zhu, Z. (2002). Preparation of polychrome silver nanoparticles in different solvents. *J Mater Chem.* 12(12), 3783–3786.

Hochella, M.F., Spencer, M.G., & Jones, K.L. (2015). Nanotechnology: nature's gift or scientists' brainchild? *Environ Sci Nanotechnol.* 2(2), 114–149.

Hussain, M.H., Abu Bakar, N.F., Mustapa, A.N., Low, K.F., Othman, N.H., & Adam, F. (2020). Synthesis of various size gold nanoparticles by chemical reduction method with different solvent polarity. *Nanoscale Res Lett.* 15(1), 1–10.

Jans, H., Liu, X., Austin, L., Maes, G., & Huo, Q. (2009). Dynamic light scattering as a powerful tool for gold nanoparticle bioconjugation and biomolecular binding studies. *Anal Chem.* 81(22), 9425–32.

Jat, S. K., Bhattacharya, J., & Sharma, M. K. (2020). Nanomaterial based gene delivery: a promising method for plant genome engineering. *J Mater Chem B*, 8, 4165–4175.

Jeevanandam, J., Barhoum, A., Chan, Y.S., Dufresne, A., & Danquah, M.K. (2018). Review on nanoparticles and nanostructured materials: history, sources, toxicity and regulations. *Beilstein J Nanotechnol.* 9(1), 1050–1074.

Johal, M.S., & Johnson, L.E. (2011). *Understanding Nanomaterials.* CRC Press, Boca Raton, FL.

Kalimuthu, K., Babu, R.S., Venkataraman, D., Bilal, M., & Gurunathan, S. (2008). Biosynthesis of silver nanocrystals by *Bacillus licheniformis*. *Colloids Surf B Biointerf.* 65(1), 150–153.

Kalishwaralal, K., Deepak, V., Pandian, S.R., Kottaisamy, M., Barath Mani Kanth, S., Kartikeyan, B., & Gurunathan, S. (2010). Biosynthesis of silver and gold nanoparticles using *Brevibacterium casei*. *Colloids Surf B Biointerf.* 77(2), 257–262.

Kaphle, A., Navya, P.N., Umapathi, A., & Daima, H.K. (2018). Nanomaterials for agriculture, food and environment: applications, toxicity and regulation, *Environ Chem Lett.* 16, 43–58.

Kaur, J., Sood, K., Bhardwaj, N., Arya, S.K., & Khatri, M. (2020). Nanomaterial loaded chitosan nanocomposite films for antimicrobial food packaging, *Mater Today Proc.* 28, 1904–1909.

Kaur, P., Duhan, J.S., & Thakur, R. (2018). Comparative pot studies of chitosan and chitosan-metal nanocomposites as nano-agrochemicals against fusarium wilt of chickpea (*Cicer arietinum* L.). *Biocat Agric Biotechnol.* 14, 466–471.

Khati, P., Bhatt, P., Kumar, R., & Sharma, A. (2018). Effect of nanozeolite and plant growth promoting rhizobacteria on maize. *3Biotech.* 8(141), 1–12.

Khati, P., Chaudhary, P., Gangola, S., & Sharma A. (2019a). Influence of nanozeolite on plant growth promotory bacterial isolates recovered from nanocompound infested agriculture field. *Environ Ecol.* 37 (2), 521–527.

Khati, P., Gangola, S., Bhatt, P., & Sharma A. (2017a). Nanochitosan induced growth of *Zea Mays* with soil health maintenance. *3 Biotech.* 7(81), 1–9.

Khati, P., Sharma A., Chaudhary, P., Singh, A K., Gangola, S., & Kumar, R. (2019b). High-throughput sequencing approach to access the impact of nanozeolite treatment on species richness and evenness of soil metagenome. *Biocat Agric Biotechnol.* 20, 101249.

Khati, P., Sharma A., Gangola, S., Kumar, R., Bhatt, P., & Kumar, G. (2017b). Impact of some agriusable nanocompounds on soil microbial activity: an indicator of soil health. *Clean Soil Air Water.* 45, 1–7.

Khodashenas, B., & Ghorbani, H.R. (2019). Synthesis of silver nanoparticles with different shapes. *Arab J Chem.* 12(8), 1823–1838.

Khot, L.A., Sankaran, S., Maja, J.M., Ehsani, R., & Schuster, E.W. (2012). Applications of nanomaterials in agricultural production and crop protection: a review. *Crop Prot.* 35, 64–70.

Kittelson, D.B. (2001). Recent measurements of nanoparticle emissions from engines. *Cur Res Diesel Exhaust Part.* 9, 451–457.

Koh, A.L., Hu, W., Wilson, R.J., Wang, S.X., & Sinclair, R. (2008). TEM analyses of synthetic anti-ferromagnetic (SAF) nanoparticles fabricated using different release layers. *Ultramicroscopy*. 108 (11), 1490–1494.

Kruis, F.E., Fissan, H., & Rellinghaus, B. (2000). Sintering and evaporation characteristics of gas-phase synthesis of size-selected PbS nanoparticles. *Mater Sci Eng B*. 69, 329–3234.

Kukreti, B., Sharma, A., Chaudhary, P., Agri, U., & Maithani, D. (2020). Influence of nanosilicon dioxide along with bioinoculants on *Zea mays* and its rhizospheric soil. *3 Biotech*. 10, 345.

Kumar, S., & Barth, A. (2010). Following enzyme activity with infrared spectroscopy. *Sensors*. 10(4), 2626–2637.

Kumari, H., Khati, P., Gangola, S., Chaudhary, P., & Sharma A. (2021). Performance of plant growth promotory rhizobacteria on maize and soil characteristics under the influence of TiO$_2$ nanoparticles. *Pant Res J*. 19(1), 28–39.

Kumari, S., Sharma A., Chaudhary, P., & Khati, P. (2020). Management of plant vigour and soil health using two agriusable nanocompounds and plant growth promontory rhizobacteria in Fenugreek. *3 Biotech*. 10(461), 1–11.

Leung, T.C., Wong, C.K., & Xie, Y. (2010). Green synthesis of silver nanoparticles using biopolymers, carboxymethylated-curdlan and fucoidan. *Mater Chem Phys*. 121(3), 402–405.

Lin, P.C., Lin, S., Wang, P.C., & Sridhar, R. (2014). Techniques for physicochemical characterization of nanomaterials. *Biotechnol Adv*. 32(4), 711–726.

Lis, D., & Cecchet, F. (2014). Localized surface plasmon resonances in nanostructures to enhance nonlinear vibrational spectroscopies: towards an astonishing molecular sensitivity. *Beilstein J Nanotechnol*. 5(1), 2275–2292.

Majhi, K.C., & Yadav, M. (2021). Synthesis of inorganic nanomaterials using carbohydrates. In *Green Sustainable Process for Chemical and Environmental Engineering and Science*. Elsevier: USA, pp. 109–135.

Mallick, K., Witcomb, M.J., & Scurrell, M.S. (2004). Polymer stabilized silver nanoparticles: a photochemical synthesis route. *J Mater Sci*. 39(14), 4459–4463.

Manna, A., Imae, T., Aoi, K., Okada, M., & Yogo, T. (2001). Synthesis of dendrimer-passivated noble metal nanoparticles in a polar medium: comparison of size between silver and gold particles. *Chem Mater*. 13(5), 1674–1681.

Maruyama, C.R. Guilger, M., Pascoli, M., & Bileshy-Jose N., Abhilash P., Fraceto L.F. & De Lima, R. (2016). Nanoparticles based on chitosan as carriers for the combined herbicides imazapic and imazapyr, *Sci Rep*. 6, 19768.

Masum, M.M.I., Siddiqa, M.M., Ali, K.A., Zhang, Y., Abdallah, Y., Ibrahim, E., Qiu, W., Yan, C. & Li, B. (2019). Biogenic synthesis of silver nanoparticles using *Phyllanthus emblica* fruit extract and its inhibitory action against the pathogen *Acidovorax oryzae* strain RS-2 of rice bacterial brown stripe. *Front Microbiol*. 10, 820.

Mukherjee, A., Majumdar, S., Servin, A.D., Pagano, L., Dhankher, O.P., & White, J. C. (2016). Carbon nanomaterials in agriculture: a critical review. *Front Plant Sci*. 7, 172.

Mullen, M.D., Wolf, D.C., Ferris, F.G., Beveridge, T.J., Flemming, C.A., & Bailey, G. (1989). Bacterial sorption of heavy metals. *Appl Environ Microbial*. 55(12), 3143–3149.

Nair, B., & Pradeep, T. (2002). Coalescence of nanoclusters and formation of submicron crystallites assisted by *Lactobacillus* strains. *Crystal Growth Des*. 2(4), 293–298.

Ning, Z., Cheung, C.S., Fu, J., Liu, M.A., & Schnell, M.A. (2006). Experimental study of environmental tobacco smoke particles under actual indoor environment. *Sci Total Environ*. 367(2–3), 822–830.

Omanovic-Miklicanina, E., & Maksimovic, M. (2016). Nanosensors applications in agriculture and food industry. *Bull Chem Technol Bosnia Herzegovina*. 47, 59–70.

Pal, S., Tak, Y.K., & Song, J.M. (2007). Does the antibacterial activity of silver nanoparticles depend on the shape of the nanoparticle? A study of the gram-negative bacterium *Escherichia coli*. *Appl Environ Microbial*. 73(6), 1712–1720.

Parisi, C., Vigani, M., & Rdríguez-Cerezo, E. (2015). Agricultural nanotechnologies: what are the current possibilities. *Nano Today*. 10, 124–127.

Poddar, K., Vijayan, J., Ray, S., & Adak T. (2018). Nanotechnology for sustainable agriculture. In: Singh, R.L., & Mondal, L. (eds.), *Biotechnology for Sustainable Agriculture Emerging Approaches and Strategies*. Elsevier, the Netherlands, pp. 281–303.

Prasad, R., Kumar, V., & Prasad, K.S. (2014). Nanotechnology in sustainable agriculture: present concerns and future aspects. *Afr J Biotechnol*. 13, 705–713.

Rastogi, A., Zivcak, M., Sytar, O., Kalaji, H.M., He, X., Mbarki, S., & Brestic, M. (2017). Impact of metal and metal oxide nanoparticles on plant: a critical review. *Front Chem*. 5, 78.

Rogers, F., Arnott, P., Zielinska, B., Sagebiel, J., Kelly, K.E., Wagner, D., Lighty. J.S., & Sarofim, A.F. (2005). Real-time measurements of jet aircraft engine exhaust. *J Air Waste Manage Assoc*. 55(5), 583–593.

Ross, M.B., Mirkin, C.A. & Schatz, G.C. (2016). Optical properties of one-, two-, and three-dimensional arrays of plasmonic nanostructures. *J Phys Chem C*. 120(2), 816–830.

Sannomiya, T., Hafner, C., & Voros, J. (2008). In situ sensing of single binding events by localized surface plasmon resonance. *Nano Lett*. 8(10), 3450–3455.

Sathiyabama, M., & Muthukumar, S. (2020). Chitosan guar nanoparticle preparation and its *in vitro* antimicrobial activity towards phytopathogens of rice. *Int J Biol Macromol*. 153, 297–304.

Sathiyabama, M., Indhumathi, M., & Amutha, T. (2020). Preparation and characterization of curcumin functionalized copper nanoparticles and their application enhances disease resistance in chickpea against wilt pathogen. *Biocat Agric Biotechnol*. 29, 101823.

Scrinis, G. & Lyons, K. (2007), The emerging nano-corporate paradigm: nanotechnology and the transformation of nature, food and agri-food systems, *Int J Soc Agric Food*. 15, 22–44.

Sekhon, B.S. (2014). Nanotechnology in agri-food production: an overview. *Nanotechnol Sci Appl*. 7, 31–53.

Servin, A., Elmer, W., Mukherjee, A., De la Torre-Roche, R., Hamdi, H., White, J.C., Bindraban, P., & Dimkpa, C. (2015). A review of the use of engineered nanomaterials to suppress plant disease and enhance crop yield. *J Nano Res*. 17, 92.

Shah, T., Xu, J., Zou, X., Cheng, Y., Zhang, X., Hussain, Q., & Gill, R.A. (2019). Impact of nanomaterials on plant economic yield and next generation. In M. Ghorbanpour & S.H. Wani (Eds.) *Advances in Phytonanotechnology*. Academic Press, pp. 203–214.

Shankar, S., & Rhim, J.W. (2015). Amino acid mediated synthesis of silver nanoparticles and preparation of antimicrobial agar/silver nanoparticles composite films. *Carb Polym*. 130, 353–363.

Sharma, V.K., Filip, J., Zboril, R., & Varma, R.S. (2015). Natural inorganic nanoparticles–formation, fate, and toxicity in the environment. *Chem Soc Rev*. 44(23), 8410–8423.

Sherkhane A., Suryawanshi, H., Mundada P. & Shinde, B. (2018). Control of bacterial blight disease of pomegranate using silver nanoparticles. *J Nanosci Nanotechnol*. 9: 1–5.

Sioutas, C., Delfino, R.J., & Singh, M. (2005). Exposure assessment for atmospheric ultrafine particles (UFPs) and implications in epidemiologic research. *Environ Health Persp*. 113(8), 947–55.

Sonkaria, S., Ahn, S.H., & Khare, V. (2012). Nanotechnology and its impact on food and nutrition: a review. *Recent Pat Food Nutr Agric*. 4, 8–18.

Spagnoletti, F.N., Spedalieri, C., Kronberg, F., & Giacometti, R. (2019). Extracellular biosynthesis of bactericidal Ag/AgCl nanoparticles for crop protection using the fungus *Macrophomina phaseolina*. *J Environ Manage*. 231, 457–466.

Stefani, D., Wardman, D., & Lambert, T. (2005). The implosion of the Calgary General Hospital: ambient air quality issues. *J Air Waste Manag Assoc*. 55(1), 52–59.

Taylor, D.A. (2002). Dust in the wind. *Environ Health Perspect*. 110(2), A80–A87.

Thakkar, K.N., Mhatre, S.S., & Parikh, R.Y. (2010). Biological synthesis of metallic nanoparticles. *Nanomed Nanotechnol Boil Medi*. 6(2), 257–262.

Tomaszewska, E., Soliwoda, K., Kadziola, K., Tkacz-Szczesna, B., Celichowski, G., Cichomski, M., Szmaja, W., & Grobelny, J. (2013). Detection limits of DLS and UV-Vis spectroscopy in characterization of polydisperse nanoparticles colloids. *J Nanomat*. 2013, 60.

Tsuji, M., Hashimoto, M., Nishizawa, Y., Kubokawa, M., & Tsuji, T. (2005). Microwave-assisted synthesis of metallic nanostructures in solution. *Chem-A Eur J*. 11(2), 440–452.

Wagner, S., Gondikas, A., Neubauer, E., Hofmann, T., & von der Kammer, F. (2014). Spot the difference: engineered and natural nanoparticles in the environment—release, behavior, and fate. *Angew Chem Int Ed*. 53(46), 12398–12419.

Wang, J.W., Grandio, E.G., Newkirk, G.M., Demirer, G.S., Butrus, S., Giraldo, J.P., & Landry, M.P. (2019). Nanoparticle mediated genetic engineering of plants. *Mol Plant*. 12, 1037–1040.

Wang, Z.L. (2000). Transmission electron microscopy of shape-controlled nanocrystals and their assemblies. *J Phys Chem B*. 104(6), 1153–1175.

Weir, A., Westerhoff, P., Fabricius, L., Hristovski, K., & Von Goetz, N. (2012). Titanium dioxide nanoparticles in food and personal care products. *Environ Sci Technol*. 46(4), 2242–2250.

Huang, X., Jain, P.K., El-Sayed, I.H., & El-Sayed. M.A. (2007). Gold nanoparticles: interesting optical properties and recent applications in cancer diagnostics and therapy. *Nanomedicine*. 2(5), 681–693.

Zhang, Q., Li, N., Goebl, J., Lu, Z., & Yin Y. (2011). A systematic study of the synthesis of silver nanoplates: is citrate a "magic" reagent? *J Am Chem Soc*. 133(46), 18931–18939.

Zhang, X.F., Liu, Z.G., Shen, W., & Gurunathan, S. (2016). Silver nanoparticles: synthesis, characterization, properties, applications, and therapeutic approaches. *Int J Mol Sci*. 17(9), 1534.

9 Role of Nanobiosensors in Agriculture Advancement

Vaibhav Dhaigude and Naliyapara HB
ICAR-NDRI

Parul Chaudhary
Graphic Era Hill University

Sami Abou Fayssal
University of Forestry
Lebanese University

Ashok Kumar Nadda
Jaypee University of Information Technology

Tatiana Minkina and Vishnu D Rajput
Southern Federal University

CONTENTS

DOI: 10.1201/9781003345565-9

9.1 INTRODUCTION

The agriculture sector is the backbone of the developing economy for the well-being of people as the economic growth of developing countries is influenced by agriculture and the food industry (i.e. India). As the world's population is drastically increasing, meeting the food demand is very critical in the current circumstances. Therefore, to ensure that advanced technologies should be implemented in the agricultural field (Agri et al., 2021; Kumari et al., 2021; Agri et al., 2022; Chaudhary et al., 2022a, b), more cutting-edge technologies are urgently needed to overcome these difficulties. Thus, nanotechnology has aided the agrotechnological revolution that is set to transform the current agricultural system while ensuring food security (Khati et al., 2017, 2018; Kukreti et al., 2020; Kumari et al., 2020). Nanotechnology is flourishing these days and has a significant impact on the progress of biosensors. A variety of nanostructures are used in their development, enhancing their efficiency and sensitivity (Sharma et al., 2022).

Arduini et al. (2017) reported that nanotechnology has conclusively demonstrated to be a crucial eco-technological invention with the help of biosensors for environmental analysis and agriculture development. Biosensors are nothing but a receptor-transducer hybrid model that detects the physicochemical characteristics of a material under the influence of a biological or organic element to recognize the presence of a particular biological substance (Sun et al., 2006). Nanobiosensors are the next innovation of biosensors that are more connected to hypersensitive elements when applied to measure selective analytes through the transducer. Nanobiosensors can detect the pH of soil, moisture and a large range of pesticides (herbicides, insecticides), pathogens and fertilizers (Ghaffar et al., 2020). Growing nutritious food is a process that is closely related to healthy life, and nanobiosensors can be useful in preventing diseases from spreading to crops, thus protecting food from toxins (Sharma et al., 2021). Adequate and managed application of nanobiosensors can aid in sustainable agriculture by increasing the crop yield. It can also help in pollution control and resource management.

The potential of nanobiosensors became more effective and consistent as it provided a wide range of substances at a low cost. To assist molecular analysis, nanobiosensors can be embedded into other technologies such as lab-on-a-chip which helps to identify different analytes such as pesticides, urea, etc. and detecting various pathogens and toxicities (Rai et al., 2012). Therefore, combining these micro- and nanotechnology-based diagnostic devices would positively benefit the food and agriculture industries while also being beneficial to the environment (Merkoci, 2021). Moreover, their mobility makes them suitable for various field applications besides laboratory work. Some heavy metal residues through pesticides, i.e. cadmium (Cd), chromium (Cr), copper (Cu), mercury (Hg), lead (Pb) and zinc (Zn), usually persist and are a significant threat to agroecosystems and human health (Singh et al., 2011). Purposely, using nanobiosensors makes the identification and measurement of agricultural contaminants and environmental pollutants an easy process (Kumar et al., 2016; Bala et al., 2017). This chapter discusses the fundamentals of different nano-based biosensors, as well as their role in agriculture and food sectors along with current developments.

9.2 NANOBIOSENSORS

The use of nanotechnology to enhance plant control over nutrient release is based on nanofertilizers and nanobiosensors, besides increasing macro- and micro-nutrient uptake efficiency by embedding nanobiosensors in a polymer. The development of high-tech biosensors through nanotechnology results in the creation of mini components called nanobiosensors, which are more effective and better assembled than conventional biosensors. Nanobiosensors are also referred to as nanosensors with immobilized bioreceptor probes (such as substrates, microorganisms, antibodies or enzymes) that specifically target analyte molecules (Vo-Dinh, 2005). Nanobiosensors are nanosensors with encapsulated biosensing probes that are specific for desired analytic molecules and could be integrated in fertilizers through biopolymer coatings to release nutrients precisely in response to signalling molecules from soil microorganisms such as rhizobium in plant root morphology (Hakeem & Pirzadah, 2020). These sensors have a similar fundamental procedure that involves the selective binding of an analyte, signal generating from the nanosensor interaction with the biomolecules and data processing into useful metrics (Foster, 2006). Due to their nano-dimension and increased surface area, aptamers can be used to model and simulate equipment that requires very small sample quantities yielding fast response times in a cost-effective process (Adam & Gopinath, 2022). Aptamers are such molecules that can be developed with unique physicochemical properties owing to high affinity and specificity to bind with any target molecules. These distinct properties of aptamers can improve the performance of sensors (Cho et al., 2020).

Nanosensors with the help of signal transducers may be capable of distinguishing between analytes and identify specific cells at the microscopic level in order to deliver desired molecules at the targeted area by monitoring the changes in the physicochemical properties (Freitas, 1999). In order to ensure crop protection and crop production improvement, commercial and potential nanobiosensors are emphasized. Therefore, nanobiosensors play an important role in the detection of pathogens, fertilizers, herbicides and insecticides (Singh, 2021). A wide variety of biosensors, including wireless nanosensors, electrochemical nanosensors, optical nanosensors and nano-barcode technology, have been designed and developed recently. In dealing with various issues related to food, agriculture and the environment, these biosensors are remarkably dependable, effective and cheap (Dar et al., 2020). Nowadays, several advanced nanobiosensors are designed in such a way that even at very low substrate exposure levels they can provide a very good signal transmission ratio and can be used in a wide range of monitoring applications (Baig et al., 2021).

9.3 PROPERTIES OF GOOD NANOBIOSENSORS

Nanosensors are essential in the area of nanotechnology for the identification of physical and chemical variations, monitoring bioactive molecules and biological changes in plant cells as well as the quantification of harmful and polluting compounds in the environment and agriculture industry (Adam & Gopinath, 2022). Their nanosized particles owing to some distinguishable properties make them ideal nanobiosensors. Some of these properties discussed by Rai et al. (2012) are outlined in Figure 9.1.

9.4 DIFFERENT PARTS OF NANOBIOSENSORS

Nano-based particles, nanomaterials made of polymers, nanoscale wires (sensitivity and selectivity to monitoring), carbon nanotubes (large surface area) and other tiny particles make up nanosensors. When a conductive carbon nanotube is revealed to specific substances, these nanosensors can identify the change in conductivity (Johnson et al., 2021). Generally, nanobiosensors are made up of three parts: probe with biologically sensitized components (bioreceptors), transducer and detector.

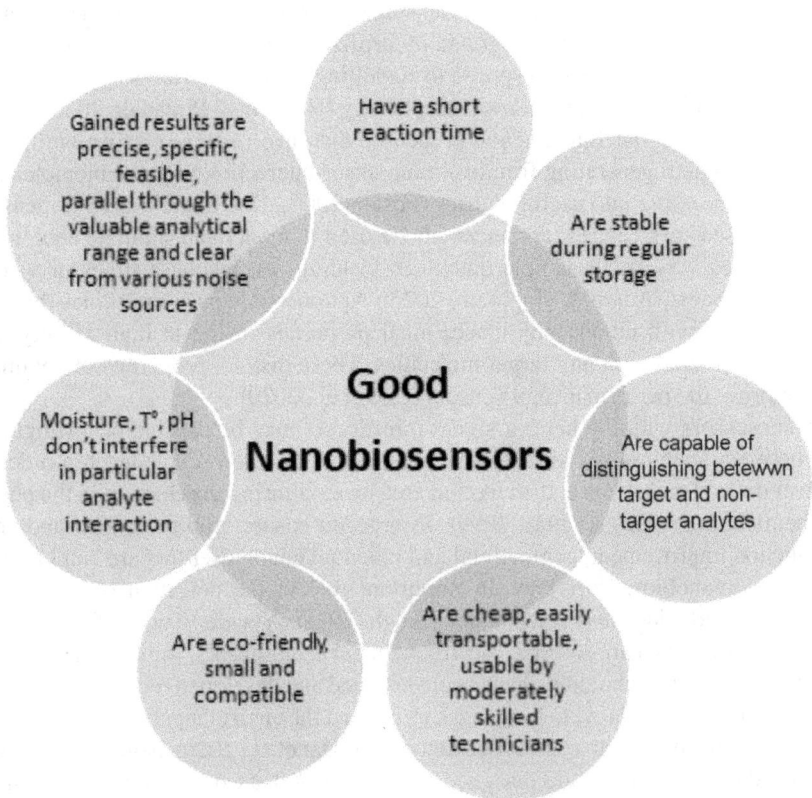

Good Nanobiosensors

- Have a short reaction time
- Are stable during regular storage
- Are capable of distinguishing betewwn target and non-target analytes
- Are cheap, easily transportable, usable by moderately skilled technicians
- Are eco-friendly, small and compatible
- Moisture, T°, pH don't interfere in particular analyte interaction
- Gained results are precise, specific, feasible, parallel through the valuable analytical range and clear from various noise sources

FIGURE 9.1 Properties of good nanobiosensors.

1. Probe-consisting receptors or bio-based components that pick up signals from the desired analytes and communicate them to the transducer (Rai et al., 2012).
2. The transducer acts as a link evaluating the physical change caused by the reaction at the desired analyte and converts that energy into detectable electrical output.
3. Whatever the signal received from the transducer, it is captured by detector elements and then amplified and analysed by a microprocessor. Then, it is monitored and stored (Cavalcanti et al., 2008).

Biosensors are managed by signal transduction, in which the transducer detects the interconnections and produces a signal. The fundamental idea behind this principle is to attach sample analytes to bioreceptors (aptamers, cell, nucleic acid, enzymes, antibodies) which changes the various physicochemical signal correlated to the binding and detects the desired data (Bhalla et al., 2016).

9.5   DIFFERENT TYPES OF NANOBIOSENSORS IN AGRICULTURE

There are numerous varieties of nanobiosensors available for different applications (Figure 9.2).

9.5.1   OPTICAL NANOBIOSENSORS

Optical nanobiosensors have been built using semiconductor quantum dots depending on fluorescence measurements (Lesiak et al., 2019). The design of optical biosensors is based on a beam of light being distributed in a closed circuit, with the

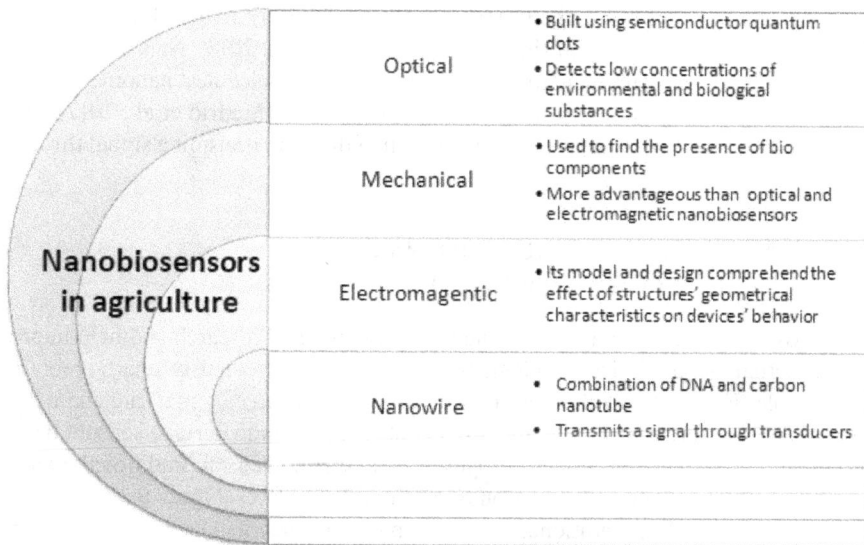

Nanobiosensors in agriculture	Optical	• Built using semiconductor quantum dots • Detects low concentrations of environmental and biological substances
	Mechanical	• Used to find the presence of bio components • More advantageous than optical and electromagnetic nanobiosensors
	Electromagentic	• Its model and design comprehend the effect of structures' geometrical characteristics on devices' behavior
	Nanowire	• Combination of DNA and carbon nanotube • Transmits a signal through transducers

FIGURE 9.2   Types of nanobiosensors and characteristics.

change in resonance frequencies caused by the analyte binding to the resonator being recorded. There is a high demand for tiny optical sensors that can specifically detect low concentrations of environmental and biological substances (Vo-Dinh, 2005). Haes and Duyne (2004) reported that triangular silver nanoparticles exhibit remarkable optical qualities and increased sensitivity to its environment.

9.5.2   MECHANICAL NANOBIOSENSORS

Mechanical nanobiosensors are tools that follow the laws of mechanics to measure physicochemical properties on a nanosized scale or to find the presence of bio components (Lim & Ramakrishna, 2006). When measuring mechanical properties at the nanoscale, mechanical nanobiosensors had a comparative advantage over optical and electromagnetic nanobiosensors (Lim, 2016). CNT-based fluidic shear-stress sensors and nanomechanical cantilever sensors are some examples of mechanical nanobiosensors (Liu et al., 2010; Chow et al., 2016).

9.5.3   ELECTROMAGNETIC NANOBIOSENSORS

The design of new types of electromagnetic biosensors as well as the amplification of the sensitivity of current sensors have both benefited from the use of electromagnetic nanoparticles and nanostructures (La Spada & Vegni, 2018). To comprehend how the dimension, shape and volume fraction affect the general behaviour of the device, it is crucial to model and design electromagnetic nanobiosensors.

9.5.4   NANOWIRE BIOSENSORS

The nanowire biosensor is a combination of single-stranded DNA (acting as the detector) and a carbon nanotube (serving as the transmitter); two molecules that are incredibly sensitive to external signals (La Spada & Vegni, 2018). Nanowire biosensors are a class of measurement tools in which biological molecules' nanometre-scale wires or fibres serve as the primary sensing components (Madrid et al., 2017). One of the main benefits of nanowire biosensors is that they can transmit a signal through transducers.

9.6   BENEFIT OF NANOBIOSENSORS OVER CONVENTIONAL BIOSENSORS

Nanobiosensors create some new avenues for fundamental research and help in providing actual bio-analytical implementations that were previously unattainable or unavailable (Fan et al., 2008; Velasco-Garcia, 2009). Due to nanomaterial characteristics that arise at the nanoscale and already absent in bulk materials, sensors based on nanomaterials have several advantages over sensors made from traditional materials in terms of accuracy and precision (Guisbiers et al., 2012) (Figure 9.3).

When compared to traditional detection methods such as chromatography and spectroscopy, nanobiosensors focus on providing real-time monitoring. These traditional methods can take days or even weeks to produce results and frequently

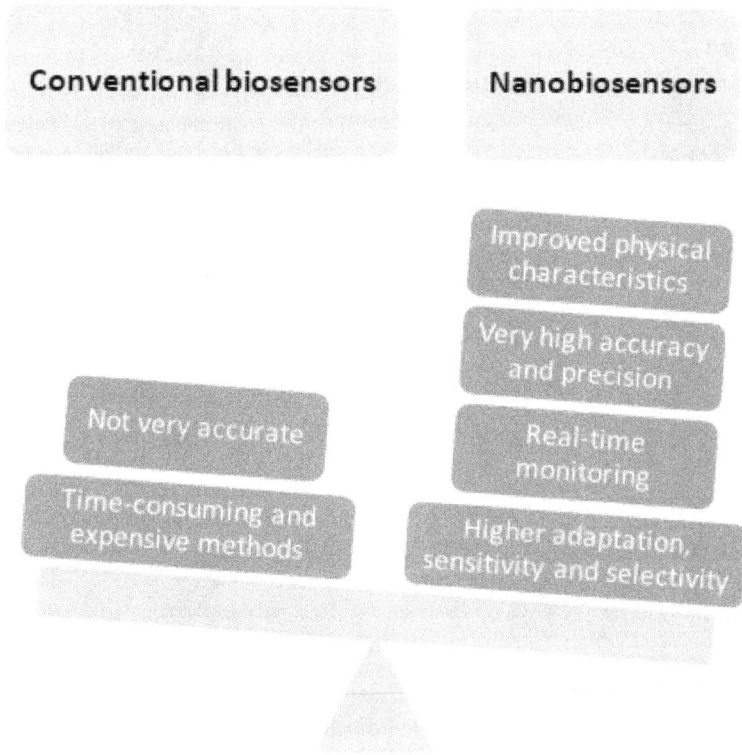

Conventional biosensors

Nanobiosensors

Improved physical characteristics

Very high accuracy and precision

Not very accurate

Real-time monitoring

Time-consuming and expensive methods

Higher adaptation, sensitivity and selectivity

FIGURE 9.3 Comparison between nanobiosensors and conventional biosensors.

necessitate some capital investments besides being time-consuming in terms of sample preparation (Callaway et al., 2017; Thangavelu et al., 2018). For instance, in the case of gas sensing (i.e. CO_2, NO_2 and SO_2), it is critical that the desired gases can be immediately detected to avoid any type of accident. Traditional techniques such as enzyme electrodes, spectrometry and high-performance liquid chromatography are used to detect gases. However, these instrumental techniques don't offer a quick analysis, while nanobiosensors can ensure an instant detection and monitoring (Kaur et al., 2019). Nanobiosensors have been also shown to be superior to conventional mechanical, electrical and optical sensors (Nandwani et al., 2019). Therefore, nanobiosensors are more advantageous compared to conventional sensors, including lower cost, higher adaptation and sensitivity and greater selectivity which will be beneficial for future aspects of growth and development in the agrifood sector.

9.7 ROLE OF NANOBIOSENSORS IN AGRICULTURE DEVELOPMENT

Through the design of diagnostic instruments and methods, nanobiosensors significantly contribute to the modernizing of agriculture field. When it comes to addressing numerous food, agricultural and environmental challenges, these sensors are

TABLE 9.1

Role of Nanobiosensors in Agricultural Practice

Nanobiosensor	Sensor Type	Applications	References
Quantum dot (Gold particles)	Fluorescence	Pathogen and contaminant detection	Thakur et al. (2022)
Plant hormone nanobiosensor	Fluorescent based	Detection of signalling molecules for plant growth	Chesterfield et al. (2020)
Smart nanobiosensor (Zinc oxide)	Electrochemical	Improvement of tomato and pepper germination	Negrete (2020)
Immobilized acetylcholinesterase on copper oxide and silver	Amperometric	Malathion and organophosphate pesticide detection	Bao et al. (2019), Zhang et al. (2019)
Acetylcholinesterase used with indophenol acetate	Colorimetric	Chlorpyrifos detection	Fu et al. (2019)
Pesticide detection (graphene based)	Electrochemical	Chlorpyrifos and chlorothalonil detection	Xu et al. (2020)
Probe based (gold NPs)	Colorimetric	Pathogen detection	Aoko et al. (2021)
Recombinase polymerase amplification (silver NPs)	Electrochemical	Plant pathogen detection	Lau et al. (2017)

precise, effective and reasonably priced. Toxicants, microbial load, heavy metal ions, pathogen and pollutant identification are a few of the uses of nanobiosensors in agriculture development. Some of the important roles of nanobiosensors in agricultural development are discussed below (Table 9.1).

9.7.1 NANOBIOSENSORS AS AN IMPORTANT APPARATUS FOR SUSTAINABLE AGRICULTURE

Nanotechnology shows a high contribution to enhancing the agriculture sector and increasing food security and productivity by making agricultural inputs more effective and providing solutions to environmental pollutions as well (Chaudhary & Sharma, 2019; Khati et al., 2019a, b; Usman et al., 2020; Chaudhary et al., 2021a, b, c, d, e; Bhatt et al., 2022). With the aid of nanobiosensors for sustainable agricultural development, nanotechnology has conclusively shown to be an essential eco-technological invention (Arduini et al., 2017).

Earlier, it was demonstrated that zinc oxide nanoparticles (ZnONPs) can penetrate the root tissue of ryegrass (Lin & Xing, 2008) while carbon nanotubes can pass through tomato seedlings (Khodakovskaya et al., 2009). This shows that new nutrient delivery methods can be created taking advantage of the nanoscale porosity structures on plant surfaces. However, nanofertilizers must demonstrate consistent nutrient release on need while inhibiting early conversion of those nutrients into toxic or gaseous forms that cannot be taken up by plants. Purposely, nanobiosensors are helpful devices enabling a controlled nitrogen release correlated to time period, environment and soil nutritional status.

9.7.2 Nanobiosensors as a Tool for Tracking Toxicants and Other Compounds

During the use of nanosensors in the detection of toxicants or in monitoring of heavy metals, one of the important factors is the response time of such detectors (how quickly it reacts to a change in signal) (Kaur et al., 2019). Nikoleli et al. (2021) reported that the use of lipid film-based nanobiosensor devices has given the chance to recreate a substantial portion of their functionality to rapid detection of food toxins. Also, different pesticides, hormones and genetically modified foods are among the analytes that can now be tracked by these nanobiosensors.

Nanobiosensors are affixed to the plant's leaves, where waves of hydrogen peroxide (H_2O_2) signalling are seen. H_2O_2 is a chemical that reacts in plants being utilized subsequently to send signals that enable leaf cells to produce substances aiming at protection from toxicants and predators, i.e. insects (Johnson et al., 2021). Many pesticides are used in agriculture to protect the plants from insects, pests and diseases, which are major causes of ground water, domestic livestock feed and water contaminations. To avoid this, novel optical methods should be incorporated for rapid pesticide tracking (detection). A recent study was done for selective tracking of diazinon, one of the most extensively applied organophosphorus pesticides. Reduced graphene quantum dots (rGQDs) and multi-walled carbon nanotubes (MWCNTs) were used to construct and create an optical apta-nanosensor that was adopted to detect diazinon pesticide with good accuracy and selectivity in contaminated areas (Talari et al., 2021).

9.7.3 Nanobiosensors as Monitoring Tool for Evaluating Soil Quality and Disease Diagnosis

With proper management of water, land, fertilizers and pesticides, crop yields can be increased with the help of nanobiosensor technology (Usman et al., 2020). Nanobiosensors have the potential to be a very efficient tool for smart targeted delivery systems, soil health promotion and disease management which play an important role in plant protection. Nanobiosensors can be a valuable tool to expand production of crops by accurately monitoring soil pH, soil quality, microbial load, humidity and so on (Table 9.2).

By reducing the deleterious consequences of excessive fertilizer application, gradual release of nanofertilizers may help in enhancing soil quality. To achieve this, nanobiosensors could be affixed to such nanofertilizers which allows a slow release of fertilizers (Rakhimol et al., 2021). There has been an increasing activity in using nanomaterial-based sensors to ameliorate the properties of contaminated soils, primarily through adsorption, reduction or chemical oxidation, due to their small particle size, high surface area and well-established reactivity (Guerra et al., 2018).

9.7.4 Nanobiosensors as a Detection Device for Sample Testing of Food and Water Quality Control

The major issue of food adulteration is explored in relation to food quality control. Purposely, nanotechnology-based detection methodologies may increase food quality and safety and access ways to tackle fraud in the food industry. As we know

TABLE 9.2

Role of Nanobiosensors to Evaluate Soil Quality and Water Quality

Nanoparticles	Material Type	Identification of Analytes from Water and Soil	References
Gold (Au) NPs	Optical colorimetric	NO_2^-, NO_3^-	Beqa et al. (2011)
Gold and Ag NPs	Optical colorimetric	Cocaine, Pb, Cu, Hg	Vilela et al. (2012)
Gold (Au) NPs	Conductive polymers such as polyaniline (PANI)	Detect highly toxic Hg^{2+} ions	Wang et al. (2011)
Bacterial biosensors	Dried paper strip	Arsenite and arsenate in potable water	Stocker et al. (2003)
Gold (Au) NPs	Fluorescence-based nanosensors	Al (III) detection	Abubaker et al. (2019)
Poly amidoamine or dendrimers	Polymer-based nanomaterials	Removal of ions such as $Cu^{2+}, Ag^+, Au^{1+}, Ni^{2+}$ from waste water	Diallo et al. (2005)
$TiO_2 + SiO_2$	Mixed oxide material	Photocatalytic degradation of methylene blue dye and industrial waste water treatment	Rasalingam et al. (2014)
Au NPs	Graphene oxide based	Detection of lead (Pb^{2+}) particles	Wang et al. (2016)
Aminophenyl and graphene oxide	Glassy carbon electrode	Simultaneous determination of Cd (II) and Cu (II)	Gupta et al. (2013)
Amphiphilic polyurethane	Polymer-based nanomaterials	Remediation of polynuclear aromatic hydrocarbons from soils	Tungittiplakorn et al. (2004)

gases and volatile substances used in feed can serve as indicators of food quality and it is possible to monitor precisely how they change during storage or adulteration. Graboski et al. (2021) reported that gas nanobiosensors are highly potential tools to be applied in food quality control, especially essential oils. Unique nanobiosensors are also reported for detecting each adulterant in various food samples, including the detection of threshold limit and their mode of operation (Singh et al., 2021).

Jebril et al. (2021) successfully developed a nanobiosensor for water quality monitoring through nanoengineering of eco-friendly silver nanoparticles using five different plant extracts. This nanobiosensor has also been used to detect environmental pollution and electrochemically determine the presence of phenol in samples of tap water, mineral water and agriculture runoff water. Furthermore, nanobiosensors can detect fruits' shelf life, viability of seeds and the quantity of nutrients required by crops (Dar et al., 2020).

9.8 LIMITATIONS IN THE APPLICATION OF NANOBIOSENSORS IN AGRICULTURE

Designing of nanobiosensors is a tough process as the synthesis methods and characterization and quantification instruments necessary for the production and deployment of nanobiosensors are relatively expensive (Kaur et al., 2019). Therefore, big efforts are needed to reduce their cost. The development of new types of electromagnetic sensors was considered costly as detectors are quite costly requiring electronic apparatus.

9.9 CONCLUSION

Greater sensitivity and lower detection limits are made possible by nanobiosensors' ability to provide an optimum surface for the target analyte in samples. Additional features and viewpoints for the development and installation of nanobiosensors have resurfaced as a result of the rapid advancement of nanotechnology in agricultural development. Nanobiosensors can be a valuable tool to expand production of crops by accurately monitoring soil pH, soil quality, microbial load, humidity, plant toxicants and so on. However, growing nutritious food is a process that is closely correlated to a healthy life. Nanobiosensors can be useful in preventing diseases from spreading to crops, thus protecting food from toxins while increasing the overall production from agricultural lands. Therefore, nanobiosensors, if used precisely and under controlled regulations, can promote sustainable development in agriculture while increasing crop productivity.

REFERENCES

Abubaker, M., Ngah, C.W.Z.C.W., Ahmad, M., & Kuswandi, B. (2019). Functionalized gold nanoparticles as optical nanosensors for determination of aluminum (III) ions in water samples. In Shukla, A.K., & Iravani, S. (eds.) *Micro and Nano Technologies, Green Synthesis, Characterization and Applications of Nanoparticles*. Elsevier, pp. 459–484.

Adam, T., & Gopinath, S.C. (2022). Nanosensors: Recent perspectives on attainments and future promise of downstream applications. *Process Biochem.* 117, 153–173.

Agri, U., Chaudhary, P., & Sharma, A. (2021). In vitro compatibility evaluation of agriusable nanochitosan on beneficial plant growth-promoting rhizobacteria and maize plant. *Nat Acad Sci Lett.* 44, 555–559.

Agri, U., Chaudhary, P., Sharma, A., & Kukreti, B. (2022). Physiological response of maize plants and its rhizospheric microbiome under the influence of potential bioinoculants and nanochitosan. *Plant Soil.* 474, 451–468.

Aoko, I.L., Ondigo, D., Kavoo, A.M., Wainaina, C., & Kiirika, L. (2021). A gold nanoparticle-based colorimetric probe for detection of gibberellic acid exuded by ralstonia solanacearum pathogen in tomato (*Solanum lycopersicum* L.). *Int J Hort Sci Technol.* 8, 203–214.

Arduini, F., Cinti, S., Scognamiglio, V., Moscone, D., & Palleschi, G. (2017). How cutting-edge technologies impact the design of electrochemical (bio) sensors for environmental analysis. A review. *Anal Chim Acta.* 959, 15–42.

Baig, N., Kammakakam, I., & Falath, W. (2021). Nanomaterials: A review of synthesis methods, properties, recent progress, and challenges. *Mater Adv.* 2(6), 1821–1871.

Bala, R., Dhingra, S., Kumar, M., Bansal, K., Mittal, S., Sharma, R.K., & Wangoo, N. (2017). Detection of organophosphorus pesticide–Malathion in environmental samples using peptide and aptamer based nanoprobes. *Chem Eng J.* 311, 111–116.

Bao, J., Huang, T., Wang, Z., Yang, H., Geng, X., Xu, G., Samalo, M., Sakinati, M., Huo, D., & Hou, C. (2019). 3D graphene/copper oxide nano-flowers based acetylcholinesterase biosensor for sensitive detection of organophosphate pesticides. *Sens Actuators B Chem.* 279, 95–101.

Beqa, L., Singh, A.K., Khan, S.A., Senapati, D., Arumugam, S.R., & Ray, P.C. (2011). Gold nanoparticle-based simple colorimetric and ultrasensitive dynamic light scattering assay for the selective detection of Pb (II) from paints, plastics, and water samples. *ACS Appl Mater Interf.* 3(3), 668–673.

Bhalla, N., Jolly, P., Formisano, N., & Estrela, P. (2016). Essays biochem. *Introd Biosens.* 60(1), 1–8.

Bhatt, P., Pandey, S.C., Joshi, S., Chaudhary, P., Pathak, V.M., Huang, Y., Wu, X., Zhou, Z., & Chen, S. (2022). Nanobioremediation: A sustainable approach for the removal of toxic pollutants from the environment. *J Hazard Mater.* 427, 128033.

Callaway, D. J., Matsui, T., Weiss, T., Stingaciu, L. R., Stanley, C. B., Heller, W. T., & Bu, Z. (2017). Controllable activation of nanoscale dynamics in a disordered protein alters binding kinetics. *J Mol Biol.* 429(7), 987–998.

Cavalcanti, A., Shirinzadeh, B., Zhang, M., & Kretly, L.C. (2008). Nanorobot hardware architecture for medical defense. *Sensors.* 8(5), 2932–2958.

Chaudhary, P., Chaudhary, A., Bhatt, P., Kumar, G., Khatoon, H., Rani, A., Kumar, S., & Sharma, A. (2022a). Assessment of soil health indicators under the influence of nano-compounds and *Bacillus* spp. in field condition. *Front Environ Sci.* 9, 769871.

Chaudhary, P., Chaudhary, A., Parveen, H., Rani, A., Kumar, G., Kumar, A., & Sharma, A. (2021e). Impact of nanophos in agriculture to improve functional bacterial community and crop productivity. *BMC Plant Biol.* 21, 519.

Chaudhary, P., Khati, P., Chaudhary, A., Gangola, S., Kumar, R., & Sharma, A. (2021a). Bioinoculation using indigenous *Bacillus* spp. improves growth and yield of *Zea mays* under the influence of nanozeolite. *3 Biotech.* 11, 11.

Chaudhary, P., Khati, P., Chaudhary, A., Maithani, D., Kumar, G., & Sharma, A. (2021d). Cultivable and metagenomic approach to study the combined impact of nanogypsum and *Pseudomonas taiwanensis* on maize plant health and its rhizospheric microbiome. *PLoS One.* 16, e0250574.

Chaudhary, P., Khati, P., Gangola, S., Kumar, A., Kumar, R., & Sharma, A. (2021c). Impact of nanochitosan and *Bacillus* spp. on health, productivity and defence response in *Zea mays* under field condition. *3 Biotech.* 11, 237.

Chaudhary, P., & Sharma, A. (2019). Response of nanogypsum on the performance of plant growth promotory bacteria recovered from nanocompound infested agriculture field. *Environ Ecol.* 37, 363–372.

Chaudhary, P., Sharma, A., Chaudhary, A., Khati, P., Gangola, S., & Maithani, D. (2021b). Illumina based high throughput analysis of microbial diversity of rhizospheric soil of maize infested with nanocompounds and *Bacillus* sp. *Appl Soil Ecol.* 159, 103836.

Chaudhary, P., Singh, S., Chaudhary, A., Sharma, A., & Kumar, G. (2022b). Overview of bio-fertilizers in crop production and stress management for sustainable agriculture. *Front Plant Sci.* 13, 930340.

Chesterfield, R.J., Whitfield, J.H., Pouvreau, B., Cao, D., Alexandrov, K., Beveridge, C.A., & Vickers, C. (2020). Rational design of novel fluorescent enzyme biosensors for direct detection of strigolactones. *ACS Synth Biol.* 9(8), 2107–2118.

Cho, I. H., Kim, D.H., & Park, S. (2020). Electrochemical biosensors: Perspective on functional nanomaterials for on-site analysis. *Biomat Res.* 24(1), 1–12.

Chow, W.W., Qu, Y., & Li, W. (2016). Carbon-nanotube-based fluidic shear-stress sensors. In T.-C. Lim (Ed.) *Nanosensors* (pp. 43–80). CRC Press, Boca Raton.

Dar, F.A., Qazi, G., & Pirzadah, T.B. (2020). Nano-biosensors: NextGen diagnostic tools in agriculture. In K. Hakeem & T. Pirzadah (Eds.) *Nanobiotechnology in Agriculture* (pp. 129–144). Cham: Springer.

Diallo, M.S., Christie, S., Swaminathan, P., Johnson, J.H., & Goddard, W.A. (2005). Dendrimer enhanced ultrafiltration. 1. Recovery of Cu (II) from aqueous solutions using PAMAM dendrimers with ethylene diamine core and terminal NH2 groups. *Environ Sci Technol.* 39(5), 1366–1377.

Fan, X., White, I. M., Shopova, S. I., Zhu, H., Suter, J. D., & Sun, Y. (2008). Sensitive optical biosensors for unlabeled targets: A review. *Anal Chim Acta.* 620(1–2), 8–26.

Foster, L.E. (2006). *Nanotechnology: Science, Innovation and Opportunity* (p. 283). Upper Saddle River, NJ: Prentice Hall.

Freitas, R.A. (1999). *Nanomedicine, Volume I: Basic Capabilities* (Vol. 1, pp. 210–219). Georgetown, TX: Landes Bioscience.

Fu, Q., Zhang, C., Xie, J., Li, Z., Qu, L., Cai, X., Ouyang, H., Song, Y., Du, D., Lin, Y. & Tang, Y. (2019). Ambient light sensor based colorimetric dipstick reader for rapid monitoring organophosphate pesticides on a smart phone. *Anal Chim Acta.* 1092, 126–131.

Ghaffar, N., Farrukh, M.A., & Naz, S. (2020). Applications of nanobiosensors in agriculture. In S. Javad (Ed.) *Nanoagronomy* (pp. 179–196). Cham: Springer.

Graboski, A.M., Paroul, N., Steffens, J., & Steffens, C. (2021). Nanosensors for food quality control especially essential oils. In S. Thomas, T.A. Nguyen, M. Ahmadi, A. Farmani, & G. Yasin (Eds.) *Nanosensors for Smart Manufacturing* (pp. 273–288). Elsevier. https://doi.org/10.1016/C2020-0-00292-6

Guerra, F. D., Attia, M. F., Whitehead, D. C., & Alexis, F. (2018). Nanotechnology for environmental remediation: Materials and applications. *Molecules.* 23(7), 1760.

Guisbiers, G., Mejía-Rosales, S., & Leonard Deepak, F. (2012). Nanomaterial properties: Size and shape dependencies. *J Nanomat.* 1–2.

Gupta, V.K., Yola, M.L., Atar, N., Ustundağ, Z., & Solak, A.O. (2013). A novel sensitive Cu (II) and Cd (II) nanosensor platform: Graphene oxide terminated p-aminophenyl modified glassy carbon surface. *Electroch Acta.* 112, 541–548.

Haes, A.J., & Duyne, R.P.V. (2004). Preliminary studies and potential applications of localized surface plasmon resonance spectroscopy in medical diagnostics. *Exp Rev Mol Diagnost.* 4(4), 527–537.

Hakeem, K.R., & Pirzadah, T.B. (2020). *Nanobiotechnology in Agriculture.* Springer International Publishing, Springer, Cham.

Jebril, S., Fdhila, A., & Dridi, C. (2021). Nanoengineering of eco-friendly silver nanoparticles using five different plant extracts and development of cost-effective phenol nanosensor. *Sci Rep.* 11(1), 1–11.

Johnson, M.S., Sajeev, S., & Nair, R.S. (2021). Role of Nanosensors in agriculture. In *2021 International Conference on Computational Intelligence and Knowledge Economy (ICCIKE)* (pp. 58–63). IEEE.

Kaur, R., Sharma, S. K., & Tripathy, S. K. (2019). Advantages and limitations of environmental nanosensors. In A. Deep & S. Kumar (Eds.) *Advances in Nanosensors for Biological and Environmental Analysis* (pp. 119–132). Elsevier. ISBN: 9780128174562

Khati, P., Bhatt, P., Kumar, R., & Sharma, A. (2018). Effect of nanozeolite and plant growth promoting rhizobacteria on maize. *3Biotech.* 8(141), 1–12.

Khati, P., Chaudhary, P., Gangola, S., & Sharma A. (2019a). Influence of nanozeolite on plant growth promotory bacterial isolates recovered from nanocompound infested agriculture field. *Environ Ecol.* 37 (2), 521–527.

Khati, P., Gangola, S., Bhatt, P., & Sharma A. (2017). Nanochitosan induced growth of *Zea Mays* with soil health maintenance. *3 Biotech.* 7(1), 81.

Khati, P., Sharma A., Chaudhary, P., Singh, A.K., Gangola, S., & Kumar, R. (2019b). High-throughput sequencing approach to access the impact of nanozeolite treatment on species richness and evenness of soil metagenome. *Biocat Agric Biotechnol.* 20, 101249.

Khodakovskaya, M., Dervishi, E., Mahmood, M., Xu, Y., Li, Z., Watanabe, F., & Biris, A. S. (2009). Carbon nanotubes are able to penetrate plant seed coat and dramatically affect seed germination and plant growth. *ACS Nano.* 3(10), 3221–3227.

Kukreti, B., Sharma, A., Chaudhary, P., Agri, U., & Maithani, D. (2020). Influence of nanosilicon dioxide along with bioinoculants on *Zea mays* and its rhizospheric soil. *3 Biotech*. 10, 345.

Kumar, N., Kumar, H., Mann, B., & Seth, R. (2016). Colorimetric determination of melamine in milk using unmodified silver nanoparticles. *Spectrochim Acta A Mol Biomol Spect*. 156, 89–97.

Kumari, H., Khati, P., Gangola, S., Chaudhary, P., & Sharma A. (2021). Performance of plant growth promotory rhizobacteria on maize and soil characteristics under the influence of TiO$_2$ nanoparticles. *Pant Res J*. 19, 28–39.

Kumari, S., Sharma, A., Chaudhary, P., & Khati, P. (2020). Management of plant vigour and soil health using two agriusable nanocompounds and plant growth promontory rhizobacteria in Fenugreek. *3 Biotech*. 10(461), 1–11.

La Spada, L., & Vegni, L. (2018). Electromagnetic nanoparticles for sensing and medical diagnostic applications. *Materials*. 11(4), 603.

Lau, H.Y., Wu, H., Wee, E.J.H., Trau, M., Wang, Y., & Botella, J.R. (2017). Specific and sensitive isothermal electrochemical biosensor for plant pathogen DNA detection with colloidal gold nanoparticles as probes. *Sci Rep*. 7, 38896.

Lesiak, A., Drzozga, K., Cabaj, J., Bański, M., Malecha, K., & Podhorodecki, A. (2019). Optical sensors based on II-VI quantum dots. *Nanomaterials*. 9(2), 192.

Lim, T.C. (ed.). (2016). *Nanosensors: Theory and Applications in Industry, Healthcare and Defense*. CRC Press. ISBN 9781439807361

Lim, T.C., & Ramakrishna, S. (2006). Conceptual review of nanosensors. *Z Naturforsch A*. 61:402–412.

Lin, D., & Xing, B. (2008). Root uptake and phytotoxicity of ZnO nanoparticles. *Environ Sci Technol*. 42(15), 5580–5585.

Liu, Y., Wang, W., & Shu, W. (2010). Nanomechanical cantilever sensors: Theory and applications. In T.-C. Lim (Ed.) *Nanosensors: Theory and Applications in Industry, Healthcare and Defence* (pp. 69–96). CRC Press. ISBN: 978-1-4398-0737-8.

Madrid, R.E., Chehín, R., Chen, T.H., & Guiseppi-Elie, A. (2017). Biosensors and nanobiosensors. In J.K.-J. Li (Ed.) *Further Understanding of the Human Machine: The Road to Bioengineering* (pp. 391–462). World Scientific.

Merkoci, A. (2021). Smart nanobiosensors in agriculture. *Nat Food*. 2(12), 920–921.

Nandwani, S.K., Chakraborty, M., & Gupta, S. (2019). Nanosensors: An improved tool for efficient reservoir characterization and oil exploration. In *EuroSciCon Conference on Advanced Nanotechnology* (p. 9838).

Negrete, J.C. (2020). Nanotechnology an option in Mexican agriculture. *J Biotechnol Bioinf Res*. 2(2), 1–3.

Nikoleli, G.P., Nikolelis, D.P., Siontorou, C.G., Nikolelis, M.T., Karapetis, S., & Bratakou, S. (2021). Nanosensors based on lipid membranes for the rapid detection of food toxicants. *Nanosens Environ Food Agric*. 1, 247–259.

Rai, V., Acharya, S., & Dey, N. (2012). Implications of nanobiosensors in agriculture. *J Biomater Nanobiotech*. 3, 2012.

Rakhimol, K.R., Thomas, S., Kalarikkal, N., & Jayachandran, K. (2021). Nanotechnology in controlled-release fertilizers. In F.B. Lewu, T. Volova, S. Thomas, & K.R. Rakhimol (Eds.) *Controlled Release Fertilizers for Sustainable Agriculture* (pp. 169–181). Academic Press. https://doi.org/10.1016/C2018-0-04238-3

Rasalingam, S., Peng, R., & Koodali, R.T. (2014). Removal of hazardous pollutants from wastewaters: Applications of TiO$_2$-SiO$_2$ mixed oxide materials. *J Nanomater*. 2014, 10.

Sharma, D., Teli, G., Gupta, K., Bansal, G., Gupta, G.D., & Chawla, P.A. (2022). Nanobiosensors from agriculture to nextgen diagnostic tools. *Curr Nanomater*. 7(2), 110–138.

Sharma, P., Pandey, V., Sharma, M. M. M., Patra, A., Singh, B., Mehta, S., & Husen, A. (2021). A review on biosensors and nanosensors application in agroecosystems. *Nanoscale Res Lett*. 16(1), 1–24.

Singh, N. A., Rai, N., & Marwal, A. (2021). Nanosensors for the detection of chemical food adulterants. In V. Kumar, P. Guleria, S. Ranjan, N. Dasgupta, & E. Lichtfouse (Eds.) *Nanotoxicology and Nanoecotoxicology* (Vol. 2, pp. 25–53). Cham: Springer.

Singh, R., Gautam, N., Mishra, A., & Gupta, R. (2011). Heavy metals and living systems: An overview. *Indian J Pharmacol.* 43(3), 246.

Singh, R.P. (2021). Recent trends, prospects, and challenges of nanobiosensors in agriculture. In Pudake, R.N., Jain, U., & Kole, C. (eds.) *Biosensors in Agriculture. Recent Trends Future Perspect* (pp. 3–13). Cham: Springer.

Stocker, J., Balluch, D., Gsell, M., Harms, H., Feliciano, J., Daunert, S., & Van der Meer, J.R. (2003). Development of a set of simple bacterial biosensors for quantitative and rapid measurements of arsenite and arsenate in potable water. *Environ Sci Technol.* 37(20), 4743–4750.

Sun, Y. P., Li, X. Q., Cao, J., Zhang, W.X., & Wang, H. P. (2006). Characterization of zero-valent iron nanoparticles. *Adv Colloid Interface Sci.* 120(1–3), 47–56.

Talari, F. F., Bozorg, A., Faridbod, F., & Vossoughi, M. (2021). A novel sensitive aptamer-based nanosensor using rGQDs and MWCNTs for rapid detection of diazinon pesticide. *J Environ Chem Eng.* 9(1), 104878.

Thakur, M., Wang, B., & Verma, M.L. (2022). Development and applications of nanobiosensors for sustainable agricultural and food industries: Recent developments, challenges and perspectives. *Environ Technol Innovat.* 26, 102371.

Thangavelu, R.M., Gunasekaran, D., Jesse, M.I., SU, M.R., Sundarajan, D., & Krishnan, K. (2018). Nanobiotechnology approach using plant rooting hormone synthesized silver nanoparticle as "nanobullets" for the dynamic applications in horticulture–an in vitro and ex vitro study. *Arab J Chem.* 11(1), 48–61.

Tungittiplakorn, W., Lion, L.W., Cohen, C., & Kim, J.Y. (2004). Engineered polymeric nanoparticles for soil remediation. *Environ Sci Technol.* 38(5), 1605–1610.

Usman, M., Farooq, M., Wakeel, A., Nawaz, A., Cheema, S.A., Rehman, H., & Sanaullah, M. (2020). Nanotechnology in agriculture: Current status, challenges and future opportunities. *Sci Total Environ.* 721, 137778.

Velasco-Garcia, M.N. (2009). Optical biosensors for probing at the cellular level: A review of recent progress and future prospects. Semin Cell Dev Biol. 2009. 20(1), 27–33.

Vilela, D., González, M.C., & Escarpa, A. (2012). Sensing colorimetric approaches based on gold and silver nanoparticles aggregation: Chemical creativity behind the assay. A review. *Anal Chimica Acta.* 751, 24–43.

Vo-Dinh, T. (2005). Optical nanosensors for detecting proteins and biomarkers in individual living cells. In Vo-Dinh, T. (ed.) *Protein Nanotechnology. Methods in Molecular Biology™* (Vol. 300, pp. 383–401). Totowa, NJ: Humana Press.

Wang, X., Jia, B., Zhang, W., Lin, B., Wang, Q., & Ding, J. (2016). Developing modified graphene oxide based sensor for lead ions detection in water. *ChemistrySelect* 1(8), 1751–1755.

Wang, X., Shen, Y., Xie, A., Li, S., Cai, Y., Wang, Y., & Shu, H. (2011). Assembly of dandelion-like Au/PANI nanocomposites and their application as SERS nanosensors. *Biosensors Bioelect.* 26(6), 3063–3067.

Xu, Y., Dhaouadi, Y., Stoodley, P., & Ren, D. (2020). Sensing the unreachable: Challenges and opportunities in biofilm detection. *Curr Opin Biotechnol.* 64, 79–84.

Zhang, P., Sun, T., Rong, S., Zeng, D., Yu, H., Zhang, Z., Chang, D., & Pan, H. (2019). A sensitive amperometric ache-biosensor for organophosphate pesticides detection based on conjugated polymer and Ag-rGO-NH$_2$ nanocomposite. *Bioelectrochem.* 127, 163–170.

10 Impact of Nanoparticles on Biotic Stress Tolerance

Shalu Priya and Ashish Kumar
Govind Ballabh Pant University of Agriculture & Technology

Viabhav Kumar Upadhayay
Dr. Rajendra Prasad Central Agricultural University

Anuj Chaudhary
Shobhit University Gangoh

Heena Parveen
National Dairy Research Institute

Govind Kumar
ICAR-Central Institute for Subtropical Horticulture

CONTENTS

DOI: 10.1201/9781003345565-10

10.1 INTRODUCTION

According to recent studies, the Food and Agriculture Organization predicts that the population will reach 9 billion in 2050, and food production must upsurge by at least 70% to meet the demand of rising population (Tyczewska et al., 2018; Mwobobia et al., 2020). In the current situation certain abiotic stresses such as heat, salinity, drought, and biotic stresses such as pests, bacteria, viruses, and fungi cause major loss of crops worldwide (Das et al., 2021; Tyagi et al., 2022). These various stresses have a significant impact on crop yield, ranging from 50% to 70% and 40% to 60% in the case of abiotic and biotic stresses (Tiwari et al., 2020). Various studies are being conducted to prevent these losses, and the use of nanoparticles (NPs) appears to be very promising in terms of promoting plant growth and conferring stress tolerance (Jeelani et al., 2020; Madzokere et al., 2021; Kumari et al., 2021). Nanoparticles are the smallest particles, with sizes ranging from 1 to 100 nm. NPs have diverse physicochemical possessions in terms of their small size and large surface area, electrical and thermal characteristics (Rastogi et al., 2019). NPs' easy solubility and transportation speeds in plants are due to their big surface area and small size (Salajegheh et al., 2020). Nanoparticles such as nanochitosan, nanozeolite, iron, silicon, and magnesium oxide play an important role in improving seed germination, increasing plant resistance to various diseases, deteriorating pesticide residues, and improving soil quality, all of which promote plant growth (Joseph et al., 2015; Cai et al., 2018; Khati et al., 2017, 2018, 2019a,b; Kukreti et al., 2020; Kang et al., 2021). There are different types of NPs known such as organic (carbon-based nanoparticles), inorganic, metallic, metal oxides, ceramics, polymeric, and lipid based which enhance seed germination and plant/soil health parameters (Khan et al., 2019; Kumari et al., 2020; Chaudhary et al., 2022a). The unique features exhibited by nanoparticles help in mitigating several abiotic and biotic stresses in plants through various mechanisms. Application of nanoparticles into the plants can improve its morphological and physiological indices such as the rate and efficiency of photosynthesis, nutrient uptake abilities, phytohormone regulation, and an enhanced plant defence system (Chaudhary & Sharma, 2019; Agri et al., 2021; Verma et al., 2022; Agri et al., 2022). The mechanism of nanoparticles on the productivity of crops under stress conditions may include a reduced oxidative stress, leading to an increased stress tolerance, maintaining ionic equilibrium according to the type of stress, enhancing various biochemical activities in plants, and synchronizing salt toxicity by reducing the deposition of malondialdehyde and water (Sarraf et al., 2022). Many different kinds of NPs have been applied to the plants to combat these stress conditions and support plant growth, such as nanoselenium and nanocopper for salinity stress (Pérez-Labrada et al., 2019; González et al., 2021) and nanosilicon for drought resistance, and all of these increase the plants' tolerance against biotic and abiotic stresses (Zahedi et al., 2020). This chapter gives an overview of the applications of nanoparticles, in particular their use to combat plant pathogens and to reduce the effects of biotic stress on plants.

10.2 BIOTIC STRESS

In plants, stress refers to the external environmental conditions that negatively affect plant development and productivity. A plant responds under stress conditions by showing some traits such as an altered pattern of gene expression and changes in growth rate and crop produce. However, plant species having exposure to stress become adapted to that particular abiotic/biotic stress (Verma et al., 2013). Plants experience biotic stress mainly caused by living organisms such as bacteria, fungi, viruses, nematodes, insects, etc. Pathogenic organisms that cause biotic stress may lead to the death of host plants (Gull et al., 2019; Chaudhary et al., 2022b). Biotic stress can be a major factor for causing pre- and post-harvest losses. There are a number of strategies plants used to counteract the effects of biotic stress; however, expression of some resistant genes helps in defence against pathogens. In fact, the climatic conditions in which plants live determine the occurrence of the type of biotic stress and also the resilience of plant species to the type of stress. Biotic stress also affects the rate of photosynthesis, as attack by chewing insects leads to a reduction in leaf surface area, and viral infections reduce photosynthesis, ultimately affecting crop yield (Gull et al., 2019; Verma et al., 2013).

10.3   USE OF NANOTECHNOLOGY FOR REDUCING EFFECTS OF BIOTIC STRESS

Existing agricultural practices are insufficient to meet the rising food demand, while pesticides and other chemicals used in agriculture have major negative consequences on the environment (Shang et al., 2019). Consequently, there is a need to devise an alternative strategy that is less destructive to the environment and could help meet food demand (Khan et al., 2021a). Environmental pollution is triggered by the usage of hazardous chemicals and pesticides, which has adverse effects on fauna and flora (Rai & Ingle, 2012). Pathogens and pests develop resistance to chemical-based fungicides and pesticides (Hirooka & Ishii, 2013). Therefore, there is a need to optimize the application of hazardous chemical pesticides and fungicides. Nanomaterials can be used to develop effective fertilizers that improve nutrient accessibility and decrease environmental pollution (Singh, 2017; Chaudhary et al., 2021a, b; Bhatt et al., 2022). Smart nanotechnology has the potential to eradicate crop diseases and increase crop yields, and can also decrease the usage of pesticides and hazardous chemicals that pollute soil, the environment, and groundwater (Khan et al., 2021). Therefore, different types of nanoparticles are used to prevent plant diseases. Nanoparticles based on silver, copper, silicon, and chitosan have shown the best antimicrobial properties against numerous plant pathogens and have also shown beneficial effects on plant growth and soil health (Worrall et al., 2018; Chaudhary et al., 2021c, d, e).

10.4   BIOTIC STRESS MITIGATION BY NANOPARTICLES

Biotic stresses affecting crop production such as pathogen infection, fungi, and herbivore attack cause a serious concern in the present world. Traditional pesticides pose a number of health and environmental risks in addition to increasing agricultural productivity. According to some studies, nanopesticides, which are pesticide formulations with effective biocidal properties, should contain NPs as an active ingredient (Adisa et al., 2019). NPs such as copper, silver, and aluminium exhibit antibacterial and pest control properties (Gogos et al., 2012; Cao et al., 2017). Under

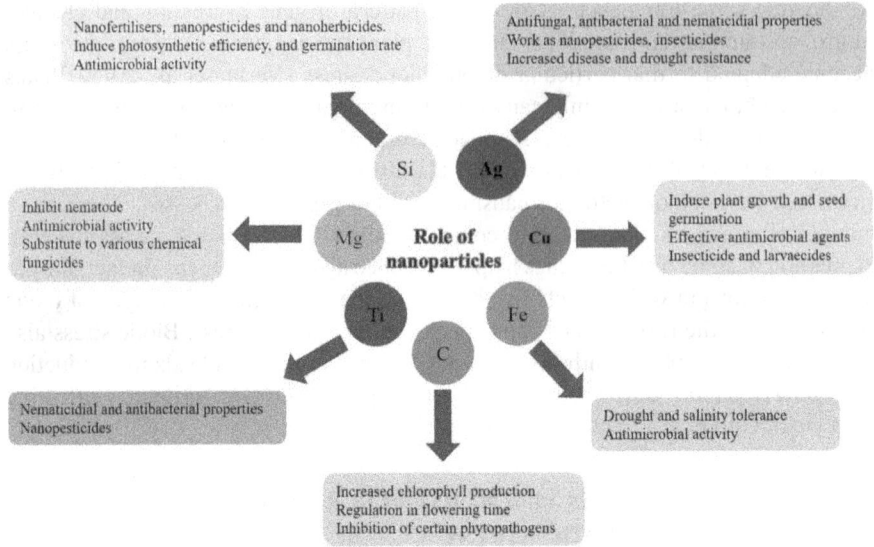

FIGURE 10.1   Role of different nanoparticles in biotic stress management.

stress, nanoparticles can mitigate the increased production of reactive oxygen species (ROS), and recently, Tripathi et al. (2022) demonstrated that nanoparticles ameliorate stress via modulating phytohormone signalling. In contrast to these nanopesticides, polymers are frequently used to encapsulate pesticides at the nanoscale level. The application of engineered nanoparticles such as cerium-, carbon- and silicon-based NPs protects plants from pathogens and insect attacks (Zhao et al., 2020). Different kinds of nanoparticles used as nanopesticides have been described below (Figure 10.1).

10.4.1   SILVER-BASED NANOPARTICLES

Silver nanoparticles (AgNPs) have received remarkable attention as significant nanopesticides in agriculture due to their wide spectrum of antibacterial activities. Silver NPs inhibit the growth of *Xanthomonas perforans* that causes bacterial spot in tomatoes and reduces yields to 10%–50% which has shown a significant reduction in its activity (Ocsoy et al., 2013). In certain studies, the effect of silver nanoparticles was studied against plant parasitic nematode *Meloidogyne incognita* and *Trachyspermum ammi* causing a huge loss of crops worldwide. The findings showed that plants inoculated with silver NPs displayed a significant increase in plant growth and defence enzyme activities such as catalase, ascorbate, and superoxide dismutase as compared to uninoculated plants (Danish et al., 2021). Several other studies suggest that silver nanoparticles possess significant antifungal properties against some fungi such as *Bipolaris sorokiniana*, *Gloeophyllum abietinum*, *Gloeophyllum trabeum*, *Chaetomium globosum*, *Phanerochaete sordid*, and *Chaetomium globosum* (Narayanan & Park, 2014; Ali et al., 2015; Zhao et al., 2020). Silver nanoparticles (AgNPs) synthesized using an extract from strawberry waste were investigated

for their effectiveness in controlling plant pathogens by evaluating their nematicidal, antibacterial, and antifungal activities towards *M. incognita, Ralstonia solanacearum,* and *Fusarium oxysporum,* respectively (Khan et al., 2021). Ocsoy et al.
(2013) developed DNA-directed silver NPs grown on graphene oxide, and these
composites efficiently reduce the extent of bacterial spot in tomato plants caused by
X. perforans. Mycogenic silver nanoparticles generated by *Trichoderma harzianum*
were evaluated against *Fusarium fujikuroi* and *Macrophomina phaseolina* and recommended as a safe way to combat fungus in Egyptian cotton (Zaki et al., 2022).
In *Arabidopsis thaliana,* bioengineered AgNPs reduced the severity of spot disease
caused by the fungus *Alternaria brassicicola* and induced immunity by modifying
the plant metabolic pathways (Kumari et al., 2020). Sabir et al. (2022) reported that
foliar spraying of 75 ppm AgNPs was more effective against yellow rust caused by
Puccinia striiformis and also improved the morpho-physiological traits of wheat.
Derbalah et al. (2022) revealed that silver oxide NP significantly inhibited the growth
of *Macophomina phaseolina,* the pathogen that causes charcoal rot, and promoted
strawberry resistance to charcoal rot by expressing a defence gene (PR-1) involved
in salicylic acid signalling pathways, as well as enhancing the growth and yield of
strawberry. El Gamal et al. (2022) showed that AgNPs exhibited virucidal activity to
treat yellow mosaic virus disease in faba beans, indicating that the use of AgNPs is
an effective tactic to prevent viral diseases in plants. Furthermore, AgNPs upregulated the pathogenesis-related gene PR-1 in treated plants and induced the production of defence-related oxidative enzymes. AgNPs embedded in greenly synthesized
chitosan-derived nanoparticles appeared to be a competitive and sustainable method
in an integrated strategic plan aimed at reducing yield losses due to wilt caused by *R.
solanacearum* (Santiago et al., 2019).

10.4.2 COPPER-BASED NANOPARTICLES

Crop losses due to biotic factors have drawn attention to the use of copper nanoparticles due to their widely recognized antimicrobial activity against phytopathogens and
their role in plant growth (Rai et al., 2018). Copper-based nanoparticles are very well
known for their antimicrobial and antifungal activities (Vinay et al., 2021). Copper
nanoparticles have been extensively utilized as antimicrobial agents against bacterial
pathogens such as *Bacillus subtilis* and *Escherichia coli,* fungal pathogens such as
F. oxysporum, Cochliobolus lunatus, Alternaria alternata, and *Phoma destructiva*
(Yoon et al., 2007; Kanhed et al., 2014). Antibacterial efficacies of copper NPs are
much better as compared with the traditional fungicide bavistin. Borgatta et al. (2018)
revealed the ability of copper NPs to suppress *F. oxysporum* which causes fungal disease in roots and observed a significant reduction (approximately 58%) in the disease
progression. The antifungal efficacy of copper sulphide nanoparticles was examined against *Gibberella fujikuroi* (bakanae disease) in *Oryza sativa.* The treatment
of these nanoparticles also altered the synthesis of salicylic acid and jasmonic acid
in the shoots and improved the defence mechanism of plant against plant pathogen
(Shang et al., 2020). Curcumin-CuNPs performed as an antifungal agent to protect
chickpea plants from the detrimental fungus *F. oxysporum* f. sp. *ciceri* and showed
the potential to induce seed germination and promote growth (Sathiyabama et al.,

2020). The biogenic CuO NPs synthesized from several *Trichoderma* species are the most effective and promising fungicides against *Alternaria* blight (Gaba et al., 2022). Therefore, biogenic CuO NPs are used to monitor the progression of the disease and its management in two different ways: (1) CuO NPs limit the growth of *Alternaria brassicae* on sensitive leaf tissues, thus preventing *Alternaria* blight; and (2) CuO NPs serve as a curative medicine that disables the pathogen by releasing *Alternaria*-inhibiting chemicals and facilitating the transportation of these chemicals to the site of infection. Besides their use as microbicides and nematicides, copper-based nanoparticles can also be used as insecticides and larvicides to control crop pests such as leafworms, mealybugs, thrips, grain pests, and mites. These nanomaterials can also be used to control a variety of mosquito vectors and insects (Kora, 2022).

10.4.3 CARBON-BASED NANOPARTICLES

Carbon-based nanoparticles have always proved to be very effective in plant stress reduction studies, and their external application to the plants causes an increase in the chlorophyll content and regulates flowering time without having any harmful consequences (Kumar et al., 2018). Studies show a remarkable antifungal activity of some single-walled, multi-walled, graphene oxide, activated carbon, and fullerene carbon-based nanoparticles' phytopathogenic fungi such as *Fusarium poae* and *Fusarium graminearum* (Wang et al., 2014). Inhibition in the water uptake activity and plasmolysis induction are described as the underlying mechanisms of the antifungal activities exhibited by carbon nanoparticles. The growth of several other plant pathogens such as *Xanthomonas* spp., *Aspergillus*, *Botrytis cinerea* has also been inhibited by using carbon nanoparticles (Hao et al., 2018). Carbon nanotubes may be a feasible option for disease management due to their powerful antimicrobial properties. González-García et al. (2021) studied the effect of multi-walled carbon nanotubes on reducing the incidence and severity of *Alternaria solani*. Furthermore, applying these nanoparticles to the leaves enhanced the antioxidant system and the photosynthetic capacity of tomato. Treatment with carbon dots (CDs) resulted in a 14.8% increase in total rice yield and an improvement in disease resistance of the rice plant (Li et al., 2018).

10.4.4 TITANIUM- AND ZINC-BASED NANOPARTICLES

The photocatalytic properties of the oxides of both titanium and zinc nanoparticles highly contribute to their antifungal and antibacterial properties, and are utilized as nanopesticides. Certain zinc and titanium nanoparticles also exhibit nematicidial properties (Elmer & White, 2018). When these NPs are exposed to light, they produce excited electrons which when combined with oxygen results in the formation of superoxide radicals (Zhao et al., 2020). It has been demonstrated that the pathogenic microbes such as *Staphylococcus aureus*, *E. coli*, and *Pseudomonas fluorescens* are inhibited by titanium and zinc oxide NPs (Azizi-Lalabadi et al., 2019). The TiO_2 NPs synthesized using leaf extract of *Moringa oleifera* were administered exogenously to wheat plants infected with the spot blotch disease-causing fungus *B. sorokiniana*, resulting in a reduction in biotic stress and an improvement in the growth and yield characteristics of wheat plants. Applications of ZnO NPs at concentrations of 50 and 100 ppm showed inhibitory activity against the plant pathogenic fungus (*Rhizoctonia solani*) and reduced disease incidence in carrot plants (Khan et al., 2022b). ZnO NPs and water melon peel waste-based hydrogel were employed to protect the pepper plant against *Fusarium* wilt (Abdelaziz et al., 2021). Parveen and Siddiqui (2021) exhibited antibacterial activity of ZnO NPs against bacterial pathogens (*Pseudomonas syringae, Xanthomonas campestris, Pectobacterium carotovorum*, and *R. solanacearum*) and fungal diseases (*F. oxysporum* and *A. solani*), and reported that foliar application of NPs resulted in highest plant growth and maximum content of proline and photosynthetic pigments. Foliar treatment of ZnO NPs in tomatoes reduced the adverse effects of tobamovirus and made the plants more disease resistant due to increased production of proline, phenolic compounds, ascorbic acid, and antioxidants (Sofy et al., 2021).

10.4.5 MAGNESIUM-BASED NANOPARTICLES

Magnesium nanoparticles being very effective against several bacteria, fungi, nematodes also promote plant growth in many other ways. Magnesium nanoparticles are known to inhibit the growth of the plant parasitic nematode, root-knot nematode (*Meloidogyne* spp.), considered as the utmost damaging species (Tauseef et al., 2021). Recently, Khan et al. (2022c) found antibacterial activity against *R. solanacearum* and nematicidal activity against the root-knot nematode *M. incognita* by green-synthesized MgO nanoparticles. The magnesium oxide nanoparticles are also preferred over other nanoparticles as a bactericide because it can be synthesized more easily, and is non-toxic (Haroon et al., 2020). Magnesium NPs exhibit a significant antibacterial activity against food borne pathogens such as *E. coli, Campylobacter*, and *Salmonella*, leading to leakage in the cell membrane, and induced oxidative stress ultimately causing bacterial cell death (He et al., 2016). Certain studies also stated the use of magnesium oxide nanoparticles as an antifungal agent against certain moulds and yeasts (Castillo et al., 2019). Imada et al. (2016) found that MgO NPs elicited resistance responses against *R. solanacearum* in the roots and hypocotyls of a tomato cultivar that was susceptible to the pathogen. Chen et al. (2020) demonstrated the antifungal activity of metallic MgO NPs against *Phytophthora nicotianae* and *Thielaviopsis basicola*, suggesting that nMgO could serve as a substitute for fungicides to control 'tobacco black shank' and 'root black rot', as well as other plant pathogens.

10.4.6 SILICA-BASED NANOPARTICLES

Silica nanoparticles (SiNPs) showed an excellent potential in agriculture in terms of their direct effect on plant growth in the form of nanopesticides, nanofertilizers, and nano-herbicides (El-Ashry et al., 2022). Because of their vast surface area, aggregation, reactivity, penetrating capacity, size, and structure, SiNPs can permeate plants and regulate their metabolic processes, and their presence on plant tissue surfaces improves pesticide delivery, photosynthetic efficiency, and germination rate (Goswami et al., 2022). The use of silica nanoparticles in the crops increases resistance to different diseases, including fungal diseases caused due to *F. oxysporum*, *Colletotrichum*, and *Aspergillus niger*; bacterial diseases caused due to *P. syringae* and *Dickeya dadantii*; and nematode diseases caused due to *M. incognita* (Khan & Siddiqui, 2020, Nisaq et al., 2021). Plant pathogens easily infiltrate the host plant by penetrating physical barriers such as wax and cell walls (Nawrath, 2006; Łaźniewska et al., 2012). Silicon nanoparticles act as physical barriers by increasing the mechanical strength of the plants by getting deposited into their leaf epidermis and stratum corneum (Buchman et al., 2019). Silica nanoparticles exhibit a great potential for molecular delivery, enhancement in release control, and insect toxicity (Athanassiou et al., 2018). Silica nanoparticles, which are safe agrochemicals, have the ability to elicit systemic acquired resistance in plants (El-Shetehy et al., 2021). They also have the potential to serve as an economically feasible, highly effective, and environmentally friendly solution for the prevention of plant diseases. Rashad et al. (2021) observed that application of SiNPs (150 ppm) significantly increased plant resistance and reduced downy mildew-infected grapevine disease severity. Similar treatment also increased the growth and yield of the plants and the quality of the berries. Silica NPs can also reduce pests by influencing peat life values such as mining values, survival, and puparia rates (Thabet et al., 2020). The study by El-Ashry et al. (2022) illustrated the combined effect of biosynthesized SiNPs and the half-recommended doses of commercial nematicides (fenamiphos, fosthiazate, and krenkel) against *M. incognita* and also improved eggplant growth. The mesoporous silica nanoparticles displayed antifungal activity against *A. solani*, promoted tomato development, and are believed to be safe for controlling tomato early blight (Derbalah et al., 2018). Abdelrhim et al. (2021) demonstrated the potential efficacy of SiO_2 NPs against *R. solani* infection, and biotic stress eradication in wheat plants was deciphered by activating a multi-layered defensive mechanism to inhibit the pathogen and neutralize the detrimental effect of ROS. Tomato plants that received a foliar spray of SiO_2 NPs (0.20 g/L) had higher plant growth characteristics, chlorophyll, carotenoid, proline, and defence enzymes against bacterial pathogens (*P. syringae*, *P. carotovorum*) and fungal pathogens (*F. oxysporum* and *A. solani*) (Parveen & Siddiqui, 2022). El-Ashry et al. (2022) studied the nematicidal action of SiNPs against egg hatching and second-stage juveniles of *M. incognita* (root-knot nematode), and their results showed that the combination of SiNPs with half the recommended dose of nematicides reduced the nematode population in the soil and improved plant growth.

10.4.7 Chitosan Nanoparticles

The strong antimicrobial potential of chitosan has drawn a lot of attention (Kheiri et al., 2017). Plants treated with chitosan have been observed to facilitate numerous physiological processes, as well as the induction of defensive responses through multiple pathways (Sarkar et al., 2022). Due to their special chemical and physical characteristics that promote cell interaction and permeability, chitosan NPs have been recognized as one of the most potent disease-suppressing agents and immunity boosters in plants (Selvaraj et al., 2022). Nanotechnology-derived chitosan-based formulations, either alone or in combination with other pesticides, exhibited significant potential against phytopathogens (Mishra et al., 2022). Nanochitosan have been reported to show antifungal potential against *M. phaseolina, A. alternata* (Divya et al., 2017), *Fusarium solani*, and *A. niger* (Xing et al., 2016). Chitosan NPs exhibited antifungal efficacy against *Fusarium* and elicited upregulation of pathogenesis-related (PR) proteins and antioxidant genes that might contribute to effective biocontrol (Chun & Chandrasekaran, 2019). Chitosan-coupled CuNPs have exhibited antimicrobial action against *R. solani*, the pathogen causing damping-off disease (Vanti et al., 2020). The application of chitosan nanoparticles protected finger millet against *Pyricularia grisea* infection by delaying blast symptom expansion on finger millet leaves, and it was concluded that induction of reactive oxygen species and the degree of peroxidase activity in finger millet leaves could be the cause of delayed symptoms (Sathiyabama & Manikandan, 2016). To achieve the goals of improved plant growth and minimizing the impact of leaf rust, chitosan nanoparticles could be used in a way that is less hazardous to the environment (Elsharkawy et al., 2022). The immunomodulatory effect of the chitosan in plants has been proven. Treating the leaves with chitosan NPs dramatically improved the plant's innate immune response and also increased levels of nitric oxide (NO), a plant defence signalling molecule (Chandra et al., 2015). Hoang et al. (2022) observed that application of chitosan NPs loaded with salicylic acid (SA) or silver reduced cassava leaf spot and improved plant growth more than conventional fungicides or nano-fungicides in net house settings. Chitosan nanoparticles have been studied for their potential to stimulate the production of antioxidants and defence-related enzymes in plants. Transcriptome assessment of plants treated with chitosan nanoparticles showed an increase in defence responses due to higher expression of genes involved in defence (Sarkar et al., 2022). The study by Narasimhamurthy et al. (2022) reported the potential of chitosan-derived nanoparticles against bacterial wilt of tomato and found that the nanomaterial significantly protected tomato plants from wilt diseases and promoted plant growth. Furthermore, this study also showed that tomato plants treated with chitosan and/or chitosan-based NPs showed a significant augmentation in upregulation of polyphenol oxidase, phenylalanine ammonia lyase, catalase, peroxidase, and β-1,3-glucanase compared to untreated plants. An experiment conducted by Ahmed et al. (2022) showed that foliar application of chitosan-iron nanocomposites (250 mg/L) significantly reduced the occurrence of bacterial leaf blight disease (67.1%) in rice via modulation of antioxidant. Furthermore, supplementation of this nanomaterial improved the efficiency of photosynthesis by improving the production of total chlorophyll, carotenoids, and the nutritional profile of rice compared to control. Research on the antiviral activity of chitosan-based nanoparticles is still in its infancy. Recently, the work of Abdelkhalek et al. (2021) demonstrated the antiviral activity of chitosan/dextran nanoparticles and reported a significant reduction in the accumulation level of alfalfa mosaic virus and a reduction in disease severity in *Nicotiana glutinosa* plants after foliar application of nanoparticles.

10.5 NANOPARTICLES-MEDIATED BIOTIC STRESS TOLERANCE

Adverse environmental factors, such as pests, diseases, toxic metals, salt, and drought, impede plant growth and yield, resulting in the demise of ecological systems. To counteract the effect of stress, plants have evolved numerous systemic signalling cascades, such as electricity, calcium, ROS, hydrodynamic waves, and plant hormones such as jasmonic acid, abscisic acid, and ethylene (Aslam et al., 2021). These systemic signal changes can be communicated to the entire system within a few minutes (Hang et al., 2021). When these systemic signals are detected in whole-body tissues of plants, systemic acquired resistance (SAR) is stimulated, allowing these tissues to withstand stress (Fichman and Mittler, 2020). Nanotechnology has incredible potential in modulating the amount of secondary metabolites, activating disease-resistant genes, activating defence enzymes, increasing the activities of antioxidant enzymes, producing ROS scavengers, delivering hormones and biomolecules, upregulation of defence-relevant phytohormones, and the production of transgenic plants that are resistant to diseases (Dong et al., 2022). The functional application of important nanoparticles against target plant pathogens and conferring stress tolerance systems in plants is illustrated in Table 10.1.

10.6 MODE OF ACTION

It is generally believed that the preponderance of nanoparticles has exhibited broad-spectrum antimicrobial effect with little or no toxicity to humans. Recently, the application of nanoparticles has been the promising choice in the agricultural sector. Nanoparticles use different ways to destroy or inactivate both cellular (bacteria, fungi) and non-cellular entities (viruses) (Upadhayay et al., 2019). The killing of microorganisms relies on several aspects, including the amount of nanoparticles, the wavelength and intensity of the light, the pH, the extent of hydroxylation, the temperature, the accessibility of oxygen and ROS (reactive oxygen species) (Markowska-Szczupak et al., 2011). Figures 10.2 and 10.3 illustrate some important mechanisms of the antimicrobial action of nanoparticles. The mechanism of action of nanoparticles has always been a focus point in research studies due to their significant results. It has been observed that silver nanoparticles when applied as a nanopesticide reduce the activity of acetylcholine resulting in the oxidative stress and cell death by affecting antioxidants and detoxifying enzymes (Gacem & Chaibi, 2022). Silver nanoparticles either up- or down-regulate the key insect genes leading to the reduction in protein synthesis and gonadotropin release, which are the major causes of reproductive and developmental damages (Nair & Choi, 2011; Jasrotia et al., 2022). Metal nanoparticles leads to the denaturation of enzymes and cell death, by binding to the sulphur and phosphorous in proteins and nucleic acids which reduce membrane permeability. Gold nanoparticles can stimulate the development and reproduction processes and act as trypsin inhibitors (Patil et al., 2016; Small et al., 2016; Lahiri et al., 2021). Alumina and silica nanoparticles act as insect dehydrating agents, by binding to the insect cuticle followed by the physical and chemical absorption of lipids and waxes (Mendes et al., 2022). Nanoparticles being efficient antimicrobial agents are well equipped with at least one of these mechanisms such as cell wall/membrane synthesis inhibitors, interrupting energy transduction, production of reactive oxygen species, photocatalysis, enzyme inhibition, and reduction in DNA production (Mishra et al., 2022; Shaikh et al., 2019; Raj et al., 2021). TiO_2 NPs can disrupt the cellular structure of microorganisms by producing a number

TABLE 10.1

Application of Nanoparticles in Controlling Plant Diseases and Mechanism of Biotic Stress Tolerance

Nanoparticles	Target Plant	Applied Method	Target Pathogen	Stress Tolerance Mechanism in Plants	References
Silica	Corn	Soil application	Fusarium oxysporum and Aspergillus niger	Augmentation in phenolic content, activation of defence-related enzymes	Suriyaprabha et al. (2014)
Chitosan	Wheat	Spraying	Puccinia triticina urediniospores	Activation of defence-related genes, increased activities of peroxidase and catalase	Elsharkawy et al. (2022)
Silver	Rice	Foliar application	Aspergillus flavus	Biochemical, osmotic, and enzymatic adjustments to tolerate biotic stress	Sultana et al. (2021)
Silver	Trachyspermum ammi	Pipetted into the holes made around 2 cm deep from the base of the stem	Meloidogyne incognita	Maximum lignin accumulation in wall, enhancement in the activities of antioxidant enzyme (SOD, POD, APX and CAT)	Danish et al. (2021)
Silicon dioxide	Beet	Foliar spray and seed soaking	Meloidogyne incognita, Pectobacterium beta-vasculorum, and Rhizoctonia solani	Improvement in growth and rate of photosynthesis, activation of defence-related enzymes	Khan and Siddiqui (2020)
Zinc oxide	Tomato (Solanum lycopersicum L.)	Foliar spray	Tobamovirus	Production of ROS scavengers, accumulation of phenolic compounds, ascorbic acid, and production of proline	Sofy et al. (2021)
Carbon nanotubes	Tomato	Foliar application	Alternaria solani	Improvement in antioxidant system, increment in the content of ascorbic acid, flavonoids, and enhancement in the activity of glutathione peroxidase	González-García et al. (2021)
C_{60} fullerenes, and carbon nanotubes	Nicotiana benthamiana	Foliar application	Tobacco mosaic virus (TMV)	Upregulation of defence-related phytohormones such as abscisic acid and salicylic acid	Adeel et al. (2021)
Fe_3O_4 NPs	Nicotiana benthamiana	Foliar application	Tobacco mosaic virus (TMV)	Activation of antioxidants, upregulation of SA synthesis, and the expression of SA-responsive PR genes	Cai et al. (2020)

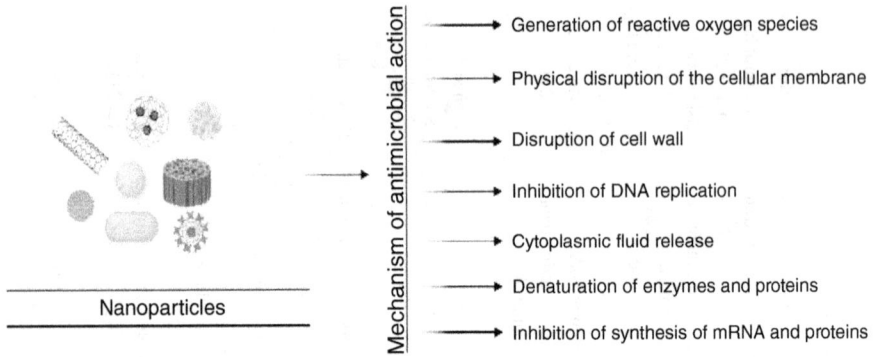

FIGURE 10.2 Mechanism of antimicrobial action of nanoparticles.

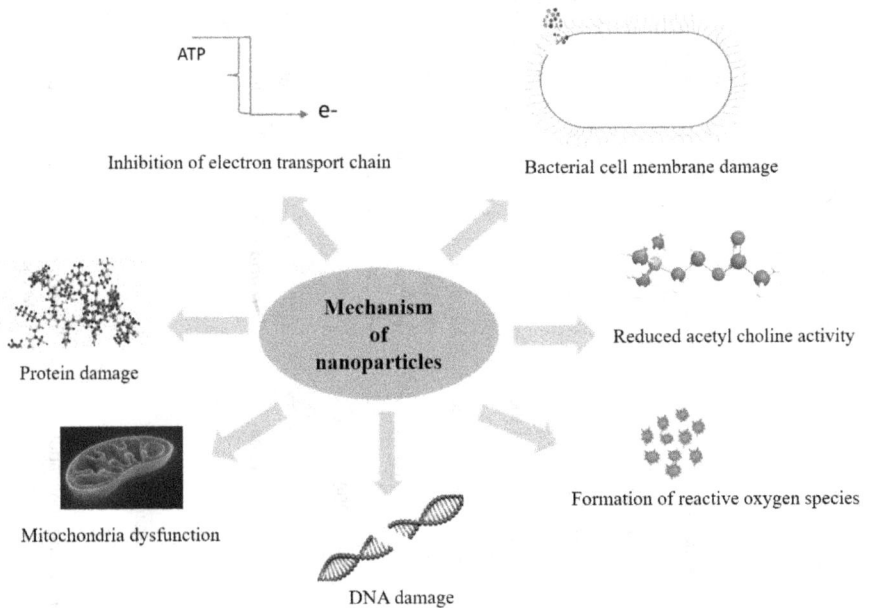

FIGURE 10.3 Mechanism of action of nanoparticles.

of ROS, including hydroxyl radicals, superoxide radicals, and hydrogen peroxide (Mammari et al., 2022; Upadhayay et al., 2019). MgO NPs are capable of destroying microbe because it generates O_2 on its surface. The inactivation of pathogenic organisms may depend on a range of factors, such as (1) the interaction of MgO NPs with microbes, (2) the alkaline condition of the particle surface, and (3) the precise surface area of particle, the length of contact, and the accessibility of O_2 on the surface (Upadhayay et al., 2019). MgO NPs produce reactive oxygen species, which attack the carbonyl group of peptide bonds and lead to the degradation of proteins (Brandelli et al., 2017; Upadhayay et al., 2019). ZnO NPs show antimicrobial properties through the production of reactive oxygen species (ROS), which induce oxidative stress by destroying nucleic acids, rupturing cellular membranes, and degrading protein molecules (Sirelkhatim et al., 2015). The most important nanoparticles, carbon nanotubes, display promising antimicrobial action through a range of mechanisms, including physical disruption of the cellular membrane, oxidative stress, as well as cytoplasmic fluid release (Azizi-Lalabadi et al., 2020). The antimicrobial effects of such NPs rely on their unique physical and chemical features, and pure single-walled nanotubes could be an effective antimicrobial agent since they have the capacity to disrupt the microbial cell membranes (Kang et al. 2007; Upadhayay et al., 2019). Chitosan has been shown to possess antimicrobial properties, making it effective against a wide range of pathogens, including bacteria and fungi. However, the antimicrobial activity of chitosan is reliant on a number of characteristics, including molecular weight, the existence of lipids and proteins that act as interferences, the extent of deacetylation or elevated ionized amino groups, the host, pH, surface charge, amount of chitosan, ionic strength of matrix, reaction period, chelating potential, and the types of microorganisms (Upadhayay et al., 2019). The surface of bacterial cells has a negative charge, which makes them more receptive to positively charged chitosan and encourages cellular leakage (Wang et al. 2017). Chitosan inhibits the growth of microorganisms and has the potential to interfere with the synthesis of mRNA and proteins after its access into cell (El Hadrami et al., 2010) (Table 10.2).

TABLE 10.2
Mode of Action of Certain Nanoparticles against Some Stress-Causing Pathogens

Nanoparticles	Host Plant	Pathogen	Response	References
Silica nanoparticles	*Zea mays*	*Fusarium oxysporum, Aspergillus niger*	Increase phenolic content, activate defence-related enzymes	Elamawi et al. (2020); Hasan et al. (2020); Nisaq et al. (2021);
	Triticum aestivum	*Meloidogyne incognita, Pectobacterium betavasculorum,* and *Rhizoctonia solani*	Promote growth, improve photosynthesis, activate defence-related enzymes	Thabet et al. (2021)
	Oryza sativa	*Fusarium fujikuroi*	Improve peroxidase activity	
	Phalaenopsis	*Harpophora maydis*	Inhibit mycelia growth	
	Solanum lycopersicum	*Spodoptera littoralis*	Induce defence-related enzymes and proteins, and genotoxicity	
	Vicia faba	*Aphis craccivora Liriomyza trifolii*		

(Continued)

TABLE 10.2 (*Continued*)
Mode of Action of Certain Nanoparticles against Some Stress-Causing Pathogens

Nanoparticles	Host Plant	Pathogen	Response	References
Zinc oxide nanoparticles	*Solanum lycopersicum*	*Tobacco mosaic virus*	Enhance plant growth, induce SAR, decrease disease severity, and transcriptional level genes were increased (POD and CHS)	Abdelkhalek and & Al-Askar (2020)
	Solanum lycopersicum	*Ralstonia solanacearum*	Antibacterial activity; stimulate plant growth; reduce bacterial soil population; decrease disease severity	Khan et al. (2021)
	Oryza sativa L. Xanthomonas oryzae pv. oryzae	*Xanthomonas oryzae*	Antibacterial effects, a reduction in the percentage of disease-affected leaves and better plant growth	Ogyunyemi et al. (2020)
Titanium oxide nanoparticles	*Beta vulgaris L.v*	*Cercospora beticola*	Enhance growth and productivity, less leaf spots, antifungal agent	Hamza et al. (2016)
	Solanum lycopersicum L.	*Meloidogyne incognita*	Reduce plant weight, root length and shows nematocidal activity	Ardakani (2013)
	Solanum lycopersicum L.	*Bactericera cockerelli*	Higher insect mortality, insecticidal effects	Gutierrez-Ramirez et al. (2021)
Magnesium oxide and magnesium hydroxide nanoparticles	*Vigna radiata*	*Fusarium solani Fusarium oxysporum*	Reduce mycelia development by preventing fungal growth	Abdallah et al. (2022)
Copper NPs	*Sphagnum*	*Meloidogyne incognita*	Cause DNA damage, disturbance of many cellular mechanisms such as synthesis of ATP, permeability of the cellular membrane, and response to oxidative stresses	Mohamed et al. (2019)

(*Continued*)

TABLE 10.2 (*Continued*)
Mode of Action of Certain Nanoparticles against Some Stress-Causing Pathogens

Nanoparticles	Host Plant	Pathogen	Response	References
	Citrullus lanatus	*Fusarium oxysporum* *Fusarium oxysporum*	Enhance release of ions, enzyme inhibition, reduction in DNA production	Borgatta et al. (2018); Buchman et al. (2019)
Cerium oxide NPs	*Solanum lycopersicum*	*Fusarium oxysporum*	Suppress disease through catalytic or other plant metabolic effects of ROS	Adisa et al. (2019)
Silver nanoparticles	*Solanum lycopersicum*	*Fusarium graminearum*	Restrict mycelial growth, inhibit spore germination, hyphal proliferation	Alvarez-Carvajal et al. (2020)

10.7 CONCLUSION

The use of nanoparticles to combat plant diseases is seen as a smart way to increase agricultural production while reducing dependence on chemical fungicides and/or pesticides. Numerous nanoparticles based on silver, silicon, copper, titanium, zinc oxide, carbon nanotubes, chitosan, etc. have been reported to alleviate the effect of biotic stress on crop. The application of nanoparticles promotes the development of disease resistance in plants through multiple mechanisms including upregulation of production of enzymes involved in plant defence, increased accumulation of secondary metabolites and proteins, scavenging of reactive oxygen species, and other related mechanisms. The usage of nanoparticles reduces the severity of disease in plants, increases photosynthetic capacity, and ultimately increases agricultural yield.

REFERENCES

Abdallah, Y., Hussien, M., Omar, M.O.A., Elashmony, R.M.S., Alkhalifah, D.H.M., & Hozzein, W.N. (2022). Mung bean (*Vigna radiata*) treated with magnesium nanoparticles and its impact on soilborne *Fusarium solani* and *Fusarium oxysporum* in clay soil. *Plants*. 11, 1514.

Abdelaziz, A.M., Dacrory, S., Hashem, A.H., Attia, M.S., Hasanin, M., Fouda, H.M., & ElSaied, H. (2021). Protective role of zinc oxide nanoparticles based hydrogel against wilt disease of pepper plant. *Biocatal Agric Biotechnol*. 35, 102083.

Abdelkhalek, A., & Al-Askar, A.A. (2020). Green synthesized ZnO nanoparticles mediated by mentha spicata extract induce plant systemic resistance against *Tobacco Mosaic Virus*. *Appl Sci*. 10, 5054.

Abdelkhalek, A., Qari, S.H., Abu-Saied, M.A.A.-R., Khalil, A.M., Younes, H.A., Nehela, Y., & Behiry, S.I. (2021). Chitosan nanoparticles inactivate alfalfa mosaic virus replication and boost innate immunity in *Nicotiana glutinosa* plants. *Plants*. 10, 2701.

Abdelrhim, A. S., Mazrou, Y. S. A., Nehela, Y., Atallah, O. O., El-Ashmony, R. M., & Dawood, M. F. A. (2021). Silicon dioxide nanoparticles induce innate immune responses and activate antioxidant machinery in wheat against *Rhizoctonia solani*. *Plants*. 10, 2758.

Adeel, M., Farooq, T., White, J.C., Hao, Y., He, Z., & Rui, Y. (2021). Carbon-based nanomaterials suppress tobacco mosaic virus (TMV) infection and induce resistance in *Nicotiana benthamiana*. *J Hazard Mater*. 404, 124167.

Adisa, I.O., Pullagurala, V.L.R., Peralta-Videa, J.R., Dimkpa, C.O., Elmer, W.H., Gardea-Torresdey, J.L., & White, J.C. (2019). Recent advances in nano-enabled fertilizers and pesticides: A critical review of mechanisms of action. *Environ Sci Nano*. 6, 2002–2030.

Agri, U., Chaudhary, P., & Sharma, A. (2021). In vitro compatibility evaluation of agriusable nanochitosan on beneficial plant growth-promoting rhizobacteria and maize plant. *Natl Acad Sci Lett*. 44, 555–559.

Agri, U., Chaudhary, P., Sharma, A., & Kukreti, B. (2022). Physiological response of maize plants and its rhizospheric microbiome under the influence of potential bioinoculants and nanochitosan. *Plant Soil*. 474, 451–468.

Ahmed, T., Noman, M., Jiang, H., Shahid, M., Ma, C., Wu, Z., & Li, B. (2022). Bioengineered chitosan-iron nanocomposite controls bacterial leaf blight disease by modulating plant defense response and nutritional status of rice (*Oryza sativa* L.). *Nano Today*. 45, 101547.

Ali, M., Kim, B., Belfield, K.D., Norman, D., Brennan, M., & Ali, G.S. (2015). Inhibition of *Phytophthora parasitica* and *P. capsici* by silver nanoparticles synthesized using aqueous extract of *Artemisia absinthium*. *Phytopathology*. 105, 1183–1190.

Alvarez-Carvajal, F., Gonzalez-Soto, T., Armenta-Calderón, A.D., Méndez Ibarra, R., Esquer Miranda, E., Juarez, J., & Encinas-Basurto, D. (2020). Silver nanoparticles coated with chitosan against *Fusarium oxysporum* causing the tomato wilt. *Biotecnia*. 22, 73–80.

Ardakani, A.S. (2013). Toxicity of silver, titanium and silicon nanoparticles on the root-knot nematode, *Meloidogyne incognita*, and growth parameters of tomato. *Nematology*. 15, 671–677.

Aslam, S., Gul, N., Mir, M.A., Asgher, M., Al-Sulami, N., Abulfaraj, A.A., & Qari, S. (2021). Role of jasmonates, calcium, and glutathione in plants to combat abiotic stresses through precise signaling cascade. *Front Plant Sci*. 12, 668029.

Athanassiou, C.G., Kavallieratos, N.G., Benelli, G., Losic, D., Rani, P.U., & Desneux, N. (2018). Nanoparticles for pest control: Current status and future perspectives. *J Pest Sci*. 91, 1–15.

Azizi-Lalabadi, M., Ehsani, A., Divband, B., & Alizadeh-Sani, M. (2019). Antimicrobial activity of titanium dioxide and zinc oxide nanoparticles supported in 4A zeolite and evaluation the morphological characteristic. *Sci Rep*. 9(1), 1–10.

Azizi-Lalabadi, M., Hashemi, H., Feng, J., & Jafari, S.M. (2020). Carbon nanomaterials against pathogens; the antimicrobial activity of carbon nanotubes, graphene/graphene oxide, fullerenes, and their nanocomposites. *Adv Colloid Interf Sci*. 284, 102250.

Bhatt, P., Pandey, S.C., Joshi, S., Chaudhary, P., Pathak, V.M., Huang, Y., Wu, X., Zhou, Z., & Chen, S. (2022). Nanobioremediation: A sustainable approach for the removal of toxic pollutants from the environment. *J Hazard Mater*. 427, 128033.

Borgatta, J., Ma, C., Hudson-Smith, N., Elmer, W., Plaza Perez, C.D., De La Torre-Roche, R., Zuverza-Mena, N., Haynes, C.L., White, J.C., & Hamers, R.J. (2018). Copper based nanomaterials suppress root fungal disease in watermelon (*Citrullus lanatus*): Role of particle morphology, composition and dissolution behavior. *ACS Sustain Chem Eng*. 6, 14847–14856.

Brandelli, A., Ritter, A.C., & Veras, F.F. (2017). Antimicrobial activities of metal nanoparticles. In M. Rai & R. Shegokar (Eds.) *Metal Nanoparticles in Pharma* (pp. 337–363). Springer, Cham.

Buchman, J.T., Elmer, W.H., Ma, C., Landy, K.M., White, J.C., & Haynes, C.L. (2019). Chitosan-coated mesoporous silica nanoparticle treatment of *Citrullus lanatus* (watermelon): Enhanced fungal disease suppression and modulated expression of stress-related genes. *ACS Sustain Chem Eng*. 7(24), 19649–19659.

Cai, L., Cai, L., Jia, H., Liu, C., Wang, D., & Sun, X. (2020). Foliar exposure of Fe_3O_4 nanoparticles on *Nicotiana benthamiana*: Evidence for nanoparticles uptake, plant growth promoter and defense response elicitor against plant virus. *J Hazard Mater*. 393, 122415.

Cai, L., Chen, J., Liu, Z., Wang, H., Yang, H., & Ding, W. (2018). Magnesium oxide nanoparticles: Effective agricultural antibacterial agent against *Ralstonia solanacearum*. *Front Microbiol*. 9, 790.

Cao, L., Zhou, Z., Niu, S., Cao, C., Li, X., Shan, Y., & Huang, Q. (2017). Positive-charge functionalized mesoporous silica nanoparticles as nanocarriers for controlled 2,4-dichlorophenoxy acetic acid sodium salt release. *J Agric Food Chem*. 66(26), 6594–6603.

Castillo, I.F., Guillén, E.G., Jesús, M., Silva, F., & Mitchell, S.G. (2019). Preventing fungal growth on heritage paper with antifungal and cellulase inhibiting magnesium oxide nanoparticles. *J Mater Chem B*. 7(41), 6412–6419.

Chandra, S., Chakraborty, N., Dasgupta, A., Sarkar, J., Panda, K., & Acharya, K. (2015). Chitosan nanoparticles: A positive modulator of innate immune responses in plants. *Sci Rep*. 5, 15195.

Chaudhary, P., Chaudhary, A., Bhatt, P., Kumar, G., Khatoon, H., Rani, A., Kumar, S., & Sharma, A. (2022a). Assessment of soil health indicators under the influence of nanocompounds and *Bacillus* spp. in field condition. *Front Environ Sci*. 9, 769871.

Chaudhary, P., Chaudhary, A., Parveen, H., Rani, A., Kumar, G., Kumar, A., & Sharma, A. (2021e). Impact of nanophos in agriculture to improve functional bacterial community and crop productivity. *BMC Plant Biol*. 21, 519.

Chaudhary, P., Khati, P., Chaudhary, A., Gangola, S., Kumar, R., & Sharma, A. (2021a). Bioinoculation using indigenous *Bacillus* spp. improves growth and yield of *Zea mays* under the influence of nanozeolite. *3 Biotech*. 11, 11.

Chaudhary, P., Khati, P., Chaudhary, A., Maithani, D., Kumar, G., & Sharma, A. (2021d). Cultivable and metagenomic approach to study the combined impact of nanogypsum and *Pseudomonas taiwanensis* on maize plant health and its rhizospheric microbiome. *PLoS One*. 16, e0250574.

Chaudhary, P., Khati, P., Gangola, S., Kumar, A., Kumar, R., & Sharma, A. (2021c). Impact of nanochitosan and *Bacillus* spp. on health, productivity and defence response in *Zea mays* under field condition. *3 Biotech*. 11, 237.

Chaudhary, P., & Sharma, A. (2019). Response of nanogypsum on the performance of plant growth promotory bacteria recovered from nanocompound infested agriculture field. *Environ Ecol*. 37, 363–372.

Chaudhary, P., Sharma, A., Chaudhary, A., Khati, P., Gangola, S., & Maithani, D. (2021b). Illumina based high throughput analysis of microbial diversity of rhizospheric soil of maize infested with nanocompounds and *Bacillus* sp. *Appl Soil Ecol*. 159, 103836.

Chaudhary, P., Singh, S., Chaudhary, A., Sharma, A., & Kumar, G. (2022b). Overview of biofertilizers in crop production and stress management for sustainable agriculture. *Front Plant Sci*. 13, 930340.

Chen, J., Wu, L., Lu, M., Lu, S., Li, Z., & Ding, W. (2020). Comparative study on the fungicidal activity of metallic MgO nanoparticles and macroscale MgO against soilborne fungal phytopathogens. *Front Microbiol*. 11, 365.

Chun, S.-C., & Chandrasekaran, M. (2019). Chitosan and chitosan nanoparticles induced expression of pathogenesis-related proteins genes enhances biotic stress tolerance in tomato. *Int J Biol Macromol*. 125, 948–954.

Danish, M., Altaf, M., Robab, M.I., Shahid, M., Manoharadas, S., Hussain, S.A., & Shaikh, H. (2021). Green synthesized silver nanoparticles mitigate biotic stress induced by *Meloidogyne incognita* in *Trachyspermum ammi* (L.) by improving growth, biochemical, and antioxidant enzyme activities. *ACS Omega*. 6(17), 11389–11403.

Das, K., Jhan, P.K., Das, S.C., Aminuzzaman, F.M., & Ayim, B.Y. (2021). Nanotechnology: Past, present and future prospects in crop protection. *Technol Agric*. 6, 30.

Derbalah, A., Essa, T., Kamel, S.M., Omara, R.I., Abdelfatah, M., Elshaer, A., & Elsharkawy, M.M. (2022). Silver oxide nanostructures as a new trend to control strawberry charcoal rot induced by *Macrophomina phaseolina*. *Pest Manag Sci.* 78(11), 4638–4648.

Derbalah, A., Shenashen, M., Hamza, A., Mohamed, A., & El Safty, S. (2018). Antifungal activity of fabricated mesoporous silica nanoparticles against early blight of tomato. *Egypt J Basic Appl Sci.* 5, 145–150.

Divya, K., Vijayan, S., George, T.K., & Jisha, M.S. (2017). Antimicrobial properties of chitosan nanoparticles: Mode of action and factors affecting activity. *Fibers Polym.* 18, 221–230.

Dong, B. R., Jiang, R., Chen, J.F., Xiao, Y., Lv, Z.Y., & Chen, W.S. (2022). Strategic nanoparticle-mediated plant disease resistance. *Crit Rev Biotechnol.* 43, 22–37.

Elamawi R.M., Tahoon A.M., Elsharnoby D.E., & El-Shafey R.A. (2020). Bio-production of silica nanoparticles from rice husk and their impact on rice bakanae disease and grain yield. *Arch Phytopathol Pflanzenschutz.* 53, 459–478

El-Ashry, R.M., El-Saadony, M.T., El-Sobki, A.E.A., El-Tahan, A.M., Al-Otaibi, S., El-Shehawi, A.M., Saad, A.M., & Elshaer, N. (2022). Biological silicon nanoparticles maximize the efficiency of nematicides against biotic stress induced by *Meloidogyne incognita* in eggplant. *Saudi J Biol Sci.* 29, 920–932.

El Gamal, A.Y., Tohamy, M.R., Abou-Zaid, M.I., Atia, M.M., El Sayed, T., & Farroh, K.Y. (2022). Silver nanoparticles as a viricidal agent to inhibit plant-infecting viruses and disrupt their acquisition and transmission by their aphid vector. *Arch Virol.* 167, 85–97.

El Hadrami, A., Adam, L.R., El Hadrami, I., & Daayf, F. (2010). Chitosan in plant protection. *Marine Drug.* 8, 968–987.

Elmer, W., White, J.C. (2018). The future of nanotechnology in plant pathology. *Annu Rev Phytopat.* 56, 111–133.

Elsharkawy, M.M., Omara, R.I., Mostafa, Y.S., Alamri, S.A., Hashem, M., Alrumman, S.A., & Ahmad, A.A. (2022). Mechanism of wheat leaf rust control using chitosan nanoparticles and salicylic acid. *J Fungi.* 8, 304.

El-Shetehy, M., Moradi, A., Maceroni, M., Reinhardt, D., Petri-Fink, A., Rothen-Rutishauser, B., Mauch, F., & Schwab, F. (2021). Silica nanoparticles enhance disease resistance in *Arabidopsis* plants. *Nat Nanotechnol.* 16, 344–353.

Fichman, Y., & Mittler, R. (2020). Rapid systemic signaling during abiotic and biotic stresses: Is the ROS wave master of all trades? *Plant J Cell Mol Biol.* 102, 887–896.

Gaba, S., Varma, A., & Goel, A. (2022). Protective and curative activity of biogenic copper oxide nanoparticles against *Alternaria* blight disease in oilseed crops: A review. *J Plant Dis Prot.* 129, 215–229.

Gacem, M.A., & Chaibi, R. (2022). Cu-based nanoparticles as pesticides: Applications and mechanism of management of insect pests. In K. Abd-Elsalam (Ed.) *Copper Nanostructures: Next-Generation of Agrochemicals for Sustainable Agroecosystems* (pp. 203–218). Elsevier.

Gogos, A., Knauer, K., & Bucheli, T.D. (2012). Nanomaterials in plant protection and fertilization: Current state, foreseen applications, and research priorities. *J Agric Food Chem.* 60, 39, 9781–9792.

González, G.Y., Cárdenas, A.C., Cadenas, P.G., Benavides, M.A., Cabrera-de-la, F.M., Sandoval, R.A., Valdés, R.J., & Juárez, M.A. (2021). Effect of three nanoparticles (Se, Si and Cu) on the bioactive compounds of bell pepper fruits under saline stress. *Plants.* 10, 217.

González-García, Y., Cadenas-Pliego, G., Alpuche-Solís, Á.G., Cabrera, R.I., & Juárez-Maldonado, A. (2021). Carbon nanotubes decrease the negative impact of *Alternaria solani* in tomato crop. *Nanomaterials.* 11, 1080.

Goswami, P., Mathur, J., & Srivastava, N. (2022). Silica nanoparticles as novel sustainable approach for plant growth and crop protection. *Heliyon.* 8, e09908.

Gull, A., Lone, A.A., & Wani, N.U.I. (2019). Biotic and abiotic stresses in plants. In A.B. de Oliveira (Ed.) *Abiotic and Biotic Stress in Plants*. IntechOpen. DOI: 10.5772/intechopen.85832.

Gutierrez-Ramirez, J.A., Betancourt-Galindo, R., Aguirre-Uribe, L.A., Cerna-Chavez, E., Sandoval-Rangel, A., Castro-del Angel, E., Chacon-Hernandez, J.C., Garcia-Lopez, J.I., & Hernandez-Juarez, A. (2021). Insecticidal effect of zinc oxide and titanium dioxide nanoparticles against *Bactericera cockerelli* Sulc. (Hemiptera: Triozidae) on tomato *Solanum lycopersicum. Agronomy*. 11, 1460.

Hamza, A., El-Mogazy, S., & Derbalah, A. (2016). Fenton reagent and titanium dioxide nanoparticles as antifungal agents to control leaf spot of sugar beet under field conditions. *J Plant Prot Res*. 56, 270–278

Hang, G., Miaojie, G., Jianbo, S., Yeye, M., & Ziqin, X. (2021). Signals in systemic acquired resistance of plants against microbial pathogens. *Mol Biol Rep*. 48, 3747–3759.

Hao, Y., Yuan, W., Ma, C., White, J.C., Zhang, Z., Adeel, M., & Xing, B. (2018). Engineered nanomaterials suppress Turnip mosaic virus infection in tobacco *(Nicotiana benthamiana). Envir Sci Nano*. 5, 1685–1693.

Haroon, A., Rai, P., Uddin, I., (2020). Synthesis, characterization and dielectric properties of BaTiO$_3$ nanoparticles. *Int J Nanosci*. 19, 1950001.

Hasan K.A., Soliman H., Baka, Z., & Shabana, Y.M. (2020). Efficacy of nano-silicon in the control of chocolate spot disease of *Vicia faba* L. caused by *Botrytis fabae. Egypt J Basic Appl Sci*. 7, 53–66.

He, Y., Ingudam, S., Reed, S., Gehring, A., Strobaugh, T.P., & Irwin, P. (2016). Study on the mechanism of antibacterial action of magnesium oxide nanoparticles against foodborne pathogens. *J Nanobiotechnol*. 14(1), 1–9.

Hirooka, T., & Ishii, H. (2013). Chemical control of plant diseases. *J General Plant Pathol*. 79(6), 390–401.

Hoang, N.H., Le Thanh, T., Thepbandit, W., Treekoon, J., Saengchan, C., Sangpueak, R., Papathoti, N.K., Kamkaew, A., & Buensanteai, N. (2022). Efficacy of chitosan nanoparticle loaded-salicylic acid and -silver on management of cassava leaf spot disease. *Polymers*. 14, 660.

Imada, K., Sakai, S., Kajihara, H., Tanaka, S., & Ito, S. (2016). Magnesium oxide nanoparticles induce systemic resistance in tomato against bacterial wilt disease. *Plant Pathol*. 65, 551–560.

Jasrotia, P., Nagpal, M., Mishra, C.N., Sharma, A.K., Kumar, S., Kamble, U., Kumar, A.B., Kashyap, P.L., Kumar, S., Singh, G.P. (2022). Nanomaterials for postharvest management of insect pests: Current state and future perspectives. *Front Nanotechnol*. 3, 811056.

Jeelani, P.G., Mulay, P., Venkat, R., & Ramalingam, C. (2020). Multifaceted application of silica nanoparticles. A review. *Silicon*. 12, 1337–1354.

Joseph, S., Anawar, H.M., Storer, P., Blackwell, P., Chia, C., Lin, Y., Munroe, P., Donne, S., Horvat, J., & Wang, J. (2015). Effects of enriched biochars containing magnetic iron nanoparticles on mycorrhizal colonisation, plant growth, nutrient uptake and soil quality improvement. *Pedosphere*. 25, 749–760.

Kang, H., Elmer, W., Shen, Y., Zuverza-Mena, N., Ma, C., Botella, P., White, J.C., & Haynes, C.L. (2021). Silica nanoparticle dissolution rate controls the suppression of *Fusarium wilt* of watermelon (*Citrullus lanatus*). *Environ Sci Technol*. 55, 13513–13522

Kang, S., Pinault, M., Pfefferle, L.D., & Elimelech, M. (2007). Single-walled carbon nanotubes exhibit strong antimicrobial activity. *Langmuir*. 23, 8670–8673.

Kanhed, P., Birla, S., Gaikwad, S., Gade, A., Seabra, A.B., Rubilar, O., Duran, N., & Rai, M. (2014). In vitro antifungal efficacy of copper nanoparticles against selected crop pathogenic fungi. *Mater Lett*. 115, 13–17.

Khan, A.U., Khan, M., Khan, A.A., Parveen, A., Ansari, S., & Alam, M. (2022a). Effect of phyto-assisted synthesis of magnesium oxide nanoparticles (MgO-NPs) on bacteria and the root-knot nematode. *Bioinorg Chem Appl.* 2022, 3973841.

Khan, A.U., Khan, M., Malik, N., Parveen, A., Sharma, P., Min, K., Gupta, M., & Alam, M. (2022b). Screening of biosynthesized zinc oxide nanoparticles for their effect on *Daucus carota* pathogen and molecular docking. *Micros Res Tech.* 85(10), 3365–3373.

Khan, I., Khalid, S., & Khan, I. (2019). Nanoparticles: Properties, applications and toxicities. *Arab J Chem.* 7, 908–931.

Khan, M., Khan, A.U., Bogdanchikova, N., & Garibo, D. (2021a). Antibacterial and anti-fungal studies of biosynthesized silver nanoparticles against plant parasitic nematode *Meloidogyne incognita*, plant pathogens *Ralstonia solanacearum* and *Fusarium oxysporum. Molecules.* 26, 2462.

Khan, M., Khan, A.U., Hasan, M.A., Yadav, K.K., Pinto, M.M., Malik, N., & Sharma, G.K. (2021b). Agro-nanotechnology as an emerging field: A novel sustainable approach for improving plant growth by reducing biotic stress. *Appl Sci.* 11(5), 2282.

Khan, M., Siddiqui, Z.A., Parveen, A., Khan, A.A., Moon, I.S., & Alam, M. (2022c). Elucidating the role of silicon dioxide and titanium dioxide nanoparticles in mitigating the disease of the eggplant caused by *Phomopsis vexans, Ralstonia solanacearum*, and root-knot nematode *Meloidogyne incognita. Nanotechnol Rev.* 11, 1

Khan, M.R., & Siddiqui, Z.A. (2020). Use of silicon dioxide nanoparticles for the management of *Meloidogyne incognita, Pectobacterium betavasculorum* and *Rhizoctonia solani* disease complex of beetroot (*Beta vulgaris* L.). *Sci Hort.* 265, 109211.

Khan, R.A.A., Tang, Y., Naz, I., Alam, S.S., Wang, W., Ahmad, M., Najeeb, S., Rao, C., Li, Y., & Xie, B. (2021c). Management of *Ralstonia solanacearum* in tomato using ZnO nanoparticles synthesized through *Matricaria chamomilla. Plant Dis.* 105, 3224–3230.

Khan, S., Naushad, M., Lima, E.C., Zhang, S., Shaheen, S.M., Rinklebe, J. (2021). Global soil pollution by toxic elements: Current status and future perspectives on the risk assessment and remediation strategies – A review, *J Hazard Mat.* 417, 126039.

Khati, P., Bhatt, P., Kumar, R., & Sharma, A. (2018). Effect of nanozeolite and plant growth promoting rhizobacteria on maize. *3Biotech.* 8, 141.

Khati, P., Chaudhary, P., Gangola, S., Bhatt, P., & Sharma, A. (2017). Nanochitosan supports growth of *Zea mays* and also maintains soil health following growth. *3 Biotech.* 7, 81.

Khati, P., Chaudhary, P., Gangola, S., & Sharma, P. (2019a). Influence of nanozeolite on plant growth promotory bacterial isolates recovered from nanocompound infested agriculture field. *Environ Ecol.* 37, 521—527.

Khati, P., Sharma, A., Chaudhary, P., Singh, A.K., Gangola, S., & Kumar R. (2019b). High-throughput sequencing approach to access the impact of nanozeolite treatment on species richness and evens of soil metagenome. *Biocatal Agric Biotechnol.* 20, 101249.

Kheiri, A., Moosawi Jorf, S. A., Malihipour, A., Saremi, H., & Nikkhah, M. (2017). Synthesis and characterization of chitosan nanoparticles and their effect on Fusarium head blight and oxidative activity in wheat. *Int J Biol Macromol.* 102, 526–538.

Kora, A.J. (2022). Copper-based nanopesticides. In K. Abd-Elsalam (Ed.) *Copper Nanostructures: Next-Generation of Agrochemicals for Sustainable Agroecosystems* (pp. 133–153). Elsevier.

Kukreti, B., Sharma, A., Chaudhary, P., Agri, U., & Maithani, D. (2020). Influence of nanosilicon dioxide along with bioinoculants on *Zea mays* and its rhizospheric soil. *3 Biotech.* 10, 345.

Kumar, A., Singh, A., Panigrahy, M., Sahoo, P.K., & Panigrahi, K.C. (2018). Carbon nanoparticle influences photomorphogenesis and flowering time in *Arabidopsis thaliana, Plant Cell Rep.* 37, 901–912.

Kumari, H., Khati, P., Gangola, S., Chaudhary, P., & Sharma, A. (2021). Performance of plant growth promotory rhizobacteria on Maize and soil characteristics under the influence of TiO$_2$ nanoparticles. *Pantnagar J Res.* 19, 28–39.

Kumari, S., Sharma, A., Chaudhary, P., & Khati, P. (2020). Management of plant vigor and soil health using two agriusable nanocompounds and plant growth promotory rhizobacteria in Fenugreek. *3 Biotech.* 10, 461.

Łaźniewska, J., Macioszek, V.K., & Kononowicz, A.K. (2012). Plant-fungus interface: The role of surface structures in plant resistance and susceptibility to pathogenic fungi. *Physiol Mol Plant Pathol.* 78, 24–30

Lahiri, D., Nag, M., Sheikh, H.L., Sarkar, T., Edinur, H.A., Pati, S., & Ray, R.R. (2021). Microbiologically-synthesized nanoparticles and their role in silencing the biofilm signaling cascade. *Front Microbio.* 12, 636588.

Li, H., Huang, J., Lu, F., Liu, Y., Song, Y., Sun, Y., Zhong, J., Huang, H., Wang, Y., Li, S., Lifshitz, Y., Lee, S.-T., & Kang, Z. (2018). Impacts of carbon dots on rice plants: Boosting the growth and improving the disease resistance. *ACS Appl Bio Mater.* 1, 663–672.

Madzokere, T.C., Murombo, L.T., & Chiririwa, H. (2021). Nano-based slow releasing fertilizers for enhanced agricultural productivity. *Mater Today Proc.* 45, 3709–3715.

Mammari, N., Lamouroux, E., Boudier, A., & Duval, R.E. (2022). Current knowledge on the oxidative-stress-mediated antimicrobial properties of metal-based nanoparticles. *Microorganisms*, 10, 437.

Markowska-Szczupak, A., Ulfig, K., & Morawski, A.W. (2011). The application of titanium dioxide for deactivation of bioparticulates: An overview. *Catal Today.* 169, 249–257.

Mendes, C.R., Dilarri, G., & Forsan, C.F. (2022). Antibacterial action and target mechanisms of zinc oxide nanoparticles against bacterial pathogens. *Sci Rep.* 12, 2658.

Mishra, K.K., Subbanna, A.R.N.S., Rajashekara, H., Paschapur, A.U., Singh, A.K., Jeevan, B., & Maharana, C. (2022). Chitosan and chitosan-based nanoparticles for eco-friendly management of plant diseases and insect pests: A concentric overview. In S. Kumar & S.V. Madihally (Eds.) *Role of Chitosan and Chitosan-Based Nanomaterials in Plant Sciences* (pp. 435–451). Elsevier. https://doi.org/10.1016/B978-0-323-85391-0.00003-4.

Mohamed, E.A., Elsharabasy, S.F., Abdulsamad, D. (2019). Evaluation of in vitro nematicidal efficiency of copper nanoparticles against root-knot nematode *Meloidogyne incognita*. *South Asian J Parasitol.* 14, 1–6.

Mwobobia, E.G., Sichangi, A.W., & Thiong'o, K.B. (2020). Characterization of wheat production using earth-based observations: A case study of Meru County, Kenya. *Model Earth Syst Environ.* 6, 13–25.

Nair, P.M.G., & Choi, J. (2011). Identification, characterization and expression profiles of *Chironomus Riparius* Glutathione S-Transferase (GST) genes in response to cadmium and silver nanoparticles exposure. *Aqua Toxicol.* 101, 550–560.

Narasimhamurthy, K., Udayashankar, A.C., De Britto, S., Lavanya, S.N., Abdelrahman, M., Soumya, K., Shetty, H.S., Srinivas, C., & Jogaiah, S. (2022). Chitosan and chitosan-derived nanoparticles modulate enhanced immune response in tomato against bacterial wilt disease. *Int J Biol Macromol.* 220, 223–237.

Narayanan, K.B., & Park, H.H. (2014). Antifungal activity of silver nanoparticles synthesized using turnip leaf extract (*Brassica rapa* L.) against wood rotting pathogens. *Eur J Plant Pathol.* 140, 185–192.

Nawrath, C. (2006). Unraveling the complex network of cuticular structure and function. *Curr Opin Plant Bio.* 9, 281–287

Nisaq, G.J., Sudarsono, S., & Sukma, D. (2021). Nano silica spray increase Phalaenopsis pulcherrima growth and resistance against Dickeya dadantii infection. In *IOP Conference Series: Earth and Environmental Science* (Vol. 694, No. 1, p. 012040). IOP Publishing.

Ocsoy, I., Paret, M.L., Ocsoy, M.A., Kunwar, S., Chen, T., You, M., & Tan, W. (2013). Nanotechnology in plant disease management: DNA-directed silver nanoparticles on graphene oxide as an antibacterial against *Xanthomonas perforans*. *ACS Nano.* 7, 8972–8980.

Parveen, A., & Siddiqui, Z.A. (2021). Zinc oxide nanoparticles affect growth, photosynthetic pigments, proline content and bacterial and fungal diseases of tomato. *Arch Phytop Plant Prot*. 54(17–18), 1519–1538.

Parveen, A., & Siddiqui, Z.A. (2022). Impact of silicon dioxide nanoparticles on growth, photosynthetic pigments, proline, activities of defense enzymes and some bacterial and fungal pathogens of tomato. *Vegetos*. 35, 83–93.

Patil, C.D., Borase, H.P., Suryawanshi, R.K., & Patil, S.V. (2016). Trypsin inactivation by latex fabricated gold nanoparticles: A new strategy towards insect control. *Enzyme Microb Technol*. 92, 18–25.

Pérez-Labrada, F., López-Vargas, E.R., Ortega-Ortiz, H., Cadenas-Pliego, G., Benavides-Mendoza, A., & Juárez-Maldonado, A. (2019). Responses of tomato plants under saline stress to foliar application of copper nanoparticles. *Plants*. 8, 151.

Rai, M., & Ingle, A. (2012). Role of nanotechnology in agriculture with special reference to management of insect pests. *Appl Microbiol Biotechnol*. 94, 287–293.

Rai, M., Ingle, A.P., Pandit, R., Paralikar, P., Shende, S., Gupta, I., Biswas, J.K., & da Silva, S.S. (2018). Copper and copper nanoparticles: Role in management of insect-pests and pathogenic microbes. *Nanotechnol Rev*. 7, 303–315.

Raj, N.B., Gowda, N.T.P., Pooja, O.S., Purushottam, B., Kumar, A.M.R., Sukrutha, S.K., Ravikumar, C.R., Nagaswarupa, H.P., Murthy, H.C.A., Bopanna, S.B. (2021). Harnessing ZnO nanoparticles for antimicrobial and photocatalytic activities. *J Photochem Photobiol*. 6, 100021.

Rashad, Y.M., El-Sharkawy, H.H.A., Belal, B.E.A., Abdel Razik, E.S., & Galilah, D.A. (2021). Silica nanoparticles as a probable anti-Oomycete compound against downy mildew, and yield and quality enhancer in grapevines: Field evaluation, molecular, physiological, ultrastructural, and toxicity investigations. *Front Plant Sci*. 12, 763365.

Rastogi, A., Tripathi, D.K., Yadav, S., Chauhan, D.K., Živčák, M., Ghorbanpour, M., El-Sheery, N.I., & Brestic, M. (2019). Application of silicon nanoparticles in agriculture. *3 Biotech*. 9, 90.

Sabir, S., Arshad, M., Ilyas, N., Naz, F., Amjad, M., Malik, N., & Chaudhari, S. (2022). Protective role of foliar application of green-synthesized silver nanoparticles against wheat stripe rust disease caused by *Puccinia striiformis*. *Green Process Synth*. 11, 29–43.

Salajegheh, M., Yavarzadeh, M., Payandeh, A., Payandeh, A., & Akbarian, M.M. (2020). Effects of titanium and silicon nanoparticles on antioxidant enzymes activity and some biochemical properties of *Cuminum cyminum* L. under drought stress. *Banat's J Biotechnol*. 11, 19–25.

Santiago, T.R., Bonatto, C.C., Rossato, M., Lopes, C.A.P., Lopes, C.A., G Mizubuti, E.S., & Silva, L.P. (2019). Green synthesis of silver nanoparticles using tomato leaf extract and their entrapment in chitosan nanoparticles to control bacterial wilt. *J Sci Food Agric*. 99, 4248–4259.

Sarkar, A., Chakraborty, N., & Acharya, K. (2022). Chitosan nanoparticles mitigate Alternaria leaf spot disease of chilli in nitric oxide dependent way. *Plant Physiol Biochem*. 180, 64–73.

Sarraf, M., Vishwakarma, K., Kumar, V., Arif, N., Das, S., Johnson, R., Janeeshma, E., Puthur, J.T., Aliniaeifard, S., & Chauhan, D.K. (2022). Metal/metalloid-based nanomaterials for plant abiotic stress tolerance: An overview of the mechanisms. *Plants*. 11, 316

Sathiyabama, M., & Manikandan, A. (2016). Chitosan nanoparticle induced defense responses in fingermillet plants against blast disease caused by *Pyricularia grisea* (Cke.) Sacc. *Carb Poly*. 154, 241–246.

Sathiyabama, M., Indhumathi, M., & Amutha, T. (2020). Preparation and characterization of curcumin functionalized copper nanoparticles and their application enhances disease resistance in chickpea against wilt pathogen. *Biocatal Agric Biotechnol*. 29, 101823.

Selvaraj, M., Ramasami, N., & Jagan, E.G. (2022). The effect of chitosan nanoparticles on immune responses in plants. In S. Kumar & S. Madihally (Eds.) *Role of Chitosan and Chitosan-Based Nanomaterials in Plant Sciences* (pp. 185–196). Elsevier. https://doi. org/10.1016/B978-0-323-85391-0.00006-X.

Shaikh, S., Nazam, N., Rizvi, S.M.D., Ahmad, K., Baig, M.H., Lee, E.J., & Choi, I. (2019). Mechanistic insights into the antimicrobial actions of metallic nanoparticles and their implications for multidrug resistance. *Int J Mol Sci*. 20, 2468.

Shang, H., Ma, C., Li, C., White, J.C., Polubesova, T., Chefetz, B., & Xing, B. (2020). Copper sulfide nanoparticles suppress *Gibberella fujikuroi* infection in rice (*Oryza sativa* L.) by multiple mechanisms: Contact-mortality, nutritional modulation and phytohormone regulation. *Env Sci Nano*. 7, 2632–2643.

Shang, Y., Hasan, M. K., Ahammed, G. J., Li, M., Yin, H., & Zhou, J. (2019). Applications of nanotechnology in plant growth and crop protection: A review. *Molecules*. 24, 2558.

Singh, M.D. (2017). Nano-fertilizers is a new way to increase nutrients use efficiency in crop production. *Int J Agric Sci*. 0975–3710.

Sirelkhatim, A., Mahmud, S., Seeni, A., Kaus, N., Ann, L.C., Bakhori, S., Hasan, H., & Mohamad, D. (2015). Review on zinc oxide nanoparticles: Antibacterial activity and toxicity mechanism. *Nano-Micro Lett*. 7, 219–242.

Small, T., Ochoa-Zapater, M.A., Gallello, G., Ribera, A., Romero, F.M., Torreblanca, A. (2016). Gold-nanoparticles ingestion disrupts reproduction and development in the German cockroach. *Sci Total Environ*. 565, 882–888.

Sofy, A.R., Sofy, M.R., Hmed, A.A., Dawoud, R.A., Alnaggar, A.E.-A.M., Soliman, A.M., & El-Dougdoug, N.K. (2021). Ameliorating the adverse effects of tomato mosaic tobamo-virus infecting tomato plants in Egypt by boosting immunity in tomato plants using zinc oxide nanoparticles. *Molecules*. 26, 1337.

Sultana, T., Javed, B., Raja, N., & Mashwani, Z. (2021). Silver nanoparticles elicited physiological, biochemical, and antioxidant modifications in rice plants to control *Aspergillus flavus*. *Green Process Synth*. 10, 314–324.

Suriyaprabha, R., Karunakaran, G., Kavitha, K., Yuvakkumar, R., Rajendran, V., & Kannan, N. (2014). Application of silica nanoparticles in maize to enhance fungal resistance. *IET Nanobiotechnol*. 8, 133–137.

Tauseef, A., Hisamuddin, K.A., Uddin, I. (2021). Role of MgO nanoparticles in the suppression of *Meloidogyne incognita*, infecting cowpea and improvement in plant growth and physiology. *Exp Parasit*. 220. 108045.

Thabet, A.F., Boraei, H.A., Galal, O.A., Samahy, M.F.M., Mousa, K.M., Zhang, Y.Z., Tuda, M., Helmy, E.A., Wen, J., & Nozaki, T. (2021). Silica nanoparticles as pesticide against insects of different feeding types and their non-target attraction of predators *Sci Rep*. 11, 14484.

Thabet, A.F., Galal, O.A., Tuda, M., El-Samahy, M.F.M., Fujita, R., & Hion, M. (2020). Silica nanoparticle effect on population parameters and gene expression of an internal feeder, American serpentine leafminer. *Preprints*. 2020100369, 1–15.

Tiwari, R.K., Lal, M.K., Naga, K.C., Kumar, R., Chourasia, K.N., Subhash, S., Kumar, D., & Sharma, S. (2020). Emerging roles of melatonin in mitigating abiotic and biotic stresses of horticultural crops. *Sci Hort*. 272, 109592.

Tripathi, D., Singh, M., & Pandey-Rai, S. (2022). Crosstalk of nanoparticles and phytohor-mones regulate plant growth and metabolism under abiotic and biotic stress. *Plant Stress*. 6, 100107.

Tyagi, J., Chaudhary, P., Mishra, A., Khatwani, M., Dey, S., & Varma, A. (2022). Role of endophytes in abiotic stress tolerance: With special emphasis on *Serendipita indica*. *Int J Environ Res*. 16, 62.

Tyczewska, A., Wozniak, E., Gracz, J., Kuczynski, J., & Twardowski, T. (2018). Towards food security: Current state and future prospects of agrobiotechnology. *Trends BioTech*. 36, 1219–1229.

Upadhayay, V.K., Khan, A., Singh, J., & Singh, A.V. (2019). Splendid role of nanoparticles as antimicrobial agents in wastewater treatment. In *Microorganisms for Sustainability* (pp. 119–136). Springer, Singapore.

Vanti, G.L., Masaphy, S., Kurjogi, M., Chakrasali, S., & Nargund, V.B. (2020). Synthesis and application of chitosan-copper nanoparticles on damping off causing plant pathogenic fungi. *Int J Biol Macromol*. 156, 1387–1395.

Verma, K.K., Song, X.-P., Joshi, A., Tian, D.-D., Rajput, V.D., Singh, M., Arora, J., Minkina, T., & Li, Y.-R. (2022). Recent trends in nanofertilizers for sustainable agriculture under climate change for global food security. *Nanomaterials*. 12, 173.

Verma, S., Nizam, S., & Verma, P. K. (2013). Biotic and abiotic stress signaling in plants. In M. Sarwat, A. Ahmad, & M. Abdin (Eds.) *Stress Signaling in Plants: Genomics and Proteomics Perspective* (Vol. 1, pp. 25–49). Springer Science, New York.

Vinay, J.U., Iliger, K.S., & Chikkannaswamy. (2021). Green nanotechnology in agriculture: Plant disease diagnosis to management. In R.K. Singh & Gopala (Eds.) *Innovative Approaches in Diagnosis and Management of Crop Diseases* (pp. 67–95). Apple Academic Press, New York.

Wang, L., Hu, C., & Shao, L. (2017). The antimicrobial activity of nanoparticles: present situation and prospects for the future. *Int J Nanomed*. 12, 1227.

Wang, X., Liu, X., Chen, J., Han, H., & Yuan, Z. (2014). Evaluation and mechanism of antifungal effects of carbon nanomaterials in controlling plant fungal pathogen. *Carbon*. 68, 798–806.

Worrall, EA., Hamid, A., Mody, K.T., Mitter, N., & Pappu, H.R. (2018). Nanotechnology for plant disease management. *Agronomy*. 8, 285.

Xing, K., Shen, X., Zhu, X., Ju, X., Miao, X., Tian, J., Feng, Z., Peng, X., Jiang, J., & Qin, S. (2016). Synthesis and in vitro antifungal efficacy of oleoyl-chitosan nanoparticles against plant pathogenic fungi. *Int J Biol Macromol*. 82, 830–836.

Yoon, K.-Y., Hoon Byeon, J., Park, J.-H., & Hwang, J. (2007). Susceptibility constants of *Escherichia coli* and *Bacillus subtilis* to silver and copper nanoparticles. *Sci Total Environ*. 373, 572–575.

Zahedi, S.M., Moharrami, F., Sarikhani, S., & Padervand, M. (2020). Selenium and silica nanostructure-based recovery of strawberry plants subjected to drought stress. *Sci Rep*. 10, 17672.

Zaki, S.A., Ouf, S. A., Abd-Elsalam, K.A., Asran, A.A., Hassan, M.M., Kalia, A., & Albarakaty, F.M. (2022). Trichogenic silver-based nanoparticles for suppression of fungi involved in damping-off of cotton seedlings. *Microorganisms*. 10, 344.

Zhao, L., Lu, L., Wang, A., Zhang, H., Huang, M., Wu, H., & Ji, R. (2020). Nanobiotechnology in agriculture: Use of nanomaterials to promote plant growth and stress tolerance. *J Agric Food Chem*. 68, 1935–1947.

11 Nanoparticles in Pest Management

Shalini Tiwari, Barkha Sharma, and Harshita Singh
Govind Ballabh Pant University of Agriculture & Technology

Pritom Biswas and Ankita Kumari
Central University of South Bihar

CONTENTS

11.1 INTRODUCTION

The use of pesticides—organic chemicals sprayed on crops, commodities, or urban environments—remains the mainstay of pest management, despite several alternative techniques. In this respect, pesticide usage has been linked to bioaccumulation, environmental pollution, and mammalian toxicity. These elements, together with the rising incidence of insect species developing resistance to many of the substances now in use, provide significant issues for agriculture and may significantly reduce the number of viable active components (Biswas & Kumar, 2021). Approximately one-third of the yearly plant crop worldwide perish due to various plant diseases caused by microbes including bacteria, fungi, and viruses. For the scientific community, managing pathogenic microorganisms that cause plant diseases is a top priority (Tyagi et al., 2022). Numerous plant diseases include fungi as a primary cause (70%) (Dutt et al., 2007). In order to solve these issues, new approaches to pest management are needed, including developing creative concepts that are resistant to pests.

DOI: 10.1201/9781003345565-11

In the near future, nanotechnology is anticipated to change the field of pest management because it is a useful tool in modern agriculture (Bhattacharya et al., 2010). In addition to being employed in new formulations of insecticides and insect repellents, nanoparticles with a diameter of 1–100 nm are efficient pesticides against weeds, plant diseases, and insect pests (Barik et al., 2008, Gajbhiye et al., 2009, Goswami et al., 2010), without the hazards posed by traditional chemical pesticides on beneficial insects (such as natural enemies) (Abbasi et al., 2020), public health, and the environment (Croissant et al., 2020). Nanocides are water-soluble (hydrophilic) as opposed to traditional hydrophobic pesticides, which increases bioactivity and uniformity of coverage (Zhang et al., 2008). Since nanocides may be sprayed in tiny amounts and are quickly absorbed by cells, their use can prevent the emergence of resistance in the pests that are the goal of the treatment (Ayoub et al., 2017).

Science-related fields such as engineering, biotechnology, analytical chemistry, and agriculture have made significant strides because of nanotechnology (Resham et al., 2015; Chaudhary et al., 2021a, b, c, d, e; Khati et al., 2017, 2018). Nanocides are usually measured between 1 and 100 nm. Over decades, advances in nanocides and nanomaterials have changed several industries, including food processing, plant productivity, soil environmental science, bioremediation, and medical, successful and secure employment (Chaudhary & Sharma, 2019; Kumari et al., 2020; Kukreti et al., 2020; Bhatt et al., 2022). However, it is thought that the application of nanocides in the agri-sector, particularly for plant production and protection, is an under-researched field (Khot et al., 2012). It has a wide range of applications, including those for conductors and semi-conductors, as well as for insecticides, coatings, sensors, and medical devices (Jordan, 2010). The use of biopesticides and biological nanomaterials, which are crucial for pest management and are currently being phased out of chemical-based agriculture in many countries, is a major problem (Bhattacharyya et al., 2010; Stadler et al., 2010; Watson et al., 2011; Gogos et al., 2012; Sahayaraj, 2014; De et al., 2014). The management of infections and pests may benefit greatly from using nanopesticides. A pesticide formulation encapsulated in nanoparticles has gradual properties as well as improved stability, solubility, transparency, and reactivity. Efficacy and toxicity of nanoparticles against pests were listed by Singh et al. (2018) up till 2013. The function of nanomaterials in pest management was examined by Chandrashekharaiah et al. (2015). In a comprehensive literature analysis, Margulis-Goshen and Magdassi (2012) shed light on how nanopesticides came to be.

Distinctive qualities of nanoparticles include uniform conductivity and unique optical characteristics. This is why there are numerous applications for these nanoparticles in biological systems and products, including drug development, medicines, pesticide delivery, hormone longevity, pheromone detection, and the detection of chemical resistance in biological systems (Bhattacharyya et al., 2016). In comparison to the bulk substance, nanoparticles have increased chemical activity and solubility. Hence, there is a consequent increase in plant or soil uptake, biological activity, as well as mobility (Sasson et al., 2007; Agri et al., 2021, 2022; Khati et al., 2019a, b; Chaudhary et al., 2022). This chapter aims on the various aspects of nanocomposites in the field of agri-sector for controlling plant diseases attacked by micro (bacteria, fungi, etc.) and macro (insects, nematodes, etc.) pathogens which can bring great economic loss globally and can lead to deterioration of soil health and yield.

11.2 TRADITIONAL STRATEGIES FOR SUSTAINABLE AGRICULTURE

Farmers have historically utilized a range of traditional methods to manage insect pests in agriculture, with "integrated pest management" (IPM) being the most well-liked strategy. These methods include crop rotation, healthier crop varieties, manipulations of sowing dates, and others. The US Academy of Sciences defined the phrase "integrated pest management" in 1969 as a means of avoiding pesticide side effects and combining the use of several pest control methods (cultural, resistant varieties, biological and chemical control). Integrated Pest Management (IPM) is therefore more challenging for producers to apply because it requires knowledge of pest dynamics, competence in pest monitoring, and producer cooperation for its application (Dhawan & Peshin, 2009; Peshin et al., 2009). When IPM was first popularized as a pest control tactic in the 1960s, there were very few IPM technologies that could be used in the field. IPM was successfully implemented in crops such as rice, cotton, sugar cane, and vegetables in the 1970s as a result of substantial study on the management of insect pests. However, it was not possible to accomplish the overly optimistic predictions that IPM adoption would result in a major reduction in pesticide use without a sizeable decline in agricultural yields (Dhawan & Peshin, 2009).

11.3 VARIOUS MICROBIAL FORMULATIONS USING NANOCIDES

11.3.1 Nano Bacterial Composites

Disease-related crop losses are considerable in crops including rice, barley, wheat, grapevine, cotton, and peanuts (Dhekney et al., 2005). Infections are treated with a wide variety of formulations. Utilizing nanoparticles (NPs) is one of the most recent developments in the fight against microbial contamination. Due to their economical and environmentally beneficial properties, use of NPs is being preferred. NPs which range in size from 0.2 to 100nm have a high surface-to-volume ratio that enhances their contact with bacteria and, consequently, their antimicrobial activity (Table 11.1). NPs' physical, chemical, and biological characteristics can be changed to suit a particular purpose (Ravishankar Rai & Jamuna, 2011). In order to avoid the antibacterial effects of antimicrobial agents, NPs have antimicrobial activity that can get around common resistance mechanisms, such as enzyme inactivation, decreased cell permeability, modification of target sites/enzymes, and increased efflux through overexpression of efflux pumps (Mulvey & Simor, 2009; Baptista et al., 2018).

Cu NPs were shown to be more susceptible when reacting with *Bacillus subtilis*, while Ag NPs were found to be less susceptible when responding with *Escherichia coli* (Yoon et al., 2007). According to Rajeshkumar et al. (2013), titanium oxide (TiO) NPs are utilized in the form of an antibacterial against gram-positive bacteria such as *B. subtilis* (3053) and gram-negative bacteria such as *Klebsiella planticola* (2727), while Au NPs made from the brown seaweed *Turbinaria conoides* are used as an effective antibacterial against *B. subtilis* and *Streptococcus* (Malarkodi et al., 2013a, b, c; Rajeshkumar et al., 2014a, b). When tested against pathogenic germs such as *Streptococcus* sp. and *Candida albicans*, zinc sulphide nanoparticles (NPs) made utilizing the sulphur-reducing bacterium *Klebsiella pneumoniae* shows strong

TABLE 11.1

Nano-PGPMs on Various Crops

Nanofertilizers	PGPMs	Effective On	References
CuO	Soil microbiota	Wheat (*Triticum aestivum*)	Kashyap et al. (2020)
SiO$_2$	*Pseudomonas stutzeri, Mesorhizobium* spp.	Land cress (*Barbarea verna*)	Boroumand et al. (2020)
TiO$_2$	*Pseudomonas fluorescens*	*Trifolium repens*	Zand et al. (2020)
ZnO	*Trichderma longibrachiatum*	*Vicia faba*	Moustafa and Taha (2021)
Ag complex	Polysaccharides of *Chlorella vulgaris*	*Triticum vulgare* and *Phaseolus vulgaris*	Ibrahim et al. (2019)
Ag complex	*Bacillus siamensis* strain C1	Rice seedlings	Afzal et al. (2021)
Nanohydroxyapatite	*Bacillus licheniformis*	Soil application	Priyam et al. (2019)
TiO$_2$	*Bacillus amyloliquefaciens*	*Brassica napus*	Palmqvist et al. (2015)

bactericidal effect (Malarkodi et al., 2014). Significant antibacterial activity against *Klebsiella aerogenes* and *Staphylococcus aureus* was found in cupric oxide nano-composites (average size 6–8 nm) produced utilizing *Tinospora cordifolia* which is a known medicinal herb (Suresh et al., 2015a). The sulphur-reducing bacteria *Serratia nematodiphila* and the marine bacterium *Enterococcus* sp. (RMAA) were used to synthesize cadmium sulphide (CdS) nanoparticles, which exhibit excellent antibacterial activity against *B. subtilis, K. planticola, E. coli*, and *Vibrio* sp. (Rajeshkumar et al., 2014a, b). Paulkumar et al. (2013) synthesized Ag NPs using leaf and stem extract of *Piper nigrum*. The nanoparticles ranged in size from 7–50 nm to 9–30 nm, respectively. Microbes such as *B. subtilis* and *K. pneumoniae* were effectively killed by Ag nanocides synthesized from the flowers of *Millingtonia hortensis*, fruit extracts of pomegranate and grapes (Gnanajobitha et al., 2013a, b), leaf extracts of *Solanum trilobatum* (Vanaja et al., 2014). *Alternanthera* dentate (active against *E. coli, P. aeruginosa, K. pneumoniae*, and *Enterococcus faecalis*) and other plant-mediated Ag NPs have antibacterial capabilities against different pathogens (Nakkala et al., 2014).

11.3.2 NANO FUNGAL AND ALGAL COMPOSITES

Titanium oxide NPs created by utilizing *Planomicrobium* sp. bacterial strain are successful in inhibiting *Aspergillus niger* fungus growth (Malarkodi et al., 2013a, b, c). Fungi such as *Botrytis cinerea* and *Penicillium expansum* can cause a number of terrible fruit diseases, such as grey and blue mould on table grapes and rotting of stored apples and pears (Spadaro et al., 2004; Cabanas et al., 2009). These two significant fruit pathogens are strongly combated by the rod-shaped ZnO NPs, which may also impair cell proliferation and result in fungal mortality (He et al., 2011). Infected tomato samples were used to identify disease-causing pathogens, and treatment to ZnO and MgO. Nanocomposites at various concentrations significantly inhibited the germination of *Mucor plumbeus*, *Rhizopus stolonifer*, and *Alternaria alternata* spores. As the concentration of NPs rose, the microorganisms that cause plant illnesses were more inhibited (Wani & Shah, 2012). This study investigated ZnO micro-particles with NPs, and the suppression of fungal growth demonstrates the size-dependent toxicity of ZnO NPs. ZnO NPs effectively combat *Fusarium graminearum*'s fungi, which have an adverse effect on crop growth (Ul Haq & Ijaz, 2019). Anthracnose, a plant disease, damages a variety of valuable plants, including mung bean, pepper, apple, cabbage, aloe, strawberry, radish, cabbage, mustard, pumpkin, cucumber, muskmelon, and watermelon (Perfect et al., 1999). Utilizing various media such as malt extract agar (MEA), potato dextrose agar (PDA), and corn meal agar (CMA), Ag nanocides with an average size of 4–8 nm were used to suppress the six *Colletotrichum* species very effectively (Chauchan et al., 2014).

In addition, ZnO NPs made from the brown marine microalga *Sargassum muticum* (Azizi et al., 2014) and other plant sources such as *Azadirachta indica* (Bhuyan et al., 2015), *Anisochilus carnosus* leaf extract (Anbuvannan et al., 2015), leaf extract of *Parthenium hysterophorus* L. (Rajiv et al., 2013), *Moringa oleifera* leaf extract (Elumalai et al., 2015), and extract of *Buchanania lanzan* (chironji) leaves (Suresh et al., 2015a) are extremely effective against microbial pathogens that cause disease, such as *K. aerogenes*, *E. coli*, and *S. aureus*. Figure 11.1 shows the mechanism of action of NPs against pathogens.

11.3.3 Various Marco Cell Formulations Using Nanocides

The accurate application of insecticides to the intended biocontrol and the specified spot in the plant is made more accessible by the use of nanoparticles. In fact, this is what the systems designed to regulate the discharge of pesticides in agricultural settings are for. By utilizing smart structures that limit and monitor pesticide release in order to steadily improve pesticide efficiency and safety, this technique seeks to minimize or eliminate the need for harmful pesticides in agriculture (Huang et al., 2018). Prasad et al. (2017) provided a list of 14 carrier substances, such as chitosan, silica, wheat gluten, etc., as well as formulation techniques, trapping, emulsion, cross-linking, and encapsulation for nanopesticides, which may be essential in the administration of pests. A nanocide formulation of the active component allows for improved water solubility, melting rate as well as dispersion ability (Muller & Keck, 2004).

In contrast to a responsive release depending on environmental factors, a simple method is made possible by the sustained delivery of nanopesticides in the plant system, delayed release. Effectiveness and efficiency of pesticide application can be increased and exact pesticide release can be achieved according to light, heating rate, humidity levels, wind velocity, and direction parameters. The rice crop affected by the blast fungus was driven to respond favourably to a chemical compound such as pyraclostrobin. This chemical alters the physiological makeup of plants and increases their ability to fend off illness. Additionally, it raises crop yields. Porous SiO_2 nano-spheres enhance the pathogen-fighting, stability, and action of pesticides (Huang et al. 2018). These essential formulations are known as "smart microcapsules" (Murugan et al. 2016).

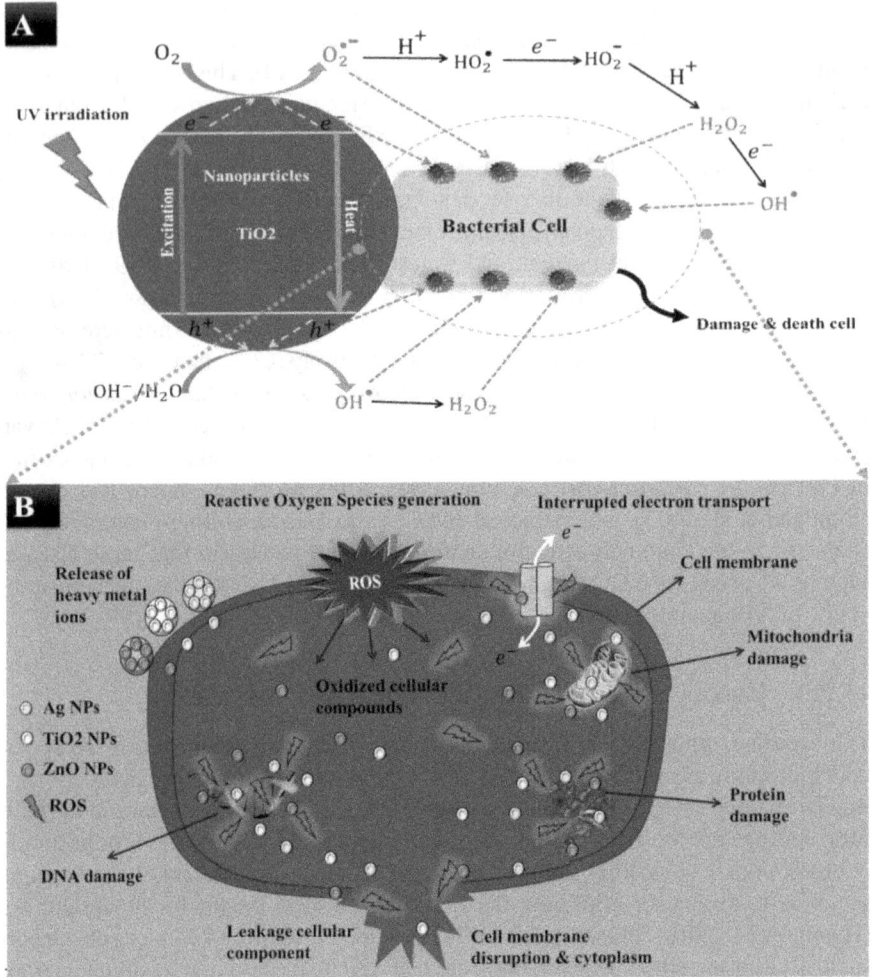

FIGURE 11.1 Mechanisms through which NPs fight infections with antimicrobial action (B). Free radicals are produced by NPs and their ions, such as those of titanium, silver, and zinc, which causes the production of oxidative stress, or ROS (A). The reactive oxygen species (ROS) that are produced can cause irreparable damage and destruction to the inner cellular components of the pathogens, ultimately leading to cell death. (Taken with permission from Khezerlou et al. (2018), Licence Number: 5382920825914.)

Broad-spectrum efficacy against pests is exhibited by nanocomposites-based pesticides such as Cu, Ag, SiO_2, and ZnO, as well as their formulations, which also significantly reduce chemical, air, water, and soil pollution. Utilizing nanopesticides has the additional benefit of allowing the addition of several agents to a single formulation, which allows for the targeting of multiple pests (Strom et al., 2001). Zn was promoted as a crop protection element (Lade et al., 2019). Being a crucial nutrient, Zn is required for plants and is rare in soils. However, further research into the interplay of nanoparticles with soil, plants, microorganisms, and humans is required before effective formulations can be created. The encased plant-based nanopesticides are more absorbable and active than synthetic nanopesticides by the pests intended to control. Since they degrade over time, plant-based nanopesticides are environmentally friendly. It seems possible that polyphagous pests and vectors will not develop resistance quickly or easily. These formulations can be used to make slow-release medications (Singh et al., 2018). It has been discovered that *F. graminearum* is resistant to the synthetic characteristics of ZnO NPs (Dimkpa et al., 2013). The aqueous extract of *Ficus racemosa* bark has been found to be efficient in treating *Culex quinquefasciatus* and *Culex gelidus* larvae (Velayutham et al., 2013). The use of nano-formulation, particularly by the fungal pathogens *Pythium* and *Fusarium* spp., in managing soft rot of ginger, is currently being examined. Due to their increased mobility, NPs can enter insect tissues and exert their active ingredient's effects (Sasson et al. 2007). The formulation of the cotton leafworm's water-soluble insecticide "Novaluron", which targets *Spodoptera littoralis* larvae, was effective against these pests. However, Elek et al. (2010) found that the utilization of chemicals was not an excellent idea for nanocomposite formulation due to its unrelenting dangerous by-products (Table 11.2).

TABLE 11.2
Various Nano-Products with Their Active Compounds for Controlling the Targeted Pathogens

Nano-Products	Active Compound	Target Organism	References
	Nanopesticides		
Chitosan/TPP-Paraquat nanoparticle	Paraquat	*Allium cepa*	Grillo et al. (2014)
PEG—essential oil of garlic nanoparticle	Essential oil (peppermint and palmarosa)	Adult beetles *Blattella germanica* L.	Yang et al. (2009b)
Graphene oxide-Cu2-xSe chlorpyrifos nanoparticle	Chlorpyrifos	*Pieris rapae*	Sharma et al. (2017)
Iron oxide nanoparticle	Iron oxide nanoparticle	Transgenic and nontransgenic Bt cotton	Van Nhan et al. (2016)
Silica nanoparticle	Silica nanoparticle	Cotton leaf worm (*Spodoptera littoralis*)	Ayoub et al. (2017)
Chitosan	Copper nanoparticles	*Curvularia* leaf spots	Choudhary et al. (2018)
Poly(*N*-isopropyl acrylamide-*co*-methacrylic acid)-methacrylic acid	Hollow mesoporous silica nanoparticles	*Nilaparvata lugens*	Gao et al. (2020)
Poly(*N*-isopropyl acrylamide-*co*-methacrylic acid)-methacrylic acid	Zein nanoparticles/neem oil	*Tetranychus urticae* and *Bemisia tabaci*	Pascoli et al. (2020)

(Continued)

TABLE 11.2 (*Continued*)
Various Nano-Products with Their Active Compounds for Controlling the Targeted Pathogens

Nano-Products	Active Compound	Target Organism	References
Poly(lactic-*co*-glycolic acid)	Nanoparticles	*Phytophthora infestans*	Fukamachi et al. (2019)
Nanoherbicide			
Microemulsion of pretilachlor	Pretilachlor	*Echinochloa crus-galli (L.) P. Beauv.*	Kumar et al. (2016b)
Nanoemulsion (*Satureja hortensis*) essential oil	Essential oil	*Amaranthus retroflexus* L.; *Chenopodium album* L.	Hazrati et al. (2017)
Guar gum nanohydrogel	Nanohydrogel	*Coronopus didymus* (L.); *Melilotus indica* (L.); *Chenopodium album* L.; *Avena sterilis ssp. ludoviciana* Dur; *Phalaris minor* Retz	Duhan et al. (2017)
Poly(ε-caprolactone)atrazine NPs	Atrazine	*Bidens pilosa* L.; *Amaranthus viridis* L.	Sousa et al. (2018)
Poly(methyl methacrylate)/ Haloxyfop-R-methyl NPs	Haloxyfop-R-methyl	*Lemna minor* L.	Chaghervand et al. (2022)
Nanonematicides			
Glycyrrhiza glabra-silver nanoparticles (GRAg NPs)	Ag NP	*Meloidogyne incognita*	Rani et al. (2022)
Colpomenia sinuosa and *Corallina mediterranea* silver nanoparticles	Silver nanoparticles	*Meloidogyne incognita*	Ghareeb et al. (2022)
PEGylated Cu, CuFe, and CuFeO$_2$ NPs	Nanoparticles	*Meloidogyne incognita* and *Meloidogyne javanica*	Ch et al. (2019)
Nanofertilizers			
Nitroxin	Ag	Potato	Davod et al. (2011)
Trichoderma sp. and *Pseudomonas* sp. NPs	Nanoclay	Rabi crops	Mukhopadhyay and De (2014)
Composted biochar	ZnO	*Triticum aestivum*	Bashir et al. (2020)
Pseudomonas fluorescens	Zero valent FeNPs	*Trifolium repens*	Daryabeigi Zand et al. (2020)
nFe$_2$O$_3$	Fe	*Glycine max* L.	Sheykhbaglou et al. (2010)
nTiO$_2$	Ti	*Cucumis sativus* L.	Servin et al. (2013)

(*Continued*)

TABLE 11.2 (*Continued*)
Various Nano-Products with Their Active Compounds for Controlling the Targeted Pathogens

Nano-Products	Active Compound	Target Organism	References
nTiO$_2$-activated carbon composite	Ti	Tomato (*Solanum lycopersicum* L.) and mung bean (*Vigna radiata* L.)	Singh et al. (2016)
Chitosan-NPK	Chitosan	*Triticum aestivum* L.	Abdel-Aziz et al. (2016)
Zein NPs	Zein	*Saccharum* spp.	Prasad et al. (2017)

11.4 NANOPARTICLES TO IMPROVE PESTICIDE FORMULATIONS

Globally, researchers have looked into the effectiveness of a variety of plant-derived NPs against arthropod pests that are significant to the economy, such as hard ticks (e.g., *Haemaphysalis bispinosa*) (Zahir & Rahuman, 2012), lice (e.g., *Pediculus humanus capitis*) (Jayaseelan et al., 2011), mosquitoes, moths (Roni et al., 2015), and louse flies (e.g., *Hippobosca maculata*) (Jayaseelan et al., 2012). The majority of pesticides currently in use are weakly water-soluble molecules, while they consist of primarily oil-in-water (O/W) emulsions, emulsifiable concentrates (ECs), or comparable compositions that are modifications of the same (Knowles, 2009). To fix the problems with ECs and O/W emulsions, based on micro- and nanoemulsions, novel formulations were created that could make NPs from 20 to 100 nm in size (Knowles, 2009; Tomlin, 2009; Song et al., 2009). Microemulsions have some restrictions even if they are more stable than nanoemulsions. Another technique of formulation based on nanodispersion or nanosuspension uses active ingredients that are 50-nm-sized nanocrystals. NPs that are either crystalline or amorphous, or both, create nanodispersions with characteristics similar to solutions, and can partially address these restrictions (Müller & Junghanns, 2006). It's interesting that this method hasn't been used very often yet; few cases (involving triclosan and novaluron) have been reported (Zhang et al., 2008; Elek et al., 2010).

In general, NPs act as "nanocarrier" transport mechanism because they can quickly enter plant cells. Since they were designed to convey a specific substance to the cell, tissue, or organism as needed, they may carry objects exactly because of this (Du et al., 2012). Despite this impressive use of nanoparticles on plants, several recent research (Slomberg & Schoenfisch, 2012; Kumari et al., 2021) concentrate on NPs' phytotoxicity and how they affect plant growth (Balalakshmi et al., 2017). On castor seeds, neither the rate of seed germination nor the growth of lepidopteran insects (*Ricinus communis* L.) was impacted by the application of silver nanoparticles (Yasur & Rani, 2015). The NPs were shown to be localized at the mitochondrial nucleoplasm in both plant tissues as well as insect tissues, demonstrating that these may be utilized for focused insecticide or fertilizer administration. This was discovered using transmission electron microscopy (TEM) (Yasur & Rani, 2013, 2015).

11.5   SYSTEM FOR DELIVERING PESTICIDES THAT USE NANOPARTICLES AS SOLID LIPID NANOCARRIERS

The idea of a "pesticide delivery system" (PDS) was developed for pest control by drawing inspiration from drug delivery techniques in the medical field, where nanocomposites have been administered to distribute drugs for medical therapy (Tsuji, 2001). Making the active ingredients available to a particular target at certain concentrations and timings is the objective that will achieve the desired effect while maintaining the highest level of biological effectiveness and minimizing any negative side effects (Tsuji, 2001). It is advantageous to employ NPs as solid lipid nanoparticles because of their increased surface area, ease of attaching single and multiple pesticide compounds, and relatively quick mass and heat transfer towards the objective, which is the body of insects. NPs also postpone the effectiveness loss brought on by degradation.

Interestingly, the essential oils of neem (Xu et al., 2010; Jerobin et al., 2012), *Artemisia arborescens* L. (Lai et al., 2006), garlic (Yang et al., 2009a), *Lippia sidoides* (Abreu et al., 2012), *Catharanthus roseus* extract (L.) G. Don (Pavunraj et al., 2017), and juniper are common examples of insecticides studied using this nanotechnology method (Athanassiou et al., 2013). Many plant extracts, such as capsaicin from chilli peppers (*Capsicum annuum*) (Chhipa, 2019) and *Lansiumamide* B extract from the seeds of *Clausena lansium* (Lour.) Skeels (Yin et al., 2012), have also been proposed as nano-delivery systems for pheromones (Bhagat et al., 2013).

NPs are crucial gene transporters in numerous plant species, as well as by modifying DNA replication and function. So, it is possible to use them to successfully combat transgenic silence (Kumar et al., 2016a). Additionally, NPs are capable of mediating multigene transformation without using the conventional complicated carrier construction process (Fu et al., 2012).

11.6   APPLICATION OF NPs IN THE CONTROL OF PESTS

For the purpose of eradicating the oriental leafworm moth, *Spodoptera litura* (F.), from *R. communis*, and the castor semilooper [*Achaea janata* (L.)], Ag NP particles stabilized with polyvinyl pyrrolidone was sprayed over the plant's leaves. Due to the physiological modifications made to the insects' bodies brought on by the presence of NPs, it was discovered that Ag NPs had a detrimental impact on both species' growth (i.e., the size and stage of the larva, the size of the pupa, and the size of the adult) (Yasur & Rani, 2015). Numerous existing paradigms have been reported for applying NPs in this situation successfully (Benelli, 2016a; Lee et al., 2018).

Recent research has shown that environmentally friendly metal nanoparticles are promising (NP) production, particularly Ag NPs, for both the in vitro and the ex vitro applications against a variety of hazardous insect species. For instance, Jayaseelan et al. (2011) showed that Ag NPs generated using the aqueous extracts of leaves of the plant *T. cordifolia* (Thunb.) Miers induced total death of *P. humanus capitis* De Geer adults when they were exposed to 25 ppm of Ag NPs for 1 hour. The majority of research on insect control has focused on the insecticidal and pupicidal efficacy of NPs with respect to the three vectors, viz., *Anopheles stephensi* Liston (the malaria vector), the arbovirus vectors *Aedes aegypti* as well as *Aedes albopictus* and *C. quinquefasciatus* (the filariasis vector). Since most of the LC_{50} values lie between 1 and 30 ppm, plant-synthesized NPs generally showed promising action against mosquito vectors in their early instar stages. The most resistant to NPs' damaging effects were the larvae and pupae of the *C. quinquefasciatus* species made from plants of all the studied species (Benelli, 2016a). Additionally, NPs produced by plants showed potential ovicide and adulticide action. According to the reports, egg hatchability of the three vectors—*A. stephensi*, *C. quinquefasciatus and A. aegypti*—was reduced by 100% after a short duration of exposure to 30 ppm of Ag NPs made from the plant *S. muticum* (Madhiyazhagan et al., 2015). The outcomes were comparable when *Anopheles subpictus* and *C. quinquefasciatus* mosquito larvae were exposed to a 20 mg/L Ag nanocide formulation for 24 hours. Additionally, *A. subpictus* and *C. quinquefasciatus* died completely as larvae after 24 hours of exposure to Ag NPs created by *Mimosa pudica* L. leaf aqueous extract (Marimuthu et al. 2011). After being applied to water reservoirs for 24, 48, and 72 hours, respectively, first–fourth instar *A. stephensi* larvae were reduced by 74.5%, 86.6%, and 97.7%, as shown by field tests conducted by Dinesh et al. (2015). Similar results were reported by Suresh et al. (2015a) who found that Ag NPs made from the leaf aqueous extract of *Phyllanthus niruri* L. caused 47.6%, 76.7%, and 100% death of *A. aegypti* larvae, respectively, 24, 48, and 72 hours after application.

Regarding mosquito adulticidal toxicity, just a few reports are known. Adults of *A. stephensi*, *A. aegypti*, and *C. quinquefasciatus* exhibited lethal effects to Ag NPs synthesized from *Feronia elephantum* leaf extract, with LD_{50} values ranging from 18.041 to 21.798 µg/mL (Veerakumar & Govindarajan, 2014). The *A. aegypti*, *C. quinquefasciatus*, and *A. stephensi* adults were tested for resistance to silver nanoparticles biosynthesized from the leaf extracts of *Heliotropium indicum*, and it was observed that *A. stephensi* showed the greatest efficacy ($LD_{50} = 26.712$ µg/mL) (Veerakumar et al., 2014). The LC_{50} values for the synthetic silver NPs synthesized from *Mimusops elengi* were 13.7 ppm for *A. stephensi* and 14.7 ppm for *A. albopictus* (Suresh et al., 2015b). The lifespan of *A. aegypti* in both sexes as well as female fertility were observed to be significantly decreased after a single exposure to concentrations of 100–500 ppm of silver nanoparticles produced from the plant *Hypnea musciformis* (Roni et al. 2015).

Houseflies, *Musca domestica*, another frequent pest linked to health problems, when exposed to *Manilkara zapota* aqueous leaf extract, which produces 10 mL/L of Ag NPs, at the adult stage, decreased entirely within 4 hours (Kamaraj et al., 2012). Since all larval instars and pupae of the cotton bollworm (*Helicoverpa armigera*) showed high mortality levels (C80%) after only 4 days of exposure in cotton leaves that had been treated with 10 ppm Ag NPs synthesized by the leaf extract of *Euphorbia hirta* plant, it was concluded that the cotton bollworm was very susceptible to Ag NPs (Devi et al., 2014). Murugan et al. (2015) reported that Ag NPs generated by an aqueous extract of the plant *Caulerpa scalpelliformis* were particularly toxic to first–fourth instar *C. quinquefasciatus* larvae, causing 80% mortality at a

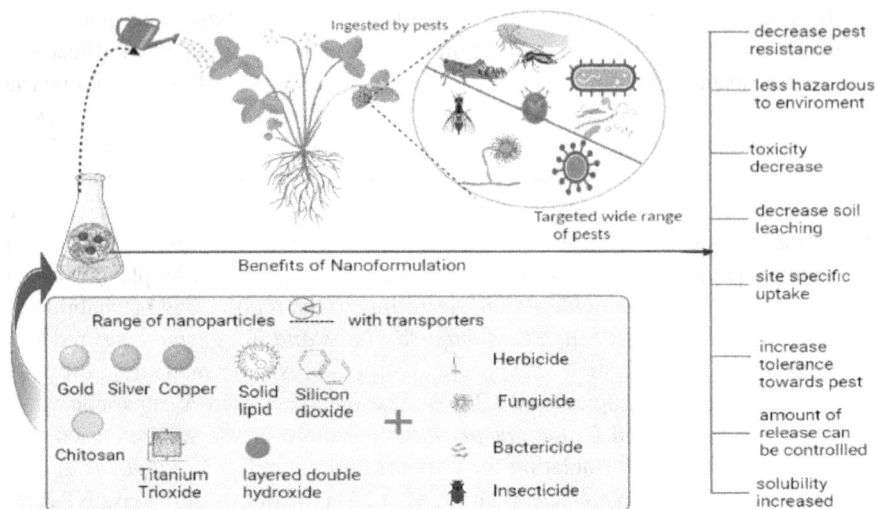

FIGURE 11.2 Application of nanoparticles in pest management—range of NPs and benefits of nano-formulations.

concentration of 10 ppm. The authors of the same paper proposed a unique biological control method against *C. quinquefasciatus* larvae using *C. scalpelliformis* Ag NPs and *Mesocyclops longisetus* (Figure 11.2).

Salunkhe et al. (2011) found that larvae of the second to fourth instars of *A. aegypti* and *A. stephensi* died after being exposed to 5 or 10 ppm of the cylindrical fungus *Cochliobolus lunatus*-produced Ag NPs for 24 hours. *Chrysosporium tropicum* is a different fungus that has been utilized to create Ag NPs and gold nanoparticles (AuNPs), which are extremely poisonous and cause mortality 100% in the second instar after 1 hour and the first instar after 24 hours, respectively (Soni & Prakash, 2012). After 24 hours of exposure, *A. aegypti* first, second, and third-instar larvae or pupae were all fatally affected. After being exposed for 24 hours, *A. aegypti* larvae or pupae in the first, second, third, or fourth instars died in 92%, 96%, or 100% of cases due to Ag NPs produced by the extracellular filtrate of the entomopathogenic fungus *Trichoderma harzianum* (Sundaravadivelan & Padmanabhan, 2014).

11.6.1 NANOHERBICIDES

To date, farmers are more concerned about the activity of weeds since they are severely damaging crop productivity (Muthuvel et al., 2020). Soil nutrients that should be supplied to crops are all consumed by weeds, thereby hindering plant growth and productivity. However, there are several reports citing the competition of *Lolium rigidum Gaudin* and *Avena fatua* with cereal crops for nitrogen, phosphorus, and potassium, leading to deficiency for the same (Lemerle et al., 1995; Angonin et al., 1996). Physical methods for removing weeds take longer and require more labour than biological methods, which are costly, time-consuming, and ineffective. Commercial herbicides used in fields have many advantages, but they also pose significant health hazards. Nanoherbicides are useful for getting rid of weeds since they are less harmful to the environment than traditional herbicides and cause less hazard (Alharby et al., 2019). It was investigated that a polymer named poly(-epsilon-caprolactone), having herbicidal activity, possesses superior physicochemical properties along with increased bioavailability and biocompatibility that was used to encapsulate atrazine to be targeted on *Brassica* sp., which was more effective in comparison to free atrazine (Pereira et al., 2014).

According to Grillo et al. (2014), polymeric poly(epsilon-caprolactone) (PCL) nanocapsule is an effective pathway responsible for the release of triazine herbicides, namely atrazine and simazine which are usually toxic in nature. Since paraquat was successfully encapsulated into chitosan/tripolyphosphate nanoparticles, its herbicidal efficacy was not reduced even after encapsulation (Grillo et al., 2014). Mahil et al. (2020) suggested the use of atrazine which was effective in managing *Striga*. Chidambaram (2016) used a nanoformulation based on rice husks that is responsible for herbicide release for a longer period of time than the commercial one; it can be used as a soil amendment to reduce herbicide leaching.

11.6.2 Nanonematicides

Among agricultural pests, root-knot nematodes (*Meloidogyne* spp.), known for repressing crops, thereby leading to annual crop losses along with reduction in production of fruits and vegetables, are one of the dangerous plant parasitic nematodes known till date (Back et al., 2002, Singh et al., 2015). To date, nematicidal control is based on synthetic organic chemical nematicides (Ntalli et al., 2011). Due to emerging effects on environment, pesticides such as synthetic carbamate and organophthalide pesticides are currently banned. Alternative nematode management methods and green nematicidal chemicals must be produced quickly in order to be less hazardous to non-target organisms, the environment, and the living organisms (Chitwood, 2002). Since nematodes are microscopic in nature, an overall crop loss of $85 billion have been estimated worldwide and annual crop losses in India are reported to be Rs 242.1 billion. Nematodes have the ability to affect agricultural crops all over the globe including tropical and subtropical regions. However, they feed on several plant parts, thereby affecting them for a long period of time.

The role of numerous nanoparticle formulations such as Ag and SiO_2 nanoparticles, in the management of root-knot nematode, has been reported (Ardakani, 2013). MgO nanoparticles have been reported to be an effective antimicrobial agent, yet they have not been experimented against the root-knot nematodes. Due to rapid synthesis as well as preparation from readily available inexpensive precursors and being non-toxic in nature, MgO is usually preferred over TiO_2, Ag, and Cu bactericides (Stoimenov et al., 2002). MgO nanoparticles are usually prepared with tremendous high surface areas (SA) possessing a crystalline structure, many boundaries, and reactive surfaces. However, they have the ability to generate reactive oxygen species, a notable property which emphasizes overall attention on the impending antibacterial treatment (Sawai et al., 1996). There have been very few references made about the antifungal properties of MgO. It has a greater inhibitory impact than nano zinc oxide at a concentration of $50\,\mu g/mL$ *in vitro* germination of conidial spores of *A. alternata, Fusarium oxysporum*, and *R. stolonifer* (Wani & Shah, 2012).

11.7 CONCLUSION

Globally researchers are interested in protecting agricultural products from pest infestations. Synthetic pesticides have been utilized for pest control since ancient times. However, environmental contamination, human health risks, and pesticide degradation have intensified the trend towards the use of natural alternatives to synthetic pesticides. In addition to the advantages of their formulation and delivery, the effective use of nanopesticides for plant protection is a rapidly developing research area that presents novel approaches to designing active ingredients in nanoscale dimensions. To increase the use of nanocomposites in the agri-sector, formulation, administration, and other cutting-edge application technologies can be developed. Additionally, biogenic nanocomposites may replace commercially available chemical insecticides and pesticides in terms of cost, sustainability, and dependability. As a result of such exhaustive research, integrated pest management strategies may increasingly employ nanoparticle-based technologies.

REFERENCES

Abbasi, A., Sufyan, M., Arif, M.J., & Sahi, S.T. (2020). Effect of silicon on tritrophic interaction of cotton, *Gossypium hirsutum* (Linnaeus), *Bemisiatabaci* (Gennadius) (Homoptera: Aleyrodidae) and the predator, *Chrysoperlacarnea* (Stephens) (Neuroptera: Chrysopidae). *Arthropod Plant Interect.* 14, 717–725.

Abdel-Aziz, H.M., Hasaneen, M.N., & Omer, A.M. (2016). Nano chitosan-NPK fertilizer enhances the growth and productivity of wheat plants grown in sandy soil. *Span J Agric Res.* 14(1), e0902.

Abreu, F.O., Oliveira, E.F., Paula, H.C., & de Paula, R.C. (2012). Chitosan/cashew gum nanogels for essential oil encapsulation. *Carb Poly.* 89(4), 1277–1282.

Afzal, S., Sharma, D., & Singh, N.K. (2021). Eco-friendly synthesis of phytochemical-capped iron oxide nanoparticles as nano-priming agent for boosting seed germination in rice (*Oryza sativa* L.). *Environ Sci Poll Res.* 28(30), 40275–40287.

Agri, U., Chaudhary, P., & Sharma, A. (2021). In vitro compatibility evaluation of agriusable nanochitosan on beneficial plant growth-promoting rhizobacteria and maize plant. *Nat Acad Sci Lett.* 44, 555–559.

Agri, U., Chaudhary, P., Sharma, A., & Kukreti, B. (2022). Physiological response of maize plants and its rhizospheric microbiome under the influence of potential bioinoculants and nanochitosan. *Plant Soil.* 474, 451–468.

Alharby, H.F., Hakeem, K.R., & Qureshi, M.I. (2019). Weed control through herbicide-loaded nanoparticles. *Nanomater Plant Potential*, 507–527.

Anbuvannan, M., Ramesh, M., Viruthagiri, G., Shanmugam, N., & Kannadasan, N. (2015). *Anisochilus carnosus* leaf extract mediated synthesis of zinc oxide nanoparticles for antibacterial and photocatalytic activities. *Mater Sci Semicond Process.* 39, 621–628.

Angonin, C., Caussanel, J.P., & Meynard, J.M. (1996). Competition between winter wheat and Veronica hederifolia: influence of weed density and the amount and timing of nitrogen application. *Weed Res.* 36(2), 175–187.

Ardakani, A.S. (2013). Toxicity of silver, titanium and silicon nanoparticles on the root-knot nematode, Meloidogyne incognita, and growth parameters of tomato. *Nematology*, 15(6), 671–677.

Athanassiou, C.G., Kavallieratos, N.G., Evergetis, E., Katsoula, A.M., & Haroutounian, S.A. (2013). Insecticidal efficacy of the enhanced silica gel with *Juniperus oxycedrus* L. ssp. oxycedrus essential oil against *Sitophilus oryzae* (L.) and *Tribolium confusum* Jacquelin du Val. *J Econ Entomol.* 106, 1902–1910.

Ayoub, H.A., Khairy, M., Rashwan, F.A., & Abdel-Hafez, H.F. (2017). Synthesis and characterization of silica nanostructures for cotton leaf worm control. *J Nanostruct Chem.* 7, 91–100.

Azizi, S., Ahmad, M.B., Namvar, F., & Mohamad, R. (2014). Green biosynthesis and characterization of zinc oxide nanoparticles using brown marine macroalga *Sargassum muticum* aqueous extract. *Mater Lett.* 116, 275–277.

Back, M.A., Haydock, P.P.J., Jenkinson, P. (2002). Disease complexes involving plant parasitic nematodes and soil borne pathogens. *Plant Pathol.* 51, 683–697.

Balalakshmi, C., Gopinath, K., Govindarajan, M., Lokesh, R., Arumugam, A., Alharbi, N. S., & Benelli, G. (2017). Green synthesis of gold nanoparticles using a cheap *Sphaeranthus indicus* extract: impact on plant cells and the aquatic crustacean *Artemia nauplii. J Photochem Photobiol B: Biol.* 173, 598–605.

Baptista, P.V., McCusker, M.P., Carvalho, A., Ferreira, D.A., Mohan, N.M., Martins, M., & Fernandes, A.R. (2018). Nano-strategies to fight multidrug resistant bacteria—"A Battle of the Titans". *Front Microbiol.* 9, 1441.

Barik, T.K., Sahu, B. & Swain, V. (2008). Nanosilica- from medicine to pest control. *Parasitol Res.* 103, 253–258.

Bashir, A., Rizwan, M., Ali, S., Adrees, M., & Qayyum, M.F. (2020). Effect of composted organic amendments and zinc oxide nanoparticles on growth and cadmium accumulation by wheat; a life cycle study. *Environ Sci Pollu Res.* 27(19), 23926–23936.

Benelli, G. (2016a). Plant-mediated biosynthesis of nanoparticles as an emerging tool against mosquitoes of medical and veterinary importance: a review. *Parasitol Res.* 115(1), 23–34.

Bhagat, D., Samanta, S.K., & Bhattacharya, S. (2013). Efficient management of fruit pests by pheromone nanogels. *Sci Rep.* 3(1), 1–8.

Bhatt, P., Pandey, S.C., Joshi, S., Chaudhary, P., Pathak, V.M., Huang, Y., Wu, X., Zhou, Z., & Chen, S. (2022). Nanobioremediation: a sustainable approach for the removal of toxic pollutants from the environment. *J Hazard Mater.* 427, 128033.

Bhattacharyya, A., Bhaumik, A., Pathipati, U., Mandel, S., & Epidi, T.T. (2010). Nanoparticles: a recent approach to insect pest control. *Afr J Biotechnol.* 9, 3489–3493.

Bhattacharyya, A., Prasad, R., Buhroo, A.A., Duraisamy, P., Yousuf, I., Umadevi, M., Bindhu, M.R., Govindarajan, M., & Khanday, A.L. (2016). One-pot fabrication and characterization of silver nanoparticles using *Solanum lycopersicum* : an eco-friendly and potent control tool against rose aphid, *Macrosiphum rosae. J Nanosci.* 2016, 4679410.

Bhuyan, T., Mishra, K., Khanuja, M., Prasad, R., & Varma, A. (2015). Biosynthesis of zinc oxide nanoparticles from *Azadirachta indica* for antibacterial and photocatalytic applications. *Mat Sci in Semiconductor Process.* 32, 55–61.

Biswas, P., & Kumar, N. (2021). Application of nanotechnology in crop improvement: an overview. In Pankaj K., Ajay K. T., Eds., *Crop Improvement* (pp. 211–224) CRC Press, Boca Raton.

Boroumand, N., Behbahani, M., & Dini, G. (2020). Combined effects of phosphate solubilizing bacteria and nanosilica on the growth of land cress plant. *J Soil Sci Plant Nutr.* 20(1), 232–243.

Cabanas, R., Abarca, M.L., Bragulat, M.R., & Cabanes, F.J. (2009). In vitro activity of imazalil against *Penicillium expansum*: comparison of the CLSI M38-A broth microdilution method with traditional techniques. *Int J Food Microbiol.* 129(1), 26–29.

Ch, S.R., Raja, A., Nadig, P., Jayaganthan, R., & Vasa, N.J. (2019). Influence of working environment and built orientation on the tensile properties of selective laser melted AlSi10Mg alloy. *Mater Sci Eng.* 750, 141–151.

Chaghervand, M.M., Torbati, M.B., Shaabanzadeh, M., Ahmadi, A., & Tafvizi, F. (2022). Hydroxyurea-loaded Fe$_3$O$_4$/SiO$_2$/chitosan-g-mPEG2000 nanoparticles; pH-dependent drug release and evaluation of cell cycle arrest and altering p53 and lincRNA-p21 genes expression. *Naunyn-Schmiedeberg's Arch Pharmacol.* 395, 51–63.

Chandrashekharaiah, M., Kandakoor, S.B., Basana Gowda, G., Kammar, V., & Chakravarthy, A.K. (2015). Nanomaterials: a review of their action and application in pest management and evaluation of DNA-tagged particles. In Chakravarthy, A. (ed.) *New Horizons in Insect Science: Towards Sustainable Pest Management* (pp. 113–126). Springer, New Delhi.

Chauchan, P., Mishra, M., & Gupta, D. (2014). Potential application of nanoparticles as Antipathogens. *Adv Mat Agr, Food, Environ Safety.* 333–367.

Chaudhary, P., Chaudhary, A., Bhatt, P., Kumar, G., Khatoon, H., Rani, A., Kumar, S., & Sharma, A. (2022). Assessment of soil health indicators under the influence of nano-compounds and *Bacillus* spp. in field condition. *Front Environ Sci.* 9, 769871.

Chaudhary, P., Chaudhary, A., Parveen, H., Rani, A., Kumar, G., Kumar, A., & Sharma, A. (2021e). Impact of nanophos in agriculture to improve functional bacterial community and crop productivity. *BMC Plant Biol.* 21, 519.

Chaudhary, P., Khati, P., Chaudhary, A., Gangola, S., Kumar, R., & Sharma, A. (2021a). Bioinoculation using indigenous *Bacillus* spp. improves growth and yield of *Zea mays* under the influence of nanozeolite. *3 Biotech.* 11, 11.

Chaudhary, P., Khati, P., Chaudhary, A., Maithani, D., Kumar, G., & Sharma, A. (2021d). Cultivable and metagenomic approach to study the combined impact of nanogypsum and *Pseudomonas taiwanensis* on maize plant health and its rhizospheric microbiome. *PLoS One.* 16, e0250574.

Chaudhary, P., Khati, P., Gangola, S., Kumar, A., Kumar, R., & Sharma, A. (2021c). Impact of nanochitosan and *Bacillus* spp. on health, productivity and defence response in *Zea mays* under field condition. *3 Biotech.* 11, 237.

Chaudhary, P., & Sharma, A. (2019). Response of nanogypsum on the performance of plant growth promotory bacteria recovered from nanocompound infested agriculture field. *Environ Ecol.* 37, 363–372.

Chaudhary, P., Sharma, A., Chaudhary, A., Khati, P., Gangola, S., & Maithani, D. (2021b). Illumina based high throughput analysis of microbial diversity of rhizospheric soil of maize infested with nanocompounds and *Bacillus* sp. *Appl Soil Ecol.* 159, 103836.

Chhipa, H. (2019). Applications of nanotechnology in agriculture. In Volker G., Andrew S. B., Sarvesh S., *Methods in Microbiology* (Vol. 46, pp. 115–142). Academic Press.

Chidambaram, R. (2016). Application of rice husk nanosorbents containing 2, 4-dichlorophenoxyacetic acid herbicide to control weeds and reduce leaching from soil. *J Taiwan Inst Chem Eng.* 63, 318–326.

Chitwood, D.J. (2002). Phytochemical based strategies for nematode control. *Ann Rev Phytopathol.* 40(1), 221–249.

Choudhary, M.K., Kataria, J., & Sharma, S. (2018). Evaluation of the kinetic and catalytic properties of biogenically synthesized silver nanoparticles. *J Cleaner Prod.* 198, 882–890.

Croissant, J.G., Butler, K.S., Zink, J.I., & Brinker, C.J. (2020). Synthetic amorphous silica nanoparticles: toxicity, biomedical and environmental implications. *Nat Rev Mat.* 5(12), 886–909.

Daryabeigi Zand, A., Tabrizi, A.M., & Heir, A.V. (2020). The influence of association of plant growth-promoting rhizobacteria and zero-valent iron nanoparticles on removal of antimony from soil by *Trifolium repens*. *Environ Sci Poll Res.* 27(34), 42815–42829.

Davod, T., Reza, Z., Ali, V.A., & Mehrdad, C. (2011). Effects of nanosilver and nitroxin biofertilizer on yield and yield components of potato minitubers. *Int J Agric Biol.* 13(6), 986–990.

De, A., Bose, R., Kumar, A., &Mozumdar, S. (2014). Management of insect pests using nanotechnology: as modern approaches. In De, A., Bose, R., Kumar, A., & Mozumdar, S. (eds.) *Targeted Delivery of Pesticides Using Biodegradable Polymeric Nanoparticles* (pp. 29–33). Springer, India.

Dhawan, A.K., & Peshin, R. (2009). Integrated pest management: concept, opportunities and challenges. In Peshin, P, & Dhawan, A.K. (eds.) *Integrated Pest Management: Innovation-Development Process* (pp. 51–81). Springer, Dordrecht, Netherlands.

Dhekney, S.A., Li, Z.T., Van Aman, M., Dutt, M., Tattersall, J., Kelley, K.T., & Gray, D.J. (2005). Genetic transformation of embryogenic cultures and recovery of transgenic plants in Vitis vinifera, Vitis rotundifolia and Vitis hybrids. In *International Symposium on Biotechnology of Temperate Fruit Crops and Tropical Species 738* (pp. 743–748).

Dimkpa, C.O., McLean, J.E., Britt, D.W., & Anderson, A.J. (2013). Antifungal activity of ZnO nanoparticles and their interactive effect with a biocontrol bacterium on growth antagonism of the plant pathogen *Fusarium graminearum. Biometals.* 26(6), 913–924.

Dinesh, D., Murugan, K., Madhiyazhagan, P., Panneerselvam, C., Mahesh Kumar, P., Nicoletti, M., & Suresh, U. (2015). Mosquitocidal and antibacterial activity of green-synthesized silver nanoparticles from Aloe vera extracts: towards an effective tool against the malaria vector *Anopheles stephensi? Parasitol Res.* 114(4), 1519–1529.

Du, L., Miao, X., Jiang, Y., Jia, H., Tian, Q., Shen, J., & Liu, Y. (2012). An effective strategy for the synthesis of biocompatible gold nanoparticles using danshensu antioxidant: prevention of cytotoxicity via attenuation of free radical formation. *Nanotoxicology.* 7(3), 294–300.

Duhan, J.S., Kumar, R., Kumar, N., Kaur, P., Nehra, K., & Duhan, S. (2017). Nanotechnology: the new perspective in precision agriculture. *Biotech Rep.* 15, 11–23.

Dutt, M., Li, Z.T., Dhekney, S.A., & Gray, D.J. (2007). Transgenic plants from shoot apical meristems of *Vitis vinifera* L. "Thompson Seedless" via *Agrobacterium*-mediated transformation. *Plant Cell Rep.* 26(12), 2101–2110.

Elek, N., Hoffman, R., Raviv, U., Resh, R, Ishaaya, I., & Magdassi, S. (2010). Novaluron nanoparticles: formation and potential use in controlling agricultural insect pests. *Colloids Surf A Physicochem Eng Asp.* 372(1–3), 66–72.

Elumalai, K., Velmurugan, S., Ravi, S., Kathiravan, V., & Kumar, A.S. (2015). RETRACTED: green synthesis of zinc oxide nanoparticles using *Moringa oleifera* leaf extract and evaluation of its antimicrobial activity. *Spectrochim Acta A Mol Biomol Spectrosc* 143, 158–164.

Fu, Y.Q., Li, L.H., Wang, P.W., Qu, J., Fu, Y.P., Wang, H., & Lü, C.L. (2012). Delivering DNA into plant cell by gene carriers of ZnS nanoparticles. *Chem Res Chin Univ.* 28(4), 672–676.

Fukamachi, K., Konishi, Y., & Nomura, T. (2019). Disease control of Phytophthora infestans using cyazofamid encapsulated in poly lactic-co-glycolic acid (PLGA) nanoparticles. *Colloids Surf A: Physicochem Eng Asp.* 577, 315–322.

Gajbhiye, M., Kesharwani, J., Ingle, A., Gade, A., & Rai, M. (2009). Fungus mediated synthesis of silver nanoparticles and its activity against pathogenic fungi in combination of fluconazole. *Nanomedicine.* 5, 382–386.

Gao, C., Lyu, F., & Yin, Y. (2020). Encapsulated metal nanoparticles for catalysis. *Chem Rev.* 121(2), 834–881.

Ghareeb, R.Y., Shams El-Din, N.G.E.D., Maghraby, D.M.E., Ibrahim, D.S., Abdel-Megeed, A., & Abdelsalam, N.R. (2022). Nematicidal activity of seaweed-synthesized silver nanoparticles and extracts against *Meloidogyne incognita* on tomato plants. *Sci Rep.* 12(1), 1–16.

Gnanajobitha, G., Paulkumar, K., Vanaja, M., Rajeshkumar, S., Malarkodi, C., Annadurai, G., & Kannan, C. (2013a). Fruit-mediated synthesis of silver nanoparticles using *Vitis vinifera* and evaluation of their antimicrobial efficacy. *J Nanostructure Chem.* 3(1), 1–6.

Gnanajobitha, G., Rajeshkumar, S., Kannan, C., & Annadurai, G. (2013b). Preparation and characterization of fruit-mediated silver nanoparticles using pomegranate extract and assessment of its antimicrobial activity. *J Environ Nanotechnol.* 2(1), 04–10.

Gogos, A., Knauer, K., & Bucheli, T.D. (2012). Nanomaterials in plant protection and fertilization: current state, foreseen applications, and research priorities. *J Agric Food Chem.* 60(39), 9781–9792.

Goswami, A., Roy, I., Sengupta, S., & Debnath, N. (2010). Novel applications of solid and liquid formulations of nanoparticles against insect pests and pathogens. *Thin Solid Films.* 519, 1252–1257.

Grillo, R., Pereira, A.E., Nishisaka, C.S., De Lima, R., Oehlke, K., Greiner, R., & Fraceto, L.F. (2014). Chitosan/tripolyphosphate nanoparticles loaded with paraquat herbicide: an environmentally safer alternative for weed control. *J Hazard Mater.* 278, 163–171.

Hazrati, H., Saharkhiz, M.J., Niakousari, M., & Moein, M. (2017). Natural herbicide activity of *Satureja hortensis* L. essential oil nanoemulsion on the seed germination and morphophysiological features of two important weed species. *Ecotoxicol Environ Saf.* 142, 423–430.

He, L., Liu, Y., Mustapha, A., & Lin, M. (2011). Antifungal activity of zinc oxide nanoparticles against *Botrytis cinerea* and *Penicillium expansum*. *Microbiol Res.* 166(3), 207–215.

Huang, B., Chen, F., Shen, Y., Qian, K., Wang, Y., Sun, C., & Cui, H. (2018). Advances in targeted pesticides with environmentally responsive controlled release by nanotechnology. *Nanomaterials.* 8(2), 102.

Ibrahim, E., Fouad, H., Zhang, M., Zhang, Y., Qiu, W., Yan, C., & Chen, J. (2019). Biosynthesis of silver nanoparticles using endophytic bacteria and their role in inhibition of rice pathogenic bacteria and plant growth promotion. *RSC Adv.* 9(50), 29293–29299.

Jayaseelan, C., Rahuman, A.A., Rajakumar, G., Kirthi, V.A., Santhoshkumar, T., Marimuthu, S., & Elango, G. (2011). Synthesis of pediculocidal and larvicidal silver nanoparticles by leaf extract from heartleaf moonseed plant, *Tinospora cordifolia* Miers. *Parasitol Res.* 109(1), 185–194.

Jayaseelan, C., Rahuman, A.A., Rajakumar, G., Santhoshkumar, T., Kirthi, A. V., Marimuthu, S., & Raveendran, S. (2012). Efficacy of plant-mediated synthesized silver nanoparticles against hematophagous parasites. *Parasitol Res.* 111(2), 921–933.

Jerobin, J., Sureshkumar, R. S., Anjali, C. H., Mukherjee, A., & Chandrasekaran, N. (2012). Biodegradable polymer based encapsulation of neem oil nanoemulsion for controlled release of Aza-A. *Carb Poly.* 90(4), 1750–1756.

Jordan, W. (2010). Nanotechnology and pesticides. Pesticide Program Dialogue Committee, April 29, 2010.

Kamaraj, C., Rajakumar, G., Rahuman, A. A., Velayutham, K., Bagavan, A., Zahir, A. A., & Elango, G. (2012). Feeding deterrent activity of synthesized silver nanoparticles using Manilkara zapota leaf extract against the house fly, *Musca domestica* (Diptera: Muscidae). *Parasitol Res.* 111(6), 2439–2448.

Kashyap, P.L., Kumar, S., Jasrotia, P., Singh, D.P., & Singh, G.P. (2020). Nanotechnology in wheat production and protection. In Nandita D., Shivendu R., Eric L., Eds., *Environmental Nanotechnology Volume 4* (pp. 165–194). Springer, Cham.

Khati, P., Bhatt, P., Kumar, R., & Sharma, A. (2018). Effect of nanozeolite and plant growth promoting rhizobacteria on maize. *3Biotech.* 8(141), 1–12.

Khati, P., Chaudhary, P., Gangola, S., & Sharma A. (2019a). Influence of nanozeolite on plant growth promotory bacterial isolates recovered from nanocompound infested agriculture field. *Environ Ecol.* 37 (2), 521–527.

Khati, P., Gangola, S., Bhatt, P., & Sharma A. (2017). Nanochitosan induced growth of *Zea Mays* with soil health maintenance. *3 Biotech.* 7(81), 1–9.

Khati, P., Sharma A., Chaudhary, P., Singh, A.K., Gangola, S., & Kumar, R. (2019b). High-throughput sequencing approach to access the impact of nanozeolite treatment on species richness and evenness of soil metagenome. *Biocat Agric Biotechnol.* 20, 101249.

Khezerlou, A., Alizadeh-Sani, M., Azizi-Lalabadi, M., &Ehsani, A. (2018). Nanoparticles and their antimicrobial properties against pathogens including bacteria, fungi, parasites and viruses. *Microbial Path.* 123, 505–526.

Khot, L.R., Sankaran, S., Maja, J.M., Ehsani, R., & Schuster, E.W. (2012). Applications of nanomaterials in agricultural production and crop protection: a review. *Crop Prot.* 35, 64–70.

Knowles, A. (2009). Global trends in pesticide formulation technology: the development of safer formulations in China. *Outlooks Pest Manag.* 20(4), 165–170.

Kukreti, B., Sharma, A., Chaudhary, P., Agri, U., & Maithani, D. (2020). Influence of nanosilicon dioxide along with bioinoculants on *Zea mays* and its rhizospheric soil. *3 Biotech.* 10, 345.

Kumar, D.R., Kumar, P.S., Gandhi, M.R., Al-Dhabi, N.A., Paulraj, M.G., & Ignacimuthu, S. (2016a). Delivery of chitosan/dsRNA nanoparticles for silencing of wing development vestigial (vg) gene in *Aedes aegypti* mosquitoes. *Int J Biol Macromol.* 86, 89–95.

Kumar, P.M., Murugan, K., Madhiyazhagan, P., Kovendan, K., Amerasan, D., Chandramohan, B., & Benelli, G. (2016b). Biosynthesis, characterization, and acute toxicity of *Berberis tinctoria*-fabricated silver nanoparticles against the Asian tiger mosquito, Aedes albopictus, and the mosquito predators *Toxorhynchites splendens* and *Mesocyclops thermocyclopoides. Parasitol Res.* 115(2), 751–759.

Kumari, H., Khati, P., Gangola, S., Chaudhary, P., & Sharma A. (2021). Performance of plant growth promotory rhizobacteria on maize and soil characteristics under the influence of TiO$_2$ nanoparticles. *Pant Res J.* 19, 28–39.

Kumari, S., Sharma A., Chaudhary, P., & Khati, P. (2020). Management of plant vigour and soil health using two agriusable nanocompounds and plant growth promontory rhizobacteria in Fenugreek. *3 Biotech.* 10(461), 1–11.

Lade, B.D., Gogle, D.P., Lade, D.B., Moon, G.M., Nandeshwar, S.B., & Kumbhare, S.D. (2019). Nanobiopesticide formulations: Application strategies today and future perspectives. In Opender K., Ed., *Nano-Biopesticides Today and Future Perspectives* (pp. 179–206). Academic Press.

Lai, F., Wissing, S.A., Müller, R.H., & Fadda, A.M. (2006). Artemisia arborescens L essential oil-loaded solid lipid nanoparticles for potential agricultural application: preparation and characterization. *Aaps Pharmscit.* 7(1), E10–E18.

Lee, J.H., Velmurugan, P., Park, J.H., Murugan, K., Lovanh, N., Park, Y.J., & Benelli, G. (2018). A novel photo-biological engineering method for *Salvia miltiorrhiza*-mediated fabrication of silver nanoparticles using LED lights sources and its effectiveness against *Aedes aegypti* mosquito larvae and microbial pathogens. *Physiol Mol Plant Pathol.* 101, 178–186.

Lemerle, D., Verbeek, B., & Coombes, N. (1995). Losses in grain yield of winter crops from Lolium rigidum competition depend on crop species, cultivar and season. *Weed Res.* 35(6), 503–509.

Madhiyazhagan, P., Murugan, K., Kumar, A.N., Nataraj, T., Dinesh, D., Panneerselvam, C., & Benelli, G. (2015). *S argassum* muticum-synthesized silver nanoparticles: an effective control tool against mosquito vectors and bacterial pathogens. *Parasitology Res.* 114(11), 4305–4317.

Mahil, E.I.T., Kumar, B.A., Babu, R., & Jones, P. (2020). Response of nanoherbicide application on striga management in sugarcane. *Int J Curr Microbiol App Sci*, 9(12), 3349–3357.

Malarkodi, C., Chitra, K., Rajeshkumar, S., Paulkumar, K., Jobitha, G.G., Vanaja, M., & Annadurai, G. (2013a). Novel eco-friendly synthesis of titanium oxide nanoparticles by using *Planomicrobium* sp., and evaluation of its antimicrobial activity. *Der Pharmacia.* 4 (3), 59–66.

Malarkodi, C., Rajeshkumar, S., Paulkumar, K., Jobitha, G.G., Vanaja, M., & Annadurai, G. (2013b). Biosynthesis of semiconductor nanoparticles by using sulfur reducing bacteria *Serratia nematodiphila. Adv Nano Res.* 1(2), 83.

Malarkodi, C., Rajeshkumar, S., Paulkumar, K., Vanaja, M., Gnanajobitha, G., & Annadurai, G. (2014). Biosynthesis and antimicrobial activity of semiconductor nanoparticles against oral pathogens. *Bioinorg Chem Appl.* 2014, 347167.

Malarkodi, C., Rajeshkumar, S., Paulkumar, K., Vanaja, M., Jobitha, G.D.G., & Annadurai, G. (2013c). Bactericidal activity of bio mediated silver nanoparticles synthesized by *Serratia nematodiphila. Drug Invent Today.* 5(2), 119–125.

Margulis-Goshen, K., & Magdassi, S. (2012). Organic nanoparticles from microemulsions: formation and applications. *Curr Opin Colloid Interface Sci.* 17(5), 290–296.

Marimuthu, S., Rahuman, A.A., Rajakumar, G., Santhoshkumar, T., Kirthi, A.V., Jayaseelan, C., & Kamaraj, C. (2011). Evaluation of green synthesized silver nanoparticles against parasites. *Parasitol. Res.* 108, 1541–1549.

Moustafa, S.M., & Taha, R.H. (2021). Mycogenic nano-complex for plant growth promotion and bio-control of *Pythium aphanidermatum. Plants.* 10(9), 1858.

Mukhopadhyay, R., & De, N. (2014). Nano clay polymer composite: synthesis, characterization, properties and application in rainfed agriculture. *Global J Bio Biotechnol.* 3(2), 133–138.

Müller, R.H., & Junghanns, J.U. (2006). Drug nanocrystals/nanosuspensions for the delivery of poorly soluble drugs. *Nanopart Drug Carr.* 1, 307–328.

Muller, R.H., & Keck, C.M. (2004). Challenges and solutions for the delivery of biotech drugs–a review of drug nanocrystal technology and lipid nanoparticles. *J. Biotechnol,* 113(1–3), 151–170.

Mulvey, M.R., & Simor, A.E. (2009). Antimicrobial resistance in hospitals: how concerned should we be? *C MAJ.* 180(4), 408–415.

Murugan, K., Dinesh, D., Kumar, P.J., Panneerselvam, C., Subramaniam, J., Madhiyazhagan, P., & Benelli, G. (2015). Datura metel-synthesized silver nanoparticles magnify predation of dragonfly nymphs against the malaria vector *Anopheles stephensi. Parasitology Res.* 114(12), 4645–4654.

Murugan, K., Panneerselvam, C., Samidoss, C.M., Madhiyazhagan, P., Suresh, U., Roni, M., & Benelli, G. (2016). In vivo and in vitro effectiveness of *Azadirachta indica*-synthesized silver nanocrystals against *Plasmodium berghei* and *Plasmodium falciparum*, and their potential against malaria mosquitoes. *Res Vet Sci.* 106, 14–22.

Muthuvel, A., Jothibas, M., Mohana, V., & Manoharan, C. (2020). Green synthesis of cerium oxide nanoparticles using Calotropis procera flower extract and their photocatalytic degradation and antibacterial activity. *Inorg Chem Commun.* 119, 108086.

Nakkala, J.R., Mata, R., Gupta, A.K., & Sadras, S.R. (2014). Biological activities of green silver nanoparticles synthesized with *Acorous calamus* rhizome extract. *Eur J Med Chem.* 85, 784–794.

Ntalli, N.G., Ferrari, F., Giannakou, I., & Menkissoglu-Spiroudi, U. (2011). Synergistic and antagonistic interactions of terpenes against Meloidogyne incognita and the nematicidal activity of essential oils from seven plants indigenous to Greece. *Pest Manag Sci,* 67(3), 341–351.

Palmqvist, N.G.M., Bejai, S., Meijer, J., Seisenbaeva, G.A., & Kessler, V.G. (2015). Nano titania aided clustering and adhesion of beneficial bacteria to plant roots to enhance crop growth and stress management. *Sci Rep.* 5(1), 1–12.

Pascoli, M., de Albuquerque, F.P., Calzavara, A.K., Tinoco-Nunes, B., Oliveira, W.H.C., Gonçalves, K.C., & Fraceto, L.F. (2020). The potential of nanobiopesticide based on zein nanoparticles and neem oil for enhanced control of agricultural pests. *J Pest Sci.* 93(2), 793–806.

Paulkumar, K., Rajeshkumar, S., Gnanajobitha, G., Vanaja, M., Malarkodi, C., & Annadurai, G. (2013). Biosynthesis of silver chloride nanoparticles using *Bacillus subtilis* MTCC 3053 and assessment of its antifungal activity. *Int Sch Res Notices.* 2013, 317963.

Pavunraj, M., Baskar, K., Duraipandiyan, V., Al-Dhabi, N.A., Rajendran, V., & Benelli, G. (2017). Toxicity of Ag nanoparticles synthesized using stearic acid from *Catharanthusroseus* leaf extract against Earias vittella and mosquito vectors (*Culex quinquefasciatus* and *Aedes aegypti*). *J Cluster Sci.* 28(5), 2477–2492.

Pereira, A.E., Grillo, R., Mello, N.F., Rosa, A.H., & Fraceto, L.F. (2014). Application of poly (epsilon-caprolactone) nanoparticles containing atrazine herbicide as an alternative technique to control weeds and reduce damage to the environment. *J Hazard Mat.* 268, 207–215.

Perfect, S.E., Hughes, H.B., O'Connell, R.J., & Green, J.R. (1999). *Colletotrichum*: a model genus for studies on pathology and fungal–plant interactions. *Fungal Genet Biol.* 27(2–3), 186–198.

Prasad, R., Bhattacharyya, A., & Nguyen, Q.D. (2017). Nanotechnology in sustainable agriculture: recent developments, challenges, and perspectives. *Front Microbiol.* 8, 1014.

Priyam, A., Das, R.K., Schultz, A., & Singh, P.P. (2019). A new method for biological synthesis of agriculturally relevant nanohydroxyapatite with elucidated effects on soil bacteria. *Sci Rep.* 9(1), 1–14.

Rajeshkumar, S., Malarkodi, C., Paulkumar, K., Vanaja, M., Gnanajobitha, G., & Annadurai, G. (2014a). Algae mediated green fabrication of silver nanoparticles and examination of its antifungal activity against clinical pathogens. *Int J Metals.* 2014, 692643.

Rajeshkumar, S., Malarkodi, C., Vanaja, M., Gnanajobitha, G., Paulkumar, K., Kannan, C., & Annadurai, G. (2013). Antibacterial activity of algae mediated synthesis of gold nanoparticles from *Turbinaria conoides*. *Der Pharma Chem.* 5(2), 224–229.

Rajeshkumar, S., Ponnanikajamideen, M., Malarkodi, C., Malini, M., & Annadurai, G. (2014b). Microbe-mediated synthesis of antimicrobial semiconductor nanoparticles by marine bacteria. *J Nanostin Chem.* 4(2), 1–7.

Rajiv, P., Rajeshwari, S., & Venckatesh, R. (2013). Bio-Fabrication of zinc oxide nanoparticles using leaf extract of *Parthenium hysterophorus* L. and its size-dependent antifungal activity against plant fungal pathogens. *Spectrochim Acta A Mol Biomol Spectrosc.* 112, 384–387.

Rani, P., Mishra, A.R., Deveci, M., & Antucheviciene, J. (2022). New complex proportional assessment approach using Einstein aggregation operators and improved score function for interval-valued Fermatean fuzzy sets. *Comput Ind Eng.* 169, 108165.

Ravishankar Rai, V., & Jamuna B.A. (2011). Nanoparticles and their potential application as antimicrobials, science against microbial pathogens: communicating current research and technological advances. In Méndez-Vilas, A., Ed., *Formatex*, Microbiology Series. No. 3 (Vol. 1, pp. 197–209). Spain.

Resham, S.M., Khalid, M., & Kazi, A.G. (2015). Nanobiotechnology in agricultural development. In Barh, D., Khan, M.S., & Davies, E. (eds.) *PlantOmics: The Omics of Plant Science* (pp. 683–698). Springer, India.

Roni, M., Murugan, K., Panneerselvam, C., Subramaniam, J., Nicoletti, M., Madhiyazhagan, P., & Benelli, G. (2015). Characterization and biotoxicity of *Hypnea musciformis*-synthesized silver nanoparticles as potential eco-friendly control tool against *Aedes aegypti* and *Plutella xylostella*. *Ecotoxicol Environ Saf.* 121, 31–38.

Sahayaraj, K. (2014). Novel biosilver nanoparticles and their biological utility and overview. *Int J Pharm.* 4, 26–39.

Salunkhe, R.B., Patil, S.V., Patil, C.D., & Salunke, B.K. (2011). Larvicidal potential of silver nanoparticles synthesized using fungus *Cochliobolus lunatus* against *Aedes aegypti*
(Linnaeus, 1762) and *Anopheles stephensi* Liston (Diptera; Culicidae). *Parasitol Res.*
109(3), 823–831.

Sasson, Y., Levy-Ruso, G., Toledano, O., & Ishaaya, I. (2007). Nanosuspensions: emerging
novel agrochemical formulations. In *Insecticides Design Using Advanced Technologies*
(pp. 1–39). Springer, Berlin, Heidelberg.

Sawai, J., Kawada, E., Kanou, F., Igarashi, H., Hashimoto, A., Kokugan, T., & Shimizu, M.
(1996). Detection of active oxygen generated from ceramic powders having antibacterial
activity. *J Chem Eng Japan*, 29(4), 627–633.

Servin, A.D., Morales, M.I., Castillo-Michel, H., Hernandez-Viezcas, J.A., Munoz, B., Zhao,
L., & Gardea-Torresdey, J.L. (2013). Synchrotron verification of TiO_2 accumulation in
cucumber fruit: a possible pathway of TiO_2 nanoparticle transfer from soil into the food
chain. *Environ Sci Technol.* 47(20), 11592–11598.

Sharma, S., Singh, S., Ganguli, A.K., & Shanmugam, V. (2017). Anti-drift nano-stickers made
of graphene oxide for targeted pesticide delivery and crop pest control. *Carbon.* 115,
781–790.

Sheykhbaglou, R., Sedghi, M., Shishevan, M.T., & Sharifi, R.S. (2010). Effects of nano-iron
oxide particles on agronomic traits of soybean. *Not Sci Biol.* 2(2), 112–113.

Singh, J.S., Koushal, S., Kumar, A., Vimal, S.R., & Gupta, V.K. (2016). Book review: microbial inoculants in sustainable agricultural productivity-Vol. II: functional application.
Front Microbiol. 7, 2105.

Singh, P., Kumari, K., Vishvakarma, V. K., Aggarwal, S., Chandra, R., & Yadav, A. (2018).
Nanotechnology and its impact on insects in agriculture. In Kumar, D., Gong, C. (eds.)
Trends in Insect Molecular Biology and Biotechnology (pp. 353–378). Springer, Cham.

Singh, S., Singh, B., & Singh, A.P. (2015). Nematodes: a threat to sustainability of agriculture.
Proced Environ Sci. 29, 215–216.

Slomberg, D.L., & Schoenfisch, M.H. (2012). Silica nanoparticle phytotoxicity to Arabidopsis
thaliana. *Environ Sci Technol.* 46(18), 10247–10254.

Song, S., Liu, X., Jiang, J., Qian, Y., Zhang, N., & Wu, Q. (2009). Stability of triazophos
in self-nanoemulsifying pesticide delivery system. *Colloids Surf A Physicochem Eng
Asp.* 350(1–3), 57–62.

Soni, N., & Prakash, S. (2012). Efficacy of fungus mediated silver and gold nanoparticles
against *Aedes aegypti* larvae. *Parasitol Res.* 110(1), 175–184.

Sousa, G.F., Gomes, D.G., Campos, E.V., Oliveira, J.L., Fraceto, L.F., Stolf-Moreira, R., &
Oliveira, H.C. (2018). Post-emergence herbicidal activity of nanoatrazine against susceptible weeds. *Front Environ Sci.* 6, 12.

Spadaro, D., Garibaldi, A., &Gullino, M.L. (2004). Control of *Penicillium expansum* and
Botrytis cinerea on apple combining a biocontrol agent with hot water dipping and
acibenzolar-S-methyl, baking soda, or ethanol application. *Postharvest Biol Technol.*
33(2), 141–151.

Stadler, T., Buteler, M., & Weaver, D.K. (2010). Novel use of nanostructured alumina as an
insecticide. *Pest Manag Sci.* 66(6), 577–579.

Stoimenov, P.K., Klinger, R.L., Marchin, G.L., Klabunde, K.J. (2002). Metal oxide nanoparticles as bactericidal agents. *Langmuir.* 18, 6679–6686.

Strom, R., Price, D., & Lubetkin, S. (2001). U.S. Patent Application No. 09/865,360.

Sundaravadivelan, C., & Padmanabhan, M.N. (2014). Effect of mycosynthesized silver
nanoparticles from filtrate of *Trichoderma harzianum* against larvae and pupa of dengue
vector *Aedes aegypti* L. *Environ Sci Poll Res.* 21(6), 4624–4633.

Suresh, D., Nethravathi, P.C., Kumar, M.P., Naika, H.R., Nagabhushana, H., & Sharma, S.C. (2015a). Chironji mediated facile green synthesis of ZnO nanoparticles and their photoluminescence, photodegradative, antimicrobial and antioxidant activities. *Mater Sci Semicond Process*. 40, 759–765.

Suresh, U., Murugan, K., Benelli, G., Nicoletti, M., Barnard, D.R., Panneerselvam, C., & Chandramohan, B. (2015b). Tackling the growing threat of dengue: Phyllanthus niruri-mediated synthesis of silver nanoparticles and their mosquitocidal properties against the dengue vector *Aedes aegypti* (Diptera: Culicidae). *Parasitol Res.*, 114, 1551–1562.

Tomlin, C.D. (2009). *The pesticide manual: a world compendium* (No. Ed. 15). British Crop Production Council.

Tsuji, K. (2001). Microencapsulation of pesticides and their improved handling safety. *J Microencapsul*. 18(2), 137–147.

Tyagi, J., Chaudhary, P., Mishra, A., Khatwani, M., Dey, S., & Varma, A. (2022). Role of endophytes in abiotic stress tolerance: with special emphasis on *Serendipita indica*. *Int J Environ Res*. 16, 62.

Ul Haq, I., & Ijaz, S. (2019). Use of metallic nanoparticles and nanoformulations as nano-fungicides for sustainable disease management in plants. *Nanobiotech Bioformulat*. 289–316, 198364985.

Van Nhan, L., Ma, C., Rui, Y., Cao, W., Deng, Y., Liu, L., & Xing, B. (2016). The effects of Fe_2O_3 nanoparticles on physiology and insecticide activity in non-transgenic and Bt-transgenic cotton. *Front Plant Sci*. 6, 1263.

Vanaja, M., Paulkumar, K., Gnanajobitha, G., Rajeshkumar, S., Malarkodi, C., & Annadurai, G. (2014). Herbal plant synthesis of antibacterial silver nanoparticles by *Solanum trilobatum* and its characterization. *Int J Met*. 2014, 692461.

Veerakumar, K., & Govindarajan, M. (2014). Adulticidal properties of synthesized silver nanoparticles using leaf extracts of *Feronia elephantum* (Rutaceae) against filariasis, malaria, and dengue vector mosquitoes. *Parasitol Res*. 113(11), 4085–4096.

Veerakumar, K., Govindarajan, M., & Hoti, S.L. (2014). Evaluation of plant-mediated synthesized silver nanoparticles against vector mosquitoes. *Parasitol Res*. 113(12), 4567–4577.

Velayutham, K., Rahuman, A.A., Rajakumar, G., Roopan, S. M., Elango, G., Kamaraj, C., & Siva, C. (2013). Larvicidal activity of green synthesized silver nanoparticles using bark aqueous extract of *Ficus racemosa* against *Culex quinquefasciatus* and *Culex gelidus*. *Asian Pacific J Trop Med*. 6(2), 95–101.

Wani, A.H., & Shah, M.A. (2012). A unique and profound effect of MgO and ZnO nanoparticles on some plant pathogenic fungi. *J Appl Pharmaceut Sci*. 2(3), 40–44.

Watson, S. B., Gergely, A., & Janus, E. R. (2011). Where is agronanotechnolgoy heading in the United States and European Union. *Nat Resour Environ*. 26, 8.

Xu, J., Fan, Q.J., Yin, Z.Q., Li, X.T., Du, Y.H., Jia, R.Y., & Shi, D.X. (2010). The preparation of neem oil microemulsion (*Azadirachta indica*) and the comparison of acaricidal time between neem oil microemulsion and other formulations in vitro. *Vet Parasitol*. 169(3–4), 399–403.

Yang, A.H., Moore, S.D., Schmidt, B.S., Klug, M., Lipson, M., & Erickson, D. (2009a). Optical manipulation of nanoparticles and biomolecules in sub-wavelength slot waveguides. *Nature*. 457(7225), 71–75.

Yang, F.L., Li, X.G., Zhu, F., & Lei, C.L. (2009b). Structural characterization of nanoparticles loaded with garlic essential oil and their insecticidal activity against *Tribolium castaneum* (Herbst)(*Coleoptera: Tenebrionidae*). *J Agric Food Chem*. 57(21), 10156–10162.

Yasui, J., & Rani, P.U. (2013). Environmental effects of nanosilver: impact on castor seed germination, seedling growth, and plant physiology. *Environ Sci Poll Res*. 20(12), 8636–8648.

Yasur, J., & Rani, P.U. (2015). Lepidopteran insect susceptibility to silver nanoparticles and measurement of changes in their growth, development and physiology. *Chemosphere.* 124, 92–102.

Yin, Y.H., Guo, Q.M., Yun, H.A.N., Wang, L.J., & Wan, S.Q. (2012). Preparation, characterization and nematicidal activity of lansiumamide B nano-capsules. *J Integr Agric.* *11*(7), 1151–1158.

Yoon, K.Y., Byeon, J.H., Park, J.H., & Hwang, J. (2007). Susceptibility constants of Escherichia coli and Bacillus subtilis to silver and copper nanoparticles. *Sci Total Environ.* 373(2–3), 572–575.

Zahir, A.A., Bagavan, A., Kamaraj, C., Elango, G., & Rahuman, A.A. (2012). Efficacy of plant-mediated synthesized silver nanoparticles against *Sitophilus oryzae.* *J Biopest.* 5, 95.

Zand, A.D., Tabrizi, A.M., & Heir, A.V. (2020). Co-application of biochar and titanium dioxide nanoparticles to promote remediation of antimony from soil by Sorghum bicolor: metal uptake and plant response. *Heliyon.* 6(8), e04669.

Zhang, H., Wang, D., Butler, R., Campbell, N. L., Long, J., Tan, B., & Rannard, S.P. (2008). Formation and enhanced biocidal activity of water-dispersable organic nanoparticles. *Nat Nanotechnol.* 3(8), 506–511.

12 Impact of Nanoparticles on Abiotic Stress Tolerance

*Geeta Bhandari, Shalu Chaudhary,
and Sanjay Gupta*
Swami Rama Himalayan University

Saurabh Gangola
Graphic Era Hill University

CONTENTS

12.1   INTRODUCTION

The population across the globe is expanding rapidly and is predicted to surpass 9.1 billion by 2050; however, agrarian productivity is hardly increasing at a comparable speed (Das & Das, 2019). Increasing agricultural production worldwide is a huge challenge, as arable land is expected to stay stable or may decline due to the enhancing restriction on land for other non-agricultural purposes. An estimate from the Food and Agriculture Organization (FAO, 2017) shows that a 70% increase in agricultural productivity by 2050 will be required in developing countries to satisfy the intense food requirement of this expanding overpopulation. However, crop productivity is limited by several factors, including biotic and abiotic stress. The environmental factors which limit plant development, viability and fertility are called abiotic stress. Abiotic stress is regarded as a prime issue in agriculture. Several kinds of abiotic stress such as water, salt, temperature, nutrient and metal stress affect plant growth and production (Tyagi et al., 2022). Additionally, environmental issues of climate change and global warming also add up to reduced crop productivity and thus abiotic stresses are regarded as a major issue for the maintenance of agricultural production. According to an estimate, about 70% of crop production is affected by abiotic stresses across the globe. Salt, water, temperature and heavy metal stresses are the most significant kinds of abiotic stresses that cause huge losses in agricultural productivity and thus threaten global food security (Tripathi et al., 2015; Shome et al., 2015; Nahar et al., 2016; Singh et al., 2017). Abiotic stresses cause severe detrimental impact on biochemical, physiological and morphological characters of the plant and minimize crop production. Therefore, there is an urgent need of technologies and genetic reserves for developing stress-tolerant crop varieties and effective methods of nutrient management and irrigation for increasing food production that satisfies the rapidly expanding population.

Agricultural production will increase globally with the production of stress-tolerant crops that can ensure food security and address the issue of growing food demand. Conventional plant breeding strategies such as interspecific or intergeneric hybridization have been found to be capable of enhancing the stress tolerability in plant varieties by a few authors (Das & Das, 2019). All conventional breeding strategies such as screening, cross-breeding, polyploidism and mutation are used for improving the characters of cultivated crops. The traditional breeding strategies are, however, restricted by the complex nature of stress-tolerant characters, lower genetic variations of developed varieties during unfavourable environment and the deficiency of effective screening methods. Therefore, it is significant to search other methods for developing stress-resistant plant varieties. Despite additional successes in the history of crop development in agricultural practices, currently yields have come to a quiescent period and there is food scarcity and deficiency in several economically weaker nations. Thus, there is a need to undertake the issue of appropriate environmental management and increase in agricultural productivity under stress conditions; the discovery of new, environmentally friendly, cost-effective, advance cultivation tools; and sustainable methods to replace current conventional method. In such cases, these approaches should be based on enhancing the adaptive ability of the plants (Kang et al., 2009).

Nanotechnology is a highly promising technology among the recent technologies and can be an emerging remedy for dealing with abiotic stresses in this era (Hatami et al., 2016). Nanotechnology incorporates employment of materials, systems and procedures operating at the nanoscale (<100 nm). It consists of nanoscale components possessing better reactivity, greater surface area, malleable pore size, variable morphology, higher surface dynamism and surface area-to-volume ratio which enhances their reactiveness and chemical reactivity (Dubchak et al., 2010). Nanotechnology possesses the capability of revamping the agricultural and food sector by inventing innovative technologies for the molecular regulation of pathogenic organisms and improving plants' capability of mineral absorption, tolerate harsh environmental conditions and reduce the detrimental impact of stress in plants (Khan et al., 2016; Khati et al., 2018) (Figure 12.1). As a result, it has the potential to increase crop productivity and nutritional value, develop better environment monitoring tools, soil remediation and improve plant tolerance mechanisms to environmental stress (Tarafdar et al., 2013; Kumari et al., 2020; Agri et al., 2021; Bhatt et al., 2022).

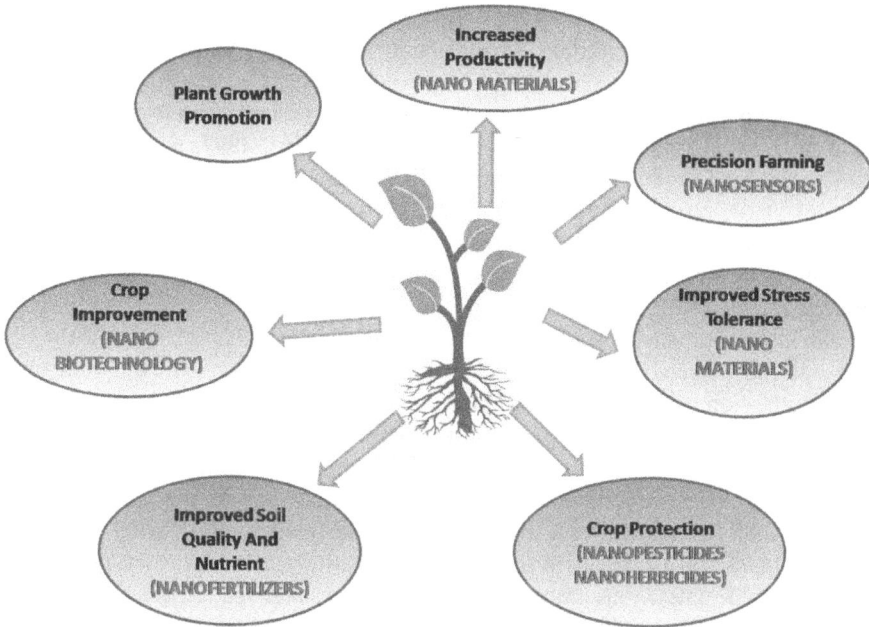

FIGURE 12.1 Application of nanotechnology in agriculture.

Nanotechnology also has the potential to furnish nanosensors and delivery tools that will greatly benefit the agricultural sector. Nanostructured catalysts may be available to improve the input utilization efficiency, allow low volumes of agricultural inputs and manage crop productivity more effectively (Liu & Lal 2015; Chaudhary et al., 2021a, b; Kukreti et al., 2020). While nanochemical pesticides are being used, some applications are mainly concerned with better preparation and less precise control of the polluted soil, greater efficiency and target-specific employment due to climate change and global warming, and the development of new pesticides is yet to come fully (Kah, 2015). In the field of agriculture, exploration and innovations in nanotechnology can intensify the production of genetically modified crops, including animal production, chemical pesticides and precise farming methods (Prasad et al., 2014). This chapter provides detailed information on the role and mechanism of nanoparticles in abiotic stress.

12.2 PLANT AND ABIOTIC STRESS

Plant growth is greatly controlled by a variety of adverse ecological factors, the most significant of which is abiotic stress, which reduces the net crop productivity by more than 50% for the majority of crops (Jalil & Ansari, 2019). Optimal environmental factors are required for maximal plant growth, and deviation in these factors is referred to as abiotic stress, which causes detrimental impact on crop development, yield and production (Bray et al., 2000). Temperature, water, irradiation, salt, heavy metal and nutrient stress are the most common abiotic stresses

(Pasala et al., 2016). Plants have the capability to acclimatize and survive with adversity by inducing cellular and molecular responses. At the onset of abiotic stresses, certain changes occur in plant physiology such as transitory amplification in cytoplasmic Ca2+ and increased intracellular secondary messenger, ROS, abscisic acid and stimulation of MAPK pathways (Hirt, 1997; Nakagami et al., 2005; Alcazar-Roman & Wente, 2008; Gill & Tuteja, 2010; Baxter et al., 2013). Various acidic conditions in the soil have a detrimental effect on soil structures, resulting in a lack of nutrients for plants that cause them to lose their normal ability to grow and develop (Rorison, 1986). Primary and long-term saline conditions are toxic to the plant cell and disrupt the osmotic balance. The loss of osmotic balance results in osmotic stress, leading to detrimental effect on plant growth (Munns & Tester, 2008). To counteract this, plants use molecular mechanisms by modifying gene expression and synthesize trehalose, inositol, sorbitol, glycerol, betaine, glycine and proline which maintain the optimal level of ions in plant cells. In response to the presence of heavy metal, plants assemble certain organic compounds such as polyphosphates, metal-chelators and acids which lead to the reduction and extraction of toxic metal ions in the cytoplasmic membrane. During advanced stage of stress regulation, various proteins are synthesized which assist in protecting cells from deleterious impact and modulating stress-related gene expression (Mahalingam & Fedoroff, 2003; Xiong et al., 2002).

Plant metabolites of secondary metabolism help the plants in combating abiotic stress through stabilizing cellular structure, protecting photosystem from oxidative damage, signalling mechanism and production of polyamines. Under unfavourable environment, the cellular wall functions as a physiological boundary to detection of stress and helps in acclimatization of plants in stress conditions (Degenhardt & Gimmler, 2000). Oxidative stress induces ROS production, phenylpropanoid accumulation, enzyme production and expression of plant defence genes (Daudi et al., 2012). Generation and destruction of ROS are regulated under favourable environmental conditions, but during abiotic stress ROS balance is disrupted by increased ROS production, resulting in the breakdown of macromolecules and membrane lipids (Foyer & Noctorc, 2005), destruction of DNA, proteins and toxicity in cells (Yadav et al., 2014), thus inhibiting growth of the plant.

12.3 NANOMATERIALS AND ABIOTIC STRESSES

Water, temperature, salt, nutrient, toxic metals and oxidative stress have become the leading source of loss in plant yield across the globe, resulting in the reduction of plants' yield by more than 50%. Plants improve their resistance to abiotic stresses at multiple stages by regulating molecular, biochemical and physiological procedures (Husen et al., 2018). Numerous reports have found the plant growth-promoting impact of nanoparticles (Bhandari, 2018). These nanoparticles affect crop physiology by regulating processes such as photosynthesis, plant pigments, photosystems, protein content, ROS and antioxidant metabolism. The application of NPs in ameliorating the detrimental impact of abiotic stress is discussed in the section below (Figure 12.2).

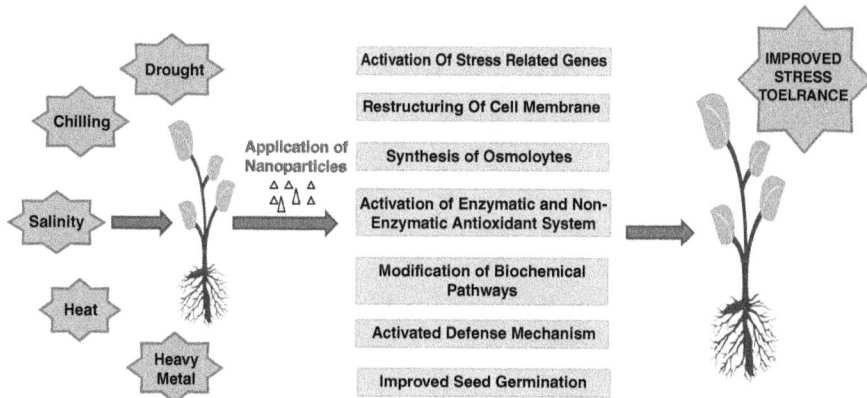

FIGURE 12.2 Application of nanoparticles for mitigation of abiotic stress in plants.

12.3.1 DROUGHT STRESS

Drought is an important abiotic problem that restricts crop productivity. Drought stress occurs during water scarcity in the rhizospheric zone or high transpiration from the leaves. These two conditions frequently coexist in semiarid and arid environments and thus these conditions significantly limit crop production in such areas. Therefore, identifying drought-tolerant plants or increasing drought tolerance mechanism in plants is always a top priority regarding sustainable agriculture. Different nanoparticles may be employed to modulate water stress by positively effecting plant development and yield through changes in their physiological processes (Table 12.1). The employment of nanosensors for monitoring and providing satellite images of drought-struck areas can help farmers in detecting such conditions at a preliminary stage (Khot et al., 2012). Additionally, nanoparticles possess porous structures and exert capillary absorption, so nanozeolites have been employed for enhancing water-retaining capacities of soil and subsequently increasing plant yield in drought-prone areas (Gururaj & Krishna, 2016). It has been found that maghemite (Fe_2O_3) alleviated drought stress while having no impact on proline, total amino acids or trace element mobilization in *Helianthus annuus*. Foliar employment of TiO_2 nanoparticles was found to alleviate detrimental impact on cell membrane and oxidative stress activated due to drought. In a similar study, silver nanoparticles decreased the adverse impact of polyethylene glycol-activated drought stress on seed germination and plant biomass of *Lens culinaris* (Hojjat & Ganjali, 2016). Shallan et al. (2016) demonstrated that application of TiO_2 nanoparticles or SiO_2 nanoparticles improved cotton plant drought stress tolerance. The use of Cu, Zn nanoparticles increased drought-tolerating capability of wheat by inducing the antioxidative metabolism and stabilization of plant pigment contents (Taran et al., 2017). In addition, Cao et al. (2017) also demonstrated that CeO_2 nanoparticles stimulate plant growth and increase water utilization efficiency and photosynthetic rate by modulating the action of Rubisco carboxylase in *Glycine max*. Sun et al. (2018) used glutathione-responsive mesoporous silica nanoparticles to successfully deliver abscisic acid to plants. The regulated release of ABA from the nanoparticles induced the translation of the ABA-inducible

TABLE 12.1

Applications of Nanoparticles in Drought and Salinity Stress Mitigation by Altering the Morphophysiological Response of Plants

Abiotic Stress	Plant	Nanoparticle Used	Physiological Response	References
Drought	*Linum usitatissimum* L.	TiO_2 (0, 10, 100, 500 mg/L)	Enhanced chlorophyll, carotenoids content, plant growth; reduced H_2O_2 and MDA content	Aghdam et al. (2016)
Drought	Cotton	TiO_2 and SiO_2 (25, 50, 100, 200 ppm) or nano-SiO_2 (400, 800, 1,600, 3,200 ppm)	Improved total phenol, total soluble proteins, total free amino acids and proline content; activation of enzymatic antioxidant metabolism	Magdy et al. (2016)
Drought	*Glycine max*	-	Improved plant growth and yield	Cao et al. (2018)
Drought	*Triticum aestivum* L.	Fe NPs (25, 50 and 100 mg/kg)	Improved plant growth; activation of antioxidant metabolism	Adrees et al. (2020)
Drought	*Triticum aestivum* L.	Si NPs (0, 25, 50 and 100 mg/kg)	Increased chlorophyll content; ameliorated oxidative stress by reduction of MDA, H_2O_2 content; electrolyte leakage	Khan et al. (2020)
Drought	*Zea mays*	Zero-valent copper NPs (3.333, 4.444 and 5.556 mg/L)	Improved plant biomass; increased anthocyanin, chlorophyll and carotenoid content; activation of antioxidant metabolism	Nguyen et al. (2021)
Drought	*Solanum melongena* L.	ZnO NPs (50 and 100 ppm)	Increased nutrient uptake and relative water content	Semida et al. (2021)
Salinity				
Salinity	*Solanum lycopersicum* L.	Nano SiO_2	Up-regulation of four salt stress genes; suppression of impact of salinity on seed germination rate, root length and fresh weight	Almutairi (2016)
Salinity	*Lupinus termis*	ZnO (20–60 mg/L)	Increased plant growth, photosynthetic pigments and antioxidant responses	Latef et al. (2017)
Salinity	*Lycopersicon esculentum*	Chitosan–PVA and Cu NPs	Alleviated the harmful effects of salinity stress	Hernandez-Hernandez et al. (2018)
Salinity	*Cucumis sativus*	SiO_2	Increased plant germination and growth characteristics	Alsaeedi et al. (2018)

(*Continued*)

TABLE 12.1 (*Continued*)
Applications of Nanoparticles in Drought and Salinity Stress Mitigation by Altering the Morphophysiological Response of Plants

Abiotic Stress	Plant	Nanoparticle Used	Physiological Response	References
Salinity	*Zea mays* L.	TiO_2 (40, 60 and 80 ppm)	Enhanced K^+ concentration, relative water, total phenolic and proline content; increased SOD, CAT and PAL activities; reduced Na^+ concentration, membrane electrolyte leakage and MDA content	Shah et al. (2021)
Salinity	*Musa acuminata*	SiO_2 (0, 50, 100, 150 mg/L)	Enhanced chlorophyll content, reduced electrolyte leakage, MDA content	Mahmoud et al. (2020)
Salinity	*Solanum tuberosum* L., Diamond cultivar	Zeolite NPs	Enhanced leaf relative water content, stomatal conductivity and chlorophyll; improved nutrient, leaf proline and gibberellic acid; activated antioxidative enzymes	Mahmoud et al. (2019)

marker gene (*AtGALK2*), which increased drought tolerance in *Arabidopsis thaliana*. The authors also reported that short-chain molecules function as gatekeepers for encapsulating biomolecules in these porous nanoparticles. The employment of ZnO nanoparticles has been reported to amplify grain nitrogen translocation and increased drought stress tolerance of sorghum by restoring total N content (Dimkpa et al., 2019).

12.3.2 SALINITY STRESS

Salt stress, also known as soil salinity, refers to the presence of a high number of salts in the soil and is recognized as a significant ecological threat resulting in reduction of plant yield across the globe. It is anticipated that greater than 20% of arable agricultural land globally is facing salinity stress, and this area is growing rapidly. Salt-affected soils have a number of issues, including higher salt content, lower porosity, waterlogging, nutrient loss and hydraulic limitations, and thus it adversely impacts a number of biochemical and physiological processes. In high-salt conditions a reduction in osmotic potential, photosynthesis, protein and lipid metabolism, mineral imbalance and increase in specific ionic toxicity have been observed. Secondary salinization in soils due to inappropriate irrigation approaches and deforestation has also been reported (Yadav et al., 2011). Plants exposed to saline environments show increased ROS production, decreased efficacy of photosystem II, stomata and photosynthetic pigment content (Khan & Upadhyaya, 2019). Various strategies such as employment of salt-resistant plant varieties, chemicals and nanoparticles have been used for managing salt-affected areas (Iqbal et al., 2020). To overcome the issue of salt stress, nanotechnology has been applied in various manners. Plants can be treated with various nanoparticles to improve their growth under salt stress. Nanomaterials or nano-remediation agents have the potential to effectively rehabilitate salt-affected soils. It is known that nano gypsum, nanochitosan, nanozeolite, nano calcium and magnesium are capable of effectively increasing the hydraulic properties and stability of the soil and thus are effective reclaimants (Mukhopadhyay & Kaur, 2016; Patra et al., 2016; Chaudhary et al., 2021c, d; Chaudhary et al., 2022) (Table 12.2). The usage of nanofertilizers can address the problem of salt toxicity and other related stress problems. According to a report by Kalteh et al. (2014), silicon nanoparticles and silicon fertilizer show potentiality for improving physiological, morphological and vegetative properties of *Ocimum basilicum* in high-salt conditions. The authors found that silicon nanoparticles and silicon fertilizer significantly enhanced growth index, pigment and proline content of the plant (Kalteh et al., 2014). Wu et al. (2018) reported that anionic CeO_2 nanoparticles were effective in scavenging high amounts of ROS such as OH radicals; regulating K^+ and nonspecific ion channels, non-selective cation channels and lower K^+ ion movements from mesophyll cells (improvement in K+ retention capability of mesophyll); and improving salt resistance in *Arabidopsis*. On the other hand, Askary et al. (2017) demonstrated that Fe_2O_3 nanoparticles improve biomass production and P, Ca, Fe, Zn and K contents in *Mentha piperita* in high-salinity conditions, but didn't impact Na content. The authors also found significant reduction in lipid peroxidation, proline content and antioxidant metabolism. Torabian et al. (2016) demonstrated increased carbon dioxide assimilation ability,

TABLE 12.2

Applications of Nanoparticles in Abiotic Stress Mitigation by Altering the Morphophysiological Responses of Plants

Abiotic Stress	Plant	Nanoparticle Used	Physiological Response	References
			Heavy Metal	
Cd	Oryza sativa L.	Fe_2O_3 (0, 25, 50 and 100 mg/kg soil)	Enhanced plant biomass and growth; ROS scavenging; down-regulation of Cd transport genes	Ahmed et al. (2021)
Cd	Triticum aestivum L.	Fe_2O_3 (25, 50 and 100 mg/kg soil)	Enhanced plant biomass and growth; ROS scavenging; down-regulation of Cd transport genes	Manzoor et al. (2021)
Cd	Zea mays	TiO_2 (0, 100 and 250 mg/L soil)	Enhanced shoot weight; decreased Cd accumulation; increased antioxidant metabolism	Zhou et al. (2020)
As	Glycine max	ZnO (0, 50 and 100 mg/L soil)	Enhanced shoot weight; decreased As accumulation; augmented antioxidant metabolism	Ahmad et al. (2020)
As	Oryza sativa L.	ZnO (10–100 mg/L)	Enhanced seedling growth; decreased As accumulation; increased phytochelatin level	Yan et al. (2021)
Pb	Vigna radiata	AgNPs (25 mg/L)	Increased photosynthetic rate, increased antioxidant metabolism	Chen et al. (2022)
Cd	Satureja hortensis L.	Si NPs (0–2.25 mM)	Enhanced growth, water, phenolic, flavonoid and essential oil content	Memari-Tabrizi et al. (2021)
Cd	Oryza sativa	FeO NPs	Increased photosynthesis, yield, Fe content; decreased Cd concentrations in tissues; alleviation of oxidative stress	Adrees et al. (2020)
Cd	Zea mays L.	Foliar TiO_2 NPs (0–250 mg/L)	Decrease shoot Cd content, increasing SOD and GST; up-regulation of various metabolic pathways	Lian et al. (2020)
Co	Zea mays	ZnO NPs	Decrease in ROS and MDA accumulation; reduced Co uptake and conferred; increased antioxidant metabolism; down-regulation of ultrastructural damage and improved photosynthetic apparatus	Salam et al. (2022)
Al	Zea mays	Nano-SiO_2	Increased enzymatic and non-enzymatic antioxidant metabolism; metal detoxification	de Sousa et al. (2019)
				(Continued)

TABLE 12.2 (Continued)
Applications of Nanoparticles in Abiotic Stress Mitigation by Altering the Morphophysiological Responses of Plants

Abiotic Stress	Plant	Nanoparticle Used	Physiological Response	References
As	*Oryza sativa* L.	*Bacillus subtilis*-synthesized iron oxide nano particles (Fe_3O_4 NP) (5 mg/L)	Alleviation of As stress and enhanced plant growth	Khan et al. (2021)
As	*Oryza sativa*	Graphene oxide, hydroxyapatite, nano-Fe_3O_4 and nano-zerovalent iron	Alleviation of As stress and enhanced plant growth	Huang et al. (2018)

substomatal carbon dioxide level, chlorophyll, Fv/Fm, Zn content and reduction in Na content of *H. annuus* on application of ZnO nanoparticles. Some researchers have also used multi-walled carbon nanotubes-induced water uptake by improving CO_2 assimilation, aquaporin transduction, modifying cell membrane of root to mitigate salt stress and increase growth of broccoli (Martınez-Ballesta et al., 2016). Abou-Zeid and Ismail (2018) found that silver nanoparticles stimulated seed germination and development of wheat during salinity stress.

12.3.3 TEMPERATURE STRESS

Too high or too low temperature conditions can also be stressful for plant growth. Temperature stress can be categorized into three types: high, freezing or chilling and imposes adverse impact on plants by affecting growth, photosynthesis and seed germination with subsequent death of the plant. Cold stress (0°C–15°C) is induced due to very low temperatures and causes injury to plant tissues without formation of ice crystals. Freezing (<0°C) causes ice crystal formation in plant tissues, resulting in the death of the plant due to frost (Hasanuzzaman et al., 2013). Freezing and chilling stress result in impairment of permeable and fluid nature of cell membranes, and ion leakage occurs from the plasma membrane that imposes negative impact on plant by affecting seed germination and plant growth (Suzuki et al., 2008).

Mohammadi et al. (2013) investigated the impact of TiO_2 nanoparticles on restoring ion leakage, membrane distortion, malondialdehyde content in low-temperature-sensitive (ILC 533) and -tolerant (Sel 11439) *Cicer arietinum* varieties. The authors reported accumulation of nanoparticles in the vacuole and chloroplast during temperature stress and found that sensitive variety showed higher nanoparticle permeability than the tolerant genotype. TiO_2 content in plants was reported to be greater during cold stress in comparison to normal conditions. Nanoparticles clearly had a positive effect on physiological parameters during the treatments and not only reduced oxidative damage in chickpea but also reduced membrane damage caused by low-temperature stress. Biogenic nanoparticles synthesized from *Tridax procumbens* were used for stimulating antioxidant metabolism, seed germination and seedling growth in wheat. The authors reported that these nanoparticles such as nanophos and nanochitosan caused significant improvement in antioxidant activities of super oxide dismutase and others (Bhati-Kushwaha et al., 2013; Chaudhary et al., 2021e; Agri et al., 2022). Application of TiO_2 nanoparticles was also reported to enhance the activity of Rubisco, translation of chlorophyll-binding gene (Hasanpour et al., 2015), antioxidant system (Mohammadi et al., 2014), plant pigment and susceptibility to low-temperature stress.

Heat stress is defined as prolonged exposure to high temperatures that result in detrimental and irreparable impact on plant growth (Jalil & Ansari, 2019). High temperature induces reactive oxygen species production and causes damage resulting in degeneration of cytomembrane lipids and ion discharge with subsequent protein denaturation (Karuppanapandian et al., 2011), as well as a lowering of photosynthetic rate and pigment amount (Prasad et al., 2011). Iqbal et al. (2017) determined the impact of silver nanoparticles on wheat growth regulation under high-temperature conditions. The authors prepared silver nanoparticles by utilizing *Moringa oleifera*

extract, which was then characterized using UV-Vis spectroscopy, X-ray diffractometry, scanning electron microscopy and atomic force microscopy (AFM). Silver
nanoparticles were given to wheat plants at three-leaf stage under high-temperature
conditions for 3 days. The results suggested that employment of the nanoparticles
at different concentrations helped in alleviating temperature stress in wheat and
enhanced plant biomass significantly. Iqbal et al. (2018) discovered that biogenic silver nanoparticles were critical in alleviating the negative impact of high-temperature
stress in wheat plants. During heat stress, AgNPs caused reduction in malondialdehyde, hydrogen peroxide content and enhanced antioxidant mechanism. Furthermore,
application of TiO_2 nanoparticles caused reduction in damaging impact of heat stress
by regulating stomatal opening (Qi et al., 2013). Plants synthesize various heat shock
molecules and molecular chaperones when exposed to high-temperature conditions
to assist other proteins for maintenance of stability in stressful conditions (Wahid,
2007). Multi-walled carbon nanotubes play a significant role in increasing the production of heat shock proteins such as HSP90 by translation of heat shock genes
(Khodakovskaya et al., 2011). Moreover, application of CeO_2 nanoparticle in maize
resulted in high amount of H_2O_2 production and up-regulation of HSP70 (Zhao et al.,
2012).

12.3.4 Heavy Metal Stress

Several heavy metals such as chromium, copper, lead, mercury, iron, arsenic and cadmium are contaminants of the environment and occur in areas of high anthropogenic
activities (Bhandari & Bhatt, 2021). Among these some metals (Cu, Mn, Co, Zn, and
Cr) are vital mineral ions for plants and are needed in minute quantities by the plants.
However, when heavy metal occurs in excessive quantities in soil, it imposes toxicity to plant tissues. Plants cultivated on heavy metal-polluted soils display metabolic
changes, impaired growth, low biomass generation and metal accumulation in plant
tissues (Table 12.2). The physiological and biochemical properties of plants are negatively impacted by the presence of heavy metals (Iqbal et al., 2020). Supplementation
of heavy metals causes increase in ROS generation, resulting in deterioration of plant
tissues by modifying cellular formation, reducing cytomembrane permeability and
degradation of proteins (Sharma et al., 2012). Additionally, plants synthesize metal
chelating organic compounds and polyphosphates which restrict the metal uptake,
leak of metal ions and the induction of antioxidative mechanism that limits ROS
generation. Stimulation of the defence mechanism is significant for developing tolerance to toxic metal ions. A study was conducted to assess the impact of magnetic
(Fe_3O_4) nanoparticles in mitigating toxic effects of Pb, Zn, Cd and Cu in wheat seedlings (Konate et al., 2017). The authors found that magnetic nanoparticles caused
significant reduction in growth inhibition and stimulation of antioxidant defence
mechanism for reducing oxidative stress and MDA content. It has been suggested
that the mitigating impacts of nano-Fe_3O_4 are related to heavy metal adsorption due
to electrostatic interaction between heavy metal ions and negatively charged adsorption sites (Konate et al., 2017).

The efficacy of Fe_3O_4 nanoparticles in mitigating the adverse impact of arsenic
toxicity in *Brassica juncea* was studied by Praveen et al. (2018). In the presence of

Fe_3O_4 nanoparticles, a reduction in plant stress-related characters was observed by limiting arsenic absorption through plant roots (Praveen et al., 2018). It has been found that ZnO nanoparticles enhance pigment content, gaseous transfer rate, antioxidative metabolism, zinc absorption and crop production in wheat during Cd stress (Hussain et al., 2018). Application of TiO_2 nanoparticles restricts toxic effects of cadmium and enhances photosynthetic rate and plant growth (Singh and Lee, 2016). Pradhan et al. (2014) reported that ROS generation caused by Cd ions in *Saccharomyces cerevisiae* is reduced by polyhydroxy fullerene (PHF). ROS production and distortion of cytomembranes can be mitigated up to 36.7% and 30.7%, respectively, by employing PHF in Cd-contaminated environment. Venkatachalam et al. (2017) found that in *Leucaena leucocephala* (Lam.), Cd and Pb stress can be mitigated by applying ZnO nanoparticles which assist in reversing the damage caused by ROS. The authors also reported an increase in total soluble protein and pigment content and reduction in lipid peroxidation of *L. leucocephala* (Lam.), on applying ZnO nanoparticles (Venkatachalam et al., 2017). A proteomics study was employed by Mustafa and Komatsu (2016) to recognize the metabolic processes impacted by the existence of metal ions and decipher the molecular mechanism. Nanoparticles have been suggested to regulate ROS, antioxidative and energy metabolism in plant tissue and therefore mitigate heavy metal-activated oxidative destruction. Various nanoparticles of selenium, iron, manganese, cerium, titanium and zinc oxides display attraction towards metal/metalloids adsorption, and thus can be employed for the development of stress tolerance and cost-effective remediation of metal-polluted soils. Martınez-Fernandez et al. (2017) demonstrated that nanoparticles mitigate the toxic effects of metal/metalloids by immobilizing them in the soil and stimulating plant growth and development by the process of phytoremediation.

12.4   NANOPARTICLE IMPACT ON THE MOLECULAR AND ANTIOXIDANT ASPECTS OF PLANTS

Nanoparticles display chemical or physical interaction with living systems such as plants due to their minute structure, larger surface area and innate catalytic property. Nanoparticle-plant interaction in abiotic stress conditions occurs due to ROS generation, where ROS function as a signal for triggering plants' defence mechanism to mitigating the cellular damage (Figure 12.3). Nanoparticles not only activate ROS but also mimic antioxidant enzymes which assist in scavenging ROS. Nanoceria (cerium oxide nanoparticles) plays a vital role in scavenging ROS and protecting photosynthetic activities from the destructive impact of ROS generation during abiotic stress (Giraldo et al., 2014). Silicon nanoparticles have been reported to increase abiotic stress tolerance and enhance uptake processes in plants through the activation of antioxidant enzymes (Tripathi et al., 2016, 2017). The use of Au nanoparticles in *B. juncea* caused a significant up-regulation of antioxidative metabolism by the synthesis of enzymes, ascorbate peroxidase, guaiacol peroxidase, catalase and glutathione reductase in addition to greater concentration of H_2O_2 and proline in the experimental plants (Gunjan et al., 2014). In a proteomic analysis, the application of Ag nanoparticles in rice revealed that nanoparticle-responsive proteins are generally connected with ROS response metabolism, Ca^{2+} regulation and signalling, transcription, protein

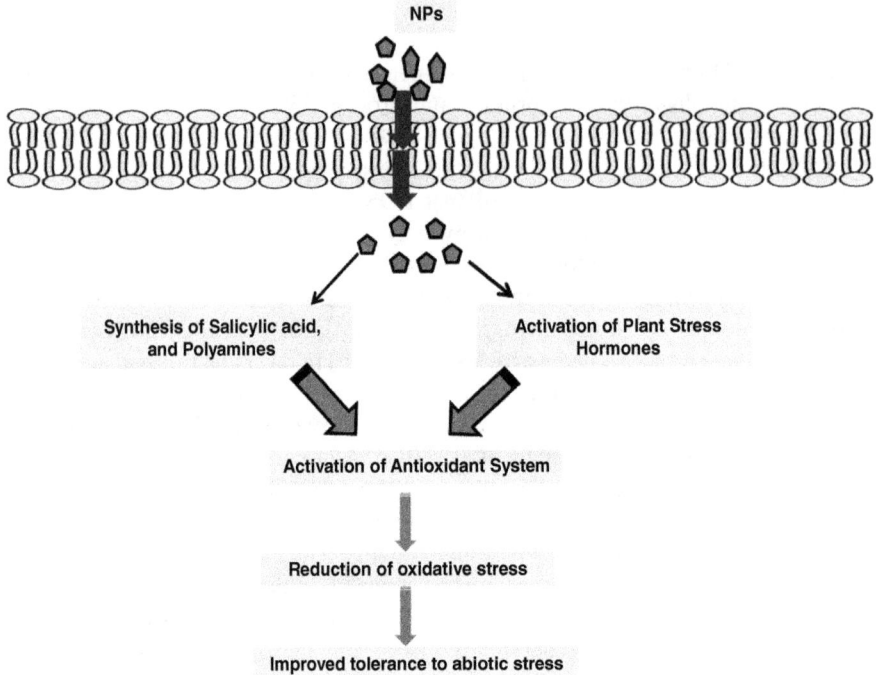

FIGURE 12.3   Mechanism of abiotic stress mitigation by nanoparticles.

degradation, cell wall synthesis, cell division and apoptosis (Mirzajani et al., 2014). In addition, it has been found that low concentration of nanoparticles sustains an active antioxidant defence mechanism which assists in regulating production of ROS in appropriate amount required for signalling process without causing detrimental impact on plant tissues (Syu et al., 2014).

12.5   CONCLUSION

Nanotechnology is an advancing area and has immense potentiality for implementation in various fields such as electronics, energy, medicine and biological sciences. Nanoscience is pre-eminent for innovation of a diverse range of cost-effective nanotech-based tools for enhancing plant growth and productivity. Nanoparticles increase stress tolerance in plants by increasing root hydraulic conductance and water absorption, and by displaying different proteins content responsible for oxidation-reduction, ROS mitigation, stress signalling and hormonal pathways. The application of nanoparticles causes changes in plant gene expression and thus subsequently affecting metabolic pathways and plant growth. However, the implementation of nanotech-based products in agriculture is still in its infancy, especially in stress conditions. Few research investigations suggesting positive impact of nanoparticles on plant growth in recent times have given confidence to agricultural scientists for applying nanotechnology for improving crop productivity under various abiotic stress conditions. However, extensive investigations are required in the field of fabricating, characterizing, standardizing nanoparticles for agricultural application and also assessing their biodegradability, environment-friendly nature and uptake and translocation by plants. Investigation for understanding the molecular mechanism of action of nanoparticles and their impact on gene expression in plants is also required. In addition, the increased utilization of nanotechnology in agrarian procedures can cause the release of nanoparticles into the environmental matrices and their impacts need to be investigated in depth for avoiding potential risks of these nanoparticles to the environment. Currently, only a few reports on detrimental effects of nanoparticles are available and thus it is vital to understand the interaction between plants and nanoparticles prior to their use in the fields, so as to minimize their toxic effects if any and develop cost-efficient methods to deal with abiotic stresses.

REFERENCES

Abou-Zeid, H., & Ismail, G. (2018). The role of priming with biosynthesized silver nanoparticles in the response of *Triticum aestivum* L. to salt stress. *Egypt J Bot*. 58, 73–85.

Adrees, M., Khan, Z.S., Ali, S., Hafeez, M., Khalid, S., Rehman, M.Z.U., Hussain, A., Hussain, K., Chatha, S.A.S., & Rizwan, M. (2020). Simultaneous mitigation of cadmium and drought stress in wheat by soil application of iron nanoparticles. *Chemosphere*. 238, 124681.

Aghdam, M.T.B., Mohammadi, H., & Ghorbanpour, M. (2016). Effects of nanoparticulate anatase titanium dioxide on physiological and biochemical performance of *Linumusitatissimum* (*Linaceae*) under well watered and drought stress conditions. *Braz J Bot*. 39,139–146.

Agri, U., Chaudhary, P., & Sharma, A. (2021). In vitro compatibility evaluation of agriusable nanochitosan on beneficial plant growth-promoting rhizobacteria and maize plant. *Natl Acad Sci Lett*. 44, 555–559.

Agri, U., Chaudhary, P., Sharma, A., & Kukreti, B. (2022). Physiological response of maize plants and its rhizospheric microbiome under the influence of potential bioinoculants and nanochitosan. *Plant Soil*. 474, 451–468.

Ahmad, P., Alyemeni, M.N., Al-Huqail, A.A., Alqahtani, M.A., Wijaya, L., Ashraf, M., Kaya, C., & Bajguz, A. (2020). Zinc oxide nanoparticles application alleviates Arsenic (As) toxicity in soybean plants by restricting the uptake of as and modulating key biochemical attributes, antioxidant enzymes, ascorbate-glutathione cycle and glyoxalase system. *Plants*. 9, 825.

Ahmed, T., Noman, M., Manzoor, N., Shahid, M., Abdullah, M., Ali, L., Wang, G., Hashem, A., Al-Arjani, A.B.F., & Alqarawi, A.A. (2021). Nanoparticle-based amelioration of drought stress and cadmium toxicity in rice via triggering the stress responsive genetic mechanisms and nutrient acquisition. *Ecotoxicol Environ Saf*. 209, 111829.

Alcazar-Roman, A., & Wente, S. (2008). Inositol polyphosphates: a new frontier for regulating gene expression. *Chromosoma*. 117, 1–13.

Almutairi, Z.M. (2016). Effect of nanosilicon application on the expression of salt tolerance genes in germinating tomato (*Solanum lycopersicum* L.) seedlings under salt stress. *Plant Omics J*. 9,106–114.

Alsaeedi, A., El-Ramady, A., Alshaal, T., El-Garawani, M., Elhawat, N., & Al-Otaibi, A. (2018). Exogenous nanosilica improves germination and growth of cucumber by maintaining K+/Na+ ratio under elevated Na+ stress. *Plant Physiol Biochem*. 125, 164–171.

Askary, M., Talebi, S.M., Amini, F., & Bangan, A.D.B. (2017). Effects of iron nanoparticles on *Menthapiperita* under salinity stress. *Biologica*. 63, 65–75.

Baxter, A., Mittler, R., & Suzuki, N. (2013). ROS as key players in plant stress signalling. *J. Exp. Bot.* 65, 1229–1240.

Bhandari, G. (2018). Environmental nanotechnology: applications of nanoparticles for bioremediation. In Prasad, R., & Aranda, E. (eds.) *Approaches in Bioremediation. Nanotechnology in the Life Sciences.* Springer, Cham.

Bhandari, G., & Bhatt, P. (2021). Concepts and application of plant–microbe interaction in remediation of heavy metals. In A. Sharma (Ed.) *Microbes and Signaling Biomolecules against Plant Stress* (pp. 55–77). Springer, Singapore.

Bhati-Kushwaha, H., Kaur, A., & Malik, C.P. (2013). The synthesis and role of biogenic nanoparticles in overcoming chilling stress. *Ind J Plant Sci.* 2, 54–62.

Bhatt, P., Pandey, S.C., Joshi, S., Chaudhary, P., Pathak, V.M., Huang, Y., Wu, X., Zhou, Z., & Chen, S. (2022). Nanobioremediation: a sustainable approach for the removal of toxic pollutants from the environment. *J Hazard Mater.* 427, 128033.

Bray, A.B., Bailey-Serres, J., & Weretilnyk, E. (2000). Responses to abiotic stress. In Buchanan, B.B., Gruissem, W., & Jones, R.L. (eds.) *Biochemistry & Molecular Biology of Plants* (pp. 1158–1203). American Society of Plant Physiologists, Rockville, MD.

Cao, Z., Rossi, L., Stowers, C., Zhang, W., Lombardini, L., & Ma, X. (2018). The impact of cerium oxide nanoparticles on the physiology of soybean (*Glycine max* (L.) Merr.) under different soil moisture conditions. *Environ Sci Pollut Res.* 25, 930–939.

Cao, Z., Stowers, C., Rossi, L., Zhang, W., Lombardini, L., & Ma, X. (2017). Physiological effects of cerium oxide nanoparticles on the photosynthesis and water use efficiency of soybean (*Glycine max* (L.) Merr.). *Environ Sci Nano.* 4, 1086–1094.

Chaudhary, P., Chaudhary, A., Bhatt, P., Kumar, G., Khatoon, H., Rani, A., Kumar, S., & Sharma, A. (2022). Assessment of soil health indicators under the influence of nano-compounds and *Bacillus* spp. in field condition. *Front Environ Sci.* 9, 769871.

Chaudhary, P., Chaudhary, A., Parveen, H., Rani, A., Kumar, G., Kumar, A., & Sharma, A. (2021e). Impact of nanophos in agriculture to improve functional bacterial community and crop productivity. *BMC Plant Biol.* 21, 519.

Chaudhary, P., Khati, P., Chaudhary, A., Gangola, S., Kumar, R., & Sharma, A. (2021a). Bioinoculation using indigenous *Bacillus* spp. improves growth and yield of *Zea mays* under the influence of nanozeolite. *3 Biotech.* 11, 11.

Chaudhary, P., Khati, P., Chaudhary, A., Maithani, D., Kumar, G., & Sharma, A. (2021d). Cultivable and metagenomic approach to study the combined impact of nanogypsum and *Pseudomonas taiwanensis* on maize plant health and its rhizospheric microbiome. *PLoS One.* 16, e0250574.

Chaudhary, P., Khati, P., Gangola, S., Kumar, A., Kumar, R., & Sharma, A. (2021c). Impact of nanochitosan and *Bacillus* spp. on health, productivity and defence response in *Zea mays* under field condition. *3 Biotech.* 11, 237.

Chaudhary, P., Sharma, A., Chaudhary, A., Khati, P., Gangola, S., & Maithani, D. (2021b). Illumina based high throughput analysis of microbial diversity of rhizospheric soil of maize infested with nanocompounds and *Bacillus* sp. *Appl Soil Ecol.* 159, 103836.

Chen, F., Aqeel, M., Maqsood, M.F., Khalid, N., Irshad, M.K., Ibrahim, M., Akhter, N., Afzaal, M., Ma, J., Hashem, M., Alamri, S., Noman, A., & Lam, S.S. (2022). Mitigation of lead toxicity in Vigna radiata genotypes by silver nanoparticles. *Environ Poll.* 308, 119606.

Das, A., & Das, B. (2019). Nanotechnology a potential tool to mitigate abiotic stress in crop plants. In *Abiotic and Biotic Stress in Plants.* DOI: http://dx.doi.org/10.5772/intechopen.83562

Daudi, A., Cheng, Z., O'Brien, J.A., Mammarella, N., Khan, S., & Ausubel, F.M. (2012). The apoplastic oxidative burst peroxidise in *Arabidopsis* is a major component of pattern-triggered immunity. *Plant Cell Online.* 24, 275–287.

de Sousa, A., Saleh, A.M., Habeeb, T.H., Hassan, Y.M., Zrieq, R., Wadaan, M.A.M., Hozzein, W.N., Selim, S., Matos, M., & AbdElgawad, H. (2019). Silicon dioxide nanoparticles ameliorate the phytotoxic hazards of aluminum in maize grown on acidic soil. *Sci Total Environ.* 693, 133636.

Degenhardt, B., & Gimmler, H. (2000). Cell wall adaptations to multiple environmental stresses in maize roots. *J Exp Bot.* 51, 595–603.

Dimkpa, C.O., Singh, U., Bindraban, P.S., Elmer, W.H., Gardea-Torresdey, J.L., & White, J.C. (2019). Zinc oxide nanoparticles alleviate drought-induced alterations in sorghum performance, nutrient acquisition, and grain fortification. *Sci Total Environ.* 688, 926–934.

Dubchak, S., Ogar, A., Mietelski, J.W., & Turnau, K. (2010). Influence of silver and titanium nanoparticles on arbuscular mycorrhiza colonization and accumulation of radiocaesium in *Helianthus annuus*. *Span J Agric Res.* 8, 103–108.

FAO (2017). *The Future of Food and Agriculture – Trends and Challenges.* FAO, Rome.

Foyer, C.H., & Noctorc, G. (2005). Oxidant and antioxidant signalling inplants: a re-evaluation of the concept of oxidative stress in a physiological context. *Plant Cell Environ.* 28, 1056–1071.

Gill, S.S., & Tuteja, N. (2010). Reactive oxygen species and antioxidant machinery in abiotic stress tolerance in crop plants. *Plant Physiol Biochem.* 48, 909–930.

Giraldo, J.P., Landry, M.P., Faltermeier, S.M., McNicholas, T.P., Iverson, N.M., Boghossian, A.A., Reuel, N.F., Hilmer, A.J., Sen, F., Brew, J.A., & Strano, M.S. (2014). Plant nanobionics approach to augment photosynthesis and biochemical sensing. *Nat Mater.* 13, 400–408.

Gunjan, B., Zaidi, M.G.H., & Sandeep, A. (2014). Impact of gold nanoparticles on physiological and biochemical characteristics of *Brassica juncea*. *J Plant Biochem Physiol.* 2, 133.

Gururaj, S.B., & Krishna, B.S.V.S.R. (2016) Water retention capacity of biochar blended soils. *J Chem Pharm Sci.* 9, 1438–144.

Hasanpour, H., Maali-Amiri, R., & Zeinali, H. (2015). Effect of TiO$_2$ nanoparticles on metabolic limitations to photosynthesis under cold in chickpea. *Russ J Plant Physiol.* 62, 779–787.

Hasanuzzaman, M., Nahar, K., & Fujita, M. (2013). Extreme temperature responses, oxidative stress and antioxidant defense in plants. In Vahdati, K., & Leslie, C. (eds.) *Abiotic Stress- Plant Responses and Applications in Agriculture* (pp. 169–205). In Tech Open Access Publisher, London, UK.

Hatami, M., Kariman, K., & Ghorbanpour, M. (2016). Engineered nanomaterial-mediated changes in the metabolism of terrestrial plants. *Sci Total Environ.* 571, 275–291.

Hernandez-Hernandez, H., Gonzalez-Morales, S., Benavides-Mendoza, A., Ortega-Ortiz, H., Cadenas-Pliego, G., & Juarez Maldonado, A. (2018). Effects of chitosan–PVA and Cu nanoparticles on the growth and antioxidant capacity of tomato under saline stress. *Molecules.* 23, 178

Hirt, H. (1997). Multiple roles of map kinases in plant signal transduction. *Trends Plant Sci.* 2, 11–15.

Hojjat, S.S., & Ganjali, A. (2016). The effect of silver nanoparticle on lentil seed germination under drought stress. *Int J Farm Allied Sci.* 5, 208–212.

Huang, Q., Liu, Q., Lin, L., Li, F., Han, Y., & Song, Z.G. (2018). Reduction of arsenic toxicity in two rice cultivar seedlings by different nanoparticles. *Ecotoxicol Environ Saf.* 159, 261–271.

Husen, A., Iqbal, M., Sohrab, S.S., & Ansari, M.K.A. (2018). Salicylic acid alleviates salinity-caused damage to foliar functions, plant growth and antioxidant system in Ethiopian mustard (*Brassica curinata* A. Br.). *Agric Food Security.* 7, 44.

Hussain, A., Ali, S., Rizwan, M., Zia Ur Rehman, M., Javed, M.R., Imran, M.S., Ali, S., Chatha, S., & Nazir, R. (2018). Zinc oxide nanoparticles alter the wheat physiological response and reduce the cadmium uptake by plants. *Environ Pollut.* 242B, 1518–1152.

Iqbal, M., Raja, N.I., Mashwani, Z.R., Hussain, M., Ejaz, M., & Yasmeen, F. (2017). Effect of silver nanoparticles on growth of wheat under heat stress. *Iran J Sci Technol Trans A Sci*. 43(2), 387–395.

Iqbal, M., Raja, N.I., Mashwani, Z.R., Wattoo, F.H., Hussain, M., Ejaz, M., & Saira, H. (2018). Assessment of AgNPs exposure on physiological and biochemical changes and antioxidative defence system in wheat (*Triticum aestivum* L) under heat stress. *IET Nanobiotechnol*. 13, 230–236.

Iqbal, S., Waheed, Z., & Naseem, A. (2020). Nanotechnology and abiotic stresses. In Javad, S. (ed.) *Nanoagronomy*, Springer, Cham. https://doi.org/10.1007/978-3-030-41275-3_3.

Jalil, S.U., & Ansari, M.I. (2019). Nanoparticles and abiotic stress tolerance in plants: synthesis, action, and signaling mechanisms. In *Plant Signaling Molecules*. DOI: https://doi.org/10.1016/B978-0-12-816451-8.00034-4

Kah, M. (2015). Nanopesticides and nanofertilizers: emerging contaminants or opportunities for risk mitigation? *Front Chem*. 3, 64.

Kalteh, M., Alipour, Z.T., Ashraf, S., Aliabadi, M.M., & Nosratabadi, A.F. (2014). Effect of silica nanoparticles on basil (*Ocimum basilicum*) under salinity stress. *J Chem Health Risks*. 4, 49–55.

Kang, Y., Khan, S., & Ma, X. (2009). Climate change impacts on crop yield, crop water productivity and food security. *Prog Nat Sci*. 19, 1665–1674.

Karuppanapandian, T., Wang, H.W., Prabakaran, N., Jeyalakshmi, K., Kwon, M., Manoharan, K., & Kim, W. (2011). 2,4-dichlorophenoxyacetic acid-induced leaf senescence in mung bean (*Vigna radiate* L.) Wilczek) and senescence inhibition by co-treatment with silver nanoparticles. *Plant Physiol Biochem*. 49,168–217.

Khan, M.N., Mobin, M., Abbas, Z.K., Almutairi, K.A., & Siddiqui, Z.H. (2016). Role of nanomaterials in plants under challenging environments. *Plant Physiol Biochem*. 110, 194–209.

Khan, S., Akhtar, N., Rehman, S.U., Shujah, S., Rha, E.S., & Jamil, M. (2021). Biosynthesized iron oxide nanoparticles (Fe$_3$O$_4$ NPs) mitigate arsenic toxicity in rice seedlings. *Toxics*. 9(1), 2.

Khan, Z., & Upadhyaya, H. (2019). Impact of nanoparticles on abiotic stress responses in plants: an overview. In *Nanomaterials in Plants, Algae, and Microorganisms*. https://doi.org/10.1016/B978-0-12-811488-9.00015-9.

Khan, Z.S., Rizwan, M., Hafeez, M., Ali, S., Adrees, M., Qayyum, M.F., Khalid, S., Rehman, M.Z.U., & Sarwar, M.A. (2020). Effects of silicon nanoparticles on growth and physiology of wheat in cadmium contaminated soil under different soil moisture levels. *Environ Sci Pollut. Res*. 27, 4958–4968.

Khati, P., Bhatt, P., Kumar, R., & Sharma, A. (2018). Effect of nanozeolite and plant growth promoting rhizobacteria on maize. *3Biotech*. 8, 141.

Khodakovskaya, M.V., de Silva, K., Nedosekin, D.A., Dervishi, E., Biris, A.S., Shashkov, E.V., Galanzha, E.I., & Zharov, V.P. (2011). Complex genetic, photothermal, and photoacoustic analysis of nanoparticle-plant interactions. *Proc Natl Acad Sci U S A*. 108, 1028–1033.

Khot, L.R., Sankaran, S., Maja, J.M., Ehsani, R., Schust, E.W. (2012). Applications of nanomaterials in agricultural production and crop protection: a review. *Crop Prot*. 35, 64–70.

Konate, A., He, X., Zhang, Z., Ma, Y., Zhang, P., Alugongo, G.M., & Rui, Y. (2017). Magnetic (Fe3O4) nanoparticles reduce heavy metals uptake and mitigate their toxicity in wheat seedling. *Sustainability*. 9, 790.

Kukreti, B., Sharma, A., Chaudhary, P., Agri, U., & Maithani, D. (2020). Influence of nanosilicon dioxide along with bioinoculants on *Zea mays* and its rhizospheric soil. *3 Biotech*. 10, 345.

Kumari, S., Sharma, A., Chaudhary, P., & Khati, P. (2020). Management of plant vigor and soil health using two agriusable nanocompounds and plant growth promotory rhizobacteria in Fenugreek. *3 Biotech*. 10, 461.

Latef, A.A.H.A., Alhmad, M.F.A., & Abdelfattah, K.E. (2017). The possible roles of priming with ZnO nanoparticles in mitigation of salinity stress in lupine (*Lupinus termis*) plants. *J Plant Growth Regul.* 36, 60–70.

Lian, J., Zhao, L., Wu, J., Xiong, H., Bao, Y., Zeb, A., Tang, J., & Liu, W. (2020). Foliar spray of TiO_2 nanoparticles prevails over root application in reducing Cd accumulation and mitigating Cd-induced phytotoxicity in maize (*Zea mays* L.). *Chemosphere.* 239, 124794.

Liu, R., & Lal, R. (2015). Potentials of engineered nanoparticles as fertilizers for increasing agronomic productions. *Sci Total Environ.* 514, 131–139.

Magdy, A.S., Hazem, M.M.H., Alia, A.M.N., & Alshaimaa, A.I. (2016). Effects of TiO_2 and SiO_2 nanoparticles on cotton plant under drought stress. *Res J Pharm Bio Chem Sci Biochem Physiol.* 7(4), 1540.

Mahalingam, R., & Fedoroff, N. (2003). Stress response, cell death, and signalling: the many faces of reactive oxygen species. *Physiol Plant.* 119, 56–68.

Mahmoud, A.W.M., Abdeldaym, E.A., Abdelaziz, S.M., El-Sawy, M.B.I., & Mottaleb, S.A. (2019). Synergetic effects of zinc, boron, silicon, and zeolite nanoparticles on confer tolerance in potato plants subjected to salinity. *Agronomy.* 10, 19

Mahmoud, L.M., Dutt, M., Shalan, A.M., El-Kady, M.E., El-Boray, M.S., Shabana, Y., & Grosser, J.W. (2020). Silicon nanoparticles mitigate oxidative stress of in vitro-derived banana (*Musa acuminata* 'Grand Nain') under simulated water deficit or salinity stress. *S Afr J Bot.* 132, 155–163.

Manzoor, N., Ahmed, T., Noman, M., Shahid, M., Nazir, M.M., Ali, L., Alnusaire, T.S., Li, B., Schulin, R., & Wang, G. (2021). Iron oxide nanoparticles ameliorated the cadmium and salinity stresses in wheat plants, facilitating photosynthetic pigments and restricting cadmium uptake. *Sci Total Environ.* 769, 145221.

Martınez-Ballesta, M.C., Zapata, L., Chalbi, N., & Carvajal, M. (2016). Multiwalled carbon nanotubes enter broccoli cells enhancing growth and water uptake of plants exposed to salinity. *J Nanobiotechnol.* 14, 42.

Martınez-Fernandez, D., Vıtková, M., Michalková, Z., & Komarek, M. (2017). Engineered nanomaterials for phytoremediation of metal/metalloid-contaminated soils: implications for plant physiology. In Ansari, A., Gill, S., Gill, R., Lanza, G.R., Newman, L. (eds.), *Phytoremediation*. Springer, Cham.

Memari-Tabrizi, E.F., Yousefpour-Dokhanieh, A., & Babashpour-Asl, M. (2021). Foliar-applied silicon nanoparticles mitigate cadmium stress through physio-chemical changes to improve growth, antioxidant capacity, and essential oil profile of summer savory (*Satureja hortensis* L.). *Plant Physiol Biochem.* 165, 71–79.

Mirzajani, F., Askari, H., Hamzelou, S., Schober, Y., Römpp, A., Ghassempour, A., & Spengler, B. (2014). Proteomics study of silver nanoparticles toxicity on *Oryza sativa* L. *Ecotoxicol Environ Saf.* 108, 335–339.

Mohammadi, R., Amiri, N.M., & Mantri, L. (2013). Effect of TiO_2 nanoparticles on oxidative damage and antioxidant defense systems in chickpea seedlings during cold stress. *Russ J Plant Physiol.* 61, 768–775.

Mohammadi, R., Maali-Amiri, R. & Mantri, N.L. (2014). Effect of TiO_2 nanoparticles on oxidative damage and antioxidant defense systems in chickpea seedlings during cold stress. *Russ J Plant Physiol.* 61, 768–775.

Mukhopadhyay, S.S., & Kaur, N. (2016). Nanotechnology in soil-plant system. In C. Kole, D. Sakthi Kumar, & M.V. Khodakovskaya (Eds.) *Plant Nanotechnology* (pp. 329–348). Springer International Publishing, Cham.

Munns, R., & Tester, M. (2008). Mechanisms of salinity tolerance. *Annu Rev Plant Biol.* 59, 651–681.

Mustafa, G., & Komatsu, S. (2016). Toxicity of heavy metals and metal-containing nanoparticles on plants. *Biochim Biophys Acta Proteins Proteom.* 1864, 932–944.

Nahar, K., Hasanuzzaman, M., Rahman, A., Alam, M.M., Mahmud, J.A., Suzuki, T., & Fujita, M. (2016). Polyamines confer salt tolerance in mung bean (*Vigna radiata* L.) by reducing sodium uptake, improving nutrient homeostasis, antioxidant defense, and methylglyoxal detoxification systems. *Front Plant Sci.* 7, 1104.

Nakagami, H., Pitzschke, A., & Hirt, H. (2005). Emerging map kinase pathways in plant stress signalling. *Trends Plant Sci.* 10, 339–346.

Nguyen, D.V., Nguyen, H.M., Le, N.T., Nguyen, K.H., Nguyen, H.T., Le, H.M., Nguyen, A.T., Dinh, N.T.T., Hoang, S.A., & Ha, C.V. (2021). Copper nanoparticle application enhances plant growth and grain yield in maize under drought stress conditions. *J Plant Growth Regul.* 1–12.

Pasala, R.K., Khan, M.I.R., Minhas, P.S., Farooq, M.A., Sultana, R., Per, T.S. (2016). Can plant bio-regulators minimize crop productivity losses caused by drought, heat and salinity stress? An integrated review. *J Appl Bot Food Qual.* 89, 113–125.

Patra, A.K., Adhikari, T., & Bhardwaj, A.K. (2016). Enhancing crop productivity in salt-affected environment by stimulating soil biological processes and remediation using nanotechnology. In J.C. Dagar, P.C. Sharma, D.K. Sharma, & A.K. Singh (Eds.) *Innovative Saline Agriculture* (pp. 83–103). Springer, New Delhi.

Pradhan, S., Patra, P., Mitra, S., Dey, K.K., Jain, S., Sarkar, S., Roy, S., Palit, P., & Goswami, A. (2014). Manganese nanoparticles: impact on non-nodulated plant as a potent enhancer in nitrogen metabolism and toxicity study both in vivo and in vitro. *J Agric Food Chem.* 62, 8777.

Prasad, P.V.V., Pisipati, S.R., Mom, I., & Ristic, Z (2011). Independent and combined effects of high temperature and drought stress during grain filling on plant yield and chloroplast EF-Tu expression in spring wheat. *J Agron Crop Sci.* 197, 430–441.

Prasad, R., Kumar, V., & Prasad, K.S. (2014). Nanotechnology in sustainable agriculture: present concern and future aspects. *Afr J Biotechnol.* 13,705–713.

Praveen, A., Khan, E., Ngiimei, D., Perwez, M., Sardar, M., & Gupt, M. (2018). Iron oxide nanoparticles as nano-adsorbents: a possible way to reduce arsenic phytotoxicity in Indian mustard plant (*Brassica juncea* L). *J Plant Growth Regul.* 37, 612–624.

Qi, M., Liu, Y., & Li, T. (2013). Nano-TiO$_2$ improves the photosynthesis of tomato leaves under mild heat stress. *Biol Trace Elem Res.* 156, 323–328.

Rorison, I.H. (1986). The response of plants to acid soils. *Experintia.* 42, 357–362.

Salam, A., Khan, A.R., Liu, L., Yang, S., Azhar, W., Ulhassan, Z., Zeeshan, M., Wu, J., Fan, X., & Gan, Y. (2022). Seed priming with zinc oxide nanoparticles downplayed ultrastructural damage and improved photosynthetic apparatus in maize under cobalt stress. *J Hazard Mater.* 423, 127021.

Semida, W.M., Abdelkhalik, A., Mohamed, G.F., Abd El-Mageed, T.A., Abd El-Mageed, S.A., Rady, M.M., & Ali, E.F. (2021). Foliar application of zinc oxide nanoparticles promotes drought stress tolerance in eggplant (*Solanum melongena* L.). *Plants.* 10, 421.

Shah, T., Latif, S., Saeed, F., Ali, I., Ullah, S., Abdullah., Alsahli, A., Jan, S., & Ahmad, P. (2021). Seed priming with titanium dioxide nanoparticles enhances seed vigor, leaf water status, and antioxidant enzyme activities in maize (*Zea mays* L.) under salinity stress. *J King Saud Univ Sci.* 33, 101207.

Shallan, A., Hassan, M.H., Namich, A.M.A., & Ibrahim, A. (2016). Biochemical and physiological effects of TiO$_2$ and SiO$_2$ nanoparticles on cotton plant under drought stress. *Res J Pharm Biol Chem Sci.* 7, 1540–1551.

Sharma, P., Jha, A.B., Dubey, R.S., & Pessarakli, M. (2012). Reactive oxygen species, oxidative damage, and antioxidant defence mechanisms in plants under stressful conditions. *J Bot.* 2012, 217037.

Shome, S., Bhattacharya, M.K., Panda, S.K., & Upadhyaya, H. (2015). PEG induced water stress alters growth and antioxidant responses in rice (*Oryza sativa* L.). *Ecobios* 8, 73–82.

Singh, J., & Lee, B.K. (2016). Influence of nano-TiO_2 particles on the bioaccumulation of Cd in soybean plants (*Glycine max*): a possible mechanism for the removal of Cd from the contaminated soil. *J Environ Manag.* 170, 88–96.

Singh, S., Tripathi, D.K., Singh, S., Sharma, S., Dubey, N.K., Chauhan, D.K., & Vaculík, M. (2017). Toxicity of aluminium on various levels of plant cells and organism: a review. *Environ Exp Bot.* 137, 177–193.

Sun, D., Hussain, H.I., Yi, Z., Rookes, J.E., Kong, L., & Cahill, D.M. (2018). Delivery of abscisic acid to plants using glutathione responsive mesoporous silica nanoparticles. *J Nanosci Nanotechnol.* 18, 1615–1625.

Suzuki, K., Nagasuga, K., & Okada, M. (2008). The chilling injury induced by high root temperature in the leaves of rice seedlings. *Plant Cell Physiol.* 49, 433–442.

Syu, Y.Y., Hung, J.H., Chen, J.C., & Chuang, H.W. (2014). Impact of size and shape of silver nanoparticles on *Arabidopsis* plant growth and gene expression. *Plant Physiol Biochem.* 83, 57–64.

Tarafdar, J.C., Sharma, S., & Raliya, R. (2013). Nanotechnology: interdisciplinary science of applications. *Afr J Biotechnol.* 12, 219–226.

Taran, N., Storozhenko, V., Svietlova, N., Batsmanova, L., Shvartau, V., & Kovalenko, M. (2017). Effect of zinc and copper nanoparticles on drought resistance of wheat seedlings. *Nanoscale Res Lett.* 12, 60.

Torabian, S., Zahedi, M., & Khoshgoftar, A.H. (2016). Effects of foliar spray of two kinds of zinc oxide on the growth and ion concentration of sunflower cultivars under salt stress. *J Plant Nutr.* 39, 172–180.

Tripathi, D.K., Singh, S., Singh, V.P., Prasad, S.M., Chauhan, D.K., & Dubey, N.K. (2016). Silicon nanoparticles more efficiently alleviate arsenate toxicity than silicon in maize cultivar and hybrid differing in arsenate tolerance. *Front Environ Sci.* 4, 46.

Tripathi, D.K., Singh, S., Singh, V.P., Prasad, S.M., Dubey, N.K., & Chauhan, D.K. (2017). Silicon nanoparticles more effectively alleviated UV-B stress than silicon in wheat (*Triticum aestivum*) seedlings. *Plant Physiol Biochem.* 110, 70–81.

Tripathi, D.K., Singh, V.P., Prasad, S.M., Chauhan, D.K., & Dubey, N.K. (2015). Silicon nanoparticles (SiNp) alleviate chromium (VI) phytotoxicity in *Pisum sativum* (L.) seedlings. *Plant Physiol Biochem.* 96, 189–198.

Tyagi, J., Chaudhary, P., Mishra, A., Khatwani, M., Dey, S., & Varma, A. (2022). Role of endophytes in abiotic stress tolerance: with special emphasis on *Serendipita indica*. *Int J Environ Res.* 16, 62.

Venkatachalam, P., Jayaraj, M., Manikandan, R., Geetha, N., Rene, E.R., Sharma, N.C., & Sahi, S.V. (2017). Zinc oxide nanoparticles (ZnONPs) alleviate heavy metal-induced toxicity in *Leucaena leucocephala* seedlings: a physiochemical analysis. *Plant Physiol Biochem.* 110, 59–69.

Wahid, A. (2007). Physiological implications of metabolites biosynthesisin net assimilation and heat stress tolerance of sugarcane (*Saccharum officinarum*) sprouts. *J Plant Res.* 120, 219–228.

Wu, H., Shabala, L., Shabala, S., & Giraldo, J.P. (2018). Hydroxyl radical scavenging byerium oxide nanoparticles improves *Arabidopsis* salinity tolerance by enhancing leaf mesophyll potassium retention. *Environ Sci Nano.* 5, 1567–1583.

Xiong, L., Schumaker, K.S., & Zhu, J.K. (2002). Cell signalling during cold, drought, and salt stress. *Plant Cell Online.* 14, S165–S183.

Yadav, S., Irfan, M., Ahmad, A., & Hayat, S. (2011). Causes of salinity and plant manifestations to salt stress: a review. *J Environ Biol.* 32, 667–685.

Yadav, T., Mungray, A.A., & Mungray, A.K. (2014). Fabricated nanoparticles: current status and potential phytotoxic threats. In Whitacre, D.M. (ed.) *Reviews of Environmental Contamination and Toxicology*. Springer International Publishing, Switzerland.

Yan, S., Wu, F., Zhou, S., Yang, J., Tang, X., & Ye, W. (2021). Zinc oxide nanoparticles allevi-
    ate the arsenic toxicity and decrease the accumulation of arsenic in rice (*Oryza sativa*
    L.). *BMC Plant Biol.* 21, 1–11.
Zhao, L., Peng, B., Hernandez-Viezcas, J.A., Rico, C., Sun, Y., Peralta-Videa, J.R., Tang, X.,
    Niu, G., Jin, L., Varela-Ramirez, A., Zhang, J., & Gardea-Torresdey, J.L. (2012). Stress
    response and tolerance of *Zea mays* to CeO_2 nanoparticles: cross talk among H_2O_2, heat
    shock protein and lipid peroxidation. *ACS Nano.* 6, 9615–9622.
Zhou, P., Adeel, M., Shakoor, N., Guo, M., Hao, Y., Azeem, I., Li, M., Liu, M., & Rui, Y.
    (2020). Application of nanoparticles alleviates heavy metals stress and promotes plant
    growth: an overview. *Nanomaterials.* 11, 26.

13 Environmental-Friendly Nanoparticles in Agriculture

Parul Chaudhary and Anuj Chaudhary
Graphic Era Hill University and Shobhit University

Vaibhav Dhaigude
ICAR-NDRI

Sami Abou Fayssal
University of Forestry
Lebanese University

Pankaj Kumar
Gurukul Kangri University

Anita Sharma
Govind Ballabh Pant University of Agriculture & Technology

Shaohua Chen
South China Agricultural University

CONTENTS

DOI: 10.1201/9781003345565-13

13.1 INTRODUCTION

India's most significant economic sector has always been agriculture, which plays a significant role in the development of the national economy. It is very necessary nowadays to meet the dietary needs of the constantly expanding world population. Nearly one-third of crops grown conventionally get damaged, mostly as a result of pest infestation, abiotic stress, microbiological assaults, natural disasters, poor soil quality and lack of available nutrients (Mittal et al., 2020; Tyagi et al., 2022; Chaudhary et al., 2022a). To solve such problems, more avant-garde technologies are immediately needed. In this way, the agrotechnological revolution that is about to change the current agricultural system while ensuring food security has been made possible in part by nanotechnology. Particles with at least one dimension less than 100 nm are referred to as nanoparticles (NPs), which are thought to be the basis of nanotechnology (Biswas & Wu, 2005). Nanotechnology has shown promising potential in promoting sustainability in the agricultural field. It plays a significant role in crop production and protection with a focus on nano-enabled remediation techniques for contaminated soils, nano-enabled fertilizers, nanopesticides and nanobiosensors (Usman et al., 2020; Bhatt et al., 2022).

However, the increased use of NPs in various sectors has increased environmental contamination. Abiotic elements, i.e. water, soil and air, that are closely linked to human health may be affected by NPs released into the environment by industrial or commercial sectors (Hashimoto et al., 2017). Some NPs are excessively utilized in consumer items; particularly, silver NPs (AgNPs) alone accounted for around 20% of all NPs' utilization, which is roughly three times that of the second most frequently used carbon-based NPs (Rai et al., 2018). Due to the use of hazardous chemicals or the production of toxic by-products, the NM synthesis by chemical methods is often not totally safe or environmentally friendly. However, NP synthesis via plants by green methods is considered both cost-effective and environmentally safe. Green methods play a vital role in the phytosynthesis and biosynthesis of NPs, making them an eco-friendly and sustainable tool (Nadaroglu et al., 2017). NPs have shown efficiency when compared to bulk particles due to the former's exceptional electrical and optical properties, enormous specific surface area, ease of functionalization, presence of active sites on the surface, insane high stability and high adsorption capacity (Wang et al., 2015; Gowda et al., 2016; Kumar et al., 2016; Xiao et al., 2016). We firmly believe that in order to achieve the goal of promoting global agriculture, a comprehensive perspective on nanotechnology is necessary. In this book chapter, we focus on the key issues surrounding the environmentally friendly NP-based agricultural research that has the potential to increase crop production and food security in the near future.

13.2 NANOPARTICLES AND ENVIRONMENT

Nanotechnology is a growing area of pharmaceutical science, where particles evolve to nanoscales and become more responsive when compared to their original counterparts. The topic of "environmental nanoparticles" research is new and expanding quickly. Before we can optimize the benefits of NPs and ensure that there are no potential negative effects on the environment and human health, there is still much work to be done. Understanding the environment in which NPs are found or used is crucial when discussing their significance in the environment. For example, iron metal NPs are being used to clean up places that have been contaminated by anthropogenic contaminants such as chlorinated hydrocarbons (Zhang, 2003).

As the size, shape and quantification of these particles to be released and consumed determine their positive impacts on the environment, their characterization via SEM, EDS and TEM should be carried out in order to understand their destiny and environmental and health implications (Burleson et al., 2004). The source of NPs is another important consideration in this context because it is still poorly understood how anthropogenic and naturally occurring NPs are transported in the environment and in living systems and whether they help or hinder the movement of chemical species (Lead & Smith, 2009). Anthropogenic NPs, which are manufactured materials, must be distinguished from naturally occurring NMs (Bakshi et al., 2015). The methods used for NPs' synthesis are also an important factor as the biological method of this process is relatively simple, cheap and environmentally friendly compared to the conventional chemical method of synthesis (Kulkarni & Muddapur, 2014). However, the causes of nanoecotoxicity and the ambient NP concentrations remain largely undefined. The clarification of the nanotoxicity of NPs in the environment and plant biosystems should be the focus of future research.

13.3 NANOPARTICLES IN AGRICULTURE

The potential uses of NMs in agriculture to increase crop productivity and boost soil health are emphasized in this section. Here, we highlight a number of advancements in the fields of nanopesticides, nanofertilizers and nanobiosensors.

13.3.1 NANOPESTICIDES

Nanotechnology is an increasingly significant aspect in the evolution of agriculture. The development of fertilizers, pesticides, genetically modified organisms and agricultural equipment is made possible by this route. Pesticides are one of the most significant areas of the agricultural field. Approximately six billion pounds of pesticides are consumed worldwide yearly. Therefore, by turning these pesticides into nano ones, the amount of yearly consumed pesticides would be decreased significantly (Sabry, 2020). The amounts of pesticides used to control pests can be reduced by adopting these nanopesticides, which would lower crop production costs and subsequently pesticide residues in food and detrimental impact on human health and the environment. The potential of nanopesticides is summarized in Figure 13.1.

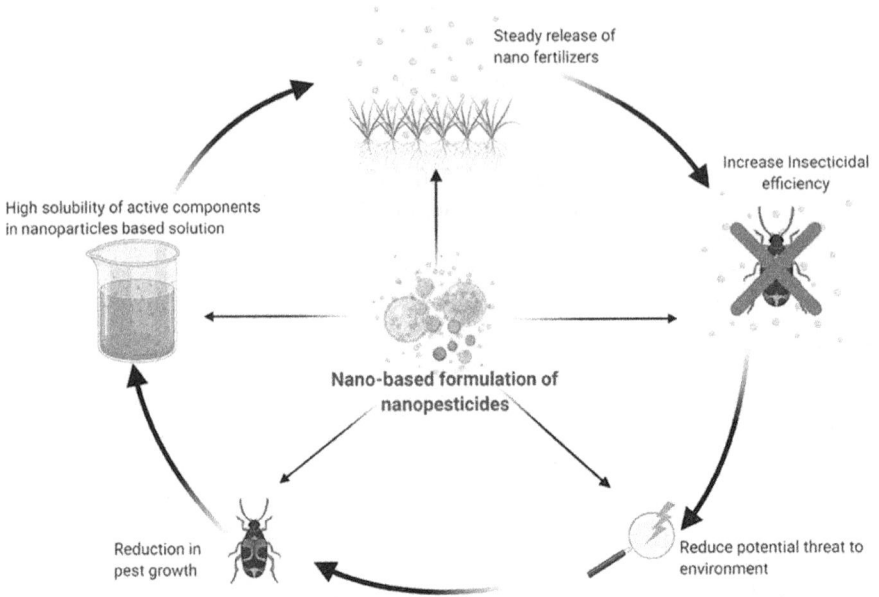

FIGURE 13.1 Nano-based formulation of nanopesticides.

The term "nanopesticides" refers to pesticides that have been particularly fixed on a hybrid substrate, enclosed in a matrix or functionalized as nanocarriers for environmental triggers or enzyme-mediated triggers as an NM (Chaud et al., 2021). Chhipa (2017) mentioned that nanopesticides such as Ag, Cu, SiO_2, ZnO and nanoformulations showed better broad-spectrum pest protection efficiency in comparison with ordinary pesticides. Plant-derived nanopesticides have been demonstrated to be more successful at reducing pest populations and plant infestation levels than bulk forms of pesticides due to their controlled and sustained toxicant release. Especially, polymer-based nanoformulations have been commercially used to encapsulate neem-derived products (Gahukar & Das, 2020). Nanopesticide formulations potentially revolutionize the management of pathogens, weeds and insects in crops by improving water solubility, bioavailability and protecting agrochemicals from environmental degradation (Yadav et al., 2020).

13.3.2 NANOFERTILIZERS

Crop yield per unit area increased as a result of the industrial and green revolutions, followed by a growth in the use of synthetic fertilizers in agriculture. Agriculture's heavy reliance on inorganic fertilizers results in a variety of health issues and irreversible environmental degradation. Traditional fertilizers are not only costly for the producer, but also harmful to humans and the environment. New agricultural approaches have been developed to lessen and eventually remove the negative impacts of synthetic fertilizers on the environment and human health

(Sharma & Chetani, 2017). Among these approaches, nanofertilizers were produced to improve agriculture productivity by crop nutrition management through delivering nutrients to crops in a controlled release manner and by improving the growth of beneficial microbes (Chaudhary & Sharma, 2019; Al-Mamun et al., 2021; Agri et al., 2022).

NFs seem to be a promising substitute to chemical fertilizers by exerting a positive role with their gradual nutrient release, improved nutrient use efficiency and reduced nutrient leaching into ground water (Zulfiqar et al., 2019). The use of NFs in agriculture provides excellent prospects to enhance plant nutrition and stress tolerance in order to obtain higher production in a climate change context. Metal oxides, carbon-based NPs and other nanoporous materials, which vary in compositional and combinational properties, are included under the NFs' concept (Liu & Lal, 2015). It is possible to develop various types of NFs, such as micronutrient NFs, macronutrient NFs, and NM-enhanced fertilizers, through years of research based on specific nutrient supplies (Duhan et al., 2017). These NFs showed improvements in terms of solubility, dispersion of insoluble nutrients, bioavailability and accessibility of definite delivery to reduce nutrient losses (Prasad et al., 2017).

13.3.3 NANOBIOSENSORS

When nanotechnology is used to develop biosensor technology, it creates an effective nanobiosensor (NBS) with a smaller structure than traditional biosensors. Numerous fertilizers, herbicides, pesticides, insecticides, diseases, moisture and soil pH may all be detected with NBSs (Rai et al., 2012). When used controllably and appropriately, NBSs can promote sustainable agriculture by increasing crop yield. By effectively monitoring soil composition in terms of pH, humidity, microbial load and other factors, NBSs can serve as an effective tool to increase production. It can also cope with resource management and pollution prevention. With multiplex and real-time sensing capabilities, electrochemical nanosensors, optical nanosensors, nano barcode technology and wireless nanosensors have revolutionized the sensing in the food and agricultural sectors (Srivastava et al., 2018).

13.4 PLANTS' INTERACTION WITH NANOPARTICLES

The interaction between NPs and plants is a core part of the risk assessment since plants are fundamentally important components of the ecosystem. NPs' phytotoxicity is a crucial prerequisite for encouraging the use of nanotechnology and preventing potential ecological concerns. Different NPs have been shown to have both enhancing and inhibiting effects on the growth of different plants (Agri et al., 2021). NPs can be involved in a variety of agricultural systems, i.e. sewage sludge addition to biosolids, soil amendment and irrigation of polluted water. In view of crop nutrients, biochemistry, molecular biology and human health, NPs must be thoroughly explored (Rai et al., 2018). NMs' absorption, bioaccumulation, biotransformation and potential hazards to food crops are still poorly investigated as only few NMs and their interaction with plant species have been studied, mostly in their early phases of development.

Interestingly, numerous NPs showed beneficial effects, such as titanium NP (TiNP) and cerium NP (CeNP) phytoaccumulation in tomato, cucumber and rice (Rico et al., 2013; Song et al., 2013; Zhao et al., 2013). Nowadays, the fate of NPs inside crops can be visualized using some advanced tools such as SP-ICP-MS, which gave a first-hand knowledge on the phytoaccumulation of gold NPs (AuNPs) in tomato plants by clarifying the particle size and concentration using the macerozyme R-10 enzyme (Dan et al., 2015). Wang et al. (2015) also found that silver NPs (AgNPs) are translocated to roots and shoots of two terrestrial agro-crops: *Vigna unguiculata* and *Triticum aestivum*. Balancing plant growth and stress responses is one of the most critical functions of agricultural application to improve plant production. Under drought stress conditions, applying copper NPs (CuNPs) to plants enhances total seed quantity and grain yield by reducing stress factors in maize (Van Nguyen et al., 2022).

13.5   IMPACT OF NANOPARTICLES ON ENVIRONMENTAL ASSESSMENT AND RISK ANALYSIS

Nanotechnological products, methods and applications are expected to significantly contribute to environmental and climate protection by saving natural resources, energy and water besides reducing greenhouse gas emissions (GHGs) and hazardous wastes. Therefore, by deploying certain advanced sensors to identify contaminants, we may help in the protection of both human health and the environment. In order to decrease waste production, GHGs and the release of harmful chemicals into water bodies, nanotechnology acts as an innovative solution (Liu & Meng, 2020). Due to their widespread and unrestricted usage, NPs have prompted researchers to think about the issues, difficulties and negative impacts on the environment demanding undoubtedly risk assessments (Gottschalk et al. 2015; Tolaymat et al. 2015). NPs have a poor ability to migrate through soil; nevertheless, their route of entry is through soil application, atmospheric deposition, rain erosion, surface runoff or other pathways. The exposure modelling also revealed that soils may be the primary source of NPs released into the environment as their concentrations found in soil are higher than those found in water or air (Gottschalk et al. 2009). Therefore, for a better sustainment of agriculture and an avoidance of any potential environmental risk, the assessment of soil quality, seed quantity and overall crop production with the help of nanotechnology is a must.

13.6   ADVANTAGES OF ENVIRONMENTAL-FRIENDLY NANOPARTICLES IN AGRICULTURE FIELD

After phytoaccumulation, the effects of NPs/NMs on agroecosystem crops might be favourable, negative or neutral. The uptake of NPs by edible plants or agro-resources is an important topic due to the implications for human health besides the effect on agricultural output. Batley et al. (2013) reported that NPs' amounts found in the environment are now far lower than the hazardous level. Recently, few advanced studies were conducted regarding the beneficial impact of NPs on agriculture crops (Table 13.1). Drought stress was regarded as more damaging to wheat plants than heat stress, whereas combined stresses are much more detrimental for plant growth.

TABLE 13.1

Environmentally Friendly NPs and Their Impacts on Crop Production/ Growth and Soil Health

NPs	Crop Studied	Effects of NPs on Crop	References
ZnNPs	*Triticum*	Reduce drought and heat stress in wheat	Azmat et al. (2022)
CuNPs	*Zea mays*	Enhance total seed quantity and grain yield	Van Nguyen et al. (2022)
Bioconjugation of AgNPs with plant extracts	Medicinal plant extraction	Decreases AgNP toxicity and enhances their medicinal effect	Majeed et al. (2022)
Melatonin-selenium nanoparticles (MT-SeNPs)	*Brassica napus* leaves	Increase antioxidant enzymes and mitigate the arsenic-induced stress	Farooq et al. (2022)
CeO_2 Cerium Oxide	*Trigonella foenum-graecum*	Used as biofertilizer, enhances crop yield	Sonali et al. (2022)
FeNPs	*Hordeum vulgare* L.	Diminish contaminants from soil and enhance bioavailability of minerals for plant	Rodríguez-Seijo et al. (2022)
AgNPs	Hayani cv.	Increase shoot and root length and leaf number	Elsayh et al. (2022)
Hydroxyapatite	*Rosmarinus officinalis* L.	Enhances plant growth and oil components	Elsayed et al. (2022)
Carbon NPs	*Zea mays* L.	Improve plant growth and soil health	Xin et al. (2022)
MgONPs	*Vigna radiata*	Promote plant growth and show antifungal activity to prevent plant from diseases	Abdallah et al. (2022)
AuNPs	*Triticum aestivum*	Develop salt tolerance and sustain nitrogen metabolism and ionic balance	Wahid et al. (2022)
PtNPs	*Cucumis sativus*	Enhance flavonoids content that boosts plant's defence system	Astafurova et al. (2017)
PdNPs	*Sinapis alba*	(Pd $(His)_2$) is formed upon exposure to PdNPs	Kińska et al. (2019)
NiONPs	*Brachychiton populneus*	Synthesis of biodiesel from non-edible oil	Dawood et al. (2021)
Co_3O_4NPs	*Brassica napus* L.	At low concentrations, it promotes plant growth and development	Jahani et al. (2020)
TiO_2	*Abelmoschus esculentus* L. Moench	Reduces the uptake of heavy metals by plant	Kumar et al. (2022)

The impact of stresses was reduced by zinc oxide NPs (ZnONPs), which improved the biochemical metabolism, while rhizobacteria provided stress resistance through increased osmoregulation and antioxidants (Azmat et al., 2022). These authors found that a co-application of ZnONPs and plants' growth-promoting rhizobacteria (PGPR) enhances the stress tolerance of wheat crop.

In another study, Van Nguyen et al. (2022) concluded that applying copper NPs (Cu-NPs) to the crop increases the total seed number and grain yield under drought stress conditions. To the best of our knowledge, although the toxicity of AgNPs has been documented to be related to the silver ion fraction in the AgNP suspension, it has been extensively used in different therapeutic applications (Beer et al., 2012). To overcome the toxic effect of silver, Majeed et al. (2022) found that the bioconjugation of AgNPs with plant extracts decreases silver toxicity to biosystems and hence enhances their effectiveness. In *Brassica napus* leaves, melatonin-selenium NPs (MT-SeNPs) increased the antioxidant enzymes, decreased the effects of arsenic-induced stress and promoted plant growth (Farooq et al., 2022).

To better understand the environmentally friendly role of such NPs, their use in controllable manner with safety margin should be followed. Purposely, more research is needed in this field to assess the benefits and risks of nanotechnology in an equitable manner.

13.6.1 Nanozeolite

With an infinite three-dimensional crystal structure, zeolites are hydrated aluminosilicates that include cations of alkaline elements, alkaline soil elements and, less frequently, other cations (Andrzejewska et al., 2017). Nanozeolites are applied in agriculture for crop yield enhancement and maintenance of soil health (Babu et al., 2022; Chaudhary et al., 2022b), in industrial application for photocatalysis (Oviedo et al., 2022a; Wang et al., 2022) and antimicrobial activity (Oviedo et al., 2022b). Clinoptilolite, a natural zeolite, is one of the most prevalent minerals in soils and sediments and frequently employed in farming operations as a soil reformer to enhance soil nitrogen uptake. It has a high cation exchange property and attracts ammonium ion (Kianfar, 2019). Chabazite is a small-pore zeolite that can be utilized in ion exchange for the removal of contaminants from effluents (Zeng et al., 2010). Izzo et al. (2022) reported that chabazite acts as a sustainable-remediation agent for wastewater contaminants' removal. Adhikari et al. (2022) used nanozeolites for the removal of lead from contaminated soils. Nanozeolite along with beneficial microbes improved soil physicochemical properties and microbial population and subsequently crop productivity (Chaudhary et al., 2021a, b, 2022b). Additionally, nanozeolite NPs also increased the population of beneficial microbes and improved the soil health through enhancing soil properties (Khati et al., 2019a, b).

13.6.2 Nanochitosan

Chitosan is the second largest renewable energy source in the world and is the most prevalent polysaccharide in marine systems (Fernandez & Ingber, 2014). Chitosan is a non-toxic, biodegradable and high-molecular-weight polymer being remarkably comparable to the plant fibre cellulose. It actually comes from fungus cell walls as well as from the crustacean shells of prawns and crabs. It is a linear copolymer made of D-glucosamine and N-acetyl-D-glucosamine units connected by (1–4) glycosidic linkages and has been widely used to create NPs. Due to its antibacterial, antioxidant and chelating properties as well as its non-toxic and biocompatible makeup, chitosan has gained popularity (Gedda et al., 2016). Due to their low toxicity, muco-adhesion and adaptable physical properties, chitosan-based NPs are especially suitable for mucosal-route drug delivery approaches (Nagpal et al., 2010). Chitosan has long been recognized for their potential as biocontrol agents (Malerba & Cerana, 2016). Many antifungal activities of nanochitosan (NCS) have been detected against phytopathogens, hence acting as an efficient biotic elicitor that causes Induced Systemic Resistance (ISR) in plants (Kalagatur et al., 2018; Al-Zahrani et al., 2021). Through biochemical and gene expression modification, Ali et al. (2021) recently examined the function of NCS in the adaptation to drought stress tolerance in *Catharanthus roseus* (L.). NCS (50 ppm) improved maize yield, growth parameters and soil health (Chaudhary et al., 2021b, c). Agri et al. (2022) detected the positive effect of putative bioinoculants and NCS on the physiological response of maize plants. Kumari et al. (2020) also reported a positive impact of NCSNPs on fenugreek crop by improving soil enzymatic activity.

13.6.3 NANOGYPSUM

Gypsum is a soluble source of calcium and sulphur that helps plants to thrive. Gypsum amendments can enhance some soils' physical characteristics (Chaudhary et al., 2022c). It encourages soil aggregation, which stops soil particles from dispersing, minimizes the formation of surface crusts, encourages seedling emergence and speeds up water infiltration rates and movement through the soil profile (Chen et al., 2005). Gypsum and nanogypsum aid in the reclamation of sodic soil by producing Ca^{2+} exchange complexes (Kumar & Thiyageshwari, 2018). Chaudhary et al. (2021d) reported an enhanced maize yield and an improved soil microbial diversity after nanogypsum implementation. Because of some ecotoxicological effects and inconsistent performance, employing nanocompounds in agriculture has not yet been practised to a great extent. Therefore, using a blend of native PGPR and nanocompounds in agricultural activities (as a bioinoculant) could be a great way to promote plant health, preserve soil health and lessen the impact of hazardous chemicals on agricultural products (Sekhon, 2014; Liu & Lal, 2015).

13.6.4 SILICON DIOXIDE NANOPARTICLES

After oxygen, silicon (Si) is the second most prevalent element in the Earth's crust. Despite being a non-essential component for higher plants, it is intricately tied to plant growth, particularly for the Gramineae species (Gaur et al., 2020). Silicon NPs (SiNPs) have several uses since they can be frequently produced at low cost, are hydrophobic, have a large surface area and pore volume and are biocompatible. For instance, SiNPs were used to address numerous agricultural issues due to their exceptional adsorption power and non-toxic nature. SiNPs can give plants resistance by creating a dual layer on the epidermal cell wall and by modulating the antioxidant defence system to lessen the damaging effects of stress (Du et al., 2022). Furthermore, SiO_2 also prevents plants from physical stresses (radiation, drought) (Kandhol et al., 2022; Zaki et al., 2022) and chemical stresses (salt and heavy metals) (Cui et al., 2022; Ding et al., 2022). Khafri et al. (2022) documented enhanced lateral roots in Myrobalan 29C (*Prunus cerasifera* L.) using chemically synthesized SiNPs, rice husk-derived biogenic SiNPs and amine-modified SiNPs. Mousavi et al. (2022) found a reduced soil lead contamination after the application of SiNPs to *Pseudomonas fluorescens.*

13.6.5   Titanium Dioxide Nanoparticles

Due to their widespread usage in coatings, cosmetics and pigments as well as their low toxicity, excellent optical qualities, potent adhesive properties and high stability, TiO_2NPs are among the most engineered NPs worldwide (Camps et al., 2017). TiO_2NPs have also been applied in nanomedicine, nanobiotechnology, solar and electrochemical cells, wastewater treatment, food, soil remediation, gas sensing, plastics, paint and paper production, hydrogen fuel generation, antiseptics and antibacterial compositions, self-cleaning devices and printing inks. TiO_2 owns an antimicrobial activity and can be extracted from *Lippia adoensis* (Kusaayee) (Gudata et al., 2022). Balaraman et al. (2022) reported the formation of TiO_2 using *Sargassum myriocystum* playing a crucial role in different dye degradation. Parveen and Siddiqui (2022) reported the impact of TiO_2 on tomato growth, development, bacterial and fungal diseases. Application of TiO_2NPs along with bacterial isolates improved maize plant growth and soil enzyme activities under maize cultivation (Kumari et al., 2021).

13.6.6   Silver Nanoparticles

A lot of attention has been paid to AgNPs due to their powerful antimicrobial properties (Bruna et al., 2021). They are frequently used in the food preservation industry, healthcare (medical equipment and medicine delivery) and textile coatings (Backx et al., 2021; Ahmed et al., 2022; Zhang et al., 2022). Silver ions are important for the development of shoots, roots and somatic embryos. Using silver, plants produce less ethylene, which lowers their risk of chlorosis and chlorophyll degradation. Additionally, it speeds up the formation of secondary metabolites, slows down ageing and enhances crop development and grain yield (Yasmeen et al., 2015). Due to their low cost, activity at lower concentrations, reduced complexity and higher reliability, biosynthesized NMs have been used to address environmental issues and promote economic sustainability (Bapat et al., 2022, Mechouche et al., 2022). Mahajan et al. (2022) stated that using AgNPs under *in vitro* conditions enhanced plant growth and production of metabolites. Jiao et al. (2022) reported the AgNPs used in nitrogen transformation enlighten the role of enzymes such as nitrite- and nitrate-reductase. Danish et al. (2022) reported improved growth, phenolic content and gas exchange parameter in *Withania somnifera* L. (Ashwagandha) treated with AgNPs.

13.7 NANOPARTICLES' EFFECT ON SOIL ENZYMES, SOIL PHYSICOCHEMICAL CHARACTERISTICS AND MICROBIAL POPULATION

The effect of NPs on plants is an indirect process, being a subsequent result of the former's impact on soil enzymes (Khati et al., 2017). There are contradictory reports whether NPs have positive or negative impacts on soil enzymes (Khati et al., 2018). Some types of NPs such as CuNPs, NiNPs and ZnNPs are considered as contaminants for soils as Cu, Ni and Zn are classified as potentially toxic elements for the environment, crops and, further, human life. On the other hand, an improved enzymatic activity in soils was acknowledged (Chaudhary et al., 2021e). Such types of NPs have a high potential to spread in soils and make structural changes to the physicochemical properties of different soil types on a long term; however, until recent time, few details were known in this context. Subsequently, changes in these properties have a direct effect on soil microbiotas that are very sensitive even to very minor fluctuations in soil nutrient availability, pH, electrical conductivity (EC) and so on.

In addition, the concentration of such NPs in soils as well as the type of soils are factors determining the enzymes' and microbiotas' sensitivity. For instance, low to high concentrations of CuNPs, NiNPs and ZnNPs had an average low reduction (14%–32%) of catalase and dehydrogenase enzymes' activity. Some bacterial populations, i.e. *Azotobacter* sp., were more sensitive (22%–53%) to NPs application to soils. Soil ecotoxicity levels of 22%, 30% and 34% were also reported for Ni, Zn and Cu, respectively (Kolesnikov et al., 2021). In the same vein, NPs are main sources of soil organic carbon responsible for soil fertility (Table 13.2). Enzymes are responsible of organic matter degradation and liberation of organic carbon. Some negative impacts of NPs may be observed on soils in the future in terms of reduced β-glucosidase, xylanase and protease activities (Tripathi et al., 2022). Another study mentioned a decrease of saline-alkali soil's total microbial population by 36.3%–48.0% when ZnONPs were implemented (You et al., 2018). However, soil type was considered as the main factor influencing and accentuating this negative impact rather than the used NPs. CuNPs had severe effects on soil microbiotas, i.e. *Bacillus subtilis* and *Pseudomonas fluorescens*, with Gram-negative ones more affected than Gram-positive ones (Sharma et al., 2021). High NPs' concentrations can also accentuate more or less these effects; a reduction of dehydrogenase activity was previously outlined as a result of excessive NP application (Josko et al., 2014). Moreover, NPs can directly increase the mobility of soil pollutants; the latter are also influenced by NP size and soil's organic matter which further alters microbial populations (Rajput et al., 2018). Bacterial population diversity and abundance can also be impacted by NPs, i.e. titanium dioxide NPs (Solanki et al., 2008). These NPs even if applied in very low concentrations triggered cascading negative effects on denitrification enzyme activity associated with a modification of the structure of bacterial populations (Simonin et al., 2016). Native soil bacteria were severely affected by CuONPs' implementation in soils (Concha-Guerrero et al., 2014). According to Khan et al. (2022), these negative results could be obtained due to the application of unrealistically high NP concentrations, being much higher than could be found in natural soils.

Low to moderate concentrations of AgNPs led to a survival of *Nitrosomonas europaea* while high AgNP concentrations were more or less lethal as a result of the big damage in the cell wall of ammonia-oxidizing bacteria under study (Wang et al., 2017). In the same vein, AgNPs acted as a bactericidal against *Actinobacteria*, *Firmicutes* and *Proteobacteria* (Chavan & Nadanathangam, 2019). A recent report outlined the high toxicity, lethality and inhibition of plant growth-promoting bacteria by high application of NPs (Ameen et al., 2021).

On the other hand, the oxidation of organic pollutants was catalysed by metal oxide NPs, i.e. CuO and Fe_3O_4. However, the type of soil and NP size played a big role in promoting more or less this impact (Ben-Moshe et al., 2013). There was very minor to no impact of CeO_2NPs on plants; however, the effect of such NPs on soil and the interaction between soil composition and microbiotas on one hand and NPs on the other was not evaluated (Servin et al., 2017). In a similar vein, AuNPs showed no impact on soil enzymatic activity and soil composition (Maliszewska, 2016). Besides that, NPs' nature (organic, inorganic) can significantly influence soil microbiotas; organic NPs showed fewer toxic impacts compared to inorganic NPs (Frenk et al., 2013). More findings were reported regarding the delay of nitrogen fixation (Huang et al., 2014), i.e. by using AgNPs (Kumar et al., 2014). Later, contradictory results were depicted by Kumar et al. (2015) and Mehta et al. (2016) who found an enhanced root nodulation in cowpea, green gam, cluster and moth bean when using low proportions of AgNPs and ZnONPs; it was a complete antagonist effect when these proportions arise. Moreover, an improved gene expression by 2.0–3.0-fold of *N. europaea* was detected when using AgNPs (Yang et al., 2013).

Maintaining a healthy soil could be promoted using NPs (Peteu et al., 2010) by decreasing nitrogen's transfer rate into the environment (Cai et al., 2014) expressed by nitrogen runoff and leaching (decrease by 21.6% and 24.5%) (Liu et al., 2016). While some reports mentioned the negative short-term effect of NPs on soil enzymatic activity, others mentioned the difficulty to notice any impact just only on a long term. These contradictory claims are obvious as the mechanisms of most NPs are still unexplored reliably (Eivazi et al., 2018).

Although some studies reported the efficiency of NPs to decrease the ability of microbial population to cope with abiotic and biotic stresses (Simonin et al., 2018), other researchers found numerous positive impacts of NPs, i.e. SiNPs, in the promotion of enzymatic activity in order to face different stresses by increasing their tolerance levels compared to no or traditional fertilizers (Figures 13.2 and 13.3). Among them are the following: drought tolerance, heat tolerance, salt tolerance, heavy metal (Al, As, Cd, Cr) tolerance and fluoride tolerance (Bhat et al., 2021).

Another study found that ZnONPs improved the structure of microbiota acting as a green chemical reagent against apple replant disease (ARD) (Pan et al., 2022). This type of NP was also reported to enhance β-glucosidase and cellulase activities, microbial biomass and soil respiration (Jiang et al., 2020). Citrate-coated AuNPs enhanced enzyme activity and altered soil microbial microbiota, while small-sized PVP-coated AuNPs improved the enzymatic activity of soils when added in low concentrations only (Asadishad et al., 2017). Chaudhary et al. (2022b) reported that NPs combined with *Bacillus* spp. improved total microbial count, NPK solubilizing bacteria, the level of soil health indicator enzymes

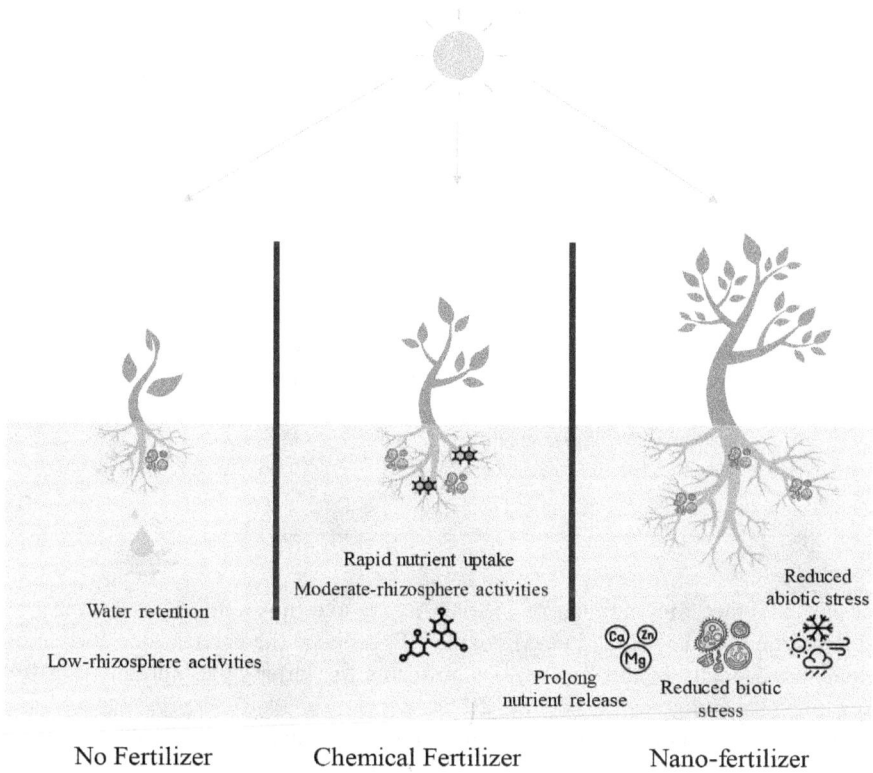

FIGURE 13.2 Comparison between control, chemical and nanofertilizers' impacts.

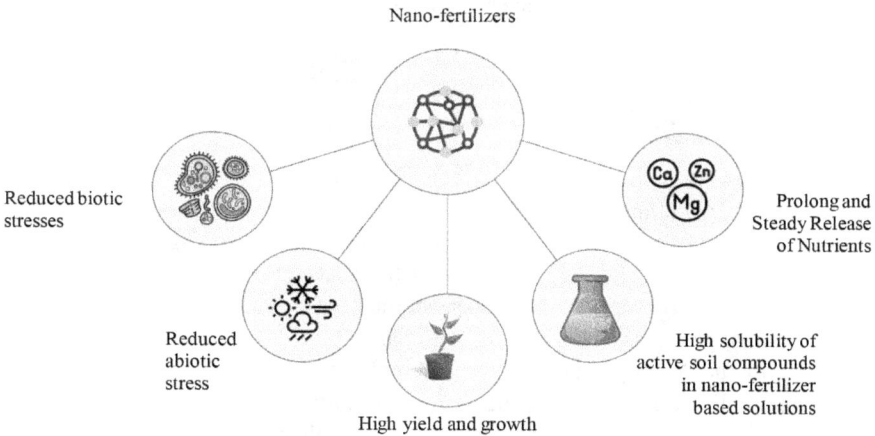

FIGURE 13.3 NPs benefits on soil physicochemical status and microbial populations.

TABLE 13.2
Effect of Nanoparticles on Soil Enzyme Activities and Microbial Population

Nanoparticles	Microbial Activities	References
Nanochitosan	Improved fluorescein (FDA) and alkaline phosphatase activities	Khati et al. (2017)
AuNPs	Improved five important soil enzyme activities	Asadishad et al. (2017)
ZnNPs	Improved soil enzymes and microbial population	Jiang et al. (2020)
Fe-Cu	Improved soil enzymes	Das et al. (2021)
SiO_2NPs	Improved FDA activity	Chai et al. (2015)
AgNPs	Increased *Proteobacteria* population	Chavan and Nadanathangam (2019)
$nCeO_2$	Decrease bacterial population of *Proteobacteria*	Kamika and Tekere (2017)
Nanosilver NPs	Decrease FDA activity	Sillen et al. (2015)
Silver NPs	Decrease dehydrogenase activity	McGee et al. (2017)
MnO_2 NPs	Decrease bacterial population such as *Cyanobacteria*	Shao et al. (2015)

besides a higher copy number of 16SrRNA gene and thus a flourishing impact on soil microbiotas. Fe-Cu and Fe-Mn oxide NPs reduced the particle size and cation exchange capacity of the soil besides buffering the latter's pH, upregulating soil key enzymes and genes (Das et al., 2021).

Thus, the complexity of soils and their initial enzymatic and microbiota diversities are main factors affecting the degree of influence and interaction with NPs' implementation into soils (Dimkpa, 2018). It is noteworthy to mention that sometimes the impact of NPs on plants and microorganisms is contradictory; it could have varied according to soil and particles' physicochemical characteristics (Tomacheski et al., 2017).

13.8   CONCLUSION

In this chapter, we outlined the effect of environmentally friendly properties of NPs on agriculture, plant and soil health. The application of nanotechnology in the agriculture sector contributes to the world economy's growth by assisting crop production and fulfilment of food shortage. Nanobiofertilizers, nanopesticides and nanobiosensors tend to raise crop yield, being good substitutes for hazardous chemical fertilizers and traditional pesticides. NPs interact with plants by either enhancing or inhibiting their growth and development; therefore, their effect should be thoroughly explored. Environmentally friendly NPs' benefits for agriculture include drought and salinity tolerance, enhanced induction of nutrients bioavailability, and increased antimicrobial activity also acting as a biocontrol agent. Therefore, nanotechnology-based assessments of soil quality, seed availability and crop yield are required for enhanced agricultural sustainability and environmental risk reduction. Further plant nanotoxicology studies should be considered in the context of evaluating current risks linked to NPs, their usage, distribution and release.

REFERENCES

Abdallah, Y., Hussien, M., Omar, M.O., Elashmony, R., Alkhalifah, D.H.M., & Hozzein, W.N. (2022). Mung bean (*Vigna radiata*) treated with magnesium nanoparticles and its impact on soilborne *Fusarium solani* and *Fusarium oxysporum* in clay soil. *Plants.* 11(11), 1514.

Adhikari, T., Dharmarajan, R., Lamb, D., & Zhang, H. (2022). Remediation of frogmore mine spoiled soil with nano enhanced materials. *Soil Sediment Contam.* 31(3), 367–385.

Agri, U., Chaudhary, P., & Sharma, A. (2021). In vitro compatibility evaluation of agriusable nanochitosan on beneficial plant growth-promoting rhizobacteria and maize plant. *Natl Acad Sci Lett.* 44, 555–559.

Agri, U., Chaudhary, P., Sharma, A., & Kukreti, B. (2022). Physiological response of maize plants and its rhizospheric microbiome under the influence of potential bioinoculants and nanochitosan. *Plant Soil.* 474, 451–468.

Ahmed, O., Sibuyi, N.R.S., Fadaka, A.O., Madiehe, M.A., Maboza, E., Meyer, M., & Geerts, G. (2022). Plant extract-synthesized silver nanoparticles for application in dental therapy. *Pharmaceutics.* 14(2), 380.

Ali, E.F., El-Shehawi, A.M., Ibrahim, O.H.M., Abdul-Hafeez, E.Y., Moussa, M.M., & Hassan, F.A.S. (2021). A vital role of chitosan nanoparticles in improvisation the drought stress tolerance in *Catharanthus roseus* (L.) through biochemical and gene expression modulation. *Plant Physiol Biochem.* 161, 166–175.

Al-Mamun, M.R., Hasan, M.R., Ahommed, M.S., Bacchu, M.S., Ali, M.R., & Khan, M.Z.H. (2021). Nanofertilizers towards sustainable agriculture and environment. *Environ Technol Innovat.* 23, 101658.

Al-Zahrani, S.S., Bora, R.S., & Al-Garni, S.M. (2021). Antimicrobial activity of chitosan nanoparticles. *Biotechnol Biotechnol Equip.* 35(1), 1874–1880.

Ameen, F., Alsamhary, K., Alabdullatif, J.A., & Nadhari, S. (2021). A review on metal-based nanoparticles and their toxicity to beneficial soil bacteria and fungi. *Ecotoxicol Environ Saf.* 213, 112027.

Andrzejewska, A., Diatta, J., Spiżewski, T., Krzesiński, W., & Smurzyńska, A. (2017). Application of zeolite and bentonite for stabilizing lead in a contaminated soil. *Ecol Eng Environ Technol.* 18(4), 1–6.

Asadishad, B., Chahal, S., Cianciarelli, V., Zhou, K., & Tufenkji, N. (2017). Effect of gold nanoparticles on extracellular nutrient-cycling enzyme activity and bacterial community in soil slurries: role of nanoparticle size and surface coating. *Environ Sci Nano.* 4(4), 907–918.

Astafurova, T. P., Burenina, A. A., Suchkova, S. A., Zotikova, A.B.P., Kulizhskiy, S. P., & Morgalev, Y. (2017). Influence of ZnO and Pt nanoparticles on cucumber yielding capacity and fruit quality. In A. Gusev, T. Dyatcheck, & A. Godymchuk (Eds.) *Nano Hybrids and Composites* (Vol. 13, pp. 142–148). Trans Tech Publications Ltd, Cham.

Azmat, A., Tanveer, Y., Yasmin, H., Hassan, M.N., Shahzad, A., Reddy, M., & Ahmad, A. (2022). Coactive role of zinc oxide nanoparticles and plant growth promoting rhizobacteria for mitigation of synchronized effects of heat and drought stress in wheat plants. *Chemosphere.* 297, 133982.

Babu, S., Singh, R., Yadav, D., Rathore, S.S., Raj, R., Avasthe, R., & Singh, V.K. (2022). Nanofertilizers for agricultural and environmental sustainability. *Chemosphere.* 292, 133451.

Backx, B.P., Dos Santos, M.S., & Dos Santos, O.A. (2021). The role of biosynthesized silver nanoparticles in antimicrobial mechanisms. *Curr Pharmaceut Biotechnol.* 22(6), 762–772.

Bakshi, S., He, Z.L., & Harris, W.G. (2015). Natural nanoparticles: implications for environment and human health. *Crit Rev Environ Sci Technol.* 45(8), 861–904.

Balaraman, P., Balasubramanian, B., Liu, W.C., Kaliannan, D., Durai, M., Kamyab, H., & Maruthupandian, A. (2022). *Sargassum myriocystum*-mediated TiO$_2$-nanoparticles and their antimicrobial, larvicidal activities and enhanced photocatalytic degradation of various dyes. *Environ Res*. 204, 112278.

Bapat, M.S., Singh, H., Shukla, S.K., Singh, P.P., Vo, D.V.N., Yadav, A., & Kumar, D. (2022). Evaluating green silver nanoparticles as prospective biopesticides: an environmental standpoint. *Chemosphere*. 286, 131761.

Batley, G.E., Kirby, J.K., & McLaughlin, M.J. (2013). Fate and risks of nanomaterials in aquatic and terrestrial environments. *Acc Chem Res*. 46(3), 854–862.

Beer, C., Foldbjerg, R., Hayashi, Y., Sutherland, D. S., & Autrup, H. (2012). Toxicity of silver nanoparticles—nanoparticle or silver ion? *Toxicol Lett*. 208(3), 286–292.

Ben-Moshe, T., Frenk, S., Dror, I., Minz, D., & Berkowitz, B. (2013). Effects of metal oxide nanoparticles on soil properties. *Chemosphere*. 90(2), 640–646.

Bhat, J.A., Rajora, N., Raturi, G., Sharma, S., Dhiman, P., Sanand, S., Shivaraj, S.M., Sonah, H., & Deshmukh, R. (2021). Silicon nanoparticles (SiNPs) in sustainable agriculture: major emphasis on the practicality, efficacy and concerns. *Nanoscale Adv*. 3(14), 4019–4028.

Bhatt, P., Pandey, S.C., Joshi, S., Chaudhary, P., Pathak, V.M., Huang, Y., Wu, X., Zhou, Z., & Chen, S. (2022). Nanobioremediation: a sustainable approach for the removal of toxic pollutants from the environment. *J Hazard Mater*. 427, 128033.

Biswas, P., & Wu, C.Y. (2005). Nanoparticles and the environment. *J Air Waste Manag Assoc*. 55(6), 708–746.

Bruna, T., Maldonado-Bravo, F., Jara, P., & Caro, N. (2021). Silver nanoparticles and their antibacterial applications. *Inter J Mol Sci*. 22(13), 7202.

Burleson, D.J., Driessen, M.D., & Penn, R.L. (2004). On the characterization of environmental nanoparticles. *J Environ Sci Health*. 39(10), 2707–2753.

Cai, D.Q., Wu, Z.Y., Jiang, J., Wu, Y.J., Feng, H.Y., Brown, I.G., Chu, P.K., Yu, Z.L. (2014). Controlling nitrogen migration through micro-nano networks. *Sci Rep*. 4, 3665.

Camps, I., Borlaf, M., Colomer, M.T., Moreno, R., Duta, L., Nita, C., & György, E. (2017). Structure-property relationships for Eu doped TiO$_2$ thin films grown by a laser assisted technique from colloidal sols. *RSC Adv*. 7(60), 37643–37653.

Chai, H., Yao, J., Sun, J., Zhang, C., Liu, W., Zhu, M., Ceccanti, B. (2015). The effect of metal oxide nanoparticles on functional bacteria and metabolic profiles in agricultural soil. *Bull Environ Contam Toxicol*. 94, 490–495.

Chaud, M., Souto, E.B., Zielinska, A., Severino, P., Batain, F., Oliveira-Junior, J., & Alves, T. (2021). Nanopesticides in agriculture: benefits and challenge in agricultural productivity, toxicological risks to human health and environment. *Toxics*. 9(6), 131.

Chaudhary, A., Chaudhary, P., Upadhyay, A., Kumar, A., & Singh, A. (2022c). Effect of gypsum on plant growth promoting rhizobacteria. *Environ Ecol*. 39(4A): 1248–1256.

Chaudhary, P., Chaudhary, A., Bhatt, P., Kumar, G., Khatoon, H., Rani, A., Kumar, S., & Sharma, A. (2022a). Assessment of soil health indicators under the influence of nanocompounds and *Bacillus* spp. in field condition. *Front Environ Sci*. 9, 769871.

Chaudhary, P., Chaudhary, A., Parveen, H., Rani, A., Kumar, G., Kumar, A., & Sharma, A. (2021e). Impact of nanophos in agriculture to improve functional bacterial community and crop productivity. *BMC Plant Biol*. 21, 519.

Chaudhary, P., Khati, P., Chaudhary, A., Gangola, S., Kumar, R., & Sharma, A. (2021a). Bioinoculation using indigenous *Bacillus* spp. improves growth and yield of *Zea mays* under the influence of nanozeolite. *3 Biotech*. 11, 11.

Chaudhary, P., Khati, P., Chaudhary, A., Maithani, D., Kumar, G., & Sharma, A. (2021d). Cultivable and metagenomic approach to study the combined impact of nanogypsum and *Pseudomonas taiwanensis* on maize plant health and its rhizospheric microbiome. *PLoS One*. 16, e0250574.

Chaudhary, P., Khati, P., Gangola, S., Kumar, A., Kumar, R., & Sharma, A. (2021c). Impact of nanochitosan and *Bacillus* spp. on health, productivity and defence response in *Zea mays* under field condition. *3 Biotech.* 11, 237.

Chaudhary, P., & Sharma, A. (2019). Response of nanogypsum on the performance of plant growth promotory bacteria recovered from nanocompound infested agriculture field. *Environ Ecol.* 37, 363–372.

Chaudhary, P., Sharma, A., Chaudhary, A., Khati, P., Gangola, S., & Maithani, D. (2021b). Illumina based high throughput analysis of microbial diversity of rhizospheric soil of maize infested with nanocompounds and *Bacillus* sp. *Appl Soil Ecol.* 159, 103836.

Chaudhary, P., Singh, S., Chaudhary, A., Sharma, A., & Kumar, G. (2022b). Overview of bio-fertilizers in crop production and stress management for sustainable agriculture. *Front Plant Sci.* 13, 930340.

Chavan, S., & Nadanathangam, V. (2019). Effects of nanoparticles on plant growth-promoting bacteria in Indian agricultural soil. *Agronomy.* 9(3), 140.

Chen, L., Lee, Y.B., Ramsier, C., Bigham, J., Slater, B., & Dick, W.A. (2005). Increased crop yield and economic return and improved soil quality due to land application of FGD-gypsum. In: *Proceedings of the World of Coal Ash.* Lexington, Ky.

Chhipa, H. (2017). Nanofertilizers and nanopesticides for agriculture. *Environ Chem Lett.* 15(1), 15–22.

Concha-Guerrero, S.I., Brito, E.M.S., & Piñón-Castillo, H.A. (2014). Effect of CuO nanoparticles over isolated bacterial strains from agricultural soil. *J Nanomater.* 2014, 13.

Cui, J., Jin, Q., Li, F., & Chen, L. (2022). Silicon reduces the uptake of cadmium in hydroponically grown rice seedlings: why nanoscale silica is more effective than silicate. *Environ Sci Nano.* 9(6), 1961–1973.

Dan, Y., Zhang, W., Xue, R., Ma, X., Stephan, C., & Shi, H. (2015). Characterization of gold nanoparticle uptake by tomato plants using enzymatic extraction followed by single-particle inductively coupled plasma–mass spectrometry analysis. *Environ Sci Technol.* 49(5), 3007–3014.

Danish, M., Shahid, M., Zeyad, M.T., Bukhari, N.A., Al-Khattaf, F.S., Hatamleh, A. A., & Ali, S. (2022). Bacillus mojavensis, a metal-tolerant plant growth-promoting bacterium, improves growth, photosynthetic attributes, gas exchange parameters, and alkalo-polyphenol contents in silver nanoparticle (Ag-NP)-treated *Withania somnifera* L. (Ashwagandha). *ACS Omega.* 7(16), 13878–13893.

Das, P., Gogoi, N., Sarkar, S., Patil, S.A., Hussain, N., Barman, S., Pratihar, S., & Bhattacharya, S.S. (2021). Nano-based soil conditioners eradicate micronutrient deficiency: soil physicochemical properties and plant molecular responses. *Environ Sci Nano.* 8, 2824–2843.

Dawood, S., Koyande, A. K., Ahmad, M., Mubashir, M., Asif, S., Klemeš, J.J., & Show, P.L. (2021). Synthesis of biodiesel from non-edible (*Brachychiton populneus*) oil in the presence of nickel oxide nanocatalyst: parametric and optimisation studies. *Chemosphere.* 278, 130469.

Dimkpa, C.O. (2018). Soil properties influence the response of terrestrial plants to metallic nanoparticles exposure. *Curr Opin Environ Sci Health.* 6, 1–8.

Ding, Z., Zhao, F., Zhu, Z., Ali, E. F., Shaheen, S. M., Rinklebe, J., & Eissa, M. A. (2022). Green nanosilica enhanced the salt-tolerance defenses and yield of Williams banana: a field trial for using saline water in low fertile arid soil. *Environ Exp Bot.* 197, 104843.

Du, J., Liu, B., Zhao, T., Xu, X., Lin, H., Ji, Y., & Ding, X. (2022). Silica nanoparticles protect rice against biotic and abiotic stresses. *J Nanobiotechnol.* 20(1), 1–18.

Duhan, J. S., Kumar, R., Kumar, N., Kaur, P., Nehra, K., & Duhan, S. (2017). Nanotechnology: the new perspective in precision agriculture. *Biotechnol Rep.* 15, 11–23.

Eivazi, F., Afrasiabi, Z., & Jose, E. (2018). Effects of silver nanoparticles on the activities of soil enzymes involved in carbon and nutrient cycling. *Pedosphere.* 28(2), 209–214.

Elsayed, A.A., Ahmed, E.G., Taha, Z.K., Farag, H.M., Hussein, M.S., & AbouAitah, K. (2022). Hydroxyapatite nanoparticles as novel nano-fertilizer for production of rosemary plants. *Sci Hort.* 295, 110851.

Elsayh, S.A., Arafa, R.N., Ali, G.A., Abdelaal, W.B., Sidky, R.A., & Ragab, T.I. (2022). Impact of silver nanoparticles on multiplication, rooting of shoots and biochemical analyses of date palm *Hayani* cv. by in vitro. *Biocatal Agric Biotechnol.* 43, 102400.

Farooq, M.A., Islam, F., Ayyaz, A., Chen, W., Noor, Y., Hu, W., & Zhou, W. (2022). Mitigation effects of exogenous melatonin-selenium nanoparticles on arsenic-induced stress in *Brassica napus. Environ Poll.* 292, 118473.

Fernandez, J.G., & Ingber, D.E. (2014). Manufacturing of large-scale functional objects using biodegradable chitosan bioplastic. *Macromol Mater Eng.* 299(8), 932–938.

Frenk, S., Ben-Moshe, T., Dror, I., Berkowitz, B., & Minz, D. (2013). Effect of metal oxide nanoparticles on microbial community structure and function in two different soil types. *PLoS One.* 8, 84441.

Gahukar, R. T., & Das, R. K. (2020). Plant-derived nanopesticides for agricultural pest control: challenges and prospects. *Nanotech Environ Eng.* 5(1), 1–9.

Gaur, S., Kumar, J., Kumar, D., Chauhan, D.K., Prasad, S.M., & Srivastava, P.K. (2020). Fascinating impact of silicon and silicon transporters in plants: a review. *Ecotoxicol Environ Saf.* 202, 110885.

Gedda, G., Lee, C.Y., Lin, Y.C., & Wu, H.F. (2016). Green synthesis of carbon dots from prawn shells for highly selective and sensitive detection of copper ions. *Sens Actuators B Chem.* 224, 396–403.

Gottschalk, F., Lassen, C., Kjoelholt, J., Christensen, F., & Nowack, B. (2015). Modeling flows and concentrations of nine engineered nanomaterials in the Danish environment. *Int J Environ Res Public Health.* 12(5), 5581–5602.

Gottschalk, F., Sonderer, T., Scholz, R.W., & Nowack, B. (2009). Modeled environmental concentrations of engineered nanomaterials (TiO_2, ZnO, Ag, CNT, fullerenes) for different regions. *Environ Sci Technol.* 43(24), 9216–9222.

Gowda, A. N., Kumar, M., Thomas, A. R., Philip, R., & Kumar, S. (2016). Self-assembly of silver and gold nanoparticles in a metal-free phthalocyanine liquid crystalline matrix: structural, thermal, electrical and nonlinear optical characterization. *Chem Select.* 1(7), 1361–1370.

Gudata, L., Saka, A., Tesfaye, J.L., Shanmugam, R., Dwarampudi, L.P., Nagaprasad, N., & Krishnaraj, R. (2022). Investigation of TiO_2 nanoparticles using leaf extracts of *Lippia adoensis* (Kusaayee) for antibacterial activity. *J Nanomater. 2022.*

Hashimoto, Y., Takeuchi, S., Mitsunobu, S., & Ok, Y. S. (2017). Chemical speciation of silver (Ag) in soils under aerobic and anaerobic conditions: Ag nanoparticles vs. ionic Ag. *J Hazard Mater.* 322, 318–324.

Huang, Y.C., Fan, R., Grusak, M.A., Sherrier, J.D., & Huang, C.P. (2014). Effects of nano ZnO on the agronomically relevant *Rhizobium*–legume symbiosis. *Sci Total Environ.* 497, 78–90.

Izzo, F., Langella, A., de Gennaro, B., Germinario, C., Grifa, C., Rispoli, C., & Mercurio, M. (2022). Chabazite from campanian ignimbrite tuff as a potential and sustainable remediation agent for the removal of emerging contaminants from water. *Sustainability.* 14(2), 725.

Jahani, M., Khavari-Nejad, R. A., Mahmoodzadeh, H., & Saadatmand, S. (2020). Effects of cobalt oxide nanoparticles (Co3O4 NPs) on ion leakage, total phenol, antioxidant enzymes activities and cobalt accumulation in *Brassica napus* L. *Not Bot Horti Agrobot Cluj Napoca.* 48(3), 1260–1275.

Jiang, H., Wang, H., Yang, B., Cao, X., & Liu, X. (2020). Effects of ZnO nanoparticles on microbial biomass and enzyme activity in soils containing cellulose and lignin. In *E3S Web of Conferences* (Vol. 194, p. 05028). EDP Sciences.

Jiao, K., Yang, B., Wang, H., Xu, W., Zhang, C., Gao, Y., & Ji, D. (2022). The individual and combined effects of polystyrene and silver nanoparticles on nitrogen transformation and bacterial communities in an agricultural soil. *Sci Total Environ*. 820, 153358.

Josko, I., Oleszczuk, P., & Futa, B. (2014). The effect of inorganic nanoparticles (ZnO, Cr_2O_3, CuO and Ni) and their bulk counterparts on enzyme activities in different soils. *Geoderma*. 232, 528–537.

Kalagatur, N. K., Nirmal Ghosh, O. S., Sundararaj, N., & Mudili, V. (2018). Antifungal activity of chitosan nanoparticles encapsulated with *Cymbopogon martinii* essential oil on plant pathogenic fungi *Fusarium graminearum*. *Front Pharmacol*. 9, 610.

Kamika, I., & Tekere, M. (2017). Impacts of cerium oxide nanoparticles on bacterial community in activated sludge. *AMB Exp*. 7, 1–11.

Kandhol, N., Jain, M., & Tripathi, D.K. (2022). Nanoparticles as potential hallmarks of drought stress tolerance in plants. *Physiol Plant*. 174(2), e13665.

Khafri, A.Z., Zarghami, R., Ma'mani, L., & Ahmadi, B. (2022). Enhanced efficiency of in vitro rootstock micro-propagation using silica-based nanoparticles and plant growth regulators in myrobalan 29C (*Prunus cerasifera* L.). *J Plant Growth Regulat*. 1–15. https://doi.org/10.1007/s00344-022-10631-3.

Khan, S.T., Adil, S.F., Shaik, M.R., Alkhathlan, H.Z., Khan, M., & Khan, M. (2022). Engineered nanomaterials in soil: their impact on soil microbiome and plant health. *Plants*. 11, 109.

Khati, P., Bhatt, P., Kumar, R., & Sharma, A. (2018). Effect of nanozeolite and plant growth promoting rhizobacteria on maize. *3Biotech*. 8, 141.

Khati, P., Chaudhary, P., Gangola, S., Bhatt, P., & Sharma, A. (2017). Nanochitosan supports growth of *Zea mays* and also maintains soil health following growth. *3 Biotech*. 7, 81.

Khati, P., Chaudhary, P., Gangola, S., & Sharma, P. (2019a). Influence of nanozeolite on plant growth promotory bacterial isolates recovered from nanocompound infested agriculture field. *Environ Ecol*. 37, 521—527.

Khati, P., Sharma, A., Chaudhary, P., Singh, A.K., Gangola, S., & Kumar R. (2019b). High-throughput sequencing approach to access the impact of nanozeolite treatment on species richness and evens of soil metagenome. *Biocatal Agric Biotechnol*. 20, 101249.

Kianfar, E. (2019). Nanozeolites: synthesized, properties, applications. *J Sol-Gel Sci Technol*. 91(2), 415–429.

Kińska, K., Bierla, K., Godin, S., Preud'Homme, H., Kowalska, J., Krasnodębska-Ostręga, B., & Szpunar, J. (2019). A chemical speciation insight into the palladium (ii) uptake and metabolism by *Sinapis alba*. exposure to Pd induces the synthesis of a Pd–histidine complex. *Metallomics*. 11(9), 1498–1505.

Kolesnikov, S., Timoshenko, A., Minnikova, T., Tsepina, N., Kazeev, K., Akimenko, Y., Zhadobin, A., Shuvaeva, V., Rajput, V.D., & Mandzhieva, S. (2021). Impact of metal-based nanoparticles on cambisol microbial functionality, enzyme activity, and plant growth. *Plants*. 10, 2080.

Kulkarni, N., & Muddapur, U. (2014). Biosynthesis of metal nanoparticles: a review. *J Nanotechnol*. 2014, 510246.

Kumar, M.S., & Thiyageshwari, S. (2018). Performance of nano-gypsum on reclamation of sodic soil. *Int Curr Microbiol Appl Sci*. 6, 56–62.

Kumar, N., Palmer, G.R., Shah, V., & Walker, V.K. (2014). The effect of silver nanoparticles on seasonal change in arctic tundra bacterial and fungal assemblages. *PLoS One*. 9, e99953.

Kumar, P., Alamri, S.A.M., Alrumman, S.A., Eid, E.M., Adelodun, B., Goala, M., Choi, K.S., & Kumar, V. (2022). Foliar use of TiO_2 nanoparticles for okra (*Abelmoschus esculentus* L. Moench) cultivation on sewage sludge amended soils: biochemical response and heavy metals accumulation. *Environ Sci Poll Res*. 29(44), 66507–66518.

Kumar, P., Burman, U., & Santra, P. (2015). Effect of nano-zinc oxide on nitrogenase activity in legumes: an interplay of concentration and exposure time. *Int Nano Lett*. 5, 191–198.

Kumar, V., Mahajan, R., Bhatnagar, D., & Kaur, I. (2016). Nanofibers synthesis of ND: PANI composite by liquid/liquid interfacial polymerization and study on the effect of NDs on growth mechanism of nanofibers. *Euro Poly J.* 83, 1–9.

Kumari, H., Khati, P., Gangola, S., Chaudhary, P., & Sharma, A. (2021). Performance of plant growth promotory rhizobacteria on Maize and soil characteristics under the influence of TiO$_2$ nanoparticles. *Pantnagar J Res.* 19, 28–39.

Kumari, S., Sharma, A., Chaudhary, P., & Khati, P. (2020). Management of plant vigor and soil health using two agriusable nanocompounds and plant growth promotory rhizobacteria in Fenugreek. *3 Biotech.* 10, 461.

Lead, J. R., & Smith, E. (eds.). (2009). *Environmental and Human Health Impacts of Nanotechnology.* John Wiley & Sons. https://www.wiley.com/en-us/Environmental+and+Human+Health+Impacts+of+Nanotechnology-p-9781444307504.

Liu, R., & Lal, R. (2015). Potentials of engineered nanoparticles as fertilizers for increasing agronomic productions. *Sci Total Environ.* 514, 131–139.

Liu, R.H., Kang, Y.H., Pei, L., Wan, S.Q., Liu, S.P., & Liu, S.H. (2016). Use of a new controlled-loss-fertilizer to reduce nitrogen losses during winter wheat cultivation in the Danjiangkou reservoir Area of China. *Commun Soil Sci Plant Anal.* 47, 1137–1147.

Liu, Y.L., & Meng, C.S. (2020). Nanominerals and their environmental effects. *Int J Nanotechnol.* 17(2–6), 237–256.

Mahajan, S., Kadam, J., Dhawal, P., Barve, S., & Kakodkar, S. (2022). Application of silver nanoparticles in in-vitro plant growth and metabolite production: revisiting its scope and feasibility. *Plant Cell Tissue Organ Cult.* 150(1), 15–39.

Majeed, M., Hakeem, K.R., & Rehman, R.U. (2022). Synergistic effect of plant extract coupled silver nanoparticles in various therapeutic applications-present insights and bottlenecks. *Chemosphere.* 288, 132527.

Malerba, M., & Cerana, R. (2016). Chitosan effects on plant systems. *Int J Mol Sci.* 17(7), 996.

Maliszewska, I. (2016). Effects of the biogenic gold nanoparticles on microbial community structure and activities. *Ann Microbiol.* 66, 785–794.

McGee, C.F., Storey, S., Clipson, N., & Doyle, E. (2017). Soil microbial community responses to contamination with silver, aluminium oxide and silicon dioxide nanoparticles. *Ecotoxicology.* 26, 449–458.

Mechouche, M. S., Merouane, F., Messaad, C. E. H., Golzadeh, N., Vasseghian, Y., & Berkani, M. (2022). Biosynthesis, characterization, and evaluation of antibacterial and photocatalytic methylene blue dye degradation activities of silver nanoparticles from *Streptomyces tuirus* strain. *Environ Res.* 204, 112360.

Mehta, C.M., Srivastava, R., Arora, S., & Sharma, A.K. (2016). Impact assessment of silver nanoparticles on plant growth and soil bacterial diversity. *3 Biotech.* 6, 254–263.

Mittal, D., Kaur, G., Singh, P., Yadav, K., & Ali, S.A. (2020). Nanoparticle-based sustainable agriculture and food science: recent advances and future outlook. *Front Nanotechnol.* 2, 10.

Mousavi, S.M., Motesharezadeh, B., Hosseini, H.M., Zolfaghari, A.A., Sedaghat, A., & Alikhani, H. (2022). Efficiency of different models for investigation of the responses of sunflower plant to Pb contaminations under SiO$_2$ nanoparticles (NPs) and *Pseudomonas fluorescens* treatments. *Arab J Geosci.* 15(14), 1–12.

Nadaroglu, H., GÜNGÖR, A.A., & Selvi, İ.N. (2017). Synthesis of nanoparticles by green synthesis method. *Int J Innov Res Rev.* 1(1), 6–9.

Nagpal, K., Singh, S.K., & Mishra, D.N. (2010). Chitosan nanoparticles: a promising system in novel drug delivery. *Chem Pharm Bull.* 58, 1423–1430.

Oviedo, L.R., Muraro, P.C.L., Chuy, G., Vizzotto, B.S., Pavoski, G., Espinosa, D.C.R., & da Silva, W.L. (2022a). Antibacterial activity of nanozeolite doped with silver and titanium nanoparticles. *J Sol-Gel Sci Technol.* 101(1), 235–243.

Oviedo, L.R., Muraro, P.C.L., Pavoski, G., Espinosa, D.C.R., Ruiz, Y.P.M., Galembeck, A., & da Silva, W.L. (2022b). Synthesis and characterization of nanozeolite from (agro) industrial waste for application in heterogeneous photocatalysis. *Environ Sci Poll Res.* 29(3), 3794–3807.

Pan, L., Zhao, L., Jiang, W., Wang, M., Chen, X., Shen, X., Yin, C., & Mao, Z. (2022). Effect of zinc oxide nanoparticles on the growth of *Malus hupehensis Rehd.* seedlings. *Front Environ Sci.* 10, 835194.

Parveen, A., & Siddiqui, Z.A. (2022). Foliar spray and seed priming of titanium dioxide nanoparticles and their impact on the growth of tomato, defense enzymes and some bacterial and fungal diseases. *Arch Phytopath Plant Prot.* 55(5), 527–548.

Peteu, S.F., Oancea, F., Sicuia, O.A., Constantinescu, F., & Dinu, S. (2010). Responsive polymers for crop protection. *Polymers.* 2, 229–251.

Prasad, R., Bhattacharyya, A., & Nguyen, Q.D., (2017). Nanotechnology in sustainable agriculture: recent developments, challenges, and perspectives. *Front Microbiol.* 8. 1–13.

Rai, P.K., Kumar, V., Lee, S., Raza, N., Kim, K. H., Ok, Y.S., & Tsang, D.C. (2018). Nanoparticle-plant interaction: implications in energy, environment, and agriculture. *Environ Int.* 119, 1–19.

Rai, V., Acharya, S., & Dey, N. (2012). Implications of nanobiosensors in agriculture. *J Biomater Nanobiotechnol.* 3, 315–324.

Rajput, V., Minkina, T., Fedorenko, A., Sushkova, S., Mandzhieva, S., Lysenko, V., Duplii, N., Fedorenko, G., Dvadnenko, K., & Ghazaryan, K. (2018). Toxicity of copper oxide nanoparticles on spring barley (*Hordeum sativum distichum*). *Sci Total Environ.* 645, 1103–1113.

Rico, C.M., Morales, M.I., Barrios, A.C., McCreary, R., Hong, J., Lee, W.Y., & Gardea-Torresdey, J.L. (2013). Effect of cerium oxide nanoparticles on the quality of rice (*Oryza sativa* L.) grains. *J Agric Food Chem.* 61(47), 11278–11285.

Rodríguez-Seijo, A., Soares, C., Ribeiro, S., Amil, B.F., Patinha, C., Cachada, A., & Pereira, R. (2022). Nano-Fe_2O_3 as a tool to restore plant growth in contaminated soils–assessment of potentially toxic elements (bio) availability and redox homeostasis in *Hordeum vulgare* L. *J Hazard Mater.* 425, 127999.

Sabry, A.K.H. (2020). Role of nanopesticides in agricultural development. In Chetan K., Ed., *Intellectual Property Issues in Nanotechnology* (pp. 107–120). CRC Press, Boca Raton.

Sekhon, B.S. (2014). Nanotechnology in agri-food production: an overview. *Nanotech Sci Appl.* 7, 31.

Servin, A.D., De la Torre-Roche, R., Castillo-Michel, H., Pagano, L., Hawthorne, J., Musante, C., Pignatello, J., Uchimiya, M., & White, J.C. (2017). Exposure of agricultural crops to nanoparticle CeO_2 in biochar-amended soil. *Plant Physiol Biochem.* 110, 147–157.

Shao, J., He, Y., Zhang, H., Chen, A., Lei, M., Chen, J., & Gu, J.D. (2015). Silica fertilization and nano-MnO_2 amendment on bacterial community composition in high arsenic paddy soils. *Appl Microbiol Biotechnol.* 100(5), 2429–2437.

Sharma, A., & Chetani, R. (2017). A review on the effect of organic and chemical fertilizers on plants. *Int J Res Appl Sci Eng Technol.* 5, 677–680.

Sharma, P., Goyal, D., Chudasama, B. (2021). Ecotoxicity of as-synthesised copper nanoparticles on soil bacteria. *IET Nanobiotechnol.* 15(2), 236–245.

Sillen, W.M., Thijs, S., Abbamondi, G.R, Janssen, J., Weyens, N., White, J.C., & Vangronsveld, J. (2015). Effects of silver nanoparticles on soil microorganisms and maize biomass are linked in the rhizosphere. *Soil Biol Biochem.* 91, 14–22.

Simonin, M., Colman, B.P., Anderson S.M., King R.S., Ruis M.T., Avellan, A., Bergemann C.M., Perrotta, B.G., Geitner, N.K., Ho, M., Barrera, B., Unrine, J.M., Lowry, G.V., Richardson, C.J., Wiesner, M.R., & Bernhardt, E.S. (2018). Engineered nanoparticles interact with nutrients to intensify eutrophication in a wetland ecosystem experiment. *Ecol Appl.* 28, 1435–1449.

Simonin, M., Richaume, A., Guyonnet, J.P., Dubost, A., Martins, J.M.F., & Pommier, T. (2016). Titanium dioxide nanoparticles strongly impact soil microbial function by affecting archaeal nitrifiers. *Sci Rep.* 6(1), 33643.

Solanki, A., John, D.K., & Ki-Bum, L. (2008). Nanotechnology for regenerative medicine: nanomaterials for stem cell imaging. *Nanomedicine.* 3, 567–578.

Sonali, J.M.I., Kavitha, R., Kumar, P. S., Rajagopal, R., Gayathri, K.V., Ghfar, A.A., & Govindaraju, S. (2022). Application of a novel nanocomposite containing micro-nutrient solubilizing bacterial strains and CeO_2 nanocomposite as bio-fertilizer. *Chemosphere.* 286, 131800.

Song, U., Shin, M., Lee, G., Roh, J., Kim, Y., & Lee, E. J. (2013). Functional analysis of TiO_2 nanoparticle toxicity in three plant species. *Biol Trace Element Res.* 155(1), 93–103.

Srivastava, A.K., Dev, A., & Karmakar, S. (2018). Nanosensors and nanobiosensors in food and agriculture. *Environ Chem Lett.* 16(1), 161–182.

Tolaymat, T., El Badawy, A., Sequeira, R., & Genaidy, A. (2015). A system-of-systems approach as a broad and integrated paradigm for sustainable engineered nanomaterials. *Sci Total Environ.* 511, 595–607.

Tomacheski, D., Pittol, M., Simões, D.N., Ribeiro, V.F., Santana, R.M.C. (2017). Impact of silver ions and silver nanoparticles on the plant growth and soil microorganisms. *Global J Environ Sci Manage.* 3(4), 341–350.

Tripathi, G.D., Javed, Z., Gattupalli, M., & Dashora, K. (2022). Impact of nanomaterials accumulation on the organic carbon associated enzymatic activities in soil. *Soil Sediment Contam.* 1–19. doi: 10.1080/15320383.2022.2105813.

Tyagi, J., Chaudhary, P., Mishra, A., Khatwani, M., Dey, S., & Varma, A. (2022). Role of endophytes in abiotic stress tolerance: with special emphasis on *Serendipita indica.* *Int J Environ Res.* 16, 62.

Usman, M., Farooq, M., Wakeel, A., Nawaz, A., Cheema, S.A., Rehman, H., & Sanaullah, M. (2020). Nanotechnology in agriculture: current status, challenges and future opportunities. *Sci Total Environ.* 721, 137778.

Van Nguyen, D., Nguyen, H.M., Le, N.T., Nguyen, K.H., Nguyen, H.T., Le, H.M., & Van Ha, C. (2022). Copper nanoparticle application enhances plant growth and grain yield in maize under drought stress conditions. *J Plant Growth Regul.* 41(1), 364–375.

Wahid, I., Rani, P., Kumari, S., Ahmad, R., Hussain, S.J., Alamri, S., & Khan, M.I.R. (2022). Biosynthesized gold nanoparticles maintained nitrogen metabolism, nitric oxide synthesis, ions balance, and stabilizes the defense systems to improve salt stress tolerance in wheat. *Chemosphere.* 287, 132142.

Wang, J., Shu, K., Zhang, L., & Si, Y. (2017). Effects of silver nanoparticles on soil microbial communities and bacterial nitrification in suburban vegetable soils. *Pedosphere.* 27(3), 482–490.

Wang, Y., Li, J., Tong, W., Shen, Z., Li, L., Zhang, Q., & Yu, J. (2022). Mesoporogen-free synthesis of single-crystalline hierarchical beta zeolites for efficient catalytic reactions. *Inorg Chem Front.* 9(11), 2470–2478.

Wang, Y.G., Mei, D., Glezakou, V.A., Li, J., & Rousseau, R. (2015). Dynamic formation of single-atom catalytic active sites on ceria-supported gold nanoparticles. *Nat Commun.* 6(1), 1–8.

Xiao, L., Cao, Y., Henderson, W.A., Sushko, M.L., Shao, Y., Xiao, J., & Liu, J. (2016). Hard carbon nanoparticles as high-capacity, high-stability anodic materials for Na-ion batteries. *Nano Energy.* 19, 279–288.

Xin, X., Nepal, J., Wright, A.L., Yang, X., & He, Z. (2022). Carbon nanoparticles improve corn (*Zea mays* L.) growth and soil quality: comparison of foliar spray and soil drench application. *J Clean Prod.* 363, 132630.

Yadav, R.K., Singh, N.B., Singh, A., Yadav, V., Bano, C., & Khare, S. (2020). Expanding the horizons of nanotechnology in agriculture: recent advances, challenges and future perspectives. *Vegetos*. 33(2), 203–221.

Yang, Y., Wang, J., Xiu, Z., & Alvarez, P.J. (2013). Impacts of silver nanoparticles on cellular and transcriptional activity of nitrogen- cycling bacteria. *Environ Toxicol Chem*. 32, 1488–1494.

Yasmeen, F., Razzaq, A., Iqbal, M.N., & Jhanzab, H.M. (2015). Effect of silver, copper and iron nanoparticles on wheat germination. *Int J Biosci*. 6(4), 112–117.

You, T., Liu, D., Chen, J., & Yang, Z. (2018). Effects of metal oxide nanoparticles on soil enzyme activities and bacterial communities in two different soil types. *J Soils Sed*. 18(1), 211–221.

Zaki, A.G., Hasanien, Y.A., & El-Sayyad, G.S. (2022). Novel fabrication of SiO_2/Ag nanocomposite by gamma irradiated *Fusarium oxysporum* to combat *Ralstonia solanacearum*. *AMB Exp*. 12(1), 1–18.

Zeng, Y., Woo, H., Lee, G., & Park, J. (2010). Removal of chromate from water using surfactant modified Pohang clinoptilolite and Haruna chabazite. *Desalination*. 257(1–3), 102–109.

Zhang, H., Ou, J., Fang, X., Lei, S., Wang, F., Li, C., & Wang, P. (2022). Robust superhydrophobic fabric via UV-accelerated atmospheric deposition of polydopamine and silver nanoparticles for solar evaporation and water/oil separation. *Chem Eng J*. 429, 132539.

Zhang, W.X. (2003). Nanoscale iron particles for environmental remediation: an overview. *J Nanop Res*. 5(3), 323–332.

Zhao, L., Sun, Y., Hernandez-Viezcas, J.A., Servin, A.D., Hong, J., Niu, G., & Gardea-Torresdey, J.L. (2013). Influence of CeO_2 and ZnO nanoparticles on cucumber physiological markers and bioaccumulation of Ce and Zn: a life cycle study. *J Agric Food Chem*. 61(49), 11945–11951.

Zulfiqar, F., Navarro, M., Ashraf, M., Akram, N.A., & Munné-Bosch, S. (2019). Nanofertilizer use for sustainable agriculture: advantages and limitations. *Plant Sci*. 289, 110270.

14 Nanotechnology in Climate Smart Agriculture
Need of the Hour

Alka Rani
Gurukul Kangri University

Tarun Kumar
Dr. Rajendra Prasad Central Agricultural University

Bhavya Trivedi
Maya Group of Colleges

Parul Chaudhary
Graphic Era Hill University

CONTENTS

14.1 INTRODUCTION

The term 'climate smart agriculture' is defined by the Food and Agricultural Organization of the United Nations. Agriculture is strongly dependent on the climate. Agriculture has been practised by human beings for many centuries. The last one century has seen mechanization in 50 years of inception while the development of various technologies such as transgenic crop production and plant breeding has taken 50 years. According to research, by 2050, the population of the world will be about 10 billion; to feed this population, we will not have enough food and so using sustainable approach we should increase crop production (United Nations, Department of Economic and Social Affairs, Population Division, 2019). We can achieve this by adopting cutting-edge technologies which have not yet been fully used in the agricultural field. Nanotechnology comes under 'smart agriculture' under which effective pesticides and next-generation biosensors to detect soil health are employed. Many biosensors are used without using nanoparticles coated with particular proteins; these work at the nanoscale level but on the other side many biosensors use small particles. A best example of this type of biosensor is a composite electrode coated with nanoparticles along with DNA or protein (Gupta et al., 2020). When modern technologies are used for the enhancement of the quality and quantity of agricultural products without worsening the environment condition, it is called smart agriculture. Smart agriculture management works on the strategies of less input of harmful chemicals which are used in the form of fertilizers and pesticides, but enhances the yield and quality by using advanced and new technologies.

Productivity in the agricultural sector can be improved by using these technologies (Chaudhary et al., 2021a, b, c). This will also reduce the infection of plant pathogens (Floreano & Wood, 2015). In the future with the help of biotechnology, farmers will be able to determine the nutritional and microbial status of the soil in minutes. The potential of the land can be fully utilized by using new technologies with low cost. Smart farming is being used well in many developed countries of world. Sensors, probes, slow-release fertilizers and pesticides are considered as key for 'smart', 'unspoilt' and 'sustainable' agriculture, where nanotechnology plays an important part in this field (Cui et al., 2001; Chaudhary & Sharma, 2019; Khati et al., 2019a). Small-sized nanostructures are highly sensitive in nature; the sensitivity of the nanoparticles is due to their small size and some special properties. It was observed that nanoscale sensor is more important than conventional sensors (Zhao et al., 2017a).

The growth and productivity of crops are directly and indirectly affected by temperature, carbon dioxide (CO_2), water availability, moisture, microflora and other factors. When CO_2 level increases in the atmosphere it not only causes global warming but also disturbs the agriculture ecosystem. Increasing levels of atmospheric CO_2 not only cause global warming but also alters the agricultural ecosystem (Panneerselvam et al., 2019). When excessive amounts of chemical-based fertilizers and pesticides are used in agriculture, it deteriorates the health of soil and agriculture ecosystem (Kumar et al., 2019). Evolution of new strategies for enhancing the crop productivity is the demand of time. By using nanotechnology, we can overcome this situation. Nanotechnology provides nanomaterial and different nanodevices such as nanobiosensors which play very important roles in agriculture. Biosensors detect the nutrient status in the soil and moisture content. Agriculture production can be increased by using the nanoparticles produced by nanotechnology (Tarafdar et al., 2012; Chaudhary et al., 2021d, e). The nutrients are controlled by nanotechnology, which is very important for agriculture (Gruère, 2012; Mukhopadhyay, 2014; Khati et al., 2019b); along with this, nanotechnology also participates in monitoring of water quality and pesticides for sustainable development. Keeping the future in mind, the role of nanotechnology in the agriculture sector is becoming an important factor for sustainable development. This technology has proven to be useful in agriculture in various ways such as helping to maintain soil fertility, drug delivery mechanisms in plants and in resources management of agricultural field (Chaudhary et al., 2022a;

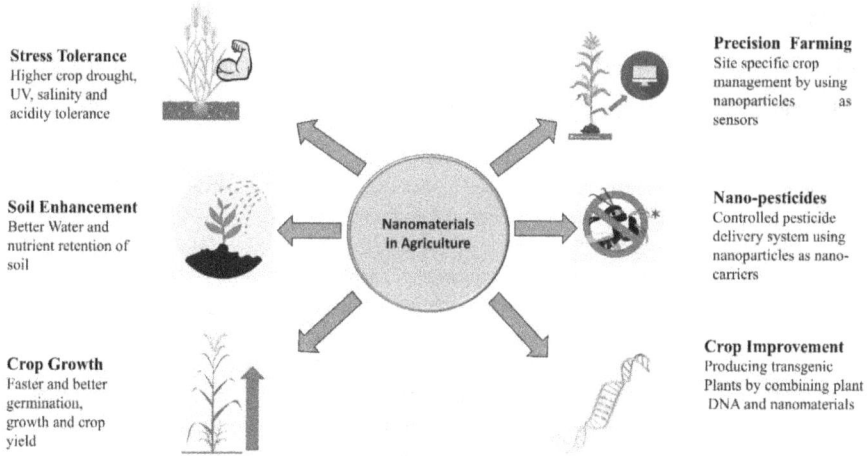

FIGURE 14.1 Application of nanomaterials in various areas of agriculture.

Agri et al., 2022). Nanotechnology has huge prospect in the agriculture field such as stress tolerance, soil enhancement, crop growth, crop improvement, nanopesticides and precision farming, as shown in Figure 14.1.

In recent decades agriculture has been receiving continuous benefits from the use of pesticides and various technological innovations such as 'hybrid' varieties, fertilizers and pesticides (Table 14.1). At present nanotechnological innovation in agriculture is strongly needed for enhancement of agricultural production (Agri et al., 2021; Zulfiqar et al., 2019; Kumar et al., 2020). The uses of nanomaterial are necessary for enhancing fertilizer efficiency, crop yields, decreasing pesticide need and smart pesticides and fertilizer delivery system (Moraru et al., 2003; Chau et al., 2007). By using nanomaterial-induced genetically improved plants farmers can improve agricultural productivity and yield (Kuzma, 2007; Scott, 2007, Naderi & Danesh-Shahraki, 2013).

14.2 TYPES OF NANOFERTILIZERS ON THE BASIS OF REQUIREMENT

Fertilizers made from nanosized particles are called nanofertilizers. Nanofertilizers release the essential nutrients in a controlled manner, which is necessary for plant growth and development. On the basis of plant requirement, these are classified as follows.

14.2.1 Macronutrient Nanofertilizer

These types of NFs are prepared from the nanosized macronutrients required by crop plants in large quantities. Macronutrients include N, P, K, Ca, Mg and S. The need of macronutrients directly depends on the growing population and changes in the climate. According to research, by 2050, the demand for macronutrients will be 263 Mt. Development of nanotechnology-based macronutrients is necessary to fulfil

TABLE 14.1

Application of Nanotechnology in Agricultural Sectors

Area of Application	Uses	References
Crop Production		
Pest control	Nanoparticles encapsulate pesticides, nanocapsule and nanoemulsions for pest control.	Anjali et al. (2012)
Nanofertilizers	Buckyball fertilizer, nanoparticles, nanocapsules and viral capsids for the absorption of essential nutrients and the distribution of nutrients in a particular place.	Anjali et al. (2012)
Precision farming	Nanosensors are globally used for the study of soil, crop growth and to pinpoint application of pesticides and fertilizer in agriculture sector.	Kalpana-Sastry et al. (2009)
Diagnostic		
Diagnostic devices (Nanosensor)	Nanosensors are used for study of the environment and plant condition.	Vamvakaki and Chaniotakis (2007)
Plant Breeding		
Gene modification	Delivery of desired DNA and RNA with nanoparticles.	Torney et al. (2007)
Nanomaterials from Plant		
Nanoparticles from plants	Production of nanofiber.	Kalpana-Sastry et al. (2009)

the needs of the populations' macronutrients. Nanofertilizer macronutrients enhance crop productivity, which is much higher than conventional fertilizers. Macronutrient nanofertilizers are produced under nanotechnology. Research conducted around the world shows that the use of macronutrient nanofertilizers has reduced the cost and increased crop productivity as compared to the conventional methods. Macronutrient nanofertilizer is produced in the agriculture sector by using nanotechnology, and it was found that there was a significant increase in the growth and productivity of crops as a result of using nanofertilizers. The use of nanofertilizer has reduced the cost of cultivation; thus, it can be an economical alternative to the existing conventional fertilizers. Table 14.2 shows the application of nanotechnology-based formulations and conventional fertilizers.

14.2.2 MICRONUTRIENT NANOFERTILIZERS

Overexploitation of soil causes deficiency of micronutrients, although plants require this micronutrient in very less amounts; the deficiency of these micronutrients causes hidden hunger. Although micronutrients are only required in small amounts (100 ppm), they are necessary for various metabolic activities of plants. The common micronutrients are iron (Fe), boron (B), manganese (Mn), zinc (Zn), copper (Cu),

TABLE 14.2

Comparison between Nanotechnology-Based Technology and Conventional Technology (Millan et al., 2008)

Properties	Nanoparticles-Based Technology	Conventional Technology
Solubility	High bioavailability and solubility	Low bioavailability and solubility because of the large particle size
Nutrient absorption ability	Nanostructure formulation enhances fertilizer efficiency and absorption ability	Due to bulky size not available for root and decreased efficiency
Regulation of release mode	In water-soluble fertilizer nutrients are regulated by encapsulation (semipermeable membranes coated with polymer and waxes)	Due to high release of fertilizers it spreads toxicity which destroys the balance and nature of soil
Nutrient releasing time	Nanoparticles prolong the effective time of nutrient delivery of fertilizer	Plants are able to use them only for delivery time; rest of the time they get converted to insoluble salts in the soil
Loss rate	Low loss rate of fertilizer nutrient	High loss rate

molybdenum (Mo), chloride (Cl), and nickel (Ni). These nanoparticles are used on crops with different types of NPK-containing composite fertilizers. Micronutrients have low environmental risks and also provide enough nutrients.

The requirement of micronutrients can be completed by the use of nanofertilizer-containing micronutrients. The agricultural efficiency of micronutrients depends on the size of the particles; by decreasing the particle size the efficiency can be increased; due to the reduction in size the surface area increases. For example, $ZnSO_4$ (1.4–2 mm) was found to be somewhat less effective than fine $ZnSO_4$ (0.8–1.2 mm), whereas granular ZnO (2.0 or 2.5 mm) has been found completely ineffective (Liscano et al., 2000).

14.3 CROP PRODUCTION

It is found that nanomaterials such as nanochitosan, SiO_2, TiO_2, ZnO and carbon nanotubes increased the growth of crop and crop quality (Kumari et al., 2020, 2021; Kukreti et al., 2020). It has been found that nanoparticles have more ability to absorb water and nutrients. Nanoparticles enhance the strength of root system and break-down of carbon-containing compounds (Harrison, 1996). Nanotubes of carbon have potential to enhance the growth of plant and seed germination percentage (Khodakovskaya et al., 2009). When the plant cell interacts with the nanoparticles, it modifies the expression of genes in the plant cell and biological pathways, which help in the growth of plants (Nair et al., 2010). Carbon nanoparticles have been found to be very useful in crop production; these particles enhance the process of photosynthesis in plants and also help to convert plants' leaves to biochemical sensor. Single-walled nanotubes use infrared fluorescence, which enables monitoring nitric oxide, and this feature converts the plant leaves to a photonic chemical sensor (Giraldo et al., 2014). Chemical fertilizer and pesticides have long been used to increase plant growth and to prevent diseases, but about 90% of these fertilizers are being wasted due to runoff and other processes and cause downstream surface and ground water pollution. Nanofertilizers are eco-friendly and highly effective as compared to chemical fertilizers (Kottegoda et al., 2011; Chaudhary et al., 2022b). The growth and productivity of crops depend on the concentration of the nanomaterial used. It was found that the photosynthesis rate of spinach was increased by 3.13 times that of TiO_2 and decreased beyond 4% of TiO_2 concentration (Zheng et al., 2005). Metal oxides such as FeO, ZnO and TiO_2 can be sprayed directly on the plants because it penetrates directly through the pores of leaves and is able to affect the growth of plants (Dhoke et al., 2013). Using multi-walled carbon nanotubes can increase the water content inside the seed because the multi-walled carbon nanotubes penetrate the seed coating and begin to affect biological activity (Tripathi et al., 2011). According to the research conducted so far, it has been found that carbon-walled, metal-based, nanochitosan and nanozeolite nanofertilizers successfully helped in increasing photosynthesis, germination and nutrient use efficiency (Khati et al., 2017, 2018).

14.4   NANOPESTICIDES FOR CROP PROTECTION

During the study of nanomaterials, the interaction between the plant and pathogen is known. By using nanosized material, we can reduce the degradation rate of fungicides and pesticides; on the other side it enhances the effectiveness of nanosized material, helps to reduce degradation of pesticides and fungicides and enhances the effectiveness of application with reduced amount. Some families such as Cucurbits are very sensitive for fungal infection (powdery mildew disease), but the fungal hyphae and the conidia can be inhibited by using nanosilver (100 ppm) (Lamsal et al., 2011). Some metal oxides such as ZnO, Cu, TiO_2 and SiO_2 combine with fungicides and pesticides and enhance the efficiency to protect plants from the pathogens and control microbial activity.

14.5   NANOSENSOR/NANOBIOSENSOR IN AGRICULTURE

Biosensor is a system by which the organophosphate pesticides and nerve agents are monitored online. Nanobiosensors are effectively used for monitoring the herbicide, fertilizer, soil pH, moisture content and pathogenic microorganisms; their regulated use can support sustainable agriculture for enhancing crop productivity (Rai et al., 2012). Using Bio-analytical nanosensor microamounts of contaminants such as pathogenic microorganisms and hazardous substances in agricultural field and food systems can be detected and quantified. Mostly conventional methods are used for conducting these toxins, although researchers are currently working on biosensor as screening tools for use in field analysis (Tothill, 2011). It is expected that nanosensors can positively influence the agriculture, food and environment sectors. Portable nanodevices play an important role in instantly detecting insects, pathogens, chemicals, diseases and contaminants and can result in faster treatment (Farrell et al., 2013).

14.6 NANOPARTICLES IN GENETIC IMPROVEMENT OF CROPS

In previous studies, it has been found that the gene distribution in plants depended on *Agrobacterium* infection and bioplastic particles, although these methods were limited to the certain plant species and are highly genotype specific. DNA enters the plant cell by the rigid cell wall, and the entry of DNA across the rigid cell wall and the stability found in DNA create obstacles in the replication inside the cell. Nanoparticles easily overcome the arising barrier in the cell wall and also have the potential to overcome the bottleneck of current transgene delivery systems. Nanoparticles and nucleic acid which has a positive charge are easily conjugated in plant cell. DNA is protected from enzymatic attack inside the cytosol, but here it releases nanoparticles into the nucleus. The proper mechanism of how transgene are released and distributed inside the plant cell is not yet well understood, but insinuations are deep. It has been observed that when nucleic acids conjugate with carbon nanotubes it can be used for transgene integration. This technique is used to overcome the species- or genotype-specific barrier normally observed in gene delivery methods (Demirer et al., 2019). In past research, it has been found that this method has successfully been used in *Nicotiana benthamiana* (tobacco), *Eruca sativa* (arugula), *Triticum aestivum* (wheat) and *Gossypium hirsutum* (cotton) to corroborate this fact. Stabilized dsDNA is useful for protecting the plant against viruses while protecting plants with the existence of unstable naked DNA was a very tough task. Bioclays (double hydroxide clay nanosheets) have the potential to successfully protect a plant from virus for a long time (Mitter et al., 2017). Similar study has been conducted by other researchers by using dsDNA coated with silica (Hajiahmadi et al., 2019). DNA has a special feature to convert itself into a solid stiff nanostructure (Zhang et al., 2019). In recent years, the study of chloroplast transformation has been an important focus for agricultural biotechnologists. It has been proven that chloroplast genome is maternally inherited. This regulates the horizontal transfer of transgene within similar wild species through pollen. Chloroplast-specific transformation is the rate-limiting step in this method. Nowadays chitosan-complexed single-walled carbon nanotubes are used for chloroplast-specific transformation of foreign gene in plants (Kwak et al., 2019). Nanoparticle-DNA conjugates are used to construct stable transgenic plants (Zhao et al., 2017b). Transgenic plants are produced without tissue culture and regeneration which is considered as the major roadblock in the field of transgenic plant production; this method is called 'pollen magnetofection'. The two methods, that is, chloroplast-specific transformation and pollen magnetofection, have done a great job in revolutionizing gene delivery effort in plants (Figure 14.2). It is very important to do research on these methods.

14.7 ROLE OF NANOTECHNOLOGY IN STRESS CONDITIONS

Nanotechnology plays a valuable role in agriculture to monitor the amount of water, insect attacks, nutrient availability and stress induced by biotic or abiotic factors (Shang et al., 2019). Plant growth and productivity are influenced by extreme abiotic stress conditions, such as heavy metals, disease, pests, salinity and drought, causing

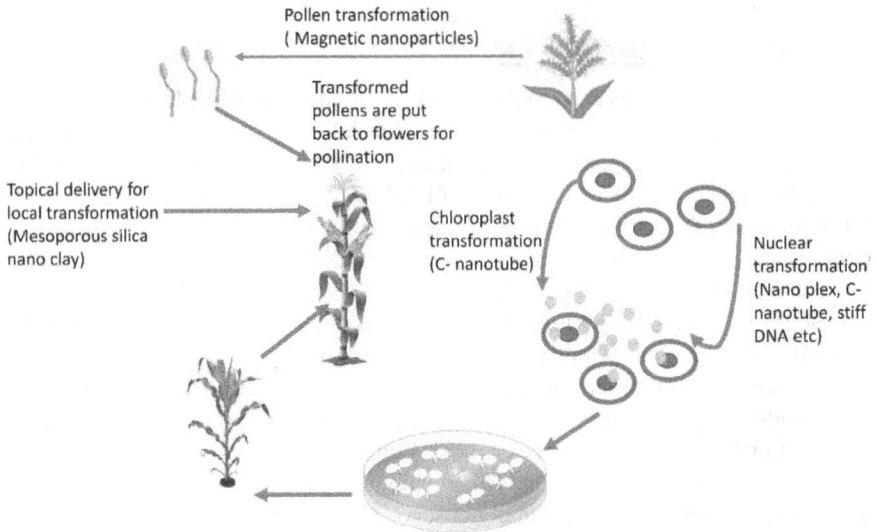

FIGURE 14.2 Methods of nanoparticle-assisted gene delivery in plants.

the collapse of whole ecosystems (Tyagi et al., 2022). In a previous study it was observed that nanoparticles improve plant growth, productivity and development by overcoming the effect of various abiotic and biotic stress of crops (Figure 14.3). On the basis of the current scenario, the loss of yield per year due to abiotic stresses is around 50%. It has been proven that drought and low soil fertility affect crop production mainly in developing countries. For the last several decades nanotechnology has proven to be helpful to agriculture, and nanotechnology has immense potential to open many avenues in the field of biotechnology. Some of them are discussed here in detail (Shang et al., 2019).

Plants have been facing water stress condition for a long time. Tall trees are the first victims when drought occurs, which are always obliged to tolerate or avoid it by adjusting their life cycle. For plant physiologists, drought is not just the lack of rain but rather it is the different types of stresses that affect the plant life, such as (1) low moisture availability, (2) high evaporative load, (3) high temperature, (4) high solar irradiance, (5) increased soil hardness, (6) lack of nutrient and (7) deposition of salt in the upper layer of soil. Copper-zinc nanoparticles enhance the activity of anti-oxidative enzymes, decrease the amount of accumulation of TBARS (Thiobarbituric Acid Reactive Substances) and enhance the content of water in leaves and stabilize the content of photosynthetic pigments (Schulze et al., 2005). It has been found that plant resistance increased against drought by using SiO_2 nanoparticles. SNPs pre-treatment increased the photosynthesis rate and stomatal conductance. It has also been proved that silver nanoparticles help to overcome the negative impact of drought stress on lentils (*Lens culinaris* Medik). In previous studies, it was found that various concentrations of nanoparticles and poly ethylene glycol significantly affects seed germination percentage, shoot length, root length and dry weight in lentil seeds (Khodakovskaya et al., 2011).

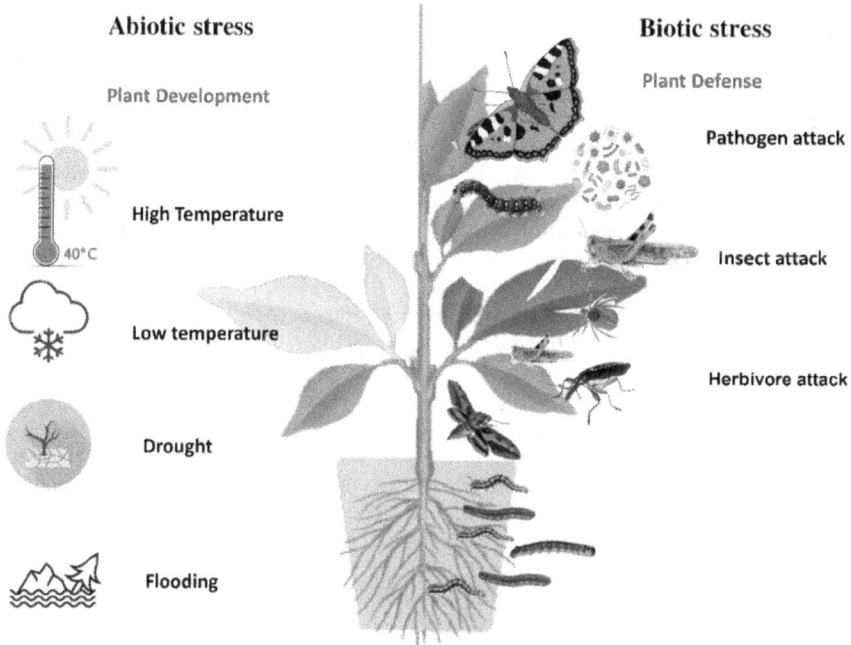

FIGURE 14.3 Various types of biotic and abiotic stresses and their effect on plants.

The heat stress negatively influences the crop productivity and grain yield as well. High thermal environments create a big problem for crop plants to survive and sustain. Heat stress causes irreparable loss in plant growth. It has been found that the low concentration of selenium nanoparticles reduces the effect of heat stress by enhancing hydration capability, plant growth and development and chlorophyll content (Zandalinas et al., 2018).

14.8   CONCLUSION

The whole world at present is facing great challenges including changes in climate, lack of water and land resources, abiotic stresses, etc. Increasing food production by reducing the impact of adverse environment is necessary. Nanomaterials play important roles in this regard. Nanotechnology has a special contribution to overcome these challenges. It has been found that nanomaterials can reduce the damage caused by various biotic and abiotic stresses. Some studies have proven that nanofertilizers are actually more effective than convention fertilizers. Nanofertilizers increase the efficiency of plant uptake with controlled release. Biosynthesis of nanoparticles use biological system, development of sensitive and effective delivery system, nanofertilizers, insect pest management using nanoformulations of insecticides, herbicides, pesticides, fungicides, etc. Under nanotechnology, nanobiosensors are used to monitor environmental attributes and diagnosis of disease. Nanotechnology is not only helping us in the present but also plays an active role in giving food security to the coming generation.

REFERENCES

Agri, U., Chaudhary, P., & Sharma, A. (2021). In vitro compatibility evaluation of agriusable nanochitosan on beneficial plant growth-promoting rhizobacteria and maize plant. *Nat Acad Sci Lett*. 44, 555–559.

Agri, U., Chaudhary, P., Sharma, A., & Kukreti, B. (2022). Physiological response of maize plants and its rhizospheric microbiome under the influence of potential bioinoculants and nanochitosan. *Plant Soil*. 474, 451–468.

Anjali, C.H., Sharma, Y., Mukherjee, A., & Chandrasekaran, N. (2012). Neem oil (*Azadirachta indica*) nanoemulsion—a potent larvicidal agent against *Culex quinquefasciatus*. *Pest Manag Sci*. 68(2), 158–163.

Chau, C.F., Wu, S.H., & Yen, G.C. (2007). The development of regulations for food nanotechnology. *Trends Food Sci Technol*. 18(5), 269–280.

Chaudhary, P., Chaudhary, A., Bhatt, P., Kumar, G., Khatoon, H., Rani, A., Kumar, S., & Sharma, A. (2022a). Assessment of soil health indicators under the influence of nanocompounds and *Bacillus* spp. in field condition. *Front Environ Sci*. 9, 769871.

Chaudhary, P., Chaudhary, A., Parveen, H., Rani, A., Kumar, G., Kumar, A., & Sharma, A. (2021e). Impact of nanophos in agriculture to improve functional bacterial community and crop productivity. *BMC Plant Biol*. 21, 519.

Chaudhary, P., Khati, P., Chaudhary, A., Gangola, S., Kumar, R., & Sharma, A. (2021a). Bioinoculation using indigenous *Bacillus* spp. improves growth and yield of *Zea mays* under the influence of nanozeolite. *3 Biotech*. 11, 11.

Chaudhary, P., Khati, P., Chaudhary, A., Maithani, D., Kumar, G., & Sharma, A. (2021d). Cultivable and metagenomic approach to study the combined impact of nanogypsum and *Pseudomonas taiwanensis* on maize plant health and its rhizospheric microbiome. *PLoS One*. 16, e0250574.

Chaudhary, P., Khati, P., Gangola, S., Kumar, A., Kumar, R., & Sharma, A. (2021c). Impact of nanochitosan and *Bacillus* spp. on health, productivity and defence response in *Zea mays* under field condition. *3 Biotech*. 11, 237.

Chaudhary, P., & Sharma, A. (2019). Response of nanogypsum on the performance of plant growth promotory bacteria recovered from nanocompound infested agriculture field. *Environ Ecol*. 37, 363–372.

Chaudhary, P., Sharma, A., Chaudhary, A., Khati, P., Gangola, S., & Maithani, D. (2021b). Illumina based high throughput analysis of microbial diversity of rhizospheric soil of maize infested with nanocompounds and *Bacillus* sp. *Appl Soil Ecol*. 159, 103836.

Chaudhary, P., Singh, S., Chaudhary, A., Sharma, A., & Kumar, G. (2022b). Overview of biofertilizers in crop production and stress management for sustainable agriculture. *Front Plant Sci*. 13, 930340.

Cui, Y., Wei, Q., Park, H. and Lieber, C.M. (2001). Nanowire nanosensors for highly sensitive and selective detection of biological and chemical species. *Science*. 293, 1289–1292.

Demirer, G.S., Zhang, H., Matos, J.L., Goh, N.S., Cunningham, F.J., Sung, Y., Chang, R., Aditham, A.J., Chio, L., Cho, M.J., & Staskawicz, B. (2019). High aspect ratio nanomaterials enable delivery of functional genetic material without DNA integration in mature plants. *Nat Nanotechnol*. 14(5), 456–464.

Dhoke, S.K., Mahajan, P., Kamble, R., & Khanna, A. (2013). Effect of nanoparticles suspension on the growth of mung (Vigna radiata) seedlings by foliar spray method. *Nanotechnol Develop*. 3(1), 1.

Farrell, D., Hoover, M., Chen, H., & Friedersdorf, L. (2013). *Overview of Resources and Support for Nanotechnology for Sensors and Sensors for Nanotechnology: Improving and Protecting Health, Safety, and the Environment*. Arlington: US National Nanotechnology Initiative.

Floreano, D., & Wood, R.J. (2015). Science, technology and the future of small autonomous drones. *Nature*. 521(7553), 460–466.

Giraldo, J.P., Landry, M.P., Faltermeier, S.M., McNicholas, T.P., Iverson, N.M., Boghossian, A.A., Reuel, N.F., Hilmer, A.J., Sen, F., Brew, J.A., & Strano, M.S. (2014). Plant nanobionics approach to augment photosynthesis and biochemical sensing. *Nat Mater*. 13, 400–408.

Gruère, G.P. (2012). Implications of nanotechnology growth in food and agriculture in OECD countries. *Food Pol*. 37(2), 191–198.

Gupta, J., Juneja, S., & Bhattacharya, J. (2020). UV Lithography-assisted fabrication of low-cost copper electrodes modified with gold nanostructures for improved analyte detection. *ACS Omega*. 5(7), 3172–3180.

Hajiahmadi, Z., Shirzadian-Khorramabad, R., Kazemzad, M., & Sohani, M.M. (2019). Enhancement of tomato resistance to Tuta absoluta using a new efficient mesoporous silica nanoparticle-mediated plant transient gene expression approach. *Sci Hort*. 243, 367–375.

Harrison, C.C. (1996). Evidence for intramineral macromolecules containing protein from plant silicas. *Phytochemistry*. 41(1), 37–42.

Kalpana-Sastry, R., Rashmi, H.B., Rao, N.H., & Ilyas, S.M. (2009). Nanotechnology and agriculture in India: the second green revolution?; Presented at the OECD conference on "Potential environmental benefits of nanotechnology: fostering safe innovation-led growth" Session 7. Agricultural nanotechnology, Paris, France.

Khati, P., Bhatt, P., Kumar, R., & Sharma, A. (2018). Effect of nanozeolite and plant growth promoting rhizobacteria on maize. *3Biotech*. 8(141), 1–12.

Khati, P., Chaudhary, P., Gangola, S., & Sharma A. (2019a). Influence of nanozeolite on plant growth promotory bacterial isolates recovered from nanocompound infested agriculture field. *Environ Ecol*. 37 (2), 521–527.

Khati, P., Gangola, S., Bhatt, P., & Sharma A. (2017). Nanochitosan induced growth of *Zea Mays* with soil health maintenance. *3 Biotech*. 7(81), 1–9.

Khati, P., Sharma A., Chaudhary, P., Singh, A.K., Gangola, S., & Kumar, R. (2019b). High-throughput sequencing approach to access the impact of nanozeolite treatment on species richness and evenness of soil metagenome. *Biocat Agric Biotechnol*. 20, 101249.

Khodakovskaya, M., Dervishi, E., Mahmood, M., Xu, Y., Li, Z., Watanabe, F., & Biris, A.S. (2009). Carbon nanotubes are able to penetrate plant seed coat and dramatically affect seed germination and plant growth. *ACS Nano*. 3(10), 3221–3227.

Khodakovskaya, M.V., de Silva, K., Nedosekin, D.A., Dervishi, E., Biris, A.S., Shashkov, E.V., Galanzha, E.I., & Zharov, V.P. (2011). Complex genetic, photothermal, and photoacoustic analysis of nanoparticle-plant interactions. *Proc Natl Acad Sci U S A*. 108(3), 1028–1033.

Kottegoda, N., Munaweera, I., Madusanka, N., & Karunaratne, V. (2011). A green slow-release fertilizer composition based on urea-modified hydroxyapatite nanoparticles encapsulated wood. *Curr Sci*. 101, 73–78

Kukreti, B., Sharma, A., Chaudhary, P., Agri, U., & Maithani, D. (2020). Influence of nanosilicon dioxide along with bioinoculants on *Zea mays* and its rhizospheric soil. *3 Biotech*. 10, 345.

Kumar, R., Kumar, R., & Prakash, O. (2019). Chapter-5 the impact of chemical fertilizers on our environment and ecosystem. *Chief Ed*. 35, 69.

Kumar, S., Nehra, M., Dilbaghi, N., Marrazza, G., Tuteja, S.K., & Kim, K.H. (2020). Nanovehicles for plant modifications towards pest-and disease-resistance traits. *Trends Plant Sci*. 25(2), 198–212.

Kumari, H., Khati, P., Gangola, S., Chaudhary, P., & Sharma A. (2021). Performance of plant growth promotory rhizobacteria on maize and soil characteristics under the influence of TiO_2 nanoparticles. *Pant Res J*. 19, 28–39.

Kumari, S., Sharma A., Chaudhary, P., & Khati, P. (2020). Management of plant vigour and soil health using two agriusable nanocompounds and plant growth promontory rhizobacteria in Fenugreek. *3 Biotech.* 10(461), 1–11.

Kuzma, J. (2007). Moving forward responsibly: oversight for the nanotechnology-biology interface. *J Nanop Res.* 9(1), 165–182.

Kwak, S.Y., Lew, T.T.S., Sweeney, C.J., Koman, V.B., Wong, M.H., Bohmert-Tatarev, K., Snell, K.D., Seo, J.S., Chua, N.H., & Strano, M.S. (2019). Chloroplast-selective gene delivery and expression in planta using chitosan-complexed single-walled carbon nanotube carriers. *Nat Nanotechnol.* 14(5), 447–455.

Lamsal, K., Kim, S.W., Jung, J.H., Kim, Y.S., Kim, K.S., & Lee, Y.S. (2011). Application of silver nanoparticles for the control of Colletotrichum species in vitro and pepper anthracnose disease in field. *Mycobiology.* 39(3), 194–199.

Liscano, J.F., Wilson, C.E., Norman-Jr, R.J. and Slaton, N.A. (2000). *Zinc Availability to Rice from Seven Granular Fertilizers* (Vol. 963). Fayetteville, CA: Arkansas Agricultural Experiment Station.

Millan, G., Agosto, F., & Vazquez, M. (2008). Use of clinoptilolite as a carrier for nitrogen fertilizers in soils of the Pampean regions of Argentina. *Cien Inv Agric.* 35, 293–302.

Mitter, N., Worrall, E.A., Robinson, K.E., Li, P., Jain, R.G., Taochy, C., Fletcher, S.J., Carroll, B.J., Lu, G.Q., & Xu, Z.P. (2017). Clay nanosheets for topical delivery of RNAi for sustained protection against plant viruses. *Nat Plants.* 3(2), 1–10.

Moraru, C.I., Panchapakesan, C.P., Huang, Q., Takhistov, P., Liu, S., & Kokini, J.L. (2003). Nanotechnology: a new frontier in food science understanding the special properties of materials of nanometer size will allow food scientists to design new, healthier, tastier, and safer foods. *Nanotechnology.* 57(12), 24–29.

Mukhopadhyay, S.S. (2014). Nanotechnology in agriculture: prospects and constraints. *Nanotechnol Sci Appl.* 7, 63.

Naderi, M.R., & Danesh-Shahraki, A. (2013). Nanofertilizers and their roles in sustainable agriculture. *Int J Agric Crop Sci.* 5(19), 2229–2232.

Nair, R., Varghese, S.H., Nair, B.G., Maekawa, T., Yoshida, Y., & Kumar, D.S. (2010). Nanoparticulate material delivery to plants. *Plant Sci.* 179(3), 154–163.

Panneerselvam, P., Senapati, A., Kumar, U., Sharma, L., Lepcha, P., Prabhukarthikeyan, S.R., Jahan, A., Parameshwaran, C., Govindharaj, G.P.P., Lenka, S., & Nayak, P.K. (2019). Antagonistic and plant-growth promoting novel *Bacillus* species from long-term organic farming soils from Sikkim, India. *3 Biotech.* 9(11), 1–12.

Rai, V., Acharya, S., & Dey, N. (2012). Implications of nanobiosensors in agriculture. *J Biomater Nanobiotechnol.* 3, 315–324.

Schulze, E.D., Beck, E., & Müller-Hohenstein, K. (2005). *Plant Ecology.* Springer, Berlin.

Scott, N.R., 2007. Nanoscience in veterinary medicine. *Vet Res Commun.* 31(1), 139–144.

Shang, Y., Hasan, M.K., Ahammed, G.J., Li. M., Yin, H., & Zhou, J. (2019). Applications of Nanotechnology in plant growth and crop protection: a review. *Molecules.* 24(14):2558.

Tarafdar, J.C., Raliya, R., & Rathore, I. (2012). Microbial synthesis of phosphorous nanoparticle from tri-calcium phosphate using *Aspergillus tubingensis* TFR-5. *J Bionanosci.* 6(2), 84–89.

Torney, F., Moeller, L., Scarpa, A., & Wang, K. (2007). Genetic engineering approaches to improve bioethanol production from maize. *Curr Opin Biotechnol.* 18, 193–199.

Tothill, I. (2011). Biosensors and nanomaterials and their application for mycotoxin determination. *World Mycotoxin J.* 4(4), 361–374.

Tripathi, S., Sonkar, S.K., & Sarkar, S. (2011). Growth stimulation of gram (*Cicer arietinum*) plant by water soluble carbon nanotubes. *Nanoscale.* 3(3), 1176–1181.

Tyagi, J., Chaudhary, P., Mishra, A., Khatwani, M., Dey, S., & Varma, A. (2022). Role of endophytes in abiotic stress tolerance: with special emphasis on *Serendipita indica. Int J Environ Res.* 16, 62.

United Nations, Department of Economic and Social Affairs, Population Division (2019). World Population Prospects 2019: Highlights (ST/ESA/SER.A/423).

Vamvakaki, V., & Chaniotakis, N.A. (2007). Pesticide detection with a liposome-based nano-biosensor. *Biosens Bioelectr.* 22(12), 2848–2853.

Zandalinas, S.I., Mittler, R., Balfagón, D., Arbona, V., & Gómez-Cadenas, A. (2018). Plant adaptations to the combination of drought and high temperatures. *Physiol Plant.* 162(1), 2–12.

Zhang, H., Demirer, G.S., Zhang, H., Ye, T., Goh, N.S., Aditham, A.J., Cunningham, F.J., Fan, C., & Landry, M.P. (2019). DNA nanostructures coordinate gene silencing in mature plants. *Proc Natl Acad Sci U S A.* 116(15), 7543–7548.

Zhao, X., Cui, H., Wang, Y., Sun, C., Cui, B. and Zeng, Z. (2017a). Development strategies and prospects of nano-based smart pesticide formulation. *J Agric Food Chem.* 66(26), 6504–6512.

Zhao, X., Meng, Z., Wang, Y., Chen, W., Sun, C., Cui, B., Cui, J., Yu, M., Zeng, Z., Guo, S., & Luo, D. (2017b). Pollen magnetofection for genetic modification with magnetic nanoparticles as gene carriers. *Nat Plant.* 3(12), 956–964.

Zheng, L., Hong, F., Lu, S., & Liu, C. (2005). Effect of nano-TiO_2 on strength of naturally aged seeds and growth of spinach. *Biol Trace Element Res.* 104(1), 83–91.

Zulfiqar, F., Navarro, M., Ashraf, M., Akram, N.A., & Munné-Bosch, S. (2019). Nanofertilizer use for sustainable agriculture: advantages and limitations. *Plant Sci.* 289, 110270.

15 Nanotextured Materials for Soil Health

Bartholomew Saanu Adeleke
Olusegun Agagu University of Science and Technology

Modupe Stella Ayilara
North-West University

Deepak Sharma
Jaipur National University

Rabiya Sultana
Gauhati University

Manas Mathur
Suresh Gyan Vihar University

Geeta Bhandari
Swami Rama Himalayan University

Anuj Chaudhary
Shobhit University

CONTENTS

DOI: 10.1201/9781003345565-15

15.1 INTRODUCTION

The continuous increase in the world's population has resulted in high food demand and anthropogenic activities, and if not properly managed, this can hamper soil health that might result in stunted plant growth and slow rate of food supply (Fischer & Connor, 2018). Other factors, such as limited nutrient supply, soil pollution with toxic compounds, disease attack and use of fertilizers, remain critical factors affecting food production (Adeleke & Babalola, 2022). The indiscriminate use of chemical fertilizers which supply micro- and macro-nutrients to the soil for plant nutrition can in turn result in soil toxicity (Erdiansyah et al., 2020). Aside from this, chemical fertilizer application can result in eutrophication, contamination of water bodies during surface run-off, reduction of soil nutrients and other parameters (Gopi et al., 2020). Also, it is more expensive with high transportation cost. Hence, the need to devise eco-friendly approaches through nanoscience to rescue the ecological threats posed by the use of chemical fertilizers becomes imperative.

Recent researches in the synthesis of nanoproducts (nanobiofertilizers, nanopesticides, nanoherbicides, etc.) have provided alternative solutions to agrochemical use, thus safeguarding the ecosystem (Chaudhary et al., 2022a). Interestingly, the incorporation of biofertilizers from microbial sources in crop production is of interest to both the researchers and the farmers for sustainable agriculture, but with certain setbacks (Chaudhary et al., 2021a; Fasusi et al., 2021). Biofertilizers help ensure a stable soil health depending on the types and functional attributes of microorganisms involved in the fixing, solubilization and mobilization of soil minerals, such as nitrogen, silicon, potassium and phosphorus (Chen, 2006; Chaudhary et al., 2021g, 2022b). The combination of biofertilizers with nanofertilizers to yield nanobiofertilizers has been referred to the microbes encapsulated in nanoparticles (Chaudhary & Sharma, 2019). Some factors mediating the activities of nanomaterials with the soil microbes in the bioremediation of metal-contaminated soils have been reported to exert positive influence on the soil health for plant survival and yield enhancement.

In recent times, nanoparticles have been widely employed in a revolutionized manner in medicine, agriculture and other sectors to ensure a sustainable healthy living, agricultural economy and maintenance of soil health (Zhu et al., 2019). Nanomaterials can be grouped based on their chemical composition and structure with the potential to synthesize agricultural nano-input for improving soil nutrient efficiency. The inorganic nanoparticles, such as copper (Cu), aluminium (Al), silica (Si), cerium (Ce), zinc oxide (ZnO) and mercury (Ag), can be advantageous to plants through the bioavailability of nutrients (Sonawane et al., 2022). For instance, the porous nature of silica helps improve soil water-holding capacity, aeration and other soil nutrients. Nevertheless, their continuous use may impose threats to the soil health by increasing heavy metal concentration and alteration in the soil pH, thus affecting microbial metabolism and the plant growth (Kumari et al., 2020).

Carbon nanotubes, polymers and lipids are examples of organic nanomaterials and their bioremediation of soil contaminated with heavy metals may enhance plant nutrition by increasing soil nutrient pool needed for plant growth (Ghormade et al., 2011). The importance of nano-input in agriculture can be linked to their size (small), large surface area, high surface tension and formulation of encapsulated nanoproducts

(Rodríguez-González et al., 2019). Nanofertilizers improve crop yield, boost plant immunity against environmental stresses and enhance soil health (Zulfiqar et al., 2019; Chaudhary et al., 2021b, c). Despite the importance of nanoformulations in agriculture, which aim at revolutionizing agriculture, there are still some challenges associated with these techniques, thus limiting their maximum use in agriculture. Therefore, employing nanotechnological approaches in agriculture and soil health enhancement can form the basis in a modern agricultural system. Furthermore, the advancement in nanoscience can be a gateway for the reclamation of the metal-polluted soils and sustaining soil health due to their potential in the bioremediation of toxic metals from the environment soils (Bhatt et al., 2022).

15.2 EFFECT OF NANOPARTICLES ON SOIL ORGANIC CARBON

The current nanotechnology-based agriculture, such as nano-based fertilizers, nanopesticides, nanofungicides, insecticides and nanoformulations, is widely accepted and proven to be very effective in improving crop protection and production (Mukherjee et al., 2016, Zhao et al., 2021). It has also been contemplated as stylish alternatives for addressing problems and challenges. The uncontrolled exploitation of nanoparticles probably has some harmful impacts on soil microbiota and soil health status. Though nanotechnology offers many advantages, such as increasing nutrient and water availability, carbon sequestration, strengthening soil structure, removing heavy metals, acting as soil conditioner (Figure 15.1), and also increasing soil microbiota and soil fertility, it also has some adverse effects, such as reduced germination, reduced enzymatic action, increased toxicity and reduced root length. In soils, natural carbon content ranged between 0.93% and 3.85%, and the unique properties of nanoparticles have been reported for carbon stabilization and its sequestration in soil to maintain soil organic carbon and soil health for enhanced soil fertility and better environment (Table 15.1). Therefore, it is important to identify the potential metal nanomaterials for attractive carbon storage through enhanced soil aggregation (Singh et al., 2017).

15.3 NANOMATERIALS FOR IMPROVED SOIL STRENGTH

The atmospheric CO_2 concentration is increasing by 2 ppm each year and is the latest universal concern (Lal, 2009). To enhance carbon management in soil airspace, soil organization and assembly are the imperative aspects studied extensively. The management practices comprise conservation tillage, addition of humic substances in soil, proper fertilization, crop rotation and nutrient management. Proper management of soil wealth with soil carbon regulation and its sequestration would provide a fundamental role in CO_2 fixation in the airspace.

Due to their exceptional properties, the application of nanoparticles has improved the earthbound carbon sink for better soil strength and healthier environment by augmenting the carbon fixation and its potential segregation in soil (Monreal et al.,

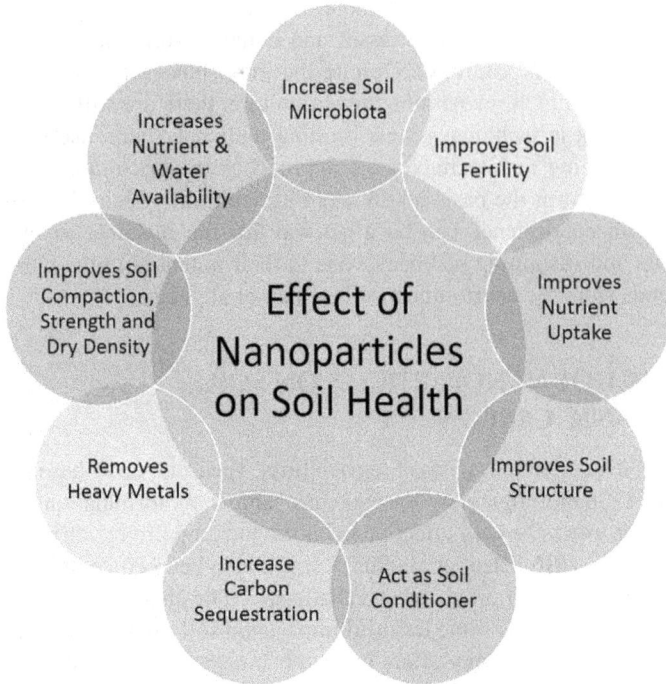

FIGURE 15.1 Positive role of nanoparticles for maintenance of soil health.

TABLE 15.1
Positive Effect of Nanoparticles on Soil Health Parameters

Nanoparticles	Effect on Microbes/Soil Enzymes	References
LiO$_2$	Increased β-glucosidase activity of soil	Avila-Arias et al. (2019)
Cu	Improved soil urease enzyme activity	Galaktionova et al. (2019)
ZnO	Improved phosphate solubilizers' bacterial population such as *Bacillus* sp.	Chavan and Nadanathangam (2019)
TiO$_2$	Enhanced secondary metabolite production in *Pseudomonas aeruginosa*	Haris and Ahmad (2017)
Ce	Improved phytohormone interaction with *Arabidopsis thaliana* and mycorrhiza	Wu et al. (2017)
Nanochitosan	Improved dehydrogenase and alkaline phosphatase enzyme activity of soil	Khati et al. (2017)
SiO$_2$	Improved FDA, dehydrogenase, alkaline phosphatase and soil physicochemical parameters	Kukreti et al. (2020)
Nanozeolite	Improved dehydrogenase and FDA activities	Kumari et al. (2020)
TiO$_2$	Improved dehydrogenase and alkaline phosphatase enzyme activities	Kumari et al. (2021)
Nanochitosan	Improved soil beneficial microbial population and soil enzyme activities	Agri et al. (2022)
Nanogypsum	Improved soil enzyme activities, soil physicochemical parameters and beneficial microbial population	Chaudhary et al. (2021e)

2010, Calabi-Floody et al., 2011, Pramanik et al., 2020). Recent studies revealed the efficacy of nanoparticles in carbon fixation by adopting proper management practices to improve soil structure, constancy of soil aggregates and associated properties (Pramanik et al., 2020). The studies documented the constructive effects of nano-ZnO, nano-Fe and nano-zeolite particles on soil strength, structure and carbon fixation in soil (Aminiyan et al., 2015).

15.4 NANOTECHNOLOGY FOR CARBON SEQUESTRATION IN SOIL AIRSPACE

Carbon sequestrations in the soil for improving soil functions and to ensure a steady carbon dioxide (CO_2) concentration in the atmosphere are able to increase by improving chemical recalcitrance of soil organic carbon (SOC), and physicochemical and biological protection (microbial decomposition) of SOC (Barre et al., 2014). Recently natural nanoparticles, such as hydrous iron (Fe) oxides, nanoclays or oxyhydroxides, have been reported for their potential impact on carbon fixation in soil (Calabi-Floody et al., 2015).

15.5 NANOMATERIALS FOR CAPTURING CO_2

Elevated concentration and emission of CO_2 in the environment frequently lead to acid rain, urban smog and health problem. The adsorption of CO_2 by solid absorbent such as active carbon, zeolite and mesoporous silica is known as a potential technology. Nanotechnology proves to be a potential alternative to develop nanoabsorbents with improved physical distinctiveness than the conservative materials and with huge distinct surface area for elevated confinement of CO_2 (Pramanik et al., 2020). Amine-functionalized mesoporous capsule base, CaO derived from nano-sized $CaCO_3$, single-walled and multi-walled carbon nanotubes (CNTs), mesoporous MgO, CuO, spinel-stabilized $CaO-MgAl_2O_4$ nanoparticles and lithium silicate nanoparticles were identified as a promising CO_2 absorber (Yang et al., 2009).

15.6 EFFECT OF NANOPARTICLES ON MICROBIAL POPULATION

Microbes are part and parcel of the basic food chain. There are several studies which reported the positive effect of NPs on soil microbes and their enzymatic activities (Kukreti et al., 2020; Kumari et al., 2021). Dehydrogenase, alkaline phosphatase activities and beneficial bacterial population in soil were improved by NPs such as nanochitosan and nanozeolite at 50 ppm, which in turn improved plant health parameters (Khati et al., 2017, 2018, 2019a, b). Gypsum, nanochitosan and nanogypsum (50 ppm) improved the beneficial bacterial population, soil microbial activity and physicochemical attributes under maize cultivation (Chaudhary et al., 2021d, e, 2022c). Nanophos, which is a combination of biofertilizers and nanoparticles, also improved soil health via improving soil microbial enzymatic activities and promoting plant productivity (Chaudhary et al., 2021f).

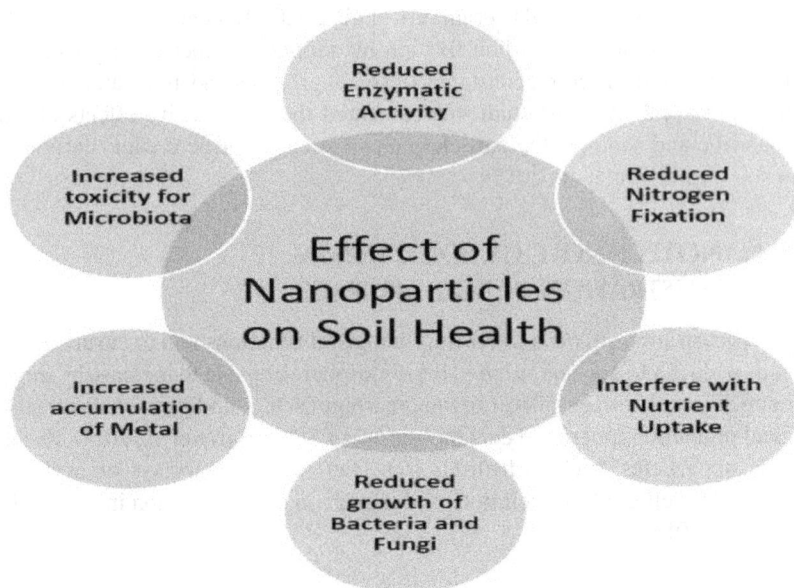

FIGURE 15.2 Toxic effects of nanoparticles on soil health.

On the other hand, the studies pertaining to nanoparticles show that they exhibit hazardous effect on the growth and abundance of the microorganisms besides environmental toxicity (Martínez et al., 2021) (Figure 15.2). Out of the total microbial population of the heterotrophic microflora, about 15% comprise bacterial population of different species. Susceptibility of bacteria varies in response to different nanoparticles depending on the cellular wall structure and composition. The Gram-positive and Gram-negative bacteria vary in their cell wall structure and composition; hence, the effect of NPs on these bacteria too varies accordingly. Reports on NPs show that ZnO-NPs, Ag-NPs, TiO$_2$-NPs, CuO-NPs and FeO-NPs which caused variable toxic effects on soil microorganisms and pure microbial strains have been investigated (Ameen et al., 2021). Although the studies of toxicity in bacterial metabolism are still in its nurturing stage, evidence of the NPs affecting the bacterial cell walls, oxidative stress, metal ions has been found from previous studies (Ameen et al., 2021, Martínez et al., 2021).

Toxic effects of NPs such as Ag and ZnO on bacteria lead to cellular disintegration together with oxidative damage (Zhang et al., 2018) and restriction of cell respiration of bacteria (Choi et al., 2008). The level of toxicity on bacteria also depends on the chemical composition, structural properties, concentration and exposure time of NPs. Prior studies on interaction of *Deinococcus radiodurans* with ZnO-NPs have revealed the morphological changes of cell, membrane damage and toxicity of gene due to the ZnO-NPs' concentration between 1 and 80 µg/mL (Martínez et al., 2021). Again, Martínez et al. estimated the outcome of different doses of TiO$_2$-NPs with bacteria wherein the result demonstrated the dose-dependent effects on the bacterial cells (Martínez et al., 2021). Likewise, Lopes et al. (2012) worked on the toxic effects in *Vibrio fischeri* by Au-NPs in which the result displayed the growth inhibition due to long exposure to NPs. A comparative study by Lin et al. (2014) on the toxic effects of five dissimilar sizes of TiO$_2$-NPs in *Escherichia coli* depicted the toxicity level based on particle properties. In some cases, the nanomaterials lead to increased rate of growth and development in organisms such as *Streptomyces*, which might be the outcome of decrease in competition among microorganisms leading to increased nutrient availability (Martínez et al., 2021).

Previous studies depicted that in arctic soils, nanoparticles of copper, silver and silicon reduce the microbial biomass to a considerable extent (Yonathan et al., 2022). Altogether, it has been reported that NPs such as zinc, silver and copper possess antimicrobial properties and hence these NPs could easily penetrate the microbial cells (Table 15.2). It has been stated that bactericidal effects together with morphological alterations occurred in *Pseudomonas stutzeri* and *Bacillus cereus* inhabiting the soils (Ameen et al., 2021). Factors such as concentration, soil texture, functionalization and exposure time also influence the effects of the Ag-NPs on the microbial community of the soil (Grün et al., 2019).

Fungi are an integral part of the microbial population and play a crucial role in the improvement of plant and soil health by protecting them from stress conditions (Nielsen et al., 2015; Tyagi et al., 2022). Soil fungi play a pioneer role in the decomposition of dead material within the soil besides being the only consumer of energy. NPs due to their size could easily penetrate the fungal hyphae, thus leading to deformation of the native morphology. The interaction between mycorrhizal fungi and varied nanoparticles resulted in both positive as well as negative aspects. It has been found that some NPs aided in the fungal colonization while some exerted a negative impact on the colonization. For example, the biological effects of comparative analysis of doses of FeO and Ag-NPs in bulk and nanoforms on AMF mycorrhizal clover

TABLE 15.2
Negative Effect of Nanoparticles on Soil Health Parameters

Nanoparticles	Target Microbes/Enzymes	References
CuO	Inhibited nitrate reductase and nitric oxide reductase activities	Zhao et al. (2019)
Polystyrene	Inhibited cellobiohydrolase and β-glucosidase	Awet et al. (2018)
ZnO	Inhibited the growth of *Rhizobium leguminosarum*	Sarabia-Castillo and Fernandez-Luqueno (2016)
Ag	Toxic effect on the growth of *Bacillus* and *Pseudomonas fluorescens*	Das et al. (2014)
Ag	*Phytophthora infestans*	Giannousi et al. (2013)
TiO$_2$ and CuO	Inhibited microbial population	Ameen et al. (2021)
Ag and ZnO	Bacterial cellular disintegration together with oxidative damage	Zhang et al. (2018)
Ag and ZnO	Restriction of cell respiration of bacteria	Choi et al. (2008)
Ag-NPs	Bactericidal effects with morphological alterations in *Pseudomonas stutzeri*	Ameen et al. (2021)
TiO$_2$-NPs	Dose-dependent toxic effects on bacterial growth	Martínez et al. (2021)

were studied by Feng et al. (2013). The end result exhibited negative effects in AMF colonization in bulk doses of NPs while significant positive impact on plant growth promotion was found in the nanorange (Ameen et al., 2021). Again, in a study, mycorrhizal colonization reduction was found in relation to Ag-NPs (Noori et al., 2017) while in another research the opposite effect was obtained with ZnO-NPs and bulk amounts of $ZnSO_4$ (Li et al., 2015). Du et al. (2020) in their experiment with aquatic fungi with ZnO-NPs found that the litter decomposition was reduced significantly by the fungi community. Lately, the transcriptome sequencing of complex mycelia of *Fusarium solani* was investigated against Ag-NPs by the aid of RNA-Seq platform in which the Ag-NPs repressed the fungal growth by obstructing with the signal transduction pathway, metabolism and genetic molecules (Shen et al., 2020).

15.7 NANOPARTICLES IN SOIL REMEDIATION

Heavy metal contaminants which include chromium, zinc, arsenic, lead, cadmium, nickel, copper, mercury, etc. enter the soil through different means, such as industrial waste water and agricultural activities, thus resulting in soil and water pollution (Tourinho et al., 2012, Chatterjee & Abraham, 2022). Heavy metal deposition into the soil and water is absorbed by plants and enters into the food chain, which has resulted in health threatening diseases, such as cancer, kidney and liver diseases, birth defects and skin lesions in humans (Chatterjee & Abraham, 2022). Over the years, nanoparticles have served as a means to bioremediate soils polluted with xenobiotic recalcitrant heavy metals, control release of pesticides and plant nutrients, etc.

About 90% of chemicals applied to the soils are usually lost due to run-offs and air (Thul & Sarangi, 2015) and this necessitates the need for more application which eventually has a negative effect on the soil's physical, chemical and biological properties; and farmer's income. In a study carried out by Xu et al. (2020), copper ion was used to chelate mesoporous silica nanoparticles to control the release of azoxystrobin. Corradini et al. (2010) reported the use of chitosan nanoparticles in the control of the release of NPK fertilizer due to its biodegradable, bactericidal, polymeric cationic and bioabsorbable nature. Also, Mushtaq et al. (2018) reported the effectiveness of silicon nanoparticles in the control of the release of NPK fertilizer and in enhancing plants to adapt to the salinity and drought stresses.

The control of soil-borne plant pathogen by nanoparticles has also been reported using magnesium oxide nanoparticles in the control of *Fusarium* wilt caused by *F. solani* and *Fusarium oxysporum* in mung bean (Abdallah et al., 2022). Furthermore, the magnesium oxide nanoparticles were able to increase the shoot length, fresh and dry shoot weight, fresh and dry root weight, and the nodule number in mung bean (Abdallah et al., 2022). Suresh et al. (2018) revealed that silver nanoparticles with *Suaeda maritima* as stabilizing and reducing agents were able to control the tobacco cutworm *Spodoptera litura* as well as the dengue vector, *Aedes aegypti*. Silver nanoparticles from the essential oil of lemon grass have been reported as biological agent in the control of lichen on rock (Mulwandari et al., 2022).

Generally, nanoparticles are used to remove pollutants from the soil (Baligah & Chesti, 2022). For instance, zero-valent iron nanoparticles have been successfully used in the bioremediation of soil pollutants (Galdames et al., 2020). Silicon oxide nanoparticles successfully helped to bioremediate calcareous soil polluted with nickel, zinc and cadmium (Naderi & Jalali, 2019). Goethite nanosphere and zero-valent iron nanoparticles helped to immobilize arsenic in polluted soils, although lower dosage of goethite nanosphere is more promising compared to others (Baragaño et al., 2020b). Magnetite nanoparticles helped to immobilize arsenic and other (poly-aromatic hydrocarbons) organic pollutants in soils (Baragaño et al., 2020a). Titanium dioxide nanoparticles assisted in the bioremediation of antimony contamination in soils, and further enhanced the growth and establishment of *Sorghum bicolor* (Zand et al., 2020).

Nanofertilizer is becoming a preferable means of increasing crop productivity because they make optimum quantity of nutrients available on specific sites. They are very soluble, efficient and affordable. Also, they increase the rate of photosynthesis, are environment-friendly, and enhance seed metabolism, germination stress tolerance, nitrogen metabolism, and synthesis of carbohydrate and protein (Figure 15.3). Although, if nanofertilizers are over-applied, nanoparticles can result in soil toxicity and reduce crop yield (Butt & Naseer, 2020). Therefore, more research is needed to ascertain the optimum quantity or dosage of nanoparticles which is to be applied to obtain optimum yield of each crop. There are majorly three approaches used in the production of nanoparticles, namely: biological, bottom-up and top-down approaches. The top-down approach involves the grinding of nanomaterials to reduce the size to form organized nanoscale assemblies. Conversely, the drawback of this method is that it is highly prone to the contamination of impurities (Zulfiqar et al., 2019). The bottom-up

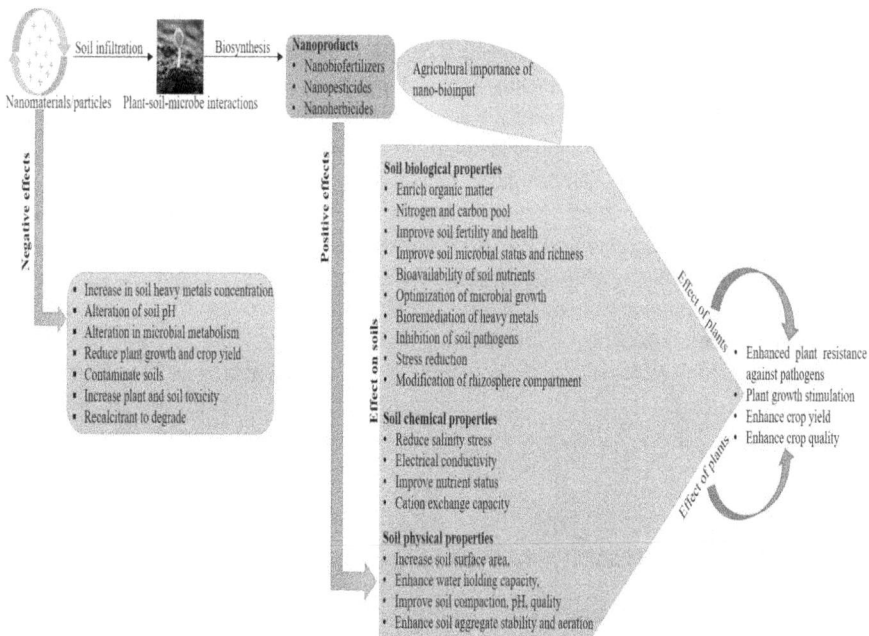

FIGURE 15.3 Effect of nanoparticles on soil and plant health status.

approach involves the build-up of nanofertilizers on an atomic or molecular scale through chemical reaction, thus reducing the presence of impurities. The biological method involves the production of nanoparticles from plants, bacteria and fungi which help to control the size of the particles and toxicity (Zulfiqar et al., 2019). Iron nanofertilizer increased the chlorophyll content and root function in *Citrus maxima* (Sharma, 2006). Equally, zinc, manganese and copper nanofertilizers have been tested on lettuce and enhanced the growth from 12% to 54% (Lu et al., 2016). Zinc oxide nanoparticles have also been reported to improve the pigment content and the immunity of coriander plant against infection (Pullagurala et al., 2018). Application of nanochitosan improved maize plant health parameters and soil microbial population (Agri et al., 2021, 2022).

The mechanisms employed by nanoparticles in bioremediating soil are complex chemical and physical reactions, which include ion-exchange, precipitation, adsorption, etc. (Zhang et al., 2021). The adsorption process can either be chemical or physical, depending on the type of molecular binding (Chen et al., 2019). For instance, Van Koetsem et al. (2016) reported the removal of cadmium pollutants from the environment by nanoparticles using adsorption reaction due to the fact that nanoparticles have higher surface area compared to other materials. Adsorption can either happen as specific adsorption which takes place in the form of amorphous oxide and hydroxides as well as crystalline nature or of non-specific adsorption which happens in the form of exchange in the soil (Chen et al., 2019). There are two types of adsorption mechanisms, which are complexation and electrostatic reaction (Xu & Chen, 2015). Ligand complexion can bioremediate heavy metals in the soil through specific surface complexation of the inner sphere and electrostatic attraction. Examples of ligands include phenolic hydroxyl, alcoholic hydroxyl and carboxyl (Zhao et al., 2018). Nanoparticles can also be dispersed in the soil to attract and convert heavy metals to a more stable form. For instance, Zhou et al. (2020) reported the use of nanoparticles to make cadmium unavailable for uptake in the soil by plants.

Redox reaction can help to remove heavy metal pollutants from the environment through the application of nanoparticle contaminants, such as rhenium, chromium, technetium, arsenic and halogenated organic matter. Nanoparticles which contain iron and sulphur are more effective because they donate electrons to the pollutants and reduce the pollutants to low-soluble and less toxic substances (Guerra et al., 2018). Reduction bioremediation process is a multistep procedure, which includes chemical sorption of pollutants on the nanoparticle surface, dissolution to reduce contaminants from high to low valence, formation between low-valence state of nanoparticles and pollutants, and precipitation, which makes the pollutants unavailable for absorption and leaching (Chen et al., 2019). More research on the mechanisms of bioremediation of heavy metal pollutants using nanoparticles will help to properly understand and fully tap into its potentials.

15.8 FACTORS AFFECTING THE BIOREMEDIATION OF SOIL BY NANOMATERIALS

15.8.1 pH OF NANOPARTICLES

The pH of nanomaterials affects the hydrolysis of pollutants, hence affecting its adsorption rate by the nanoparticles (Zhang et al., 2021). In the presence of a high pH, cadmium is hydrolysed and forms cadmium hydroxide, which enhances the affinity between the soil sites and cadmium (Zhang et al., 2021). Equally, adding nanomagnesium to the soil helps to increase its pH content. It has been reported that a linear negative relationship exists between exchangeable cadmium and pH in alkaline and acidic soils (Zhang et al., 2021).

15.8.2 CATION EXCHANGE CAPACITY

Cation exchange capacity is another factor which affects bioremediation. An increase in the cation exchange capacity as well as the clay content in soil can result in a decrease in the exchangeable cadmium content in soil, thus leading to a greater adsorption and heavy metal fixation in soil (Pu et al., 2008). High cation exchange capacity in soil leads to a greater cadmium ion adhesion and alters the permeability parameters, as well as the hydraulic conductivity of the soil (Covelo et al., 2007).

15.8.3 MOISTURE CONTENT

The water content in the soil affects the bioremediation of heavy metal pollutants by nanomaterials. For instance, Adrees et al. (2020) performed a research and observed that there was an increase in the concentration of cadmium in a water-deficient condition, which may indicate a lower bioremediation process.

15.9 DIFFERENT SOURCES OF NANOPARTICLES FOR SOIL BIOREMEDIATION

Nanoparticles which are used in bioremediation of soil can be from organic sources which include plant and microbial sources. These organic sources contain sugars, amino acids, functional groups (carboxylic acid ketones, aldehydes, amines and alcohols), proteins, citric acid, membrane proteins, phenolics and tartaric acid (Al-Shnani et al., 2017, Selvan et al., 2018). For example, green tea extract nanomaterials which are used as a means for bioremediation can convert iron (II) to iron (III) with their dispersive abilities (Wang et al., 2019). Similarly, the fruits classified in the Combretaceae family are used to produce gold, iron, lead and palladium nanoparticles, which are used to bioremediate chromium VI heavy metal in the soil. These fruits contain some chemicals such as chebulic acid, gallate esters, tannins, chebulic ellagitannins and gallic acids, which are responsible for bioremediation abilities (Kumar et al., 2013, Gopinath et al., 2014). On the other hand, microbes, such as bacteria, fungi and viruses, have been used to produce nanoparticles, which was used to bioremediate metal-contaminating soils. For instance, *Lactobacillus* spp. has been reported to produce titanium oxide nanoparticle, used for soil bioremediation (Jha et al., 2009).

15.10 CONCLUSION

The impact of microbial nanoproducts on the health of the soil has greatly benefited the soil ecosystems. The agricultural significance of nano-bioinput in the bioremediation of hazardous compounds and enhancement of soil nutrient profiling are beneficial to plant and soil health for yield enhancement. Therefore, the approaches into the biosynthesis of nanoproducts still need to be focused on for maximum exploration and application in ensuring a stable soil ecology, which can serve as potential biological agents in the bioremediation and reclamation of soils polluted with heavy metals. An in-depth study is required for proteomic and genomic analysis for future endeavours. Furthermore, research on the nanoparticles' concentration and their microbial interaction should be focused in detail. Proper strategies are needed for disposal of nanoparticles in soil ecosystem and safety measures are required to be developed for reducing the contact with soil microorganisms. Additionally, to assess the positive and harmful aspects of the nanoparticles on the soil physicochemical properties detailed investigation is of utmost importance. Altogether, gene level study of microorganisms to combat toxicity of nanoparticles is yet another area for thorough research and development. To summarize, the utilization of nanoparticles should be controlled and proper evaluation for environmental safety should be implemented.

REFERENCES

Abdallah, Y., Hussien, M., Omar, M.O., Elashmony, R., Alkhalifah, D.H.M., & Hozzein, W.N. (2022). Mung bean (*Vigna radiata*) treated with magnesium nanoparticles and its impact on soilborne *Fusarium solani* and *Fusarium oxysporum* in clay soil. *Plants*. 11, 1514.

Adeleke, B.S., & Babalola, O.O. (2022). Meta-omics of endophytic microbes in agricultural biotechnology. *Biocatal Agric Biotechnol*. 42, 102332.

Adrees, M., Khan, Z.S., Ali, S., Hafeez, M., Khalid, S., Rehman, M.Z., Hussain A., Hussain, K., Chatha, S.A.S., & Rizwan, M. (2020). Simultaneous mitigation of cadmium and drought stress in wheat by soil application of iron nanoparticles. *Chemosphere*. 238, 124681.

Agri, U., Chaudhary, P., & Sharma, A. (2021). In vitro compatibility evaluation of agriusable nanochitosan on beneficial plant growth-promoting rhizobacteria and maize plant. *Natl Acad Sci Lett*. 44, 555–559.

Agri, U., Chaudhary, P., Sharma, A., & Kukreti, B. (2022). Physiological response of maize plants and its rhizospheric microbiome under the influence of potential bioinoculants and nanochitosan. *Plant Soil*. 474, 451–468.

Al-Shnani, F., Al-Haddad, T., Karabet, F., & Allaf, A.W. (2017). Chitosan loaded with silver nanoparticles, CS-AgNPs, using thymus syriacus, wild mint, and rosemary essential oil extracts as reducing and capping agents. *J Phys Org Chem*. 30, e3680.

Ameen F., Alsamhary K., Alabdullatif J.A., & ALNadhari S. (2021). A review on metal-based nanoparticles and their toxicity to beneficial soil bacteria and fungi. *Ecotoxicol Environ Saf*. 213, 112027.

Aminiyan, M.M., Safari Sinegani, A.A., & Sheklabadi, M. (2015). Aggregation stability and organic carbon fraction in a soil amended with some plant residues, nanozeolite, and natural zeolite. *Int J Recycl Org Waste Agric*. 4, 11–22.

Avila-Arias, H., Nies, L.F., Gray, M.B., & Turco, R.F. (2019). Impacts of molybdenum-, nickel-, and lithium- oxide nanomaterials on soil activity and microbial community structure, *Sci Tot Environ*. 652, 202–211.

Awet, T.T., Kohl, Y., Meier, F., Straskraba, S., Grün, A.-L., Ruf, T., Jost, C., Drexel, R., Tunc, E., & Emmerling, C. (2018). Effects of polystyrene nanoparticles on the microbiota and functional diversity of enzymes in soil. *Environ Sci Eur* 30, 11.

Baligah, H.U., & Chesti, M. (2022). Fungal nanoparticles for soil health and bioremediation. In J.A. Malik & M.R. Goyal (Eds.) *Bioremediation and Phytoremediation Technologies in Sustainable Soil Management*. Apple Academic Press, pp. 101–118.

Baragaño, D., Alonso, J., Gallego, J., Lobo, M., & Gil-Díaz, M. (2020a). Magnetite nanoparticles for the remediation of soils co-contaminated with As and PAHs. *Chem Eng J*. 399, 125809.

Baragaño, D., Alonso, J., Gallego, J., Lobo, M., & Gil-Díaz, M. (2020b). Zero valent iron and goethite nanoparticles as new promising remediation techniques for As-polluted soils. *Chemosphere*. 238, 124624.

Barre, P., Fernandez-Ugalde, O., Virto. I., Velde, B., & Chenu, C. (2014). Impact of phyllo-silicate mineralogy on organic carbon stabilization in soils: incomplete knowledge and exciting prospects. *Geoderma*. 235–236, 382–395.

Bhatt, P., Pandey, S.C., Joshi, S., Chaudhary, P., Pathak, V.M., Huang, Y., Wu, X., Zhou, Z., & Chen, S. (2022). Nanobioremediation: a sustainable approach for the removal of toxic pollutants from the environment. *J Hazard Mater*. 427, 128033.

Butt, B.Z., & Naseer, I. (2020). Nanofertilizers. In S. Javad (Ed.) *Nanoagronomy*. Springer, Cham, pp. 125–152.

Calabi-Floody, M., Bendall, J.S., Jara, A.A., Welland, M.E., Theng, B.K.G., Rumpel, C., & Mora, M.L. (2011). Nanoclays from an Andisol: extraction, properties and carbon stabilization. *Geoderma*. 161, 159–167.

Calabi-Floody, M., Rumpel, C., Velasquez, G., Violante, A., Bol, R., Condron, L.M., & Mora, M.L. (2015). Role of Nanoclays in carbon stabilization in Andisols and Cambisols. *J Soil Sci Plant Nute*. 15, 587–604.

Chatterjee, A., & Abraham, J. (2022). Heavy metal remediation through nanoparticles. In R.N. Bharagava, S. Mishra, G.D. Saratale, R.G. Saratale, L.F.R. Ferreira (Eds.) *Bioremediation*. CRC Press, Boca Raton, pp. 235–247.

Chaudhary, A., Chaudhary, P., Upadhyay, A., Kumar, A., & Singh, A. (2022c). Effect of gypsum on plant growth promoting rhizobacteria. *Environ Ecol*. 39(4A), 1248–1256.

Chaudhary, A., Parveen, H., Chaudhary, P., Bhatt, P., & Khatoon, H. (2021g). Rhizospheric microbes and their mechanism. In Bhatt, P., Gangola, S., Udayanga, D., & Kumar, G. (eds.) *Microbial Technology for Sustainable Environment*. Springer, Singapore.

Chaudhary, P., Chaudhary, A., Agri, U., Khatoon, H., & Singh, A. (2022c). Recent trends and advancements for agro-environmental sustainability at higher altitudes. In Goel, R., Soni, R., Suyal, D.C., & Khan, M. (eds.) *Survival Strategies in Cold-Adapted Microorganisms*. Springer, Singapore.

Chaudhary, P., Chaudhary, A., Bhatt, P., Kumar, G., Khatoon, H., Rani, A., Kumar, S., & Sharma, A. (2022a). Assessment of soil health indicators under the influence of nano-compounds and *Bacillus* spp. in field condition. *Front Environ Sci*. 9, 769871.

Chaudhary, P., Chaudhary, A., Parveen, H., Rani, A., Kumar, G., Kumar, A., & Sharma, A. (2021f). Impact of nanophos in agriculture to improve functional bacterial community and crop productivity. *BMC Plant Biol*. 21, 519.

Chaudhary, P., Khati, P., Chaudhary, A., Gangola, S., Kumar, R., & Sharma, A. (2021b). Bioinoculation using indigenous *Bacillus* spp. improves growth and yield of *Zea mays* under the influence of nanozeolite. *3 Biotech*. 11, 11.

Chaudhary, P., Khati, P., Chaudhary, A., Maithani, D., Kumar, G., & Sharma, A. (2021e). Cultivable and metagenomic approach to study the combined impact of nanogypsum and *Pseudomonas taiwanensis* on maize plant health and its rhizospheric microbiome. *PLoS One*. 16, e0250574.

Chaudhary, P., Khati, P., Gangola, S., Kumar, A., Kumar, R., & Sharma, A. (2021d). Impact of nanochitosan and *Bacillus* spp. on health, productivity and defence response in *Zea mays* under field condition. *3 Biotech*. 11, 237.

Chaudhary, P., Parveen, H., Gangola, S., Kumar, G., Bhatt, P., & Chaudhary, A. (2021a). Plant growth promoting rhizobacteria and their application in sustainable crop production. In Bhatt P., Gangola S., Udayanga D., Kumar G. (eds). *Microbial Technology for Sustainable Environment*. Springer, Singapore.

Chaudhary, P., & Sharma, A. (2019). Response of nanogypsum on the performance of plant growth promotory bacteria recovered from nanocompound infested agriculture field. *Environ Ecol*. 37, 363–372.

Chaudhary, P., Sharma, A., Chaudhary, A., Khati, P., Gangola, S., & Maithani, D. (2021c). Illumina based high throughput analysis of microbial diversity of rhizospheric soil of maize infested with nanocompounds and *Bacillus* sp. *Appl Soil Ecol*. 159, 103836.

Chaudhary, P., Singh, S., Chaudhary, A., Sharma, A., & Kumar, G. (2022b). Overview of biofertilizers in crop production and stress management for sustainable agriculture. *Front Plant Sci*. 13, 930340.

Chavan, S., Nadanathangam, V. (2019). Effects of nanoparticles on plant growth-promoting bacteria in Indian agricultural soil. *Agronomy*, 9, 140.

Chen, J.-H. (2006). The combined use of chemical and organic fertilizers and/or biofertilizer for crop growth and soil fertility. In *International Workshop on Sustained Management of the Soil-Rhizosphere System for Efficient Crop Production and Fertilizer Use*. Land Development Department, Bangkok Thailand, Vol. 16, pp. 1–11.

Chen, Y., Liang, W., Li, Y., Wu, Y., Chen, Y., Xiao, W., Zhao, L., Zhang J., & Li, H. (2019). Modification, application and reaction mechanisms of nano-sized iron sulfide particles for pollutant removal from soil and water: a review. *Chem Eng J*. 362, 144–159.

Choi, O., Deng, K.K., Kim, N.J., Ross, L., Surampalli, R.Y., & Hu, Z. (2008). The inhibitory effects of silver nanoparticles, silver ions, and silver chloride colloids on microbial growth. *Water Res*. 42, 3066–3074.

Corradini, E., De Moura, M., & Mattoso, L. (2010). A preliminary study of the incorporation of NPK fertilizer into chitosan nanoparticles. *Express Polym Lett*. 4(8), 509–515.

Covelo, E., Vega, F., & Andrade, M. (2007). Competitive sorption and desorption of heavy metals by individual soil components. *J Hazard Mater*. 140, 308–315.

Das, V.L., Thomas, R., Varghese, R.T., Soniya E.V., Mathew, J., & Radhakrishnan, E.K. (2014). Extracellular synthesis of silver nanoparticles by the Bacillus strain CS 11 isolated from industrialized area. *3 Biotech* 4, 121–126.

Du, J., Zhang, Y., Yin, Y., Zhang, J., Ma, H., Li, K., & Wan, N. (2020). Do environmental concentrations of zinc oxide nanoparticle pose ecotoxicological risk to aquatic fungi associated with leaf litter decomposition? *Water Res*. 178, 115840

Erdiansyah, I., Sari, V., Pratama, A., & Wiharto, K. (2020). Utilization of Rhizobium spp as substitution agent of nitrogen chemical fertilizer on soybean cultivation. In *IOP Conference Series: Earth and Environmental Science*. IOP Publishing, Vol. 411, p. 012065.

Fasusi, O.A., Cruz, C., & Babalola, O.O. (2021). Agricultural sustainability: microbial biofertilizers in rhizosphere management. *Agriculture*. 11, 163.

Feng, Y.Z., Cui, X.C., He, S.Y., Dong, G., Chen, M., Wang, J.H., & Lin, X. (2013). The role of metal nanoparticles in influencing arbuscular mycorrhizal fungi effects on plant growth. *Environ Sci Technol*. 47, 9496–9504.

Fischer, R.A., & Connor, D.J. (2018). Issues for cropping and agricultural science in the next 20 years. *Field Crops Res*. 222, 121–142.

Galaktionova, L., Gavrish, I. & Lebedev, S. (2019). Bioeffects of Zn and Cu Nanoparticles in Soil Systems. *Toxicol Environ Health Sci*. 11, 259–270.

Galdames, A., Ruiz-Rubio, L., Orueta, M., Sánchez-Arzalluz, M., & Vilas-Vilela, J.L. (2020). Zero-valent iron nanoparticles for soil and groundwater remediation. *Int J Environ Res Public Health*. 17, 5817.

Ghormade, V., Deshpande, M.V., & Paknikar, K.M. (2011). Perspectives for nano-biotechnology enabled protection and nutrition of plants. *Biotechnol Adv*. 29, 792–803.

Giannousi, K., Avramidisb, I., & Dendrinou-Samara, C. (2013). Synthesis, characterization and evaluation of copper based nanoparticles as agrochemicals against *Phytophthora infestans*. *RSC Adv.*, 3, 21743.

Gopi, G.K., Meenakumari, K., Anith, K., Nysanth, N., & Subha, P. (2020). Application of liquid formulation of a mixture of plant growth promoting rhizobacteria helps reduce the use of chemical fertilizers in Amaranthus (*Amaranthus tricolor* L.). *Rhizosphere*. 15, 100212.

Gopinath, K., Gowri, S., Karthika, V., & Arumugam, A. (2014). Green synthesis of gold nanoparticles from fruit extract of *Terminalia arjuna*, for the enhanced seed germination activity of *Gloriosa superba*. *J Nanostructure Chem*. 4, 1–11.

Grün, AL., Manz, W., Kohl, Y.L., Meier, F., Staskraba, S., Jost, C., Drexel, R., & Emmerling, C. (2019). Impact of silver nanoparticles (AgNP) on soil microbial community depending on functionalization, concentration, exposure time, and soil texture. *Environ Sci Eur*. 31, 15.

Guerra, F.D., Attia, M.F., Whitehead, D.C., & Alexis, F. (2018). Nanotechnology for environmental remediation: materials and applications. *Molecules*. 23(7), 1760.

Haris, Z., & Ahmad, I. (2017). Impact of Metal Oxide Nanoparticles on Beneficial Soil Microorganisms and their Secondary Metabolites. *Int J Life Sci Scienti. Res.*, 3(3), 1020–1030.

Jha, A.K., Prasad, K., & Kulkarni, A. (2009). Synthesis of TiO$_2$ nanoparticles using microorganisms. *Colloids Surf B Biointerf*. 71, 226–229.

Khati, P., Bhatt, P., Kumar, R., & Sharma, A. (2018). Effect of nanozeolite and plant growth promoting rhizobacteria on maize. *3Biotech*. 8, 141.

Khati, P., Chaudhary, P., Gangola, S., Bhatt, P., & Sharma, A. (2017). Nanochitosan supports growth of *Zea mays* and also maintains soil health following growth. *3 Biotech*. 7, 81.

Khati, P., Chaudhary, P., Gangola, S., & Sharma, P. (2019a). Influence of nanozeolite on plant growth promotory bacterial isolates recovered from nanocompound infested agriculture field. *Environ Ecol*. 37, 521–527.

Khati, P., Sharma, A., Chaudhary, P., Singh, A.K., Gangola, S., & Kumar R. (2019b). High-throughput sequencing approach to access the impact of nanozeolite treatment on species richness and evens of soil metagenome. *Biocatal Agric Biotechnol*. 20, 101249.

Kukreti, B., Sharma, A., Chaudhary, P., Agri, U., & Maithani, D. (2020). Influence of nanosilicon dioxide along with bioinoculants on *Zea mays* and its rhizospheric soil. *3 Biotech*. 10, 345.

Kumar, K.M., Mandal, B.K., Kumar, K.S., Reddy, P.S., & Sreedhar, B. (2013). Biobased green method to synthesise palladium and iron nanoparticles using *Terminalia chebula* aqueous extract. *Spectroch Acta A Mol Biomol Spect*. 102, 128–133.

Kumari, H., Khati, P., Gangola, S., Chaudhary, P., & Sharma, A. (2021). Performance of plant growth promotory rhizobacteria on Maize and soil characteristics under the influence of TiO$_2$ nanoparticles. *Pantnagar J Res*. 19, 28–39.

Kumari, S., Sharma, A., Chaudhary, P., & Khati, P. (2020). Management of plant vigor and soil health using two agriusable nanocompounds and plant growth promotory rhizobacteria in Fenugreek. *3 Biotech*. 10, 461.

Lal, R. (2009). Challenges and opportunities in soil organic matter research. *Eur J Soil Sci*. https://doi.org/10.1111/j.1365-2389.2008.01114.x

Li, S., Liu, X.Q., Wang, F.Y., & Miao Y.F. (2015). Effects of ZnO nanoparticles, ZnSO$_4$ and arbuscular mycorrhizal fungus on the growth of maize. *Huanjing Kexue/Environ Sci*. 36(12), 4615–4622.

Lin, X., Li, J., Ma, S., Liu, G., Yang, K., Tong, M., & Lin, D. (2014). Toxicity of TiO$_2$ nanoparticles to *Escherichia coli*: effects of particle size, crystal phase and water chemistry. *PLoS One*. 9(10), e110247.

Lopes, I., Ribeiro, R., Antunes, F.E., Rocha-Santos, T.A., Rasteiro, M.G., Soares, A.M., Gonçalves, F., & Pereira, R. (2012). Toxicity and genotoxicity of organic and inorganic nanoparticles to the bacteria *Vibrio fischeri* and *Salmonella typhimurium*. *Ecotoxicol*. 21, 637–648

Lu, S., Feng, C., Gao, C., Wang, X., Xu, X., Bai, X., Gao, N., & Liu M. (2016). Multifunctional environmental smart fertilizer based on L-aspartic acid for sustained nutrient release. *J Agric Food Chem.* 64, 4965–4974.

Martínez, G., Merinero, M., Pérez-Aranda, M., Pérez-Soriano, E.M., Ortiz, T., Begines, B., & Alcudia, A. (2021). Environmental impact of nanoparticles' application as an emerging technology: a review. *Materials.* 14(7), 166.

Monreal, C.M., Sultan, Y., & Schnitzer, M (2010). Soil organic matter in nano-scale structures of a cultivated black Chernozem. *Geoderma.* 159, 237–242.

Mukherjee, A., Majumdar, S., Servin, A.D., Pagano, L., Dhankher, O.P., & White, J. C. (2016). Carbon nanomaterials in agriculture: a critical review. *Front in Plant Sci.* 7, 172.

Mulwandari, M., Asysyafiiyah, L., Sirajuddin, M.I., & Cahyandaru, N. (2022). Direct synthesis of lemongrass (*Cymbopogon citratus* L.) essential oil-silver nanoparticles (EO-AgNPs) as biopesticides and application for lichen inhibition on stones. *Heliyon.* 8, 1–11.

Mushtaq, A., Jamil, N., Rizwan, S., Mandokhel, F., Riaz, M., Hornyak, G., Malghani N.M., & Shahwani, N.M. (2018). Engineered silica nanoparticles and silica nanoparticles containing controlled release fertilizer for drought and saline areas. In *IOP Conference Series: Materials Science and Engineering.* IOP Publishing, p. 012029.

Naderi Peikam, E., & Jalali, M. (2019). Application of three nanoparticles (Al_2O_3, SiO_2 and TiO_2) for metal-contaminated soil remediation (measuring and modeling). *Int J Environ Sci Technol.* 16, 7207–7220.

Nielsen, U.N., Wall, D.H., & Six, J. (2015). Soil biodiversity and the environment. *Annu Rev Environ Res.* 40, 63–90.

Noori, A., White, J.C., & Newman, L.A. (2017). Mycorrhizal fungi influence on silver uptake and membrane protein gene expression following silver nanoparticle exposure. *J Nanopart Res.* 19, 66.

Pramanik, P., Ray, P., Maity, A., Das, S., Ramakrishnan, S., & Dixit, P. (2020). *Carbon Management in Tropical and Sub-Tropical Terrestrial Systems.* Springer, Singapore, pp. 403–415.

Pu, J., Fu, J., & Zhan, M. (2008). Effects of soil properties on the bioavailability of added cadmium and lead in paddy soils. *Ecol Econ.* 17, 2253–2258.

Pullagurala, V.L.R., Adisa, I.O., Rawat, S., Kalagara, S., Hernandez-Viezcas, J.A., Peralta-Videa, J.R., & Gardea-Torresdey, J.L. (2018). ZnO nanoparticles increase photosynthetic pigments and decrease lipid peroxidation in soil grown cilantro (*Coriandrum sativum*). 132, 120–127.

Rodríguez-González, V., Terashima, C., & Fujishima, A. (2019). Applications of photocatalytic titanium dioxide-based nanomaterials in sustainable agriculture. *J Photochem Photobiol C Photochem Rev.* 40, 49–67.

Sarabia-Castillo, C.R. & Fernández-Luqueño, F. (2016). TiO2, ZnO, and Fe2O3 nanoparticles effect on Bradyrhizobium japonicum-Glycine max [L.] Merr. symbiosis. 3rd Biotechnology Summit 2016, Ciudad Obregón, Sonora, Mexico, 24–28 October 2016. 160-165 ref.13.

Selvan, D.A., Mahendiran, D., Kumar, R.S., Rahiman, A.K.J.J.P., & Biology, P.B. (2018). Garlic, green tea and turmeric extracts-mediated green synthesis of silver nanoparticles: phytochemical, antioxidant and in vitro cytotoxicity studies. *J Photochem Photobiol B Biol.* 180, 243–252.

Sharma, C.P. (2006). *Plant Micronut.* CRC Press, Boca Raton.

Shen, T., Wang, Q., Li, C., Zhou, B., Li, Y., & Liu, Y. (2020). Transcriptome sequencing analysis reveals silver nanoparticles antifungal molecular mechanism of the soil fungi *Fusarium solani* species complex. *J Hazard Mater.* 388, 122063.

Singh, M., Sarkar, B., & Sarkar, S. (2017). Stabilization of soil organic carbon as influenced by clay mineralogy. *Adv Agron.* 148, 33–84.

Sonawane, H., Shelke, D., Chambhare, M., Dixit, N., Math, S., Sen, S., Borah, S.N., Islam, N.F., Joshi, S.J., & Yousaf, B. (2022). Fungi-derived agriculturally important nanoparticles and their application in crop stress management–prospects and environmental risks. *Environ Res.* 212, 113543.

Suresh, U., Murugan, K., Panneerselvam, C., Rajaganesh, R., Roni, M., & Al-Aoh, H.A.N. (2018). *Suaeda maritima*-based herbal coils and green nanoparticles as potential biopesticides against the dengue vector *Aedes aegypti* and the tobacco cutworm *Spodoptera litura*. *Physiol Mol Plant Pathol.* 101, 225–235.

Thul, S.T., & Sarangi, B.K. (2015). Implications of nanotechnology on plant productivity and its rhizospheric environment. In M.H. Siddiqui, M.H. Al-Whaibi, & F. Mohammad (Eds.) *Nanotechnology and Plant Sciences*. Springer, Cham, pp. 37–53.

Tourinho, P.S., Van Gestel, C.A., Lofts, S., Svendsen, C., Soares, A.M., & Loureiro, S. (2012). Metal-based nanoparticles in soil: fate, behavior, and effects on soil invertebrates. *Environ Toxicol Chem.* 31, 1679–1692.

Tyagi, J., Chaudhary, P., Mishra, A., Khatwani, M., Dey, S., & Varma, A. (2022). Role of endophytes in abiotic stress tolerance: with special emphasis on *Serendipita indica*. *Int J Environ Res.* 16, 62.

Van Koetsem, F., Van Havere, L., & Du Laing, G. (2016). Impact of carboxymethyl cellulose coating on iron sulphide nanoparticles stability, transport, and mobilization potential of trace metals present in soils and sediment. *J Environ Manag.* 168, 210–218.

Wang, Y., O'Connor, D., Shen, Z., Lo, I.M., Tsang, D.C., & Pehkonen, S., et al. (2019). Green synthesis of nanoparticles for the remediation of contaminated waters and soils: constituents, synthesizing methods, and influencing factors. *J Clean Prod.* 226, 540–549.

Wu, H., Tito, N., & Giraldo, J. P. (2017). Anionic cerium oxide nanoparticles protect plant photosynthesis from abiotic stress by scavenging reactive oxygen species. *ACS Nanomat.* 11, 11283–11297.

Xu, C., Shan, Y., Bilal, M., Xu, B., Cao, L., & Huang, Q. (2020). Copper ions chelated mesoporous silica nanoparticles via dopamine chemistry for controlled pesticide release regulated by coordination bonding. *Chem Eng J.* 395, 125093.

Xu, Y. & Chen, B. (2015). Organic carbon and inorganic silicon speciation in rice-bran-derived biochars affect its capacity to adsorb cadmium in solution. *J Soils Sediments* **15**, 60–70.

Yang, Z., Zhao, M., Florin, N.H., & Harris, A.T. (2009). Synthesis and characterization of CaO nanopods for high temperature CO_2 capture. *Ind Eng Chem Res.* 48, 10765–10770

Yonathan, K., Mann, R., Mahbub, K.R., & Gunawan, C. (2022). The impact of silver nanoparticles on microbial communities and antibiotic resistance determinants in the environment. *Environ Poll.* 293, 118506.

Zand, A.D., Tabrizi, A.M., & Heir, A.V. (2020). Co-application of biochar and titanium dioxide nanoparticles to promote remediation of antimony from soil by *Sorghum bicolor*: metal uptake and plant response. *Heliyon.* 6, 1–11.

Zhang, L., Wu, L., Si, Y., & Shu, K. (2018). Size-dependent cytotoxicity of silver nanoparticles to *Azotobacter vinelandii*: growth inhibition, cell injury, oxidative stress and internalization. *PLoS One.* 13 (12), e0209020.

Zhang, Y., Zhang, Y., Akakuru, O.U., Xu, X., & Wu, A. (2021). Research progress and mechanism of nanomaterials-mediated in-situ remediation of cadmium-contaminated soil: a critical review. *J Environ Sci.* 104, 351–364.

Zhao, F., Xin, X., Cao, Y., Su, D., Ji, P., Zhu, Z., & He, Z. (2021). Use of carbon nanoparticles to improve soil fertility, crop growth and nutrient uptake by corn (*Zea mays* L.). *Nanomaterials.* 11, 2717.

Zhao, G., Zhao, Y., Lou, W., Su, J., Wei, S., Yang, X., & Shen, W. (2019). Nitrate reductase-dependent nitric oxide is crucial for multi-walled carbon nanotube-induced plant tolerance against salinity. *Nanoscale.* 11, 10511–10523.

Zhao, Z., Nie, T., Yang, Z., & Zhou, W. (2018). The role of soil components in the sorption of tetracycline and heavy metals in soils. *RSC Adv*. 8(56), 32178–32187.

Zhou, P., Adeel, M., Shakoor, N., Guo, M., Hao, Y., Azeem, I., Li, M., Liu, M., & Rui, Y. (2020). Application of nanoparticles alleviates heavy metals stress and promotes plant growth: an overview. *Nanomaterials*. 11(1), 26.

Zhu, Y., Xu, F., Liu, Q., Chen, M., Liu, X., Wang, Y., Sun, Y., & Zhang, L. (2019). Nanomaterials and plants: positive effects, toxicity and the remediation of metal and metalloid pollution in soil. *Sci Total Environ*. 662, 414–421.

Zulfiqar, F., Navarro, M., Ashraf, M., Akram, N.A., & Munné-Bosch, S. (2019). Nanofertilizer use for sustainable agriculture: advantages and limitations. *Plant Sci*. 289, 110270.

16 Nanotechnology for Crop Improvement and Sustainable Agriculture

Damini Maithani
IFTM University

Anita Sharma
Govind Ballabh Pant University of Agriculture & Technology

Hemant Dasila
Eternal University

Anjali Tiwari
G.B. Pant National Institute of Himalayan Environment

Viabhav Kumar Upadhayay
Dr. Rajendra Prasad Central Agricultural University

CONTENTS

16.1 INTRODUCTION

More than 60% of the population depends on agriculture for their livelihood in the majority of developing countries. In the current scenario, changing climatic conditions, limited land and water resources, pest resistance, accumulation of toxic pollutants due to fertilizer overuse and increasing food demands are some challenges

DOI: 10.1201/9781003345565-16

the agriculture sector is facing. These challenges are predicted to intensify by 2050 (Naderi & Danesh Shahraki, 2011). Conventional agricultural practices are insufficient to resolve the persisting challenges and feed the burgeoning population. To get maximum output with limited inputs, modern agricultural practices that may help in reshaping the agricultural productivity in a sustainable manner are needed. To overcome these massive threats, different attempts have been made, including use of high-yielding varieties with desirable traits, hybrid seeds, use of synthetic fertilizers and pesticides and developing transgenic crops with higher yielding and disease-resistant traits. However, there are many concerns and risks associated with these innovations. High dosage of fertilizers and pesticides may cause accumulation of toxic chemicals in soil and water bodies causing eutrophication and biomagnification of toxic compounds in food chain. Transgenic crops do not have worldwide acceptance, and the risk of pest resistance is also associated with their use (Pandey et al., 2018).

Nanotechnology is emerging as an efficient, cost-effective and labour-saving method of improving crop productivity and promising food security of the world. Use of nanoparticles in agriculture can give better crop yield, help in disease management and diagnosis, and can also help in developing disease-resistant varieties of economically important crops, thus minimizing the crop loss caused by phytopathogens (Banerjee et al., 2021).

Nanotechnology primarily deals with manipulation of matter at the nano scale, designing and characterization of nanoparticles (NPs) with suitable shape, structure and biological activity. Nanotechnology is gaining momentum in the area of food production and agriculture sector and is considered the sixth most revolutionary technology presently as it provides sustainable solution to issues including climate change, food insecurity and decreased soil fertility (Knell, 2010). Thus, it can prove to be a revolutionary approach in the field of plant production system and help in the development of smart crops (Sugunan & Dutta, 2008).

So far, engineered nanoparticles (ENPs) have been exploited in various commercial products including electronics and cosmetics; these are also used for drug delivery and in the field of medicine (nanomedicine). Use of nanotechnology in plant systems (phyto-nanotechnology) has not received much attention yet. Application of NPs in agriculture gives better yield with low inputs and allows delivery of fertilizers for plant nutrition, smart packaging and monitoring nutrient deficiencies and diseases in plants (Khati et al., 2017a, b, 2018, 2019a, b; Chaudhary et al., 2021a, b, c, d; Kumari et al., 2020, 2021). NPs help in improving yield and health of plants by enhancing nutrient acquisition; these can also be used to improve plant health by transferring disease-resistant genes in host plant that provide protection against target pathogen. Besides, disease and pollutants can be detected in early phase using nanosensors which help in early detection and eradication of pathogen and removal of pollutants from the environment (Wang et al., 2016).

Crop improvement refers to the science of developing crop varieties which are better than pre-existing cultivars in terms of agronomic traits and improved yield. For crop improvement conventional breeding and biotechnology tools play a vital role. Combination of these approaches helps to improve desirable traits including disease resistance, stress tolerance, improved yield, quality and uniformity of the produce (Adlak et al., 2019). Conventional and non-conventional methods are used for introducing desired traits in plants. Although conventional methods have been used for a long time, due to certain limitations and for achieving crop improvement in a short duration, non-conventional methods are gaining popularity. Introduction of new genes and engineering of plant genome involve various physical and biological methods which have their own limitations. Use of nanotechnology enables cost-effective and easy way of delivering new genes in plants for obtaining desirable phenotype. NP-based delivery of biomolecules is advantageous over conventional methods due to their unique architecture, biocompatible and non-cytotoxic nature. Their use in animal system is prominent in comparison to plant cells because of complex multilayered cell wall which interferes with internalization of particles. However, due to the progress made in the field of nanomaterial designing, delivery of biomolecules across the cell wall has been made easier and does not require any special equipment (Jat et al., 2020).

Although nanotechnology is revolutionizing the field of agriculture utmost care must be taken to ensure safe design and use. Commercialization of NPs and their product increases their exposure to humankind; thus, before using these products, toxicity assessment needs to be performed (Xia et al., 2009). This chapter discusses about the benefits of nanotechnology in agriculture and crop improvement.

16.2 CROP IMPROVEMENT: AIMS AND STRATEGIES

The science of crop improvement aims at developing varieties with superior traits in terms of quality and quantity of produce (Passricha et al., 2020). Crop improvement strategies lead to improved traits in plants to meet the requirement of consumers, improved nutritional quality, yield traits and introduction of uniform growth and maturity in economically important crops. Besides, it also leads to improved resistance in plants to better withstand diseases and pest attack. Superior traits in plants can be introduced to withstand vagaries of climatic conditions. Plant breeding approaches can be divided into two fundamental categories: conventional and non-conventional, based on the instruments and procedures used, which are subject to change with advancements in science and technology. Although traditional plant breeding has been used since long, there are certain disadvantages associated with the process, as it takes a long time to achieve the desired results and exchange of genes is limited to closely related species. Besides, genes other than the targeted ones can also be transferred. On the other hand, non-conventional methods allow gene transfer in a short period of time and exchange of genes can be done in distantly related species also. As selection process in non-conventional methods is based on molecular markers, it is more reliable than traditional breeding where selection is based on phenotypic traits. Non-conventional approaches also have certain limitations as it requires extra production cost and equipment. The aim and scope of crop improvement are briefly discussed in Figure 16.1.

16.3 STRATEGIES OF CROP IMPROVEMENT

To meet the global food demand and develop plant species that can withstand the persisting challenges, crop improvement is required. Crop improvement strategies include conventional and non-conventional methods which have been discussed in this section. Plant breeding is the science and art of genetically modifying and enhancing

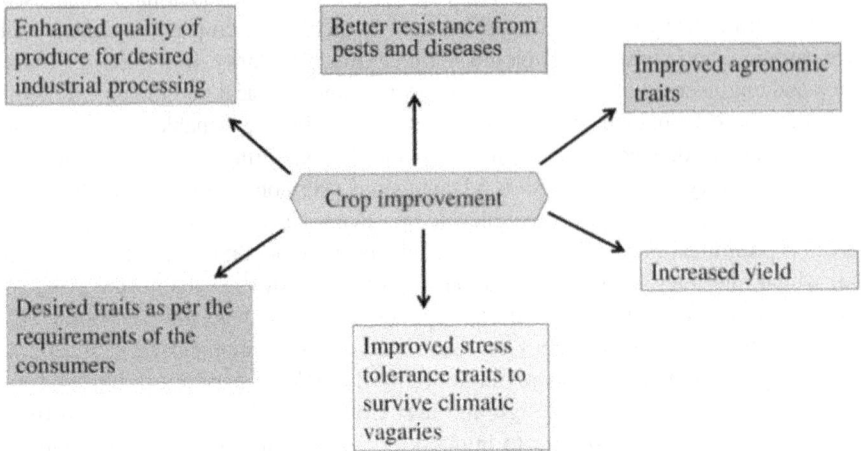

FIGURE 16.1   Aims and scope of crop improvement.

plants for human benefit. It is one of the most common methods of crop improvement (Singh et al., 2002). Plant breeding refers to a conscious human attempt to develop particular characteristics of plants in order to fulfil new functions or enhance those that already exist, i.e. different from existing varieties (Acquaah, 2012). Plant breeding can be seen as a form of co-evolution between people and food plants. People altered the agricultural plants, and as a result, new plant species facilitated shifts in human population patterns. Plant breeders improve crops by identifying sources of genetic variation for the characteristics of interest. Plant genetic materials in each species are highly variable, even within and among closely related species (Weber et al., 2012). The main objective of plant breeding is to effect genetic improvement in morphology, physiology and biochemical aspects of crops according to consumer needs. The core of plant breeding is to improve yield and quality, resistance to biotic and abiotic stress and adaptability characteristics of crop. Each of these agronomic or nutritional value factors can be broken down into a variety of distinct features, each with a unique range of variation. It is rather simple to manipulate a single feature while ignoring all others, but this is unlikely to produce a useful variety. The major limitation of plant breeding is to concurrently improve all of the desired features.

This effort is complicated by genetic correlations between various traits, which may result from pleiotropic genes, physical chromosomal links between genes or population genetic structure (Hartl & Clark, 1997). Identification of plants with desired traits and the development of plans to combine these traits to create superior varieties make up the ongoing, cyclical process of plant breeding (Acquaah, 2015).

16.4   CONVENTIONAL BREEDING

The practice of conventional breeding has developed over time, resulting in an efficient framework that has enabled the development of foods that are both safe and nutrient-rich and also having improved crop performance. Making judgements on which parents to select, which parents to cross-pollinate and which offspring is of prime importance to promote breeding of plants. In conventional breeding new varieties of plants are developed by using conventional tools and natural processes. Breeders create new cultivars through the process of crossing (hybridization), combining desirable features from various but typically related plants (Acquaah, 2012). As new genes are not added during conventional breeding, the resultant plants simply accentuate the target qualities that already exist in the species' genetic potential (Jakowitsch et al., 1999). In order to produce suitable individuals who meet the breeding objectives, breeders must first create or assemble diversity (Fehr, 1987).

The most popular traditional/conventional breeding technique for acquiring new parental material for creating cultivars of current crops is plant introduction. According to Allard and Hansche (1964) plant introduction is the oldest and rapid method for crop improvement. The first step in plant breeding for any trait is the collection and evaluation of the available germplasm, which allows genetic variation to be assembled. If a species or location lacks the desired variability, one option is to introduce exotic germplasm. The main purpose of plant introduction is prevention of genetic uniformity, substantial yield increases through the introgression of desired genes, the introduction of novel quality features, the introduction of disease- and insect-resistant genes, and the introduction of abiotic stress resistance genes (Begna, 2021).

Another important process leading to crop improvement is selection. The process that favours the survival and continued reproduction of plants with more favourable characteristics than others is known as selection. It is the most fundamental techniques used by scientists as well as the farmers (Briggs & Allard, 1953). There are various types of selection procedures, such as pure line selection, progeny selection, clonal selection, recurrent selection, and clonal and disruptive selection. The third important method is hybridization. Hybridization is a natural or artificial process that leads to the development of a hybrid. Breeders utilize a variety of methods to manage pollination in order to promote hybridization for the creation of hybrid cultivars with desired characters (Acquaah, 2015).

Crop improvement is also possible through polyploidy. A polyploid is an organism or individual with more than two basic sets of chromosomes, and this situation is referred to as polyploidy. The basic way to cause polyploidy is by using colchicine. *Colchium atumnale* is a plant whose seeds contain an alkaloid that prevents the growth of spindle fibres. Polyploidy is an essential factor in crop improvement (Touchell et al., 2020). Mutation breeding is another method for enhancing plant types through conventional breeding. By using mutagens (agents of mutation, either chemical or physical), plant breeders can develop new genes or novel sources of diversity for breeding (Broertjes & van Harten, 1988). In crop improvement induced mutation plays an important role in the development of improved varieties, induction of male sterility, production of haploids, creation of genetic variability, overcoming self-incompatibility and improvement in adaptation.

16.5 NON-CONVENTIONAL BREEDING

Non-traditional plant breeding permits the direct transfer of one or a small number of genes, between species that are closely or distantly related. Restriction enzymes,

Non-conventional methods of crop improvement

Tissue culture	Transgenic crops	Molecular breeding
(a) Micropropagation	Gene transfer	(a) Marker assisted selection
(b) Protoplast culture	(a) Agrobacterium mediated gene transfer	(b) Identification of breeding line
(c) Embryo culture	(b) Microinjection	(c) Purity of lines
(d) Haploid production	(c) Particle bombardment	(d) Germplasm characterization
(e) Anther culture	(d) Electroporation	(e) Heterosis prediction

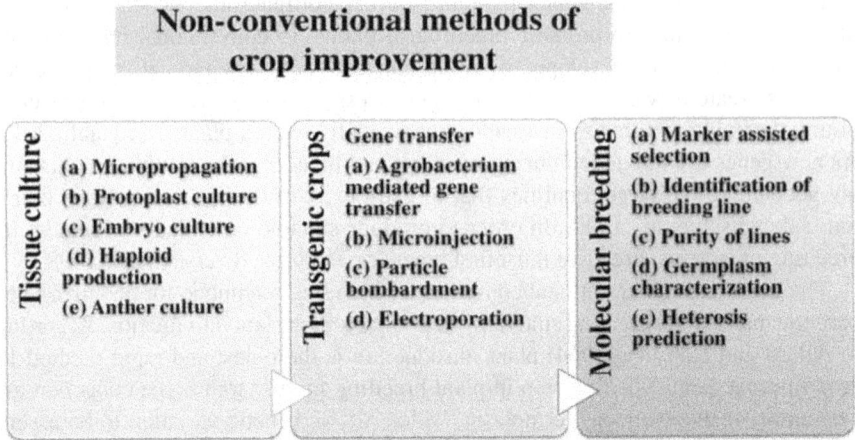

FIGURE 16.2   An overview of non-conventional methods of crop improvement.

biomarkers, molecular glue (ligases), transcription and post-translational modification machinery are some of the genomic tools used in non-conventional breeding methods (Huang, 2016). Tissue culture, transgenic approach and molecular breeding are the major approaches dealing with crop improvement in non-conventional breeding techniques which are shown in Figure 16.2.

Non-conventional methods such as plant tissue culture technique refer to the growth of living plant tissue, cells or organ in a suitable nutrient media. Plant tissue culture techniques include micropropagation, somatic embryogenesis, somaclonal variation, meristem culture, anther culture, embryo culture, protoplast culture, cryopreservation and production of secondary metabolites, depending on the plant part used as explants (Adlak et al., 2019). Micropropagation describes the process of regenerating plants from isolated meristematic cells, tissue or somatic cells. Crop plants that are difficult to reproduce sexually or vegetatively propagated species whose rate of multiplication is slow can be multiplied in large numbers using this technique. This technology has been used to develop disease-free, high-quality and super-elite planting material for further seed production. On the other hand, embryo culture is the process of regenerating an entire plant from an embryo in a culture medium. The seed is extracted, and the embryos are transferred to culture media at the proper time. Plant cells without a cell wall are referred to as protoplasts. These are able to regenerate an entire plant in tissue culture labs when the appropriate artificial media and environmental conditions are present. Protoplast culture is the process of cultivating isolated protoplasts from any plant part, such as a root, stem, leaf or embryo, in an artificial medium that encourages cell division and plant regrowth. Anthers or pollen culture is the regeneration of a full plant from anthers or pollen in a culture medium. The plant is harvested for anthers at the proper period where ideal stage varies depending on the species. It is used in the development of haploid plant.

Production of genetically modified (GM) crops is through transgenic approach on a large scale by inserting one or more genes that code for desirable features through the process of genetic engineering (GE) (Adlak et al., 2019). For genetic analysis and direct DNA modification, transgenic technology involves the transfer of genes from similar or unrelated species to the desired agricultural plant species. The importance of transgenic technique in crop improvement is to develop new genotype, genetic improvement in autogamous and allogamous crop plants and develop resistance against disease, insects and herbicides. Several crops including tomatoes, wheat, maize, brinjal, cotton, etc. have been genetically modified for their enhanced resistance to stress stimuli and improved nutrient content (Passricha et al., 2020).

In molecular breeding technique different types of markers are tightly linked to phenotypic traits to assist in a selection scheme for a particular breeding objective. The markers used in molecular breeding are morphological markers, cytological markers, and biochemical and DNA markers. Morphological markers are related to shape, size, colour and surface of various plants. They are used for varietal identification. Cytological markers are related to chromosome morphology whereas biochemical markers are related to variations in protein and amino acid banding pattern. DNA markers are related to variation in DNA fragments generated by restriction endonuclease enzyme (Amiteye, 2021). In molecular breeding restriction fragment length polymorphism (RFLP), amplified fragment length polymorphism (AFLP), random amplified polymorphic DNA (RAPDs), cleaved amplified polymorphic sequence (CAPS), etc. are used. RFLP refers to the change within a species in the length of DNA fragments that are generated by a specific endonuclease. It is used in comparative and synteny mapping. In crop improvement, RFLPs are used for gene mapping, germplasm characterization and marker-assisted selection (Amiteye, 2021). To achieve improved agronomic traits and stress tolerance attributes in plants, conventional plant breeding and advanced genetic (transgenic and non-transgenic) approaches are used. Moreover, high-throughput techniques (proteomics, genomics and metabolomics) have improved the crop. Metabolomic-assisted crop improvement is gaining popularity as metabolomic quantitative trait loci (mQTLs) and metabolomics-based genome-wide association studies (mGWAs) assist in improving selection process of breeding lines improving marker-assisted breeding of crops (Agarrwal & Nair, 2020). Besides genetic engineering, advanced molecular techniques such as genetic engineering, transcription activator such as effector nuclease, Zinc-finger nucleases and clustered regularly interspaced short palindromic repeat, etc., were able to compensate the constraints of conventional breeding methods (Passricha et al., 2020).

16.6 INTERVENTIONS OF NANOTECHNOLOGY IN CROP IMPROVEMENT

Gene introduction for quality and yield improvement, disease resistance and stress tolerance traits in plants has been used as a strategy in crop improvement. Nanotechnology can serve in genetic manipulation, delivery of DNA and growth hormones in plants. Delivery of biomolecules including DNA across intact plant cell wall is difficult. Protoplasts are generally used for gene delivery in plant cells but maintenance and regeneration of plantlets become difficult from protoplast (Jat et al., 2020). For delivery of new genes in plants various methods are used including physical methods such as electroporation, chemical fusogen (PEG), microinjection, particle bombardment where high-velocity metal particles are used for DNA transfer, biological methods including *Agrobacterium*-mediated gene transfer and transfer of nucleic acid molecules using viral vehicles (Sanford, 1990; Gelvin, 2017; Adlak et al., 2019). All these methods have some limitations, for instance, bombardment of metal particles may cause tissue damage, non-targeted integration, low penetration, etc. Electroporation can be used only for protoplast and is limited to certain species and efficiency of microinjection is very low. Besides, it is applicable only for larger cells. Biological methods of gene transfer using either *Agrobacterium* or virus as a vehicle for gene delivery also have certain limitations. *Agrobacterium*-mediated transfer is only applicable to dicot plants and efficiency of transfer is also low. Viral particles on the other hand have low DNA loading capacity, restricted in certain species, and used in case of transient gene expression (Zaidi & Mansoor, 2017; Adlak et al., 2019). Thus, due to tissue damage, low efficiency, limited cargo supply, low success rate and high operational cost, ENPs emerge as a better choice for gene transfer in plants. ENPs have been used successfully for delivering drugs and other molecules in medicine and therapeutics. Their use in plant system is also gaining momentum and thus overcome the limitations of persisting techniques and offer tremendous advantage in crop improvement (Jat et al., 2020). Nanotechnology has revolutionized the field of crop improvement through genetic engineering. In a study conducted by Naqvi et al. (2012), calcium phosphate nanoparticles were used for delivery of a reporter gene and efficiency of transfer was found to be 80.7% which was greater than *Agrobacterium*-mediated transfer (54.4%) and naked DNA transfer (8%). This finding supports the fact that use of NPs in gene transfer is a better and efficient way of gene transfer. Calcium phosphate NPs have been included in GRAS list by the FDA which thereby encourages its use for eco-friendly and non-toxic nature.

ENPs enable genome editing in plants and help to introduce desirable genes for improved yield, stress tolerance, disease resistance, etc. Nanoparticles on the basis of their nature can be organic (liposomes, polymers, micelles, etc.), inorganic (metal NPs, quantum dots) and carbon based (carbon nanotubes, fullerenes and carbon nanofibres) (Khan et al., 2019). NPs even enable unaided delivery of biomolecules without using gene gun, vortexing or any other external device (Chang et al., 2013; Kwak et al., 2019; Zhang et al., 2019). For instance, Zhang et al. (2019) demonstrated the use of DNA nanostructure-mediated delivery of siRNA in leaves of *Nicotiana benthamiana* for gene silencing purpose. Another report suggested delivery of NPs in the presence of external magnetic field and the process is known as magneto-fection (Zhao et al., 2017). In another study, dsRNA was delivered in plants using cationic-layered double hydroxide (BioClay) nanosheets which helped in protecting plants against viral infection via RNA interference. Clay nanosheets offered systemic protection to the leaves of *Nicotiana tabacum* against CMV (Mitter et al., 2017).

Zhang et al. (2019) studied the effect of cerium oxide NPs on spinach plants and found induced metabolic reprogramming in the root and leaves of plants. Various reports suggest use of NPs in plant protection from bacterial and fungal diseases due to their non-toxic and strong antimicrobial activities (Ocsoy et al., 2013; Paret et al., 2013; Mishra et al., 2014; Buchman et al., 2019). NP-mediated gene delivery allows delivery of genetic material (DNA) in active state without any damage and provides protection from nucleases. In a study conducted by Fu et al. (2015), silica

NPs modified with positively charged poly-L-lysine were used for the delivery of DNA molecule preventing its cleavage from DNase I. Besides DNA could integrate into genome of tobacco and transformation efficiency was 47.11% ± 0.4%.

NPs synthesized using *Aspergillus ochraceus* when conjugated with carbon NPs were able to deliver DNA to dicot (Tobacco) and monocot (*Oryza sativa*) plants without any damage to DNA molecule and promised efficient delivery (Okuzaki et al., 2013; Gad et al., 2020). Hajiahmadi and co-workers demonstrated transformation of tomato plants by injecting functionalized mesoporous silica NPs conjugated with pDNA and injected into tomato fruits before ripening. pDNA-MSNs cross the nuclear pore and pDNA are directly released into the nucleus where it integrates with plant genome possibly via homologous recombination (Hajiahmadi et al., 2019). There are reports scattered across literature where NPs including metal NPs (AgNPs, TiO_2, cerium and CuNPs), carbon nanotubes, calcium phosphate NPs, mesoporous silica, clay nanosheets and magnetic NPs have been used for delivery of biomolecules in plants and crop improvement (Table 16.1).

16.7 NANOTECHNOLOGY FOR SUSTAINABLE AGRICULTURE

In order to increase crop production under persisting challenges, countries that are primarily dependent on agriculture need to opt for some advanced technologies that are cost-effective and labour saving (War et al., 2020). NPs are particulate substances having one of their dimensions less than 100 nm. Due to their small size, high surface area-to-volume ratio and high reactivity, these particles have higher biological activities in comparison to their macro-scale counterparts (Wang et al., 2016). All these characteristics make their use popular in different fields. These particles can be engineered through physical, chemical and biological methods. These engineered nanoparticles (ENPs) show unique properties including biological activity, stability, structure, chemical properties, etc. These have been exploited so far in the field of medicine, therapeutics and diagnostics but their use is still at the budding stage. Some important nanocompound-based products are present in the market for their use in agriculture (Pestovsky & Martínez-Antonio, 2017). Nanofertilizers, nanopesticides and nanosensors are some products offered by nanotechnology that benefit the agricultural sector in one way or the other. Fertilizers (nanofertilizers) improve crop nutrition, and nanopesticides provide protection to economically important plants from diseases, pests and weeds. NPs can also be used for monitoring the quality of produce after harvest, as nanosensors, for early detection of diseases for protection of crops at early stages, etc. (Park et al., 2006; Tarafdar et al., 2014; Rameshaiah et al., 2015; Abobatta, 2018). Currently, NPs that have been synthesized using plant extracts and microorganisms are gaining popularity due to their biodegradable and eco-friendly nature. Plant nutrition is important for crop quality and food production which is generally achieved through application of agrochemicals. Nano-based products also offer smart packing technologies and also help in monitoring nutrient deficiencies in plants. Nanofertilizers make it easier for nutrients to be released gradually, which lowers nutrient loss, increases nutrient use efficiency and reduces environmental protection expenses, thereby improving agricultural production with low inputs and minimize the use of toxic chemicals.

TABLE 16.1

Nanoparticles for Delivery of Biomolecules and Crop Improvement

Nanoparticle Used	Host Crop	Purpose	Reference
Calcium phosphate NPs	*Brassica juncea*	Delivery of pCambia 1301 plasmid	Naqvi et al. (2012)
Mesoporous silica NPs	Tobacco protoplast	DNA	Torney et al. (2007)
SWCNTs (single-walled carbon nanotubes) and MWCNTs (multiwalled carbon nanotubes)	*Nicotiana tabacum* L.	Plasmid construct pGreen 0029	Burlaka et al. (2015)
DNA nanostructure	*Nicotiana benthamiana*	Delivery of siRNA (gene silencing)	Zhang et al. (2019)
Calcium phosphate NPs (25–55 nm)	Tobacco	Delivery of pBI121 plasmid DNA	Ardekani et al. (2014)
Calcium phosphate NPs	*Chicorium intybus*	Delivery of HMG Co-A reductase gene (results in enhanced esculin biosynthesis and improved antioxidant activity)	Ohadi et al. (2013)
Clay nanosheets	*Nicotiana tabacum*	Delivery of dsRNA for protection from virus	Mitter et al. (2017)
Chitosan-coated mesoporous silica nanoparticles (CT-MSN)	*Citrullus lanatus*	Suppression of *Fusarium* wilt	Buchman et al. (2019)
Ag@dsDNA@GO (DNA-directed silver NPs grown on graphene oxide)	Tomato	Protection from *Xanthomonas perforans*	Ocsoy et al. (2013)
Biofabricated silver nanoparticle	Wheat	Reduction of spot blotch disease (caused by *Bipolaris sorokiniana*) of wheat	Mishra et al. (2014)
TiO_2 doped with zinc	Tomato	Reduction of disease caused by *Xanthomonas perforans*	Paret et al. (2013)
Magnesium oxide (MgO) NPs	Tomato	Induction of systemic resistance	Imada et al. (2015)
AgNPs	Tomato	Reduction in incidence of bacterial wilt (causative agent: *Ralstonia solanacearum*)	Santiago et al. (2019)
DNA loaded with magnetic NP	Cotton	Delivery of magnetic NP into pollen for transgenic plant development	Zhao et al. (2017)
SiNPs charged with poly-L-Lysine	Tobacco	Stable genetic transformation	Fu et al. (2015)

Nanotechnology can benefit agriculture by increasing crop productivity/yield, promising efficient resource utilization, reducing precipitation as well as loss of nutrients from soil pool, minimizing the use of toxic agrochemicals, promising food security for the future, and monitoring plant growth and development and agriculture waste management (Chang et al., 2013; Pirzadah et al., 2019). Nanomaterials in agriculture embrace properties including controlled release of pesticides and fertilizers, improved targeted activity in a safe and eco-friendly manner. Combining nanotechnology with agriculture and food production is gaining the interest of researchers which is evident from a number of research works that indicate positive influence of NPs on growth and yield of economically important crops. Besides, NPs such as carbon nanotubes have been reported to enhance water uptake efficiency of seeds, thereby improving their germination and growth (Khodakovskaya et al., 2009). CNTs have been reported to improve photosynthetic efficiency and vegetative growth in various plants; they also assist in water consumption in plants under water limiting conditions (Safdar et al., 2022).

The role of NPs has attracted the attention of scientists in the field of plant protection, and was found to be promising against plant pathogens causing economic losses. In a study conducted by Mishra et al. (2014), AgNPs synthesized using *Serratia* sp. showed complete inhibition of conidial germination of *Bipolaris sorokiniana*. In a study conducted by Raliya et al. (2015), TiO_2NPs developed using *Aspergillus flavus* TFR7 were shown to improve shoot length, root length, root area, chlorophyll content in mung bean plants after foliar spray (10 mg/L). In another study, ZnONPs were reported to protect strawberry plants from *Botrytis cinerea* and the protection was comparable with conventional chemical fungicide fenhexamid (Luksiene et al., 2020). There are reports where NP-mediated abiotic stress tolerance in plants has been reported. In a study conducted by Wu et al. (2018), application of CeO_2NPs improved salinity stress tolerance in *Arabidopsis thaliana* via hydroxyl radical scavenging activity and increased potassium retention in leaves. Application of NPs also neutralizes the damaging effects of ROS by stimulating antioxidant machinery in plants (Mujtaba et al., 2021).

16.8 CHALLENGES IN USE OF NANOTECHNOLOGY IN AGRICULTURE

While promoting the use of nanotechnology in agriculture, one cannot neglect the constraints of phyto-nanotechnology which limits its use. Despite the fact that nanotechnology has revolutionized the world and gained a rapid momentum in every sector, at the same time there are ethical issues and safety measures that need to be addressed. Toxicity concerns among consumers and researchers all over the world have led to the development of the science of nano-toxicology which deals with evaluating toxicity of nano-based products and thereby promoting safe design and use for safety of consumers (Table 16.2). Devising quality control checkpoints for checking toxic effects of nano-based products must be compulsory for safe consumption and ensuring eco-friendly nature of the developed product (Pirzadah et al., 2019). Although no serious human ailments have been reported till date by the use of nanoparticles, much more is needed to be

TABLE 16.2

Advantages and Disadvantages of Different Carrier-Based Nanoparticles Used for Gene Targeting

Types of Nanoparticles	Advantages	Disadvantages	References
Polymer-based nanoparticle	Small size, high stability against enzymatic degradation and low toxicity	Low efficacy	Lee et al. (2013); Lu et al. (2007)
Inorganic-based nanoparticle	Diverse functionality, easy preparation, accurate and targeted delivery	Sometimes non-biocompatible	Nitta and Numata (2013)
Lipid nanoparticle	Low immunogenicity and easy preparation	Low transformation efficiency	Liu and Zhang (2011)
Hybrid nanoparticle	Precise targeted delivery and prevent loss of DNA during targeting	At high dose sometimes toxic in nature	Nitta and Numata (2013)

learned in this aspect. Ultrafine particles (UFPs) that have aerodynamic dimension of less than 100 nm show similarity with NPs. UFPs are reported to have adverse health outcomes due to their small size, deep penetration and long retention in lungs. NPs may also pose similar hazards due to their small size and high surface area-to-volume ratio (Xia et al., 2009). Nanoparticles in edible plant parts may lead to toxicity in biological systems and environment. Besides, at higher concentration, nanomaterial may lead to damage of tissues and DNA via generation of reactive oxygen species (ROS) and free radicals (Dekkers et al., 2016). Considering these facts before using NPs in agriculture, intelligent designing and optimization of dose is required for their safe use. Before commercialization, toxicological studies, NP evaluation and safety issues must be considered sincerely (Mujtaba et al., 2021).

16.9 CONCLUSION

Excessive use of synthetic agrochemicals to increase food productivity is definitely not a reliable choice for the future in light of the fact that these are deleterious for environment and soil health. Nanotechnology is a green and sustainable approach in this context which is emerging as a major economic force. Prospects of nanotechnology are very bright but the science still requires detailed understanding about interaction between plants and nanoparticles, their effect on plants' physiological processes, on consumers (animals and human beings) and on the environment. Nanotechnology provides real solution to all the persisting problems in agriculture and help in crop improvement, enhance nutrient use efficiency, provide protection and help to monitor diseases and nutrient deficiencies in plants. Besides current gene delivery methods for crop improvement and genetic transformation have low efficiency, limited cargo delivery, non-specific gene transfer, specific host range, etc. Thus, conventional practices and protocols need to be replaced by more efficient solution. Science of nanotechnology offers tremendous advantage for crop improvement by allowing efficient genetic transformations and delivery of biomolecules in plants in a more efficient way. NPs also improve plant vegetative growth, photosynthetic rates and water consumption efficiency in a safe and eco-friendly manner. NPs of various classes can easily deliver biomolecules in plant cells despite their intricate cell wall, thereby making the process of genetic transformation easier in economically important crops. Use of nanotechnology in agriculture can benefit human health and environment by reducing overuse of fertilizers and pesticides and improve plant health in a sustainable manner. Following all the safety measures, promoting green synthesis of NPs, after proper toxicological analysis and approval by some authorized agencies NPs could revolutionize the agriculture sector.

REFERENCES

Abobatta, W.F. (2018). Nanotechnology application in agriculture. *Acta Sci Agric.* 2(6), 99–102.

Acquaah, G. (2012). *Principles of Plant Genetics and Breeding*, 2nd edn. Wiley-Blackwell, Oxford.

Acquaah, G. (2015). Conventional Plant breeding principles and techniques. In Al-Khayri, J.M., Jain, S.M., & Johnson, D.V. (eds.) *Advances in Plant Breeding Strategies: Breeding, Biotechnology and Molecular Tools* (pp. 115–158). Springer, Cham, Switzerland.

Adlak, T., Tiwari, S., Tripathi, M.K., Gupta, N., Sahu, V.K., Bhawar, P., & Kandalkar, V.S (2019). Biotechnology: an advanced tool for crop improvement. *Curr J Appl Sci Technol.* 33(1), 1–11.

Agarrwal, R., & Nair, S. (2020). Metabolomics-assisted crop improvement. In N. Tuteja, R. Tuteja, N. Passricha, & S.K. Saifi (Eds.) *Advancement in Crop Improvement Techniques*. Woodhead, Duxford, UK.

Allard, R.W., & Hansche, P.E. (1964). Some parameters of population variability and their implications in plant breeding. *Adv Agron.* 16, 281–325.

Amiteye, S. (2021). Basic concepts and methodologies of DNA marker systems in plant molecular breeding. *Heliyon,* 30(10), e08093.

Ardekani, M.R.S., Abdin, M.Z., Nasrullah, N., & Samim, M. (2014). Calcium phosphate nanoparticles a novel non-viral gene delivery system for genetic transformation of tobacco. *Int J Pharm Pharm Sci.* 6(6), 605–609.

Banerjee, A., Sarkar, A., Acharya, K., & Chakraborty, N. (2021). Nanotechnology: an emerging hope in crop improvement. *Lett Appl Nano Bio Sci.* 10(4), 2784–2803.

Begna, T. (2021). Role and economic importance of crop genetic diversity in food security. *Int J Agric Sci Food Technol.* 7(1), 164–169.

Briggs, F.N., & Allard, R.W. (1953). The current status of the backcross method of plant breeding. *Agronomy J.* 45(4), 131–138.

Broertjes, C., & van Harten, A.M (1988). *Applied Mutation Breeding for Vegetatively Propagated Crops*. Elsevier, Amsterdam.

Buchman, J.T., Elmer, W.H., Ma, C., Landy, K.M., White, J.C., & Haynes, C.L (2019). Chitosan-Coated mesoporous silica nanoparticle treatment of *Citrullus lanatus* (watermelon): enhanced fungal disease suppression and modulated expression of stress-related genes. *ACS Sustain Chem Eng.* 7(24), 19649–19659.

Burlaka, O.M., Pirko, Y.V., Yemets, A.I., & Blume, Y.B. (2015). Plant genetic transformation using carbon nanotubes for DNA delivery. *Cytol Genet.* 49(6), 349–357.

Chang, Y.C., Yang, C.Y., Sun, R.L., Cheng, Y.F., Kao, W.C., & Yang, P.C. (2013). Rapid single cell detection of *Staphylococcus aureus* by aptamer-conjugated gold nanoparticles. *Sci Rep.* 3, 1863.

Chaudhary, P., Khati, P., Chaudhary, A., Gangola, S., Kumar, R., & Sharma, A. (2021d). Bioinoculation using indigenous *Bacillus spp.* improves growth and yield of *Zea Mays* under the influence of nanozeolite. *3 Biotech.* 11, 11.

Chaudhary, P., Khati, P., Chaudhary, A., Maithani, D., Kumar, G., & Sharma, A. (2021a). Cultivable and metagenomic approach to study the combined impact of nanogypsum and *Pseudomonas taiwanensis* on maize plant health and its rhizospheric microbiome. *PLoS One.* 16(4), e0250574.

Chaudhary, P., Khati, P., Gangola, S., Kumar, A., Kumar, R., & Sharma, A. (2021b). Impact of nanochitosan and *Bacillus* spp. on health, productivity and defence response in *Zea mays* under field condition. *3 Biotech.* 11(5), 237.

Chaudhary, P., Sharma, A., Chaudhary, A., Khati, P., Gangola, S., & Maithani, D. (2021c). Illumina based high throughput analysis of microbial diversity of maize rhizosphere treated with nanocompounds and *Bacillus* sp. *Appl Soil Ecol.* 159, 103386.

Dekkers, S., Oomen, A.G., Bleeker, E.A., Vandebriel, R.J., Micheletti, C., & Cabellos, J. (2016). Towards a nanospecific approach for risk assessment. *Regul Toxicol Pharmacol.* 80, 46–59.

Fehr, W.R. (1987) *Principles of Cultivar Development, Vol 1, Theory and Technique.* Macmillan, New York.

Fu, Y., Li, L., Wang, H., Jiang, Y., Liu, H., Cui, X., & Lü, C. (2015). Silica nanoparticles-mediated stable genetic transformation in *Nicotiana tabacum.* *Chem Res Chin Univ.* 31(6), 976–981.

Gad, M.A., Li, M.J., Ahmed, F.K., & Almoammar, H. (2020). Nanomaterials for gene delivery and editing in plants: challenges and future perspective. In K.A. Abd-Elsalam (Ed.) *Multifunctional Hybrid Nanomaterials for Sustainable Agri-Food and Ecosystems* (pp. 135–153). Elsevier Inc., Amsterdam.

Gelvin, S.B. (2017). Integration of *Agrobacterium* T-DNA into the plant genome. *Annu Rev Genet.* 51, 195–217.

Hajiahmadi, Z., Shirzadian-Khorramabad, R., Kazemzad, M., & Sohani, M.M. (2019). Enhancement of tomato resistance to Tutaabsoluta using a new efficient mesoporous silica nanoparticle-mediated plant transient gene expression approach. *Sci Hort.* 243, 367–375.

Hartl, D.L., & Clark, A.G. (1997). *Principles of Population Genetics* (Vol. 116, p. 542). Sinauer Associates, Sunderland, MA.

Huang, X. (2016). Genetic mapping to molecular breeding: genomics have paved the highway. *Mol Plants.* 9(7), 959–960.

Imada, K., Sakai, S., Kajihara, H., Tanaka, S., & Ito, S. (2015). Magnesium oxide nanoparticles induce systemic resistance in tomato against bacterial wilt disease. *Plant Pathol.* 65(4), 551–560.

Jakowitsch, J., Mette, M.F., van der Winden, J., Matzke, M.A., & Matzke, A.J.M. (1999). Integrated pararetroviral sequences definea unique class of dispersed repetitive DNA in plants. *Proc Natl Acad Sci U S A.* 96(23), 13241–13246.

Jat, S. K., Bhattacharya, J., & Sharma, M.K. (2020). Nanomaterial based gene delivery: a promising method for plant genome engineering. *J Mater Chem.* 8(19), 4165–4175.

Khan, I., Saeed, K., & Khan, I. (2019). Nanoparticles: properties, applications and toxicities. *Arab J Chem.* 12(7), 908–931.

Khati, P., Bhatt, P., Kumar, R., & Sharma, A. (2018). Effect of nanozeolite and plant growth promoting rhizobacteria on maize. *3Biotech.* 8(141), 1–12.

Khati, P., Chaudhary, P., Gangola, S., & Sharma, A. (2019a). Influence of nanozeolite on plant growth promotory bacterial isolates recovered from nanocompound infested agriculture field. *Environ Ecol.* 37 (2), 521—527

Khati, P., Gangola, S., Bhatt, P., & Sharma, A. (2017a). Nanochitosan induced growth of *Zea Mays* with soil health maintenance. *3 Biotech.* 7(81), 1–9.

Khati, P., Sharma, A., Chaudhary, P., Singh, A K., Gangola, S., & Kumar, R. (2019b). High-throughput sequencing approach to access the impact of nanozeolite treatment on species richness and evenness of soil metagenome. *Biocatal Agric Biotechnol.* 20, 101249.

Khati, P., Sharma, A., Gangola, S., Kumar, R., Bhatt, P., & Kumar, G. (2017b). Impact of some agriusable nanocompounds on soil microbial activity: an indicator of soil health. *Clean Soil Air Water.* 45 (5), 1–7.

Khodakovskaya, M., Dervishi, E., Mahmood, M., Xu, Y., Li, Z., Watanabe, F., & Biris, A. S. (2009). Carbon nanotubes are able to penetrate plant seed coat and dramatically affect seed germination and plant growth. *ACS Nano.* 3(10), 3221–3227.

Knell, M. (2010). Nanotechnology and the sixth technological revolution. In S. Cozzens & J. Wetmore (Eds.) *Nanotechnology and the Challenges of Equity, Equality and Development.* Springer, Dordrecht.

Kumari, H., Khati, P., Gangola, S., Chaudhary, P., & Sharma A. (2021). Performance of plant growth promotoryrhizobacteria on maize and soil characteristics under the influence of TiO$_2$ nanoparticles. *Pant Res J.* 19(1), 28–39.

Kumari, S., Sharma A., Chaudhary, P., & Khati, P. (2020). Management of plant vigour and soil health using two agriusable nanocompounds and plant growth promontory rhizobacteria in Fenugreek. *3 Biotech.* 10(461), 1–11.

Kwak, S.Y., Lee, T.Y., Jung, W.H., Hur, J.W., Bae, D., Hwang, W.J. (2019). Immediate and sustained positive effects of meditation on resilience are mediated by changes in the resting brain. *Front Human Neurosci.* 13, 101.

Lee, J.M., Yoon, T.J., & Cho, Y.S. (2013). Recent developments in nanoparticle-based siRNA delivery for cancer therapy. *BioMed Res Int.* 2013, 782041.

Liu, C., & Zhang, N. (2011). Nanoparticles in gene therapy: principles, prospects, and challenges. *Prog Mol Biol Trans Sci.* 104, 509–562.

Lu, A.H., Salabas, E.E., & Schüth, F. (2007). Magnetic nanoparticles: synthesis, protection, functionalization, and application. *Angew Chem Int Ed.* 46(8), 1222–1244.

Luksiene, Z., Rasiukeviciute, N., Zudyte, B., & Uselis, N. (2020). Innovative approach to sunlight activated biofungicides for strawberry crop protection: ZnO nanoparticles. *J Photochem Photobiol B.* 203, 111656.

Mishra, S., Singh, B.R., Singh, A., Keswani, C., Naqvi, A.H., & Singh, H.B. (2014). Biofabricated silver nanoparticles act as a strong fungicide against *Bipolaris sorokiniana* causing spot blotch disease in wheat. *PLoS One.* 9(5), e97881.

Mitter, N., Worrall, E. A., Robinson, K. E., Li, P., & Jain, R.G. (2017). Clay nanosheets for topical delivery of RNAi for sustained protection against plant viruses. *Nat Plant.* 3(2), 1–10.

Mujtaba, A., Akhter, M. H., Alam, M., Ali, M., & Hussain, A. (2021). An updated review on therapeutic potential and recent advances in drug delivery of berberine: current status and future prospect. *Curr Pharmacol Biotechnol.* 23, 60–71.

Naderi, M.R., & Danesh Shahraki, A. (2011). Nano-fertilizers and their role in agriculture. *Int J Agric Crop Sci.* 86, 66–73.

Naqvi, S., Maitra, A. N., Abdin, M. Z., Akmal, M. D., Arora, I., & Samim, M. D. (2012). Calcium phosphate nanoparticle mediated genetic transformation in plants. *J Mater Chem A.* 22(8), 3500–3507.

Nitta, S.K. & Numata, K. (2013). Biopolymer-based nanoparticles for drug/gene delivery and tissue engineering. *Int J Mol Sci.* 14(1), 1629–1654.

Ocsoy, I., Paret, M. L., Ocsoy, M. A., Kunwar, S., Chen, T., You, M., & Tan, W. (2013). Nanotechnology in plant disease management: DNA-directed silver nanoparticles on graphene oxide as an antibacterial against *Xanthomonas perforans. Am Chem Soc Nano.* 7(10), 8972–8980.

Ohadi, R.M, Alvari, A., Samim, M., & Abdin, M.Z. (2013). Plant bio-transformable HMG-CoA reductase gene loaded calcium phosphate nanoparticle: in vitro characterization and stability study. *Curr Drug Dis Technol.* 10(1), 25–34.

Okuzaki, A., Kida, S., Watanabe, J., Hirasawa, I., & Tabei, Y. (2013). Efficient plastid transformation in tobacco using small gold particles (0.07–0.3 µm). *Plant Biotechnol J.* 30(1), 65–72.

Pandey, S., Giri, K., Kumar, R., Mishra, G., & Raja, R.R. (2018). Nanopesticides: opportunities in crop protection and associated environmental risks. *Proc Natl Acad Sci India Sect B Biol Sci.* 88(4), 1287–1308.

Paret, M.L., Vallad, G.E., Averett, D.R., Jones, J.B., & Olson, S.M. (2013). Photocatalysis: effect of light-activated nanoscale formulations of TiO_2 on *Xanthomonas perforans* and control of bacterial spot of tomato. *J Phytopathol.* 103(3), 228–236.

Park, H.J., Kim, S.H., Kim, H.J., & Choi, S.H. (2006). A new composition of nanosized silica-silver for control of various plant diseases. *Plant Pathol J.* 22(3), 295–302.

Passricha, N., Saifi, S.K., Negi, H., Tuteja, R., & Tuteja, N. (2020). Transgenic approach in crop improvement. In N. Tuteja, R. Tuteja, N. Passricha, & S.K. Saifi (Eds.) *Advancement in Crop Improvement Techniques* (pp. 329–350). Woodhead Publishing, USA.

Pestovsky, Y.S., & Martínez-Antonio, A. (2017). The use of nanoparticles and nanoformulations in agriculture. *J Nanosci Nanotechnol.* 17(12), 8699–8730.

Pirzadah, T.B., Malik, B., Maqbool, T., & Rehman, R.U. (2019). Development of nano-bioformulations of nutrients for sustainable agriculture. In Prasad, R., Kumar, V., Kumar, M., & Choudhary, D. (eds.) *Nanobiotechnology in Bioformulations. Nanotechnology in the Life Sciences*, Springer, Cham.

Raliya, R., Biswas, P., & Tarafdar, J.C. (2015). TiO_2 nanoparticle biosynthesis and its physiological effect on mung bean (*Vigna radiata* L.). *Biotechnol Rep.* 5, 22–26.

Rameshaiah, G.N., Pallavi, J., & Shabnam, S. (2015). Nano fertilizers and nano sensors–an attempt for developing smart agriculture. *Int J Eng Res Gen Sci.* 3(1), 314–320.

Safdar, M., Kim, W., Park, S., Gwon, Y., Kim, Y.O., & Kim, J. (2022). Engineering plants with carbon nanotubes: a sustainable agriculture approach. *J Nanobiotechnol.* 20(1), 1–30.

Sanford, J.C. (1990). Biolistic plant transformation. *Physiol Plant.* 79(1), 206–209.

Santiago, T.R., Bonatto, C.C., Rossato, M., Lopes, C.A., Lopes, C.A., Mizubuti, E.S., & Silva, L.P. (2019). Green synthesis of silver nanoparticles using tomato leaf extract and their entrapment in chitosan nanoparticles to control bacterial wilt. *J Sci Food Agric.* 99(9), 4248–4259.

Singh, H.P., Uma, S., Sathiamoorthy, S., & Dayarani, M. (2002). Crop improvement in Musa-Evaluation of germplasm for male and female fertility. *Ind J Plant Gen Res.* 15(2), 137–139.

Sugunan, A., & Dutta, J. (2008). Pollution treatment, remediation and sensing. *Nanotechnology.* 3, 125–143.

Tarafdar, J.C., Raliya, R., Mahawar, H., & Rathore, I. (2014). Development of zinc nanofertilizer to enhance crop production in pearl millet (*Pennisetum americanum*). *Agric Res.* 3(3), 257–262.

Torney, F., Trewyn, B. G., Lin, V., & Wang, K. (2007). Mesoporous silica nanoparticles deliver DNA and chemicals into plants. *Nat Nanotechnol.* 2(5), 295–300.

Touchell, D.H., Palmer, I. E., & Ranney, T.G. (2020). In vitro ploidy manipulation for crop improvement. *Front Plant Sci.* 11, 722.

Wang, P., Lombi, E., Zhao, F.J., & Kopittke, P.M. (2016). Nanotechnology: a new opportunity in plant sciences. *Trends Plant Sci.* 21(8), 699–712.

War, J.M., Fazili, M.A., Mushtaq, W., Wani, A.H., & Bhat, M.Y. (2020). Role of nanotechnology in crop improvement. In *Nanobiotechnology in Agriculture*, Springer Nature, Switzerland

Waseda, Y., Matsubara, E. and Shinoda, K. 2011. *X-ray diffraction crystallography: introduction, examples and solved problems.* Springer Science & Business Media.

Weber, N., Halpin, C., Hannah, L.C., Jez, J.M., Kough, J., & Parrott, W. (2012). Editor's choice: crop genome plasticity and its relevance to food and feed safety of genetically engineered breeding stacks. *Plant Physiol.* 160(4), 1842–1853.

Wu, H., Shabala, L., Shabala, S., & Giraldo, J.P. (2018). Hydroxyl radical scavenging by 820 cerium oxide nanoparticles improves *Arabidopsis* salinity tolerance by enhancing 821 leaf mesophyll potassium retention. *Environ Sci Nano J.* 5, 1567–1583.

Xia, T., Li, N., & Nel, A. (2009). Potential health impact of nanoparticles. *Annu Rev Public Health.* 30, 137–150.

Zaidi, S.S.E.A., & Mansoor, S. (2017). Viral vectors for plant genome engineering. *Fron Plant Sci.* 8, 539.

Zhang, H., Demirer, G.S., Zhang, H., Ye, T., & Goh, N.S. (2019). DNA nanostructures coordinate gene silencing in mature plants. *Proc Natl Acad Sci U S A.* 116(15), 7543–7548.

Zhao, X., Meng, Z., Wang, Y., Chen, W., Sun, C., & Cui, B. (2017). Pollen magnetofection for genetic modification with magnetic nanoparticles as gene carriers. *Nat Plants.* 3(12), 956–964.

17 Impact of Nanoparticles on Human Health and Environment

Sadhna Chauhan
Govind Ballabh Pant University of Agriculture & Technology

Payal Chakraborty
Banaras Hindu University

Chandan Maharana
ICAR-Central Potato Research Institute

Neha Singh
Dr. Yashwant Singh Parmar University
of Horticulture and Forestry

Asha Kumari
ICAR-Indian Agricultural Research Institute

CONTENTS

DOI: 10.1201/9781003345565-17

17.1 INTRODUCTION

Nanotechnology is at the forefront of technology, having usage in economic, technical, and therapeutic applications, and has transformed our way of life. The nano-compounds and nanocomposites are being widely used in the field of agriculture owing to their targeted and precise action (Khati et al., 2017a, b, 2018, 2019a, b; Chaudhary & Sharma, 2019; Kumari et al., 2021). However, research related to widespread use of nanoparticles clearly indicates that these can have detrimental impacts on environment and life forms (Gajewicz et al., 2012). Additionally, the efficient manufacture and use of modified nanoparticles will result in their release into the surrounding environment, thereby augmenting the exposure and interactions of the biotic as well as abiotic components of the ecosystem with them. Nanomaterials have many useful applications, but their presence and extended exposure in the ecosystem can be dangerous and have toxic biological impacts. Therefore, understanding potential ecotoxicological consequences of how released nanoparticles interact with the environment and their impact on human health, as well as hazard identification and risk assessment is necessary in order to facilitate nanoparticles reliably (Bhatt & Tripathi, 2011). Nanoparticles are particles with size varying between 1 and 100 nm that are imperceptible to the human eye and exhibit certain unique and novel physical and chemical properties as compared to their larger equivalents. Nanoparticles are present naturally within the atmosphere, can be created as a by-product of combustion reactions, or sometimes produced tenaciously through engineering. It may also be produced from carbohydrates having different dimensions. Due to high surface energy, spatial confinement, high surface atom fraction, and reduced imperfections, they stand apart from bulk matter. Nanoparticles are essentially a transitional state between macroscopic solids and atomic-molecular systems.

17.2 BASIS FOR TOXICITY OF NANOPARTICLES

Understanding the components and traits that significantly increase NPs toxicity is the most crucial stage before assessing their toxicity. These particles' effects will be defined by how they interact with the environment and with people. In this part, we'll discuss some of the characteristics of NPs that make them hazardous (Table 17.1; Figure 17.1).

17.2.1 SIZE AND SHAPE

The surface-to-volume ratio rises exponentially as NP size decreases, making these particles more reactive and hazardous. The size of NPs greatly influences their capability to cross through various cellular barriers. For instance, brain-blood pathway can be crossed by particles smaller than 35 nm, cell nuclei can be accessed by particles smaller than 40 nm, and cell membranes can be crossed by particles smaller than 100 nm to enter cells (Oberdörster et al., 2008; Dawson et al., 2009). Silver NPs were discovered to produce cytotoxicity which was size-dependent in human lung tissues because of the release of silver in the cellular plasma (Gliga et al., 2014). The *in vitro* studies show that Ag nanoparticles generate an array of toxicities depending on the size, including genotoxicity, inflammation, developmental toxicity, and cytotoxicity (Park et al., 2011). Catalytic and adsorption activities of proteins are influenced by NP size. *Tetrahymena thermophila* is 10–20 times more vulnerable

TABLE 17.1
The Basis of Toxicity of Nanoparticles

Basis	Role of Basis	References
Size and Shape	Smaller the size more is the reactivity and hence toxicity	Carlson et al. (2008); Oberdörster et al. (2008); Dawson et al. (2009); Nangia and Suresh kumar (2012)
Reactivity and stability of compound	More reactive and hence more toxic	Gaiser et al. (2012); Nel et al. (2006); Deng et al. (2013)
Surface chemistry	The charges in the surface make the nanoparticle/compounds more reactive and hence more toxic	Zhu et al. (2010c); Goodman et al. (2004); Sarma et al. (2015)
Aggregation	The aggregating nature reduce the absorption and thus the toxic effects are also reduced	Albanese and Chan (2011)
Medium	NPs become less toxic if they are stored in the medium for longer durations	Kittler et al. (2010); Yue et al. (2014)

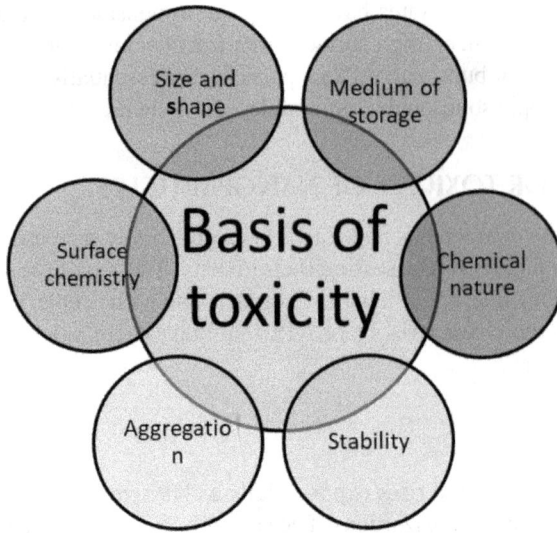

FIGURE 17.1   Basis of toxicity of nanoparticles.

to copper oxide (CuO) nanoparticles than to bulk CuO (Mortimer et al., 2010). On the basis of size, AuNPs adjuvant with non-lethal groups exhibited acute as well as chronic toxicity (Bozich et al., 2014). When AgNPs were 10 nm in size, human lung cells exhibited cytotoxicity that was unaffected by the coating process (Carlson et al., 2008). Smaller AgNPs (15 nm) resulted in more damage in the cells by forming ten times higher ROS as compared to bigger AgNP particles (30 and 55 nm) (Gliga et al., 2014). Similarly, cytotoxicity resulting from small AuNPs (1.4 nm) was 60–100 times greater than that from larger nanoparticles (15 nm) (Pan et al., 2007). Because there are no universally accepted toxicity protocols, it is very difficult to generalize the size ranges of various NPs that can cause increased toxicity. So, it can be concluded from the above that smaller NPs are more dangerous than their larger equivalents.

According to computerized simulation-based studies, nanoparticle's shape and charge can speed up their efflux mechanism through cellular membrane up to 60-fold (Nangia & Suresh kumar, 2012). It has been shown that different AgNP shapes have distinct effects on cells (Stoehr et al., 2011). Human lung epithelial cells (A549) are more harmful to ZnO nanorods than to spherical NPs of ZnO (Hsiao & Huang, 2011). Additionally, gold NPs with comparable function demonstrate that spherical particles lead more to toxicity than rod ones, which can be ascertained to the functional ligands being released from spherical surface quickly when in contact with cells (Tarantola et al., 2011).

17.2.2   Nature and Reactivity

The nature and composition of NPs is also one of the primary contributors to ecotoxicity. The damaging impacts of nano-sized particles vary from negligible to extremely high. This notion is supported by numerous studies. For instance, in a number of toxicity assessment processes, Ag nanoparticles were found to be lethal than CeO_2 particles (Gaiser et al., 2012).

NPs' extraordinarily high reactivity when compared to other materials is one of their key characteristics, and it is what made them useful for catalysis. ROS are produced within cells as a result of catalytic processes that are sparked by reactive NPs. In biological systems, a chain of reactions is to be expected because the particles do not degrade during catalytic processes. Because of their peculiar reactivity, NPs are very harmful (Stark, 2011). NPs with charge are more reactive towards cells and proteins than NPs without charge (Deng et al., 2013). Particle-induced damage works via mechanism of the production of ROS (Nel et al., 2006).

17.2.3 Mobility and Stability

NPs have high levels of mobility and are tiny in size because of which they can easily enter into plant and mammalian cells. Particle diffusion rises with decreasing particle size. By bringing NPs close to one another, mobility is another feature that aids in the aggregation of nanoparticles. NPs type, size, and concentration decide their stability within a biological system. System pH is crucial for NP stability. The majority of inorganic and organic NPs dissolve in biological systems' pH. NPs solubility can be a crucial factor determining the stability and toxicity of NPs in a given environment.

17.2.4 Surface Chemistry and Charge

Similar to this, nanomaterials are synthesized and functionalized with different organic components for application in many fields. AuNPs with functionalized cationic chains were shown to be somewhat hazardous, while those with functional anionic chains were found to be non-lethal (Goodman et al., 2004). The sort of organic groups that are surface-coated on AgNPs has a significant impact on their toxicity, destiny, and stability in any media (Sarma et al., 2015). The absorption, distribution, and toxicity of AuNPs are modified with hydrophilic and hydrophobic monolayers in medaka fish. While hydrophobic particles travelled to the fish's various organs and caused fish mortality in less than 24 hours, hydrophilic particles were discovered in the fish bowels but had no clear negative impact on health (Zhu et al., 2010c).

17.2.5 Agglomeration/Aggregation

Aggregation has a significant impact on cellular absorption and toxicity. HeLa and A 549 cells' cellular absorption of transferrin-coated AuNPs was reduced by 25% upon aggregation, but increased by two times corresponding to MDA-MB 435 cells (Albanese & Chan, 2011).

17.2.6 Medium and Storage Time

The medium in which nanoparticles are created or maintained influences how toxic they are as well. In a study, RTgill-W1 (*Oncorhynchus mykiss*) cell line exhibited solvent-dependent toxicity towards citrate-capped AgNPs (Yue et al., 2014). According to studies, NPs become less hazardous as they are kept in a solvent for a longer duration. Ag ions were discovered to be the reason behind the enhanced lethal effects of NP solution that contains dispersed Ag ions towards mesenchymal stem cells from humans over several weeks of storage (Kittler et al., 2010).

17.3   TOXICOLOGICAL IMPACT ON HUMAN HEALTH

Humans and the environment are exposed to nanoparticles which can be explained well with various methods. Workers (including engineers, scientists, and technicians) are mostly exposed at work while creating goods using nanomaterials on a commercial scale and for research. The primary cause of this exposure is the manipulation of raw materials while performing reactions using the apparatus. Other potential sources of this kind of exposure include the characterization of the resultant material, packing, and transportation. Consumers are exposed to certain nanomaterials during usage, most specifically in the second stage of synthesis that can have direct detrimental effects on them (Tsuji et al., 2006).

Above a certain level, nearly all substances from tiny heavy metals to organic complex macromolecules are lethal to cells, plants, and animals. The process of NPs' distribution into living systems, as well as their cell uptake, mode of dispersion, and possible health implications, is the subject of investigation on a global scale (Table 17.2).

17.3.1   PULMONARY EXPOSURE AND CARDIOVASCULAR TOXICITY

Studies have been done to show that RBCs can absorb various nanomaterials cellularly and to further understand the actual uptake mechanism (Wang et al., 2012; Geiser et al., 2005). For instance, the nanoplastic having entry routes into the body pass through the primary tissue barriers, and then move through all of the blood system to the secondary organs. Red blood cells (RBCs) can adsorb and be penetrated by carboxylated polystyrene nanoparticles, according to in vitro experiments (Chambers & Mitragotri, 2004; Anselmo et al., 2013). The polystyrene nanoparticles can prolong their circulation time by avoiding the liver and spleen's fast clearance processes through a process known as cellular hitchhiking. The placental barrier and the blood-brain barrier are secondary barriers that can be crossed by blood flow (BBB). RBCs are highly specialized cells that may be identified by their endocytotic uptake method and lack of a cell nucleus. RBCs were shown to take up nano-polystyrene particles (size $< 200\,nm$). This finding suggested that the absorption mechanism for this particular cell type was passive diffusion rather than endocytosis or actin-based uptake (Rothen-Rutishauser et al., 2006).

17.3.2   DERMAL EXPOSURE AND CELLULAR TOXICITY

Nanoparticles can spread widely across the different tissues and organs of biological systems after entering the body through a variety of routes. Nanoparticles that can

TABLE 17.2
Toxicological Impact on Human Health

System Affected	Problem Associated	References
Respiratory	NP aerosols when inhaled cause severe damage to respiratory system	Chambers and Mitragotri (2004); Wang et al. (2012); Geiser et al. (2005); Anselmo et al. (2013)
Skin	Inflammation and fibrosis during dermal exposure of NPs	Oberdörster et al. (2005); Zhu et al. (2006); Nowack and Bucheli (2007)
CNS	NPs can pass the blood to brain barrier and hence can cause neurotoxicity and cerebral inflammation	Hussain et al. (2006); Trickler et al. (2010)
Gastrointestine	Entry of nanoplastics into the digestive system exposes the human with high risk of nanotoxicity	Mattsson et al. (2014); Cole and Galloway (2015); Li et al. (2016); Van Pomeren et al. (2017); Pitt et al. (2018)
Muscles	AgNPs may cause N-acetyl-B-D-glucosaminidase levels to increase and creatine level to decrease	Rosenman et al. (1979); Korani et al. (2013)
Renal	AgNPs in higher dose gets deposited in the kidneys	Kim and colleagues (2008); Kim et al. (2008); Korani et al. (2013)
Reproductive system	Some NPs can severely affect male fertility	Asharani et al. (2008); Osborne et al. (2013)
Immune system	The ROS generated through NPs cause a huge damage to immune system	Brown et al. (2001); Elsaesser and Howard (2012); Fuchs et al. (2016)
Cardiovascular	Absorption of NPs by RBCs	Wang et al. (2012); Geiser et al. (2005)

affect biological systems are primarily found in wastewater. Wastewater may include nanoparticle-associated pharmaceutical residues that are later discharged into the environment. Dust from construction sites, industrial activities, and the environment is an additional source of nanoparticles (Buzea et al., 2007). On the cellular level, antioxidant activity, oxidative stress and cytotoxicity have an impact; however, at the organismal level, the main effects are inflammation and fibrosis (Oberdörster et al., 2005). Root elongation has also been discovered in hydroponic experiments using aluminium oxide (Yang & Watts, 2005). Endocytosis and phagocytosis are processes that protists and metazoan use to process the entry of nano- or micro-sized particles (Moore, 2006). Furthermore, it's been observed that unicellular protozoa localize the nanoparticles into the cell mitochondria (Zhu et al., 2006; Nowack & Bucheli, 2007). Environmental health can be potentially impacted by NPs as: (1) direct impacts on microbes, and other living things (both chordates and non-chordates); (2) damage to abiotic components; and (3) reacting with pollutants and nutrient elements that hamper the availability of those nutrients.

There are a number of potential methods a cell could use to take in nanoparticles (Krug & Wick, 2011). Although endocytosis is the most common mechanism used by cells, passive diffusion channel or transport-protein-mediated absorption has also been reported. There have been many different endocytosis mechanisms discovered, such as phagocytosis, micropinocytosis, and endocytosis mediated by caveolae and clathrin (Khalil et al., 2006; Doherty et al., 2009; Ziello & Huang, 2010).

Cellular intake of polystyrene particles was by human fibroblasts (colon cells) and bovine oviductal cells (Fiorentino et al., 2015). Uptake inhibition assays showed that the nanoparticles of polystyrene were mostly internalized through clathrin-independent mechanism. In another study, HeLa cells (human cervical), 1N1 (glial astrocytoma), and A549 lines (lung epithelial cells), as well as those used by Forte et al. (2016), were used to determine the transport inhibitors and cellular uptake of 40–200 nm carboxylated polystyrene nanoparticles (Dos Santos et al., 2011). The results clearly demonstrated that polystyrene nanoparticles get entry into cells mediated by active mechanisms only.

17.3.3 Central Nervous System Toxicity

Additionally, there is evidence that SNPs (silver nanoparticles) might cause neurotoxicity and cerebral inflammation (Hussain et al., 2006; Trickler et al., 2010). Exposure to silver might cause permanent neurotoxicity, which can ultimately result in death. High concentrations of Ag have been diagnosed in erythrocytes, cerebrospinal fluid, plasma, epileptic convulsions, and coma after regular ingestion of colloidal equivalent. Recent and updated knowledge of the effects of NP on the CNS is lacking and additional research and clinical evidence are required.

17.3.4 Inhalation and Respiratory System Toxicity

Research should pay specific attention to the airways and lungs in relation to the relative significance of potential workplace and environmental exposure. Research on carbon nanoparticles created from combustion and some inferences from studies on particulate pollution suggest that these particles may cause lung inflammation. This corroborates the idea that toxicity is determined by particle chemistry, surface properties, size, and dose (Khan, 2013). It was hypothesized that salt aerosol movement from sea can lead to human exposure to nanoplastics through inhalation. Such aerosols may form as a result of wave action and contain suitable-sized polymer particles, which the wind may subsequently carry to coastal areas (Wright & Kelly, 2017). As an illustration, CNT (carbon nanotubes) toxicity is more harmful when inhaled, leading to serious inflammation, in contrast to cutaneous or via mouth exposure, that has minor consequences (Foldvari & Bagonluri 2008). They interact with cells and different proteins causing structural changes or may simply go through metabolism (Pichardo et al., 2012), at this point translocation to other body organs takes place via blood transport. This depends on their physiochemical properties (Sharifi et al., 2012).

17.3.5 Ingestion and Gastrointestinal Toxicity

Numerous aquatic creatures, including fish, mussels, and birds, have been shown to ingest and accumulate nanoplastic particles in their GI tracts (Mattsson et al., 2014; Cole & Galloway, 2015; Li et al., 2016; Van Pomeren et al., 2017; Pitt et al., 2018).

Humans may purposely consume plastics, putting them at risk for the rare gastrointestinal obstruction known as a plastic bezoar (Yaka et al., 2015), which primarily affects people with psychiatric disorders, or inadvertently through the food chain by consuming contaminated food or beverages. In packaged materials, food products are at high risk of getting contaminated with nanoplastics.

17.3.6 Muscle Toxicity

During an occupation trial, people were subjected to Ag nanoparticles. While creatinine clearance has been observed to decrease, N-acetyl-B-D-glucosaminidase levels were observed to increase throughout studies (Rosenman et al, 1979). Colloidal AgNPs may cause a severe reaction in this organ that is dose-dependent. Utilizing transmission electron microscopy and X-ray diffraction, it was determined that AgNPs of both small and large sizes may disperse in the muscles and result in muscle toxicity (Korani et al., 2013).

17.3.7 Renal Toxicity

As the primary organ for drug disposal, the kidneys may be potential targets of NPs toxicity. For instance, AgNPs have been known to get deposited in the kidneys in a dose-dependent manner (Kim et al., 2008). Some of the NP-containing biomedical devices have the potential to gradually release silver ions (Ag+), which can build up in the kidneys via the bloodstream (Tang et al., 2009). Histological evidence in mammalian cells shows that exposure to AgNPs can thicken capsular membrane, damage PCT (proximal convoluted tubule degeneration), and cause abnormality in mesangial cells and adhesion to Bowman's capsule (Kim et al., 2008; Korani et al., 2013).

17.3.8 Reproductive and Developmental Toxicity

The possible threat of NPs due to prenatal exposure and to pregnant women is another concern. Male fertility may be compromised because of NPs' high precision to these chemicals during spermatogenesis. Zebrafish embryos' brain, heart, yolk, and blood included SNPs that led to a multitude of developmental abnormalities, which decreased the likelihood that zebrafish embryos would survive (Asharani et al., 2008; Osborne et al., 2013). Almost little relevant details are available on the adverse impact of AgNPs on the reproduction system. Grafmueller and colleagues showed that polystyrene particle (240nm) uptake by cells of the placental barrier is possible, which is then assimilated from the foetal to the maternal blood system (Grafmueller et al., 2015).

17.3.9 Immune System Toxicity

Living organisms are challenged with the adverse effects of nanoparticles which produce ROS leading to inflammation and cytotoxicity (Elsaesser & Howard, 2012). Researchers carried out studies using human cell lines and suggested that polymer nanoparticles may cause oxidative burst, activate the innate immune system, or even cause inflammation. *In vitro* experiments showed the pro-inflammatory effects of polystyrene particles on human A549 lung cells. They discovered that the expression of IL-8 was higher than what 64-nm-diameter polystyrene nanoparticles revealed (Brown et al., 2001). According to studies by Fuchs et al., polystyrene particles with 120 nm carboxylates and amino modifications can polarize human macrophages into the M1 or M2 phenotype (Fuchs et al., 2016). Because there are so few *in vivo* researches on nano-immune interactions, there is a higher need for prudence when prescribing NPs for use in biomedical applications.

17.3.10 Genotoxicity and Carcinogenicity

There have been reports of AgNPs' genotoxic effects on the DNA of the calf thymus. The findings showed that genotoxicity was caused by the interaction of AgNPs with detergent (Chi et al., 2009). These were able to generate genotoxic activity in *Drosophila* and interfere with DNA molecule replication, which resulted in mutations (Yang et al., 2009; Demir et al., 2011). SNPs are intriguingly more detrimental to cancer cells as compared to healthy ones (Foldbjerg et al., 2011; Jiang et al., 2013; Guo et al., 2013, 2014), which might hold significance for future research on anticancer drug discovery and development.

17.4 TOXICOLOGICAL IMPACT ON ENVIRONMENT

The natural world is made up of numerous intricate ecosystems that include atmospheric, terrestrial, freshwater, and marine components. In contrast to human exposure, there are a lot of species that could be at risk from nanoparticles. Risks related to exposure to nanoparticles may affect an individual or a population, as well as the overall structure and function of the ecosystem (Figure 17.2).

17.4.1 Effects in Groundwater and Soils

There is a need to comprehend the fate and function of such nanoparticles on different grounds and groundwater. It is also important to take into account how soil flora and fauna will react to and absorb nanoparticles. Once the sources of these nanoparticles in the sewer have been identified, it may also be necessary to take into account their environmental fate and toxicity to bacteria in sewage treatment facilities (Chen et al., 1998).

17.4.2 Effects on the Aquatic Environment

Water bodies are more prone to contamination as a result of the rise in nanoparticles' amount in different skincare products and cosmetics (Popov et al., 2005). Nanoparticles are hazardous to unicellular creatures as well as aquatic species such as fish and *Daphnia* (Nowack & Bucheli, 2007). According to Tjälve and Henriksson (1999), the designed NPs are extremely poisonous and have a selective route mechanism for fish and rats. According to Oberdörster et al. (2008), aquatic animals can consume NPs or have them enter through their gills. Results showed that nickel,

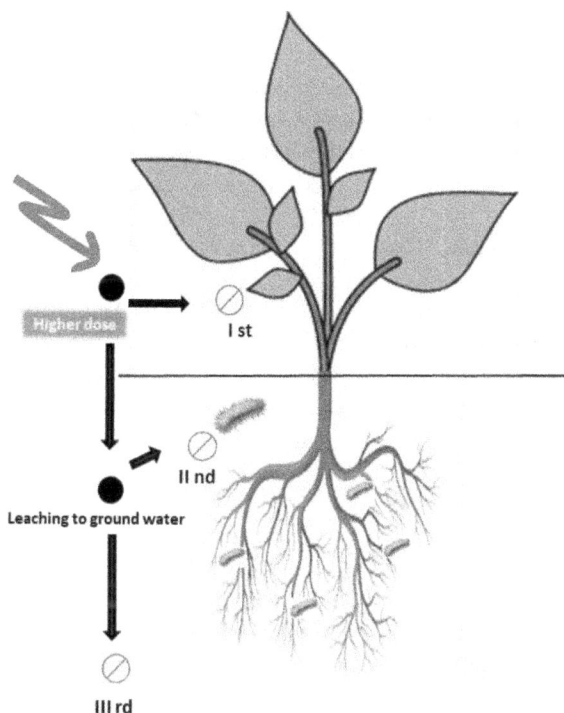

FIGURE 17.2 Toxicological impact on environment.

copper, silver, and aluminium ENPs had harmful impacts on all of the studied creatures, including zebrafish, dolphins, and algal species (Griffitt et al., 2008). Although the severity depends on size, type, charge of the nanoparticle, and the organism being exposed, NPs are harmful to aquatic life. In the same species, some organs are more impacted than others. The major problem with NP transmission arises in the food chain.

Sun et al. (2018) demonstrated the toxicity studies of nanoplastics and suggested that particles (size: 50 nm) at a dose of 80 mg/L caused damage via inducing oxidative stress in bacterium used to saline water habitat (*Halomonas alkaliphila*) in contradiction to microplastic particles (size: 1 µm). Exposure to polystyrene nanoparticles may have a lethal effect on oyster planktonic stages as doses ranging from 0.1 to 25 g/mL were more damaging to both gametes and embryos of oysters (*Crassostrea gigas*) when exposed to polystyrene (size: 50 nm) nanoparticles (Tallec et al., 2018). Nanoplastics have been shown to be proactively consumed and adsorbed by various aquatic organisms such as mussels, algae, and zooplanktons (Ward & Kach, 2009; Bhattacharya et al., 2010; Besseling et al., 2014; Booth et al., 2016).

17.4.3 Effects on Plants and Soil Microbial Communities

Nanotechnology is advancing rapidly, and its application in agriculture makes the interactions among nanoparticles in flora, fauna, or microbes (Chaudhary et al.,

2022). Some NPs such as nanochitosan, nanozeolite, nanophos, and nanogypsum at low concentrations have positive impact on soil microbes and improve the beneficial bacterial microbes which improved plant health and productivity (Chaudhary et al., 2021a, b, c, d, e). Silicon dioxide and nanochitosan NPs improved soil enzyme activities, soil physicochemical properties, and also improve plant physiological functions (Kumari et al., 2020; Kukreti et al., 2020; Agri et al., 2021, 2022).

Shrivastava et al. (2019) identified that the amount of silver in *Lolium multiflorum* exposed to gum arabic (GA) overlaid with AgNPs (size: 6–25 nm) enhanced as the nanoparticle concentration continued to rise; it was also noted that the plants' roots and shoots were inhibiting the growth of seedlings. AgNPs were found to have detrimental impacts on mitosis, germination, and nuclear abnormalities in *Allium cepa* root cells (Scherer et al., 2019). These studies highlighted the need for modified forms of these nanoparticles in potential toxic research by showing that the toxic effects on plants vary depending on the initial nanoparticle composition. Short-term range of dose of AgNPs (0.01–1 mg/kg) has no appreciable impact on soil microbiomes (Grün & Emmerling, 2018). Similar to this, ZnONPs were more readily taken up than bulk ZnO and $ZnCl_2$ and had detrimental effects on earthworm fecundity (García-Gómez et al., 2014). Nanoparticles of zinc oxide and chloride salt both exhibited harmful long-term effects on the development as well as fertility in *Daphnia magna* (Adam et al., 2014). The process of denitrification in soil samples was found to be affected by caesium NPs, and it depended on the species being studied, particle size, and dosage (Dahle & Arai, 2014).

17.4.4 WIDER EFFECTS ASSOCIATED WITH UNINTENTIONAL RELEASE

It is unclear how nanoparticles are absorbed and behave in bacteria, invertebrates, vertebrates, and plants. Subsequently, there is a dearth of information pertaining to cellular uptake, localization, and toxicity. It is crucial to consider the effects at low doses and over long time periods. It's important to look into how nanomaterials interact with different types of contaminants, such as metals and organics. Ecotoxicology may also benefit from information from mammalian toxicology. Evidence from toxicological studies shows that toxicological pathways may have adverse effects on common biological phenomenon (Khan, 2013).

17.5 MECHANISM OF TOXICITY OF NANOPARTICLES

17.5.1 DAMAGE TO CELL MEMBRANE

Toxic effects are produced by the interaction of silver nanoparticles with cell membranes. NaCl content and pH both affect how the bacterial membrane reacts. Surface charge is crucial for this interaction; however, *in vitro* research has shown that silver nanoparticles harm mammalian cell membranes (Danielle Mcshan, 2014). According to evidence from transmission electron microscopy, silver nanoparticles can stick to and enter *Escherichia coli* cells, and they can also cause pits in cell membrane (Sondi & Salopek-Sondi, 2004; Raffi et al., 2008; Choi et al., 2008). Other bacteria such as *Pseudomonas aeruginosa*, *Salmonella typhi*, and *Vibrio cholerae* have also

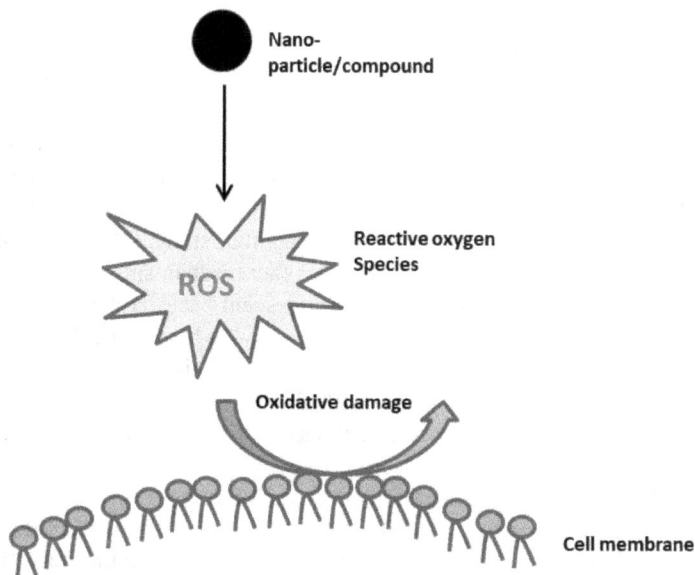

FIGURE 17.3   Mechanism of toxicity of nanoparticles.

been observed to accumulate silver nanoparticles on their cell membranes and take them up internally. Uncertainty exists regarding the specific process by which AgNPs interact with cell membranes and get entry into the cells. In bacteria, membranes are negatively charged in contrast to nanoparticles being positively charged which results in interaction between them through electrostatic attraction (Raffi et al., 2008). It has been hypothesized that cells in membrane and AgNP are more readily connected with proteins that essentially contain elemental sulphur, in a way analogous to silver attaching with thiol side branches in respiratory and membrane proteins to disrupt their normal functioning (Morones et al., 2005). Another explanation of Ag nanoparticles damaging *E. coli* membranes is metal depletion, which leads to the formation of pits with erratic forms in the membrane and alterations in permeability brought on by the release of LPS molecules in membranes (Amro et al., 2000; Sondi & Salopek-Sondi, 2004) (Figure 17.3).

17.5.2   PRODUCTION OF ROS (REACTIVE OXYGEN SPECIES)

ROS play a key role in normal body functioning; however, its excessive generation can harm proteins, lipids, and DNA, as well as the antioxidant defence system (Ahmadi, 2012; Sriram, 2012; Duran et al., 2015). The antioxidant systems in mitochondria are activated by silver nanoparticles, according to research on the potentially detrimental effects of silver NPs in fibroblast and liver tissues. The ROS that mitochondria release cause oxidative stress, disrupt ATP production, damage DNA, and cause apoptosis (Gurunathan, 2015). In eukaryotes, excessive ROS production leads to oxidative stress, but low levels can be balanced out by the antioxidant defence, such as the GSH (glutathione)/GSSG (glutathione disulphide) ratio (Nel et al., 2006). Silver nanoparticles affect GSH via direct mechanism and combine with GSH reductase to breach the antioxidant defence (Carlson et al., 2008). Increased free radical generation has the potential to compromise mitochondrial and membrane function, attack membrane lipids, and harm DNA (Mendis et al., 2005). In prokaryotes, Ag ions are reported to trigger ROS generation directly by interaction with the thiol groups in the respiratory enzymes or superoxide dismutases which is as free radical scavenger (Park et al., 2009). The emergence of free radicals from silver NPs was proved by spin resonance imaging (Kim et al., 2007). In another study, DNA damage, or cell membrane damage in bacterial cells, as well as other mechanisms of Ag nanoparticle toxicity were examined. Recombinant bioluminescent bacteria were employed in a panel. When silver nanoparticles were added to growing bacterial cultures, they produced superoxide radicals. It was concluded that toxicity is result of oxidative stress and membrane protein damage (Hwang et al., 2008).

ROS production and oxidative stress serve as the primary function of toxicity, even at low doses (Mohd Javed et al., 2019; Lojk et al., 2020). ROS appear in acidic surroundings such as lysosome aqueous medium and oxidation organelles. Media comprising superoxides, hydroxyl groups (OH), and hydrogen peroxide form ROS that interact with cell compartments, causing toxicity depending on the amount formed and the response to oxidative stress (Chang et al., 2012). ROS production can be induced by interaction with immune cells such as phagocytic cells or alveolar macrophages using the NADPH oxidase complex (Manke et al., 2013). Furthermore, excessive usage of it alters biomolecules in a way that could cause cell mortality. Silica nanoparticles at a conc. of 50 g/mL were shown to have no negative effects; however, when the concentration was increased to 400 g/mL, the damage significantly enhanced (Akhtar et al., 2010). Therefore, a rise in ROS has the ability to alter DNA, leading to the formation of links between chains that, if not rectified, can cause genetic alterations and ultimately cell death (Chang et al., 2012; Guo & Chen, 2015).

17.6 FACTORS AFFECTING THE TOXICITY OF NANOPARTICLES

Nanotoxicity is directly connected with their surface characteristics, such as their shell, ligand, and surface modifications, as well as their diameter, the type of toxicity assay employed, and the duration of exposure (Oh et al., 2016).

Nanoparticle size and the amount of surface area it has are two of the primary factors that influence the specific process by which a nanoparticle interacts with living systems. The enormous specific surface area of NPs is the root cause of their extraordinarily high catalytic activity and reaction capacity. Nanoparticles with a size of 15 nm are 60 times safer for epithelial cells, fibroblasts, macrophages, and melanoma cells than NPs (Huo et al., 2014). Necrosis of the cell is caused by 1.4-nm nanoparticles, but apoptosis is mostly caused by 1.2-nm nanoparticles (Pan et al., 2007). Nanoparticles with size <5 nm are able to breach the cell wall in a manner that is not selective, in contrast to larger particles, which can enter cells via phagocytosis, macropinocytosis, and both selective and nonspecific transport pathways (Zhang et al., 2015).

Spherical nanoparticles are more prone to be absorbed in cells through endocytosis as compared to nanofibers and nanotubes (Champion & Mitragotri, 2006). The toxicity is determined by the geometry of the particles, which gives them the ability to physically damage the cell membrane. On the other hand, research has shown that the amount of foetal calf serum present in the culture medium is inversely proportional to the level of toxicity exhibited by the organisms. This is further explained by the great ability of graphene oxide nanoparticles to adsorb protein molecules, which coat the nanoparticle surface, causing a change in the form of the nanoparticle, and helping to avoid cell membrane disruption in part (Hu et al., 2011).

Both the chemical make-up and the crystal structure of nanoparticles have a substantial impact on their hazardous potential. Comparing the adverse effects of ZnO and SiO_2 nanoparticles (with a size of 20 nm) on rat fibroblasts revealed different paths of toxicity. ZnO nanoparticles are responsible for producing oxidative stress, whereas SiO_2 nanoparticles cause the change in the structure of DNA (Yang et al., 2009). Ecosystem parameters such as hydrogen ion power and ionic strength can affect the degree to which nanoparticles deplete. The majority of the time, the deleterious effect that nanoparticles have when they interact is caused by the leaking of metallic ions from the core. While Zn and Fe and other metal ions have biological roles, they are very toxic and can harm cellular circuits when present in high amounts. However, by covering the nanoparticle cores in silica layers, or gold shells and thick polymer shells, this harmful effect can be lessened (Soenen et al., 2015).

When using human bronchial epithelial cell line and TiO_2 nanoparticles with a variety of crystal lattices, one can investigate the link between crystal structure and toxicity. The nanoparticles' crystal structures also play a role in determining their hazardous potential. It has been demonstrated that nanoparticles with anatase-like crystal lattice of the same size do not produce any harm, in contrast to those of rutile-like crystal structures (e.g. prism-shaped TiO_2 crystals) that cause damage to DNA and lipids. This is the case even though both types of nanoparticles have the same size (Gurr et al., 2005). Crystalline structure of nanoparticles is prone to change if they are exposed to certain environmental factors, such as biological fluids, interactions with water, and other dispersion media. Surface properties of nanoparticles have a significant effect on the toxicity of the particles because of the control they have over the manner in which nanoparticles interact with living organisms (El et al., 2011). By grafting with polymers that have varying charges, nanoparticle surfaces and their charges can be altered. The intracellular absorption of the nanoparticles and their capacity to target particular cells are enhanced by the addition of PEG and folic acid. Dextran, methotrexate, and polyethylene are utilized to change the charge on the surface of nanoparticles (Zhang et al., 2002). The reason behind nanoparticles with a positive charge being more dangerous than neutral and negative charged particles is that they get entry into the cells more readily.

Shells form the basis in nanoparticle toxicity, as it must be applied to the surface in order to modify their optical and electromagnetic properties and also to enhance their compatibility and dissolution in water and other liquids by lowering their tendency for agglomeration. In essence, shell lessens the toxicity of nanoparticles while enriching them with the ability to engage specifically with different kinds of other cellular and biological components (Xia et al., 2009). However, shell can be effectively used to lessen or completely eliminate this adverse effect of free radicals. It also aids in stabilizing nanoparticles by improving their environmental resistance and decreasing the chances of the emission of toxic compounds from them (Peng et al., 2013). Both organic and inorganic substances, such as silicon, polyglycolic acid, polyethylene glycol, polylactic acid, proteins, and lipids, are used to modify the surface of nanoparticles.

Research on NPs has advanced significantly over the last 20 years, and as a result, nanotoxicology, a brand-new field of study concentrated on the potential toxicity of nanoparticles on bioecological systems, was established. The main focus of nanotoxicology is to discover safe nanoparticles while simultaneously researching the toxic properties of nanoparticles and the ways in which these properties influence different cells and tissues. *In vitro* investigations, involving the use of cell line model, and *in vivo* techniques, which involve the use of laboratory animals, are two typical methods for researching the potentially dangerous impacts that nanoparticles have on a variety of living systems. The pathways and processes of nanoparticles' hazardous effects are not sufficiently understood enough for computer models to anticipate interactions between nanoparticles and living matter with adequate reliability, hence the third approach for evaluating nanoparticles' toxicity is not used (Sukhanova et al., 2018). This is because the computer models cannot predict the appropriate outcomes of reactions among nanoparticles and living beings in a large range (Sukhanova et al., 2018). Both cell culture and animal experimental approaches come with their own set of advantages as well as disadvantages when it comes to researching and investigating the toxicity. Inside the body transit of NPs into various cells and patterns of dispersion are not taken into account. However, the hampered impact of nanoparticle action *in vivo* can be approximated thanks to research on the toxicity of nanoparticles in animal trials. Furthermore, it is impossible to determine which of the broad patterns of toxicity manifestations the root cause of the experimental effects is and which of them its effects are.

17.7 HAZARD IDENTIFICATION, RISK ASSESSMENT, AND ANALYSIS

At every milestone of the development of nanotechnology, significant potential dangers must be taken into account. As a result, there is an urgent need to conduct a risk assessment on NPs. The major aspects to be addressed and examined are namely hazards associated with short- and long-term exposure to the environment, humans health, toxicological examination, and risks related to their safe disposal.

The identification of hazards and the evaluation of risks associated with synthesized particles can be described further. The physicochemical characterization of the nanoparticles is the foremost step. The main danger is thus determined to be the nanoparticle's structure or content. The nanoparticles will then be examined for their ability to produce ROS in a biological setting. Additionally, the nanoparticles must undergo *in vitro* testing for immunotoxicity, skin and eye toxicity, liver and kidney toxicity. *In vivo* testing is performed for immunotoxicity, organ toxicity, genotoxicity, and reproductive toxicity after *in vitro* investigations. The carcinogenicity and unfavourable impacts on reproductive behaviour of the nanoparticles exhibiting positive genotoxicity and mutagenicity will be examined. The evaluation of risk at various exposure levels, the determination of exposure levels, and other regulatory restrictions are all included in the risk assessment, which also includes hazard iden-tification and risk characterization. In order to manage nanoparticles effectively and apply them safely, it is necessary to combine knowledge about experimental nanoparticle exposure levels and harmful effects caused by nanoparticles (Jamuna & Ravishankar, 2014).

17.7.1   TOXICOLOGICAL ANALYSIS

One of the major challenges in nanotechnology is identifying and characterizing nanomaterials and their agglomerates (Hassellöv et al., 2008). Physicochemical char-acteristics of materials change dramatically from macroscopic to microscopic sizes. Understanding their qualities is necessary for assessing toxicological risks and envi-ronmental exposure (Hristozov & Malsch, 2009). Physical and chemical parameters such as structure, melting and boiling temperatures, water stability, solubility, molec-ular weight, composition, reactivity, and vapour pressure must all be accurately and thoroughly characterized. However, it's possible that the aforementioned criteria aren't sufficient for identifying NPs due to the unavailability of relevant procedures and database for exposure and hazard evaluation. Developing well-characterized nanoma-terials requires extensive study and more research (Hristozov & Malsch, 2009).

17.7.2   IN VITRO AND IN VIVO TOXICITY ASSAYS

Nanoparticle toxicity can be studied *in vitro* and *in vivo*. NP mechanics, pathways, and entry points in complex multicellular organisms can only be understood by *in vivo* studies. This is important for both the design of nanomaterials intended for human exposure, such as nanomedicines, and the usage of nanomaterials in indus-trial processes where exposure may occur through the surroundings. It is now well-established that skin, lungs, and digestive tract can be good entry ways for NPs. It has also been proven, albeit in small quantities, that these cells can translocate to secondary organs. Besides *in vitro* investigations, checking the biological response of nanoparticles in model animals is necessary for exploring their potential applica-tion (Schrand et al., 2010). Researchers can utilize zebrafish as a model to examine the dangerous consequences of nanoparticles because of the fish's fast embryonic growth and transparent body structure. Nanoparticle uptake, transport, and effect on organogenesis and morphological development may all be observed in real time and without damaging the organism (Fent et al., 2010).

In order to better characterize chemical (or particle) mode of action, determine human relevance, and identify and assess prospective hazards in support of human health risk assessment, genomic tools such as gene profiling are considered as a potentially quick and economically justified alternative to other time-consuming conventional techniques (Thomas et al., 2007).

17.7.3   EPIDEMIOLOGICAL ANALYSIS

An epidemiological study is a preview of the factors governing health in the gen-eral public. Epidemiological research should be carried out in-depth to study the interactions between natural, engineered, and accidental nanoparticles as observed in their degradation through stress and carcinogenic reactions (Knox, 2005; Grigoratos & Martini, 2014).

The chance of acquiring lung cancer is thought to be raised by exposure to produced nanoparticles, but there is not enough epidemiological data to support this (Roller, 2009). An epidemiological survey of workforces prone to carbon and TiO_2 (size: 1–500 nm) was carried out and summarized so as to work the relation between exposure and risk associated with cancer. Except for a substantial group of carbon workers in a study (where a doubling of the cancer risk was reported), the findings were contrasting. Inadequate studies on humans for the carcinogenicity was determined with the International Agency for Research on Cancer (IARC) (Roller, 2009). The contradictory results necessitate additional study.

17.7.4 Life Cycle Exposure Pathway

The threat of environmental exposure to nanomaterials rises with increased development and use, which heightens the potential for negative effects on the human health and environment (Lee et al., 2009). At any point in their life cycle, manufactured and processed nanomaterials have the potential to enter the environment directly or indirectly. This can happen when making materials and chemicals, when constructing using nanomaterials, when individuals utilize nanomaterials, and when disposing of nanomaterials. Certain chemical, physiological, and physical changes may occur to nanoparticles once they are discharged into the atmosphere, which may cause them to take on new properties (Lee et al., 2010).

17.8 WORKING ENVIRONMENT AND PERSONAL SAFETY

Possessing the potential to be reactive materials, irritants, and interacting substances, NPs can affect people, employees, and the environment (Nel et al., 2006). Men dealing with aluminium smelters are prone to exposure to high quantities of the NPs and ultrafine particles which are synthesized during milling process. These procedures can result in measurable NP concentrations in the workplace (Debia et al., 2013). Another difficulty with NP disposal is that care must be taken when discarding materials that contain NPs. The use of protective equipment and correct knowledge of hazardous compounds are the two most crucial factors in limiting worker exposure to NPs at work. More extensive research is required to consider hazardous limits for these particles because there is currently little literature to address this topic. To get the exposure to NPs under control, certain general strategies can be used. One study aimed at assessing the airborne NPs during nanocomposite production found that risk associated with NP exposure is affected by ventilation, feeder type and strategy, and the characteristics and nature of the nanoparticles (Tsai et al., 2012). Titanium dioxide and black carbon NPs can be brought to safe levels with the use of ventilation systems and N95 respiratory masks (Ling et al., 2011). In the same vein, fewer people and the advancement of automated manufacturing methods might lessen exposure. Use of protective gear, such as gloves, masks, and full body covering, can greatly minimize the level of exposure. According to studies, handling NPs in a glove box can help prevent exposure (Debia et al., 2013). Inhalation of NPs can be decreased or stopped by personal safety devices. Such tactics can also be used to prevent cutaneous and gastrointestinal exposure.

Some NMs can cause negative biological reactions at the cellular, membranous, proteins, tissues, and even organ levels, according to experimental results. The unique physicochemical characteristics of NMs as they go closer to the sub-100 nm length scale are what give rise to the possibility of biological harm. Quantum effects are more prominent at this size. Since biological molecules such as DNA, proteins, and membranes are also nanoscale structures, NMs with a higher surface reactivity may interact with them more readily. Interactions at the nano-biomolecule contact may cause conformational alterations, membrane permeability modifications, oxidation injury, ionic exchanges, mutational change, biocatalytic changes, signalling cascade effects, and enzyme degradation. Some of the most significant parameters of NPs in such interactions involve their form, size, surface charge, hydrophobic nature, purity, agglomeration criteria, physio morphology, crystallinity, electrostatic behaviour, and reactive oxygen species (ROS) forming capability. These characteristics might be connected to biological results.

For safe design of nanoparticles, the link between material properties and toxicological results can also be employed. The physicochemical properties that hinder cellular uptake or bioavailability, inhibit dissemination, or lessen harmful biocatalytic effects could be altered or modified as improvements. The coating of NP surfaces with polymers, ligands, and detergents to produce steric hindrance is one example of how these improvements might be made (Xia et al., 2009). The time needed for both particle wrapping and cellular absorption is significantly influenced by the particle size. Fluorescence and other imaging methods can be used to study particle ingestion, and a set of mathematical equations can be used to represent wrapping time in terms of quantity (Chithrani & Chan, 2007).

17.9 LEGISLATION AND REGULATION

NPs' human-made emissions are divided into two groups. Unintentionally generated NPs from combustion, automotive exhaust, and mechanics' shops cohabit with nanowires, nanotubes, and nanofibers (Wu et al., 2008). Breathing polluted air exposes us to NPs (Kumar et al., 2011b). The Clean Air Act, cleaner air for Europe and the EU Directive (2008/50/EC) on ambient air quality come under legislations for environmental safety and air quality maintenance. The amount of particle emissions to the ambient atmosphere is restricted by these rules. The numerical concentration of the NPs is needed to be considered in future laws regarding the emission of nanoparticles (Heal et al., 2012). Vehicle emission requirements for Euro 5 and Euro 6 are the special laws pertaining to NPs. This is the first set of regulations to limit NPs' emissions of fine materials greater than 23 nm. The lower threshold of these standards allows not to include more than 30% tiniest NPs (Heal et al., 2012). Also, regulatory designs and frameworks must take into consideration the decreased particle spectrum. The development of innovative approaches for the measurement and detection of NPs in the surroundings requires significant focus.

For ENP regulatory actions, accurate physical and chemical identification and function of exposure response time are, however, required. The US Environmental Protection Agency (USEPA) looks after nanoparticles under their Toxic Substances Control Act (TSCA) and has subsequently studied a broad range of nanoscale substances under the same, including silica, aluminium nanoparticles, and single and multiwalled carbon nanotubes, for prolonged inhalation. Additionally, the USEPA is creating the "Significant New Use Rule" (SNUR) to ensure the regulatory examination of nanoscale particles. Manufacturers must obtain approval at least 90 days in advance of planning to create new nanoscale particles using the compounds mentioned under the TSCA. According to the "Information collection regulation," manufacturers of nanoscale materials are required to disclose information regarding production volume, synthesis technique, and any data on the material's safety and potential health risks.

Under the Canadian Environmental Protection Act (CEPA), NPs are evaluated by Environment Canada (EC) (1999). Similar to how it regulates regular chemical compounds, the European Commission (EC) is regulating artificial nanoparticles under "REACH" programme (Morimoto & Kobayashi 2010; Kumar et al., 2011b). A new law on the categorization, labelling, and packaging of novel materials has been devised by the European Union (EU). Organizations ISO (International Organization for Standardization) and OECD (Organisation for Economic Co-operation and Development) are international bodies dealing with issues that arise because of nanotechnology. Currently, ISO is preparing taxonomy, meteorology, apparatus standards and terminology for assessment and science-based safety and health processes including nanomaterials. Likewise, the Working Party on Manufactured Nanomaterials (WPMN) which is founded by OECD is working for encouraging environmental safety and health research data set, exposure measurement techniques, collaborative effort on environmentally sustainable nanotechnology use, and ENP testing guidelines. The US-based organization, i.e. National Institute for Occupational Safety and Health (NIOSH), has formulated strict criteria for the individual safety, evaluation of exposure, and preventative measures at workplace and potential risks to health.

In 2006, the OECD established the Advisory Committee for Manufactured Nanomaterials (WPMN) to make sure that initiatives to exposure, associated hazards, and risk evaluation for synthesized NPs are of superior quality, scientifically founded, and globally uniform (Paranavithana & Mohajerani, 2006). According to Rasmussen et al. (2016), the OECD is recognized as a prominent player in the governance of nanotechnology concerns, particularly through the subsidiary WPMN. The efforts of the OECD are supplemented by National Coordinators of the Test Guidelines Programme.

The ISO/TC 229 technical committee was founded by the International Organization for Standardization (ISO) in the year 2005 and it supports OECD relevant works along with addressing standardization of the nanotechnology science. It has also made available standard framework for nomenclature, terminology, instrumentation, metrology, testing methodologies, modelling, and standards for reference materials, simulation models, health and environmental safety (Rauscher et al., 2017).

17.10 TOXICOLOGICAL STUDIES OF NANOPARTICLES: RECENT STATUS, WEAKNESSES, AND FUTURE CHALLENGES

The updated and revised literature makes it abundantly evident that there is a rising necessity for the development of relevant approaches to estimate the ecotoxicity of NPs synthesized by us. The synthesis of nanoparticles has increased in the past few decades, especially utilized in consumer products; therefore, the need of the hour is to study the associations, translocation, assimilation, and destiny of NPs in both biological entities and the environment (Batista et al., 2017). The investigation of the ecotoxicity of NPs has advanced significantly, but the field is still young, and the literature demonstrates high levels of inter-study heterogeneity and even contradicting findings.

Fuel combustion and deliberate advancement in nanotechnology use have increased exposure of anthropogenic NPs to humans and the environment over the past few decades (Pipal et al., 2014). There are several aspects as to how some environmental NPs are currently being studied. The most important areas to focus on are the evaluation of post-production fate of NPs regarding their entry pathways and environment channels, as well as potential impacts on ecosystems (Garner & Keller, 2014). Some of the most visible phenomena that occur during the transfer of NPs within the environment are aggregation, deposition, chemical changes in properties, oxidation, dissolution, interaction with organic molecules, and colloids' surface coating (Quik et al., 2011; Arvidsson et al., 2011; Levard et al., 2012). Because of their antibacterial nature, NPs with metal-origin have found place in insecticidal sprays, aerosol propellants, toothpaste, cosmetics, washing machines, purifiers, and other items, thereby raising worries about environment safety (Dos Santos et al., 2014). Studies show that NPs are diffusely and indirectly released into the environment on a global scale, at a rate of about 8,300 metric tonnes per year (Keller & Lazareva, 2014). Oxidation and photocatalysis convert NPs into degraded products in the environment (Tiwari & Marr, 2010). Understanding NPs' features, such as their destiny, transport, and toxicological consequences, is crucial due to these alterations (Kumar et al., 2011a; Meesters et al., 2013). Although precise measurements of NPs' deposition times and agglomeration rates are currently impossible due to shortage of appropriate tools and the NPs' complicated features in the atmosphere, estimates suggest that about one-thirtieth of NPs persist in the lower atmosphere while the rest settle down. However, there is considerable debate on NPs' ability to remain in the environment for very long (Gottschalk et al., 2010; Quik et al., 2014). The fate of engineered NPs once they are formed and subsequently released into the environment holds utmost importance (Bello et al., 2008; Curwin & Bertke, 2011; Lee et al., 2012). Although there are a majority of particles that were agglomerated, their concentration was between 10 nm and 1 m and it was rather stable in the environment for a long time period (Brouwer, 2010; Curwin & Bertke, 2011). Dry or wet deposition is used to get rid of these NPs in suspension. Dumped NPs make their way into water systems, including those used for drinking, irrigation, and, most significantly, the marine environment. The environmental toxicity of the NPs life cycle requires more thorough investigation (Gottschalk & Nowack, 2011; Sánchez et al., 2011; Love et al., 2012; Praetorius et al., 2012; Sharifi et al., 2012; Etheridge et al., 2013).

Drawing any straightforward conclusions from the research in this area is a very laborious task due to its huge dispersion and dispersion. Toxicity experiments on different NPs have been attempted in which mechanism of action is viewed from various angles. Additionally, the different ways that NPs are prepared have made it challenging to compare the results of various toxicity studies. Although many different cell lines and animals have been used in NP research, there is currently little proof of a direct connection between NPs and human health. In these circumstances, it is advised that a set of industry-recognized criteria be developed to evaluate the cytotoxicity of ENPs. The interaction of NPs and soil is being studied by researchers as a result of a recent opinion that has been published. NPs' impacts on soil macro- and microorganisms, and their association with other soil pollutants can reveal some unanticipated negative effects (Bakshi et al., 2014).

Water forming on the surface of nanoparticles after release is anticipated to be a prime entry pathway for the particles into the surroundings, where it can aid in their dispersal and forge a link between silver NPs and environmental components such as soil and life forms (Peijnenburg et al., 2015; Baun et al., 2017). Nanoparticles undergo agglomeration, dissolution, sedimentation, uptake by biota, sulfidation and translocation that govern their ultimate fate and toxicity (Jung et al., 2018; Baun et al., 2017). Trying to determine the ecotoxicity of nanomaterials is complicated by the fact that they undergo substantial transformations after being discharged into the environment from their factory of origin (Nowack et al., 2012).

Recent significant research has emphasized how changes in biodiversity would arise from nanoparticle contamination. After being exposed to metal-based synthetic nanoparticles, the make-up of microbial communities in soil and water drastically changed. Essential criteria to consider are the toxicity repercussions of produced nanoparticles in their natural environments. There is a need for accurate models to be built in order to evaluate the consequences of nanoparticle interactions with biological organisms. This calls for studying older nanoparticles and exposing subjects to micromolar doses over extended periods of time (Pradas del Real et al., 2017). *In vitro*, many toxicity studies have been carried out because of practical constraints, lagging behind the transmission of realistic information to genuine, real environments (Conie et al., 2017). Therefore, no reliable comparisons can be drawn from the studies because of the unavailability of accurate NP characterization within real experimental circumstances (Hjorth et al., 2017). Environmental conditions may complicate characterization of AgNPs due to the presence of entities that interferes with the characterization techniques. For practical purposes, characterizing developed nanoparticles necessitates a thorough re-evaluation of existing methods. Ingestion and deposition of designed nanoparticles in terrestrial and aquatic organisms has been documented via several different routes for solubilized substances (Skjolding et al., 2016). For instance, once internalized, silver NPs have the ability to release Ag+ ions and encourage the production of ROS. The transformation, internalization, and bioaccumulation of nanoparticles in aquatic and terrestrial species are still not well understood. According to established test protocols, nanoecotoxicological investigations should be planned taking into account the regulatory environment. Due to the possibility of designed nanomaterials behaving differently from traditional soluble compounds, this poses another challenge for ecotoxicity assessments. The documented resilience of few strains of bacteria of medicinal importance may be generalized to soil or aquatic microbes, although further research is required to determine this.

Nanoparticles are difficult to detect and characterize. Only a small number of particles at a time can be detected and quantified using quantitative and qualitative methods (Chen et al., 2011). Additionally, nanoparticles typically aggregate in soil and water systems, making it crucial but challenging to discern between the characteristics of individual particles and aggregates.

17.11 CONCLUSION

Exposure to nanoparticles, assessment methods for such exposure, and the resulting environmental and human health implications are still largely unknown. Nanoparticles' unique features, which boost their application in home and industrial operations, also enhance their toxicity. When these particles are taken up by cells, they may release ROS (reactive oxygen species) that can cause cell mortality. NPs' surface characteristics such as form, dimensions, nature, surface charge, formulation medium, preservation period, agglomeration, durability, mobility, and reactivity may affect their toxicity. Cosmetics with antimicrobial NPs are the most common intentional exposure method because they are applied directly to the skin. Through inhalation, these particles can travel to the lungs and then move on to various organs. NPs found in food and medication are directly ingested, enter the digestive tract, and interact with lymphatic cell tissues.

In order to characterize the NPs and comprehend the negative effects, it is also necessary to integrate various fields, including physics, biology, geology, mathematics, chemistry, and engineering, among others. In order to draw solid findings with scientific backing, toxicity of naturally occurring NPs to a wide range of organisms needs to be assessed. There needs to be more clarity on the connections between the impacts of NPs' exposure on human and environmental health. To better understand the ingress, mobility, and modifications of NPs under different conditions and media, as well as to find safety measures against the hazards caused by NPs, new methodologies have recently been introduced and implemented for environmental monitoring. Additional facets of monitoring include physical characterization; monitoring of source; release; translocation, consequences; filter frameworks that can identify, deactivate, and eliminate pollutants. The toxicity of engineered nanoparticles depends on environmental factors. Controlling engineering equipment and the workplace will help reduce exposure to NPs. Using personal safety gear and getting regular check-ups can help reduce exposure risks. While there is some consensus that NPs can be harmful, the body of evidence is limited. Based on the available literature, no firm conclusions can be reached. To investigate the toxicity of all types of NPs, standard techniques must be developed.

REFERENCES

Adam, N., Schmitt, C., & Galceran, J. (2014). The chronic toxicity of ZnO nanoparticles and ZnCl$_2$ to *Daphnia magna* and the use of different methods to assess nanoparticle aggregation and dissolution. *Nanotoxicology*. 8, 709–717.

Agri, U., Chaudhary, P., & Sharma, A. (2021). In vitro compatibility evaluation of agriusable nanochitosan on beneficial plant growth-promoting rhizobacteria and maize plant. *Natl Acad Sci Lett*. 44, 555–559.

Agri, U., Chaudhary, P., Sharma, A., & Kukreti, B. (2022). Physiological response of maize plants and its rhizospheric microbiome under the influence of potential bioinoculants and nanochitosan. *Plant Soil*. 474, 451–468.

Ahmadi, F.I. (2012). Impact of different levels of silver nanoparticles on performance oxidative enzyme and blood parameter in broiler chicks. *Pak Vet J*. 32, 325–328.

Akhtar, M.J., Ahamed, M., Kumar, S., Siddiqui, H., Patil, G., Ashquin, M., & Ahmad, I. (2010). Nanotoxicity of pure silica mediated through oxidant generation rather than glutathione depletion in human lung epithelial cells. *Toxicology*. 276, 95–102.

Albanese, A., & Chan, W.C.W. (2011). Effect of gold nanoparticle aggregation on cell uptake and toxicity. *ACS Nanotechnol*. 5, 5478–5489.

Amro, N., Kotra, L., Wadu-Mesthrige, K., Bulychev, A., Mobashery, S., & Liu, G. (2000). High-resolution atomic force microscopy studies of the *Escherichia coli* outer membrane: structural basis for permeability. *Langmuir*. 16, 2789–2796.

Anselmo, A.C., Gupta, V., Zern, B.J., Pan, D., Zakrewsky, M., Muzykantov, V., & Mitragotri, S. (2013). Delivering nanoparticles to lungs while avoiding liver and spleen through adsorption on red blood cells. *ACS Nanotechnol*. 7(12), 11129–11137.

Arvidsson, R., Molander, S., Sandén, B.A., & Hassellöv, M. (2011). Challenges in exposure modeling of nanoparticles in aquatic environments. *Human Ecol Risk Assess*. 17(1), 245–262.

Asharani PV, Lian Wu Y, Gong Z, & Valiyaveettil S. (2008). Toxicity of silver nanoparticles in zebrafish models. *Nanotechnology*, 19, 255102.

Bakshi, M., Singh, H.B., & Abhilash, P.C. (2014). The unseen impact of nanoparticles: more or less? *Curr Sci*. 106, 1–3.

Batista, D., Pascoal, C., & Cássio, F. (2017). How do physicochemical properties influence the toxicity of silver nanoparticles on freshwater decomposers of plant litter in streams? *Ecotoxicol Environ Safe*. 140, 148–155.

Baun, P., Sayre, P., Steinhäuser, K.G., & Rose, J. (2017). Regulatory relevant and reliable methods and data for determining the environmental fate of manufactured nanomaterials, *Nano Impact*. 8, 1–10.

Bello, D., Wardle, B.L., & Yamamoto, N. (2008). Exposure to nanoscale particles and fibers during machining of hybrid advanced composites containing carbon nanotubes. *J Nanoparticle Res*. 11, 231–249.

Besseling, E., Wang, B., Lürling, M., & Koelmans, A.A. (2014). Nanoplastic affects growth of *S. obliquus* and reproduction of *D. magna*. *J Sci Technol*. 48 (20), 12336–12343.

Bhatt, I., & Tripathi, B.N. (2011). Interaction of engineered nanoparticles with various components of the environment and possible strategies for their risk assessment. *Chemosphere*. 82, 308–317.

Bhattacharya, P., Lin, S., Turner, J.P., & Ke, P.C. (2010). Physical adsorption of charged plastic nanoparticles affects algal photosynthesis. *J Phys Chem*. 114(39), 16556–16561.

Booth, A.M., Hansen, B.H., Frenzel, M., Johnsen, H., & Altin, D. (2016). Uptake and toxicity of methylmethacrylate-based nanoplastic particles in aquatic organisms. *Environ Toxicol Chem*. 35(7), 1641–1649.

Bozich, J.S., Lohse, S.E., Torelli, & M.D. (2014). Surface chemistry, charge and ligand type impact the toxicity of gold nanoparticles to *Daphnia magna*. *Environ Sci Nanotechnol*. 1, 260.

Brouwer, D. (2010). Exposure to manufactured nanoparticles in different workplaces. *Toxicology*. 269, 120–127.

Brown, D.M., Wilson, M.R., MacNee, W., Stone, V., & Donaldson, K. (2001). Size-dependent proinflammatory effects of ultrafine polystyrene particles: a role for surface area and oxidative stress in the enhanced activity of ultrafines. *Toxicol App Pharm*. 175(3), 191–199.

Buzea, C., Pacheco, I., & Robbie, K. (2007). Nanomaterials and nanoparticles: sources and toxicity. *Biointerphases*. 2, 17–71.

Carlson, C., Hussain, S.M., Schrand, A.M., Braydich-Stolle, L.K., Hess, K.L., Jones, R.L., & Schlager, J.J. (2008). Unique cellular interaction of silver nanoparticles: size-dependent generation of reactive oxygen species. *J Phys Chem B*. 112, 13608–13619.

Chambers, E., & Mitragotri, S. (2004). Prolonged circulation of large polymeric nanoparticles by non covalent adsorption on erythrocytes. *J Control Release*. 100(1), 111–119.

Champion, J.A., & Mitragotri, S. (2006). Role of target geometry in phagocytosis. *Proc Natl Acad Sci U S A*. 103(13), 4930–4934.

Chang, Y.N., Zhang, M., Xia, L., Zhang, J., & Xing, G. (2012). The toxic effects and mechanisms of CuO and ZnO nanoparticles. *Nature*. 5, 2850–2871.

Chaudhary, P., Chaudhary, A., Bhatt, P., Kumar, G., Khatoon, H., Rani, A., Kumar, S., & Sharma, A. (2022). Assessment of soil health indicators under the influence of nanocompounds and *Bacillus* spp. in field condition. *Front Environ Sci*. 9, 769871.

Chaudhary, P., Chaudhary, A., Parveen, H., Rani, A., Kumar, G., Kumar, A., & Sharma, A. (2021e). Impact of nanophos in agriculture to improve functional bacterial community and crop productivity. *BMC Plant Biol*. 21, 519.

Chaudhary, P., Khati, P., Chaudhary, A., Gangola, S., Kumar, R., & Sharma, A. (2021a). Bioinoculation using indigenous *Bacillus* spp. improves growth and yield of *Zea mays* under the influence of nanozeolite. *3 Biotech*. 11, 11.

Chaudhary, P., Khati, P., Chaudhary, A., Maithani, D., Kumar, G., & Sharma, A. (2021d). Cultivable and metagenomic approach to study the combined impact of nanogypsum and *Pseudomonas taiwanensis* on maize plant health and its rhizospheric microbiome. *PLoS One*. 16, e0250574.

Chaudhary, P., Khati, P., Gangola, S., Kumar, A., Kumar, R., & Sharma, A. (2021c). Impact of nanochitosan and *Bacillus* spp. on health, productivity and defence response in *Zea mays* under field condition. *3 Biotech*. 11, 237.

Chaudhary, P., & Sharma, A. (2019). Response of nanogypsum on the performance of plant growth promotory bacteria recovered from nanocompound infested agriculture field. *Environ Ecol*. 37, 363–372.

Chaudhary, P., Sharma, A., Chaudhary, A., Khati, P., Gangola, S., & Maithani, D. (2021b). Illumina based high throughput analysis of microbial diversity of rhizospheric soil of maize infested with nanocompounds and *Bacillus* sp. *Appl Soil Ecol*. 159, 103836.

Chen, H., Chien, C., Petibois, C., Wang, C., Chu, Y.S., Lai, S.F., Hua, T.E., Chen, Y.Y., Cai, X., Kempson, I.M., Hwu, Y., & Margaritondo, G (2011). Quantitative analysis of nanoparticle internalization in mammalian cells by high resolution X-ray microscopy. *J Biotechnol*. 9(14), 1–15.

Chen, H.H.C., Yu, C., Ueng T.H., Chen, S., Huang, K.J., & Chiang, L.Y. (1998). Acute and subacute toxicity Study of water-soluble poly alkyl sulfonated C 60 in rats. *Toxicol Pathol*. 26(1), 143–151.

Chi, Z., Liu, R., Zhao, L., Qin, P., Pan, X., & Sun, F. (2009). A new strategy to probe the genotoxicity of silver nanoparticles combined with cetylpyridine bromide. *Spectrochim Acta A Mol Biomol Spectrosc*, 72, 577–581.

Chithrani, B.D., & Chan, W.C.W. (2007). Elucidating the mechanism of cellular uptake and removal of proteincoated gold nanoparticles of different sizes and shapes. *Nano Lett*. 7, 1542–50.

Choi, O., Deng, K., Kim, N., Ross, L., Surampalli, R., & Hu, Z. (2008). The inhibitory effects of silver nanoparticles, silver ions, and silver chloride colloids on microbial growth. *Water Res*. 42, 3066–3074.

Cole, M., & Galloway, T.S. (2015). Ingestion of nanoplastics and microplastics by pacific oyster larvae. *Environ Sci Technol*. 49(24), 14625–14632.

Conie, A.L., Rearick, D.C., Xenopoulos, M.A., & Frost, P.C. (2017). Variable silver nanoparticle toxicity to *Daphnia* in boreal lakes, *Aqu Toxicol*. 192, 1–6.

Curwin, B., & Bertke, S. (2011). Exposure characterization of metal oxide nanoparticles in the workplace. *J Occup Environ Hyg*. 8, 580–587.

Dahle, J.T., & Arai, Y. (2014). Effects of Ce (III) and CeO_2 nanoparticles on soil-denitrification kinetics. *Arch Environ Cont Toxicol*. 67, 474–482.

Danielle Mcshan, P.C. (2014). Molecular toxicity of mechanism of nanosilver. *J Food Drug Anal*. 22, 116–127.

Dawson, K.A., Salvati, A., & Lynch, I. (2009). Nanotoxicology: nanoparticles reconstruct lipids. *Nat Nanotechnol*. 4, 84–85.

Debia, M., Beaudry C., & Weichenthal S. (2013). Characterization and control of occupational exposure to nanoparticles and ultrafine particles.

Demir, E., Vales, G., Kaya, B., Creus A., & Marcos, R. (2011). Genotoxic analysis of silver nanoparticles in *Drosophila. Nanotoxicology*. 5, 417–24.

Deng, Z.J., Liang, M., & Toth, I. (2013). Plasma protein binding of positively and negatively charged polymer-coated gold nanoparticles elicits different biological responses. *Nanotoxicology*. 7, 314–322.

Doherty, G.J., & McMahon, H.T. (2009). Mechanisms of endocytosis. *Annu Rev Biochem* 2009, 78, 857–902.

Dos Santos, C.A., Seckler, M.M., & Ingle, A.P. (2014). Silver nanoparticles: therapeutical uses, toxicity, and safety issues. *J Pharm Sci*. 103, 1931–1944.

Dos Santos, T., Varela, J., Lynch, I., Salvati, A., & Dawson, K.A. (2011). Effects of transport inhibitors on the cellular uptake of carboxylated polystyrene nanoparticles in different cell lines. *PLoS One*. 6 (9), e24438.

Duran, N., Duran, M., de Jesus, M.D., Seabra, A.B., Favaro, W.J., & Nakajato, G. (2015). Silver Nanoparticles: a new vive on mechanistic aspects on antimicrobial activity. *Nanomedicine*. 12, 789–799.

El Badawy, A.M., Silva, R.G., Morris, B., Scheckel K.G., Suidan M.T., & Tolaymat T.M. (2011). Surface charge-dependent toxicity of silver nanoparticles. *Environ Sci Technol*. 45(1), 283–7.

Elsaesser, A., & Howard, C.V. (2012). Toxicology of nanoparticles. *Adv Drug Delivery Rev*. 64(2), 129–137.

Etheridge, M.L., Campbell, S.A., & Erdman, A.G. (2013). The big picture on nanomedicine: the state of investigational and approved nanomedicine products. *Nanomedicine*. 9, 1–14.

Fent, K., Weisbord, C.J., Wirth-Heller, A., & Pieles, U. (2010). Assessment of uptake and toxicity of fluorescent silica nanoparticles in zebrafish (*Danio reria*) early life stages. *Aqua Toxicol*. 100, 218–228.

Fiorentino, I., Gualtieri, R., Barbato, V., Mollo, V., Braun, S., Angrisani, A., Turano, M., Furia, M., Netti, P. A., & Guarnieri, D. (2015). Energy independent uptake and release of polystyrene nanoparticles in primary mammalian cell cultures. *Exp Cell Res*. 330(2), 240–247.

Foldbjerg, R., Dang, D.A., & Autrup, H. (2011). Cytotoxicity and genotoxicity of silver nanoparticles in the human lung cancer cell line, A549. *Arch Toxicol*. 85, 743–50.

Foldvari, M., & Bagonluri, M. (2008). Carbon nanotubes as functional excipients for nanomedicines: II. Drug delivery and biocompatibility issues. *Nanomedicine*. 4, 183–200.

Forte, M., Iachetta, G., Tussellino, M., Carotenuto, R., Prisco, M., De Falco, M., Laforgia, V., & Valiante, S. (2016). Polystyrene nanoparticles internalization in human gastric adenocarcinoma cells. *Toxicol Vitro*. 31, 126–136.

Fuchs, A.K., Syrovets, T., Haas, K. A., Loos, C., Musyanovych, A., Mailänder, V., Landfester, K., & Simmet, T. (2016). Carboxyl- and amino-functionalized polystyrene nanoparticles differentially affect the polarization profile of M1 and M2 macrophage subsets. *Biomaterials*. 85, 78–87.

Gaiser, B.K., Fernandes, T.F., & Jepson, M.A. (2012). Interspecies comparisons on the uptake and toxicity of silver and cerium dioxide nanoparticles. *Environ Toxicol Chem.* 31, 144–154.

Gajewicz, A., Rasulev, B., Dinadayalane, T.C., Urbaszek, P., Puzyn, T., Leszczynska, D., & Leszczynski, J. (2012). Advancing risk assessment of engineered nanomaterials: application of computational approaches. *Adv Drug Del Rev.* 64, 663–693.

García-Gómez, C., Babin, M., & Obrador, A. (2014). Toxicity of ZnO nanoparticles, ZnO bulk, and $ZnCl_2$ on earthworms in a spiked natural soil and toxicological effects of leachates on aquatic organisms. *Arch Environ Contam Toxicol.* 67(4), 465–473.

Garner, K.L., & Keller, A.A. (2014). Emerging patterns for engineered nanomaterials in the environment: a review of fate and toxicity studies. *J Nano Res.* 16, 2503.

Geiser, M., Rothen-Rutishauser, B., Kapp, N., Schürch, S., Kreyling, W., Schulz, H., Semmler, M., Im Hof, V., Heyder, J., & Gehr, P. (2005). Ultrafine particles cross cellular membranes by nonphagocytic mechanisms in lungs and in cultured cells. *Environ Health Persp.* 113(11), 1555–1560.

Gliga, A.R., Skoglund, S., & Wallinder, I.O. (2014). Size-dependent cytotoxicity of silver nanoparticles in human lung cells: the role of cellular uptake, agglomeration and Ag release. *Par Fibre Toxicol.* 11, 11.

Goodman, C.M., McCusker, C.D., Yilmaz, T., & Rotello, V.M. (2004). Toxicity of gold nanoparticles functionalized with cationic and anionic side chains. *Bioconjugate Chem.* 15, 897–900.

Gottschalk, F., & Nowack, B. (2011). The release of engineered nanomaterials to the environment. *J Environ Mon.* 13, 1145–1155.

Gottschalk, F., Scholz, R.W., & Nowack, B. (2010). Probabilistic material flow modeling for assessing the environmental exposure to compounds: methodology and an application to engineered nano-TiO_2 particles. *Environ Model Software.* 25, 320–332.

Grafmueller, S., Manser, P., Diener, L., Diener, P.A., Maeder-Althaus, X., Maurizi, L.; Jochum, W., Krug, H.F., Buerki-Thurnherr, T., & von Mandach, U. (2015). Bidirectional transfer study of polystyrene nanoparticles across the placental barrier in an ex vivo human placental perfusion model. *Environ Health Pers.* 123 (12), 1280–1286.

Griffitt, R.J., Luo, J., & Gao, J. (2008). Effects of particle composition and species on toxicity of metallic nanomaterials in aquatic organisms. *Environ Toxicol Chem.* 27, 1972–1978.

Grigoratos, T., & Martini, G. (2014). Non-Exhaust Trace Related Emissions: Brake and Tyre Wear PM, JRC Science and Policy Reports. Joint Research Centre.

Grün, A.L., & Emmerling, C. (2018). Long-term effects of environmentally relevant concentrations of silver nanoparticles on major soil bacterial phyla of a loamy soil. *Environ Sci Eur.* 30, 1–13.

Guo, D., Zhao, Y., Zhang, Y., Zhou, H., Ge, Y., & Ma, W. (2014). The cellular uptake and cytotoxic effect of silver nanoparticles on chronic myeloid leukemia cells. *J Biomed Nanotechnol.* 10, 669–678.

Guo, D., Zhu, L., Huang, Z., Zhou, H., Ge, Y., & Ma, W. (2013). Anti-leukemia activity of PVP-coated silver nanoparticles via generation of reactive oxygen species and release of silver ions. *Biomaterials.* 34, 7884–7894.

Guo, X., & Chen, T. (2015). Progress in genotoxicity evaluation of engineered nanomaterials. In Larramendy, M.L. (ed.) *Nanomaterials-Toxicity and Risk Assessment*, 1st ed.; Londres: London, UK. doi: 10.5772/61013

Gurr, J.R., Wang, A.S., Chen, C.H., & Jan, K.Y. (2005). Ultrafine titanium dioxide particles in the absence of photoactivation can induce oxidative damage to human bronchial epithelial cells. *Toxicology.* 213(1–2), 66–73.

Gurunathan, S.P.J. (2015). Comparative assessment of the synthesised by *Bacillus lequilensis* and *calocybeindica* in MDA-MB-231 human breast cells: targeting Ps3 for anticancer therapy. *Int J Nanomed.* 10, 4203–4223.

Hassellöv, M., Readman, J.W., Ranville, J.F., & Tiede, K. (2008). Nanoparticle analysis and characterization methodologies in environmental risk assessment of engineered nanoparticles. *Ecotoxicology.* 17, 344–361.

Heal, M.R., Kumar, P., & Harrison, R.M. (2012). Particles, air quality, policy and health. *Chem Soc Rev.* 41, 6606–6630.

Hjorth, R., Skjolding, L.M., Sørensen, S.N., & Baun, A. (2017). Regulatory adequacy of aquatic ecotoxicity testing of nanomaterials. *Nano Impact.* 8, 28–37.

Hristozov, D., & Malsch, I. (2009). Hazards and risks of engineered nanoparticles for the environment and human health. *Sustainability.* 1, 1161–1194.

Hsiao, I.L., & Huang Y.-J. (2011). Effects of various physicochemical characteristics on the toxicities of ZnO and TiO nanoparticles toward human lung epithelial cells. *Sci Total Environ.* 409, 1219–1228.

Hu, W., Peng, C., Lv, M., Li, X., Zhang, Y., Chen, N., Fan, C., & Huang, Q. (2011). Protein corona-mediated mitigation of cytotoxicity of graphene oxide. *ACS Nano.* 5(5), 3693–700.

Huo, S., Jin, S., Ma, X., Xue, X., Yang, K., Kumar, A., Wang, P.C., Zhang J., Hu, Z., & Liang. X.J. (2014). Ultrasmall gold nanoparticles as carriers for nucleus-based gene therapy due to size-dependent nuclear entry. *ACS Nano.* 8(6), 5852–62.

Hussain, S.M., Javorina, A.K., Schrand, A.M., Duhart, H.M., Ali, S.F., & Schlager, J.J. (2006). The interaction of manganese nanoparticles with PC-12 cells induces dopamine depletion. *Toxicol Sci.* 92, 456–63.

Hwang, E., Lee, J., Chae, Y., Kim, Y., Kim, B., Sang, B., & Gu, M. (2008). Analysis of the toxic mode of action of silver nanoparticles using stress-specific bioluminescent bacteria. *Small.* 4,746– 750.

Jamuna, B.A., & Ravishankar, R.V. (2014). Environmental risk, human health, and toxic effects of nanoparticles. In *Nanomaterials for Environmental Protection.* Wiley-Blackwell, Hoboken, NJ, pp. 523–535.

Jiang, X., Foldbjerg, R., Miclaus, T., Wang, L., Singh, R., & Hayashi, Y. (2013). Multi-platform genotoxicity analysis of silver nanoparticles in the model cell line CHO-K1. *Toxicol Lett.* 222, 55–63.

Jung, Y., Metreveli, G., Park, C.B., Baik, S., & Schaumann, G.E. (2018). Implications of Pony Lake fulvic acid for the aggregation and dissolution of oppositely charged surface-coated silver nanoparticles and their ecotoxicological effects on *Daphnia magna. Environ Sci Technol.* 52, 436–445.

Keller, A.A., & Lazareva, A. (2014). Predicted releases of engineered nanomaterials: from global to regional to local. *Environ Sci Technol Lett.* 1, 65–70.

Khalil, I., Kogure, K., Akita, H., & Harashima, H. (2006). Uptake pathways and subsequent intracellular trafficking in nonviral gene delivery. *Pharmacol Rev.* 58 (1), 32– 45.

Khan, F.H. (2013). Chemical hazards of nanoparticles to human and environment (a review). *Oriental J Chem.* 29(4), 1399.

Khati, P., Bhatt, P., Kumar, R., & Sharma, A. (2018). Effect of nanozeolite and plant growth promoting rhizobacteria on maize. *3Biotech.* 8(141), 1–12.

Khati, P., Chaudhary, P., Gangola, S., Bhatt, P., & Sharma, A. (2017a). Nanochitosan induced growth of *Zea Mays* with soil health maintenance. *3 Biotech.* 7(81), 1–9.

Khati, P., Chaudhary, P., Gangola, S., & Sharma A. (2019a). Influence of nanozeolite on plant growth promotory bacterial isolates recovered from nanocompound infested agriculture field. *Environ Ecol.* 37 (2), 521—527

Khati, P., Sharma, A., Chaudhary, P., Singh, A K., Gangola, S., & Kumar, R. (2019b). High-throughput sequencing approach to access the impact of nanozeolite treatment on species richness and evenness of soil metagenome. *Biocatal Agric Biotechnol.* 20, 101249.

Khati, P., Sharma, A., Gangola, S., Kumar, R., Bhatt, P., & Kumar, G. (2017b). Impact of some agriusablenanocompounds on soil microbial activity: an indicator of soil health. *Clean Soil Air Water*. 45 (5), 1–7.

Kim, J., Kuk, E., Yu, K, Kim, J., Park, S., Lee, H., Kim, S., Park, Y., Hwang, C., Kim, Y., Lee, Y., Jeong, D., & Cho, M. (2007). Antimicrobial effects of silver nanoparticles. *Nanomed Nanotechnol*. 3, 95–101.

Kim, Y.S., Kim, J.S., Cho, H.S., Rha, D.S., Kim, J.M., & Park, J.D. (2008). Twentyeight-day oral toxicity, genotoxicity, and gender-related tissue distribution of silver nanoparticles in Sprague-Dawley rats. *Inhalation Toxicol*. 20, 575–83.

Kittler, S., Greulich, C., & Diendorf, J. (2010). Toxicity of silver nanoparticles increases during storage because of slow dissolution under release of silver ions. *Chem Mater*. 22, 4548–4554.

Knox, A. (2005). *An Overview of Incineration and EFW Technology as Applied to the Management of Municipal Solid Waste (MSW)*. ONEIA Energy Subcommittee, University of Western Ontario.

Korani, M., Rezayat, S.M., & ArbabiBidgoli, S. (2013). Sub-chronic dermal toxicity of silver nanoparticles in guinea pig: special emphasis to heart, bone and kidney toxicities. *Iran J Pharma Res*. 12, 511–519.

Krug, H.F. & Wick, P. (2011). Nanotoxicology: an interdisciplinary challenge. *Angew Chem Int Ed*. 50(6), 1260–1278.

Kukreti, B., Sharma, A., Chaudhary, P., Agri, U., & Maithani, D. (2020). Influence of nanosilicon dioxide along with bioinoculants on *Zea mays* and its rhizospheric soil. *3 Biotech*. 10, 345.

Kumar, P., Ketzel, M., & Vardoulakis, S. (2011a). Dynamics and dispersion modelling of nanoparticles from road traffic in the urban atmospheric environment—a review. *J Aerosol Sci*. 42, 580–603.

Kumar, P., Robins, A., Vardoulakis, S., & Quincey, P. (2011b). Technical challenges in tackling regulatory concerns for urban atmospheric nanoparticles. *Particuology*. 9, 566–571.

Kumari, H., Khati, P., Gangola, S., Chaudhary, P., & Sharma, A. (2021). Performance of plant growth promotory rhizobacteria on maize and soil characteristics under the influence of TiO$_2$ nanoparticles. *Pant Res J*. 19(1), 28–39.

Kumari, S., Sharma A., Chaudhary, P., & Khati, P. (2020). Management of plant vigour and soil health using two agriusable nanocompounds and plant growth promontory rhizobacteria in fenugreek. *3 Biotech*. 10(461), 1–11.

Lee, J., Mahendra, S., Alvarez, P., Bittnar, Z., Bartos, P.J., & Němeček, J. (2009). *Nanotechnology in Construction 3: Proceedings of the NICOM3*. Springer, Prague, Czech Republic. pp. 1–435.

Lee, J., Mahendra, S., & Alvarez, P.J.J. (2010). Nanomaterials in the construction industry: a review of their applications and environmental health and safety considerations. *ACS Nano*. 4, 3580–3590.

Lee, J.H., Ahn, K., & Kim, S.M. (2012). Continuous 3-day exposure assessment of workplace manufacturing silver nanoparticles. *J Nano Res*. 14, 1134.

Levard, C., Hotze, E.M., Lowry, G.V., & Brown, G.E. (2012). Environmental transformations of silver nanoparticles: impact on stability and toxicity. *Environ Sci Technol*. 46, 6900–6914.

Li, W.C., Tse, H.F., & Fok, L. (2016). Plastic waste in the marine environment: a review of sources, occurrence and effects. *Sci Total Environ*. 566–567, 333–349.

Ling, M-P., Chio, C-P., & Chou, W.C. (2011). Assessing the potential exposure risk and control for airborne titanium dioxide and carbon black nanoparticles in the workplace. *Envirno Sci Pollut Res Int*. 18, 877–889.

Lojk, J., Repas, J., Veraniˇc, P., Bregar, V.B., & Pavlin, M. (2020). Toxicity mechanisms of selected engineered nanoparticles on human neural cells in vitro. *Toxicology*, 432, 152364.

Love, S.A., Maurer-Jones M.A., & Thompson, J.W. (2012). Assessing nanoparticle toxicity. *Annu Rev Anal Chem.* 5, 181–205.

Manke, A., Wang, L., & Rojanasakul, Y. (2013). Mechanisms of nanoparticle-induced oxidative stress and toxicity. *Biomed Res Int.* 1, 1–15.

Mattsson, K., Ekvall, M.T., Hansson, L.A., Linse, S., Malmendal, A., & Cedervall, T. (2014). Altered behavior, physiology, and metabolism in fish exposed to polystyrene nanoparticles. *Environ Sci Technol.* 49(1), 553–561.

Meesters, J.A.J., Veltman, K., Hendriks, A.J., & van de Meent, D. (2013). Environmental exposure assessment of engineered nanoparticles: environmental science pollutants research, why REACH needs adjustment. *Integr Environ Assess Manag.* 9, e15–e26.

Mendis, E., Rajapakse, N., Byun, H., & Kim, S. (2005). Investigation of jumbo squid (*Dosidicus gigas*) skin gelatin peptides for their in vitro antioxidant effects. *Life Sci.* 77, 2166–2178.

Mohd Javed, A., Maqusood, A., Hisham, A., & Salman, A. (2019). Toxicity mechanism of gadolinium oxide nanoparticles and gadolinium ions in human breast cancer cells. *Curr. Drug Metab*, 20, 907–917.

Moore, M.N. (2006). Do nanoparticles present ecotoxicologiocal risks for the health of the aquatic environment? *Environ Int.* 32, 967–976.

Morimoto, Y., & Kobayashi, N. (2010). Hazard assessments of manufactured nanomaterials. *J Occup Health.* 52, 325–334.

Morones, J., Elechiguerra, J., Camacho, A., Holt, K., Kouri, J., Ramirez, J., & Yacaman M. (2005). The bactericidal effect of silver nanoparticles. *Nanotechnology.* 16, 2346–2353.

Mortimer, M., Kasemets, K., & Kahru, A. (2010). Toxicity of ZnO and CuO nanoparticles to ciliated protozoa *Tetrahymena thermophila.* *Toxicology.* 269, 182–189.

Nangia, S., & Suresh kumar, R. (2012). Effects of nanoparticle charge and shape anisotropy on translocation through cell membranes. *Langmuir.* 28, 17666–17671.

Nel, A., Xia, T., Madler, L., & Li, N. (2006). Toxic potential of materials at the nanolevel. *Science.* 311, 622–627.

Nowack, B., & Bucheli, T.D. (2007). Occurrence, behavior and effects of nanoparticles in the environment. *Environ Pollu.* 150, 5–22.

Nowack, J.F., Ranville, S., Diamond, J.A., Gallego-Urrea, C., Metcalfe, J., Rose, N., Horne, A.A., Koelmans, S.J., & Klaine (2012). Potential scenarios for nanomaterial release and subsequent alteration in the environment, *Environ Toxicol Chem.* 31(1), 50–59.

Oberdörster, G., Oberdörster, E., & Oberdörster, J. (2005). Nanotoxicology: an emerging discipline evolving from studies of ultrafine particles. *Environ Health Pers.* 113, 823–839.

Oberdörster, G., Sharp, Z., Atudorei, V., Elder, A., Gelein, R., Kreyling, W., & Cox, C. (2004). Translocation of inhaled ultrafine particles to the brain. *Inhalat Toxicol.* 16(6–7), 437–445.

Oh, E., Liu, R., Nel, A., Gemill, K.B., Bilal, M., Cohen, Y., & Medintz, I.L. (2016). Meta-analysis of cellular toxicity for cadmium-containing quantum dots. *Nat Nanotechnol.* 11(5), 479–486.

Osborne, O.J., Johnston, B.D., Moger, J., Balousha, M., Lead, J.R., & Kudoh, T. (2013). Effects of particle size and coating on nanoscale Ag and TiO_2 exposure in zebrafish (*Danio rerio*) embryos. *Nanotoxicology.* 7, 1315–1324.

Pan, Y., Neuss, S., Leifert, A., Fischler, M., Wen, F., Simon, U., Schmid, G., Brandau W., & Jahnen-Dechent, W. (2007). Size-dependent cytotoxicity of gold nanoparticles. *Small.* 3(11), 1941–1949.

Paranavithana, S., & Mohajerani, A. (2006). Effects of recycled concrete aggregates on properties of asphalt concrete. *Res Con Recycle.* 48, 1–12.

Park, H., Kim, J., Lee, J., Hahn, J., Gu, M., & Yoon, J. (2009). Silver-ion-mediated reactive oxygen species generation affecting bactericidal activity. *Water Res*. 43, 1027–1032.

Park, M.V.D.Z., Neigh, A.M., & Vermeulen, J.P. (2011). The effect of particle size on the cytotoxicity, inflammation, developmental toxicity and genotoxicity of silver nanoparticles. *Biomaterials*. 32, 9810–9817.

Peijnenburg, W.J.G.M., Baalousha, M., Chen, J., Chaudry, Q., Kammer, Von der kammer, F., Kuhlbusch, T.A.J., Lead, J., Nickel, C., Quik, J.T.K., Renker, M., Wang, Z., & Koelmans, A.A. (2015). A review of the properties and processes determining the fate of engineered nanomaterials in the aquatic environment, *Crit Rev Environ Sci Technol*. 45, 2084–2134.

Peng, L., He, M., Chen, B., Wu, Q., Zhang, Z., Pang, D., Zhu, Y., & Hu, B. (2013). Cellular uptake, elimination and toxicity of CdSe/ZnS quantum dots in HepG2 cells. *Biomaterials*. 34(37), 9545–9558.

Pichardo, S., Gutiérrez-Praena, D., & Puerto, M. (2012). Oxidative stress responses to carboxylic acid functionalized single wall carbon nanotubes on the human intestinal cell line Caco-2. *Toxicol Vitro*. 26, 672–677.

Pipal, A.S., Taneja, A., & Jaiswar, G. (2014). Risk assessment and toxic effects of exposure to nanoparticles associated with natural and anthropogenic sources. In Gupta Bhowon, M., Jhaumeer-Laulloo, S., Li Kam Wah, H., & Ramasami, P. (eds.) *Chemistry: The Key to Our Sustainable Future*. Springer, Netherlands, Dordrecht, pp. 93–103.

Pitt, J.A., Kozal, J.S., Jayasundara, N., Massarsky, A., Trevisan, R., Geitner, N., Wiesner, M., Levin, E.D., & Di Giulio, R.T. (2018). Uptake, tissue distribution, and toxicity of polystyrene nanoparticles in developing zebrafish (*Danio rerio*). *Aqua Toxicol*. 194, 185–194.

Popov, A.P., Priezzhev, A.V., Lademann, J., & Myllylä, R. (2005). TiO$_2$ nanoparticles as an effective UV-B radiation skin-protective compound in sunscreens. *J Phys D Appl Phys*. 38, 2564–2570.

Pradas del Real, E., Vidal, V., Carrière, M., Castillo-Michel, H., Levard, C.M., Chaurand, P., & Sarret, G. (2017). Silver nanoparticles and wheat roots: a complex interplay. *Environ Sci Technol*. 51, 5774–5782.

Praetorius A., Scheringer M., & Hungerbühler, K. (2012). Development of environmental fate models for engineered nanoparticles—a case study of TiO$_2$ nanoparticles in the Rhine River. *Environ Sci Technol*. 46, 6705–6713.

Quik, J.T.K., Velzeboer, I., & Wouterse, M. (2014). Heteroaggregation and sedimentation rates for nanomaterials in natural waters. *Water Res*. 48, 69–279.

Quik, J.T.K., Vonk, J.A., & Hansen S.F. (2011). How to assess exposure of aquatic organisms to manufactured nanoparticles? *Environ Int*. 37, 1068–1077.

Raffi, M., Hussain, F., Bhatti, T., Akhter, J., Hameed, A., & Hasan, M. (2008). Antibacterial characterization of silver nanoparticles against *E. coli* ATCC-15224. *J Mater Sci Technol*. 24, 192–196.

Rasmussen, K., González, M., Kearns, P., Sintes, J.R., Rossi, F., & Sayre, P. (2016). Review of achievements of the OECD Working Party on manufactured nanomaterials' testing and assessment programme. From exploratory testing to test guidelines. *Reg Toxicol Pharmacol*. 74, 147–160.

Rauscher, H., Rasmussen, K., & Sokull-Klüttgen, B. (2017). Regulatory aspects of nanomaterials in the EU. *Chem. Ing. Tech*. 89, 224–231.

Roller, M. (2009). Carcinogenicity of inhaled nanoparticles. *Inhalation Toxicol*. 21, 144–157.

Rosenman, K.D., Moss, A., & Argyria, K.S. (1979). Clinical implications of exposure to silver nitrate and silver oxide. *J Occup Med*. 21, 430–435.

Rothen-Rutishauser, B.M., Schürch, S., Haenni, B., Kapp, N., & Gehr, P. (2006). Interaction of fine particles and nanoparticles with red blood cells visualized with advanced microscopic techniques. *Environ Sci Technol*. 40 (14), 4353–4359.

Sánchez, A., Recillas, S., & Font, X. (2011). Ecotoxicity of, and remediation with, engineered inorganic nanoparticles in the environment. *Trends Anal Chem*. 30, 507–516.

Sarma, S.J., Bhattacharya, I., & Brar, S.K. (2015). Carbon nanotube bioaccumulation and Recent advances in environmental monitoring. *Crit Rev Environ Sci Technol*. 45(9), 905–938.

Scherer, M.D., Sposito, J.C.V., Falco, W.F., Grisolia, A.B., Andrade, L.H.C., Lima S.M.A.R.L., Machado, G., Nascimento, V.A., Gonçalves, D.A., Wender, H., Oliveira, & Caires, S.L. (2019). Cytotoxic and genotoxic effects of silver nanoparticles on meristematic cells of *Allium cepa* roots: a close analysis of particle size dependence. *Sci Total Environ*. 660, 459–467.

Schrand, A.M., Rahman, M.F., Hussain, S.M., Schlager, J.J., Smith, D.A., & Syed, A.F. (2010). Metal-based nanoparticles and their toxicity assessment. *Nanomed Nanobiotechnol*. 2:544–568.

Sharifi, S., Behzadi, S., & Laurent, S. (2012). Toxicity of nanomaterials. *Chem Soc Rev*. 41, 2323–2343.

Shrivastava, M., Srivastav, A., Gandhi, S., Rao, S., Roychoudhury, A., Kumar, A., Singhal, R.K., Jha, S.K., & Singh, S.D. (2019). Monitoring of engineered nanoparticles in soil-plant system: a review. *Environ Nanotechnol Monit Manag*. 11, 100218.

Skjolding, L.M., Sørensen, S.N., Hartmann, N.B., Hjorth, R., Hansen, S.F., & Baun. A. (2016). Aquatic ecotoxicity testing of nanoparticles-the quest to disclose nanoparticle effects. *Chem Int*. 55, 15224–15239.

Soenen, S.J., Parak, W.J., Rejman, J., & Manshian, B. (2015). (Intra) cellular stability of inorganic nanoparticles: effects on cytotoxicity, particle functionality, and biomedical applications. *Chem Rev*. 115(5), 2109–2135.

Sondi, I., & Salopek-Sondi, B. (2004). Silver nanoparticles as antimicrobial agent: a case study on *E. coli* as a model for Gram-negative bacteria. *J Colloid Inter Sci*. 275, 177–182.

Sriram, M.I.K.K. (2012). Size based cytotoxicity of silver nanoparticles in bovine retinal endothelial cells. *Nano Sci Meth*. 1, 1, 56–77.

Stark, W.J. (2011). Nanoparticles in biological systems. *Angew Chem Int Ed Engl*. 50, 1242–1258.

Stoehr, L.C., Gonzalez, E., & Stampfl, A. (2011). Shape matters: effects of silver nanospheres and wires on human alveolar epithelial cells. *Particle Fibre Toxicol*. 8, 36.

Sukhanova, A., Bozrova, S., Sokolov, P., Berestovoy, M., Karaulov, A., & Nabiev, I. (2018). Dependence of nanoparticle toxicity on their physical and chemical properties. *Nanoscale Res Lett*. 13(1), 1–21.

Sun, X., Chen, B., Li, Q., Liu, N., Xia, B., Zhu, L., & Qu, K. (2018). Toxicities of polystyrene nano- and microplastics toward marine bacterium *Halomonas alkaliphila*. *Sci Total Environ*. 642, 1378–1385.

Tallec, K., Huvet, A., Di Poi, C., González-Fernández, C., Lambert, C., Petton, B., Le Goïc, N., Berchel, M., Soudant, P., & Paul-Pont, I. (2018). Nanoplastics impaired oyster free living stages, gametes and embryos. *Environ Poll*. 242(Pt B), 1226–1235.

Tang, J., Xiong, L., Wang, S., Wang, J., Liu, L., & Li, J. (2009). Distribution, translocation and accumulation of silver nanoparticles in rats. *J Nanosci Nanotechnol*. 9:4924–4932.

Tarantola, M., Pietuch, A., & Schneider, D. (2011). Toxicity of gold nanoparticles: synergistic effects of shape and surface functionalization on micromotility of epithelial cells. *Nanotoxicology*. 5(2), 254–268.

Thomas, R.S., Allen, B.C., Nong, A., Yang, L., Bermudez, E., Clewell, H.J., & Andersen, M.E. (2007). A method to integrate benchmark dose estimates with genomic data to assess the functional effects of chemical exposure. *Toxicol Sci*. 98, 240–248.

Tiwari, A.J., & Marr, L.C. (2010). The role of atmospheric transformations in determining environmental impacts of carbonaceous nanoparticles. *J Environ Quality*. 39, 1883–1895.

Tjälve, H., & Henriksson, J. (1999). Uptake of metals in the brain via olfactory pathways. *Neurotoxicology*. 20, 181–195.

Trickler, W.J., Lantz, S.M., Murdock, R.C., Schrand, A.M., Robinson, B.L., & Newport, G.D. (2010). Silver nanoparticle induced bloodbrain barrier inflammation and increased permeability in primary rat brain microvessel endothelial cells. *Toxicol Sci.* 118, 160–70.

Tsai, CS-J., White, D., & Rodriguez, H. (2012). Exposure assessment and engineering control strategies for airborne nanoparticles: an application to emissions from nanocomposite compounding processes. *J Nano Res.* 14, 989.

Tsuji, J. S., Maynard, A.D., & Howard, P.C. (2006). Research strategies for safety evaluation of nanomaterials, part IV: risk assessment of nanoparticles. *Toxicol Sci.* 89, 42–50.

Van Pomeren, M., Brun, N.R., Peijnenburg, W.J.G.M., & Vijver, M.G. (2017). Exploring uptake and biodistribution of polystyrene nanoparticles in zebrafish embryos at different developmental stages. *Aqua Toxicol.* 190, 40–45.

Wang, T., Bai, J., Jiang, X., & Nienhaus, G.U. (2012). Cellular uptake of nanoparticles by membrane penetration: a study combining confocal microscopy with FTIR spectroelectrochemistry. *ACS Nano.* 6 (2), 1251–1259.

Ward, J.E. & Kach, D.J. (2009). Marine aggregates facilitate ingestion of nanoparticles by suspension-feeding bivalves. *Marine Environ Res.* 68 (3), 137– 142.

Wright, S.L., & Kelly, F.J. (2017). Plastic and human health: a micro issue? *Environ Sci Technol.* 51 (12), 6634–6647.

Wu, Z., Hu, M., & Lin, P. (2008). Particle number size distribution in the urban atmosphere of Beijing, China. *Atmos Environ.* 42, 7967–7980.

Xia, T., Li, N., & Nel, A. E. (2009). Potential health impact of nanoparticles. *Annu Rev Pub Health.* 30(1), 137–150.

Yaka, M., Ehirchiou, A., Alkandry, T.T.S., & Sair, K. (2015). Huge plastic bezoar: a rare cause of gastrointestinal obstruction. *Pan Afr Med J.* 21 (286).

Yang, L., & Watts, D.J. (2005). Particle surface characteristics may play an important role in phytotoxicity of alumina nanoparticles. *Toxicol Lett.* 158, 122–132.

Yang, W., Shen, C., Ji, Q., An, H., Wang, J., & Liu, Q. (2009). Food storage material silver nanoparticles interfere with DNA replication fidelity and bind with DNA. *Nanotechnology.* 20, 085102.

Yue, Y., Behra, R., & Sigg, L. (2014). Toxicity of silver nanoparticles to a fish gill cell line: role of medium composition. *Nanotoxicology.* 9(1), 54–63.

Zhang, S., Gao, H., & Bao, G. (2015). Physical principles of nanoparticle cellular endocytosis. *ACS Nano.* 9(9), 8655–8671.

Zhang, Y., Kohler, N., & Zhang, M. (2002). Surface modification of superparamagnetic magnetite nanoparticles and their intracellular uptake. *Biomaterials.* 23(7), 1553–1561.

Zhu, Y., Zhao, Q., Li, Y., Cai, X., & Li, W. (2006). The interaction and toxicity of multi-walled carbon nanotubes with *Stylonychiamytilus*. *J Nanosci Nanotechnol.* 6, 1357–1364.

Zhu, Z.J., Carboni, R., & Quercio, M.J. (2010c). Surface properties dictate uptake, distribution, excretion, and toxicity of nanoparticles in fish. *Small.* 6, 2261–2265.

Ziello, J., & Huang, Y. (2010). Cellular endocytosis and gene delivery. *Mol Med.* 16(5–6), 1.

18 Commercially Important Nanocompounds Available in the Market for Agriculture Applications

Jyotsana Maura
DRDO

Priyanka Khati and Swati Lohani
ICAR-Vivekananda Parvatiya Krishi Anusandhan Sansthan

CONTENTS

18.1 INTRODUCTION

Since it provides food and raw materials to other sectors, agriculture is the most valuable and dependable sector of the economy. It is regarded as the backbone of the economy in the majority of emerging nations since it is a crucial component of progress and development. Scientists and engineers are currently working with novel strategies to increase agricultural production as a result of the world's expanding population and the ensuing need for more food. It will take 2–4 billion tons of grains, or a 50% increase in yearly production, to meet the needs of the 7.9 billion people who will be on the planet by 2050. To achieve the productivity goal several ways can be followed, e.g. innovative farming (Laing et al., 2017; Tyagi et al., 2022),

DOI: 10.1201/9781003345565-18

cyclic bio economies using ag-technologies, precision farming, and creative nano-interventions for sustainable and ecologically friendly farming (Lokko et al., 2018; Wreford et al., 2019). The agricultural revolution is a necessary occurrence if we are to end poverty and hunger in today's society. In this instance, reputable lenders can be found in remote areas where economic growth hasn't been as successful and where people are living in poverty. As a result, agriculture must develop dramatically. All things considered, the Green Revolution and current farming techniques dramatically improved the availability of food worldwide. As a result of the unanticipated and negative effects on the environment and ecosystem services, more sustainable agriculture practices are needed.

It is well recognized that improper use of pesticides and fertilizers in both underground and surface rivers causes an increase in nutrients and pollutants, increased expenditures for health care and water purification, and decreased opportunities for fishing and leisure activities. Furthermore, excessive use of synthetic fertilizers has had detrimental effects on humans as well as eutrophication, water contamination, greenhouse gas emissions, and nutrient leaching (Chanakya et al., 2012, 2013; Li et al., 2007). In this context, organic nutritive sources from inedible and underutilized waste sources with overall nutrient composition may be viable options as resilient nutrient sources (Mahapatra et al., 2022). The best approaches to solving the issues may use nanotechnology. Due to its many applications, nanotechnology has recently attracted a lot of attention. In addition, there will be a host of additional advantages in the future, such as enhanced quality of food and protection, decrease in agrarian intake, enrichment through the absorption of nanoscale nutrients from the soil, etc. (Chaudhary et al., 2022a). The United States Department of Agriculture (USDA) created the first "roadmap" for integrating nanotechnology into agriculture and food in December 2002. At Cornell University (New York, USA), a sizeable gathering of policymakers, officials from land grant universities, and scientists from private corporations gathered to discuss how they saw the future of agriculture employing nanoscale technologies.

Nanopesticides and nanofertilizers, for instance, can help produce results while also defending against a few insect infestations and microbiological problems without decontaminating soils and water. Thus, nutrient-rich organic waste from various bioremediation processes/environmental pollutants can be nanoscale and used as nano-biofertilizers to promote global agroecology (Bhatt et al., 2022). The next generation of fertilizers will include secondary nutrients, biostimulants such as plant growth promoters, and algae generated from wastewater with appropriate NPK (Haddad et al., 2017; Mahapatra & Murthy, 2021).

Agrochemical formulations for application of biopesticides and fertilizers for crop production, use of nanosensors in crops with disease identification, and nanodevices for crop genetic modification and management post-harvest, among others, are some of the ways nanotechnology can benefit agriculture (Chaud et al., 2021). Although there is a lot of potential for nanotechnology in agriculture, there are still several difficulties that need to be resolved, including how to scale up manufacturing processes while reducing prices and how to handle risk assessment issues. Of particular interest in this respect is that nanoparticles made from biopolymer substances such as carbohydrates and proteins have little impact on the environment and human health.

18.2   NANOCOMPOUNDS AS BIOFERTILIZERS AND BIOPESTICIDES

The use of synthetic fertilizer has gradually given way to a variety of bio-based fertilizing techniques. Because biofertilizers have high moisture content and are expensive to transport over long distances, it is necessary to produce them in large quantities domestically (Chittora et al., 2020). These organic fertilizers may come from microbial sources, plant or animal sources, or both (Lee et al., 2018; Chaudhary et al., 2022b). According to Bhardwaj et al. (2014), these microbial organisms perform a wide range of tasks including soil conditioning, nutrient replenishment, and improvement. As a result, the crops receive the best nutrients possible for increased yields and productivity. They are often administered to the soil, roots, and seeds as living or latent cells for absorption (Figure 18.1).

By improving the soil system and applying fertilizer in a novel way that uses nanoparticles, plant development characteristics have been promoted (Chaudhary et al., 2021a, b). As the main determining factor, slow nutrient release mechanism increases fertilizer efficiency. The use of nanoparticles also accelerates the activities of plant roots. The use of nanofertilizers has improved the physiological properties of agricultural crops in every imaginable way, from synthesis to metabolism. The slow and continuous release of fertilizers can greatly increase the nutrient availability to the assimilation apparatus of the crop system, limiting nutrient loss

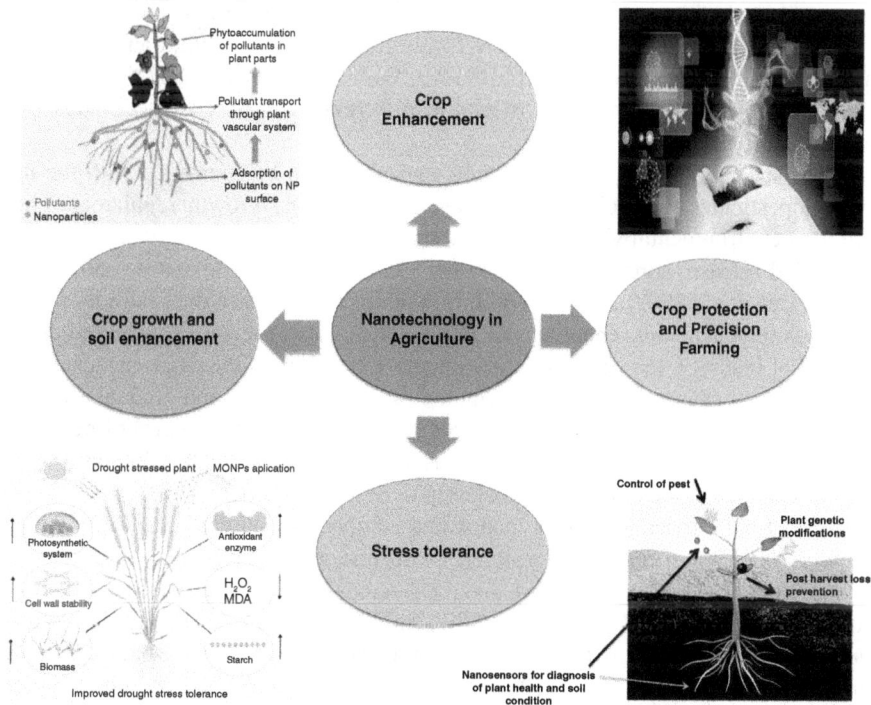

FIGURE 18.1   Application of nanotechnology in agriculture.

by leaching. Nanofortification using nanolayers on top of commonly used fertilizers can be used for this purpose. These coats also avert unwanted nutrient-microbe interactions and ensure its exterior is shielded. Studies have fully demonstrated the benefits of combining non-materials with biofertilizers such as *Bacillus* spp. and *Pseudomonas* spp., increasing the growth and development of cultures when used in vitro (Karunakaran & Behera, 2016; Agri et al., 2021). This has restored the usefulness of nanoforms as potential biofertilizers. The application of chitosan nanonutrients as NPK fertilizers has led to a significant improvement in the nutritional composition of wheat, including its high sugar content (Rashed et al., 2016).

With the new notion of precision farming that uses techniques such as delayed release, quick release, particular release, moisture release, heat release, pH release, ultrasonic release, etc., it made effective practices to enhance the agriculture system. Entire farming system has been covered in pesticide use to maximize output potential. Chemicals known as pesticides are widely employed to get rid of and control dangerous organisms that hinder agriculture production economically (Nicolopoulou-Stamati et al., 2016). Once more, these harmful substances are to blame for a wide range of health problems, including neurological impacts, respiratory conditions, cancer, Parkinson's disease, foetal ailments, infertility, and genetic disorders (Mohan et al., 2015). In light of the growing global demand for pesticides to reduce the effects of pathogens and pests, action should be taken to limit the overuse of agrochemicals by identifying suitable substitutes. The widespread use of conventional pesticides raises serious concerns about bioaccumulation, which results from the biomagnification of constant organic contaminants and the evolution of pest-target defiance (Awwad & Ibrahim, 2015). As a result, there is already a trend to reduce the overuse of pesticides. We can simply get over these restrictions by using nanotechnology to create different nano-based insecticides (An et al., 2014).

Pests such as rodents, insects, weeds, and mites can be controlled with the use of nanopesticides, fungicides, herbicides, miticides, and growth regulators. They are also environmentally benign since they enhance formulation qualities such as controlled release, adhesion, permeability, chemical solidity, and water dispersion (Vermeiren et al., 2002). Nano-based formulations can enhance the solubility rate of poor soluble chemicals, enabling the active component to release slowly and effectively. For instance, the production of nanoparticles, such as the nano-encapsulation of pesticides or insecticides, can extend the rate of absorption and gradual effective delivery of various agrochemicals to a particular host plant for control of pests. By the use of water-soluble insecticide, porous silica loaded for a delayed-release capability, nanoparticles can be used efficiently in pest management. Insect pests have a cuticular lipid barrier, which the nanosilica component can obstruct. *Rhyzopertha dominica* (F.) and *Sitophilus oryzae* (L.), two insect pests, have been successfully controlled with nanoscale alumina, and these nanoaluminium particles have demonstrated great mortality (Gajjar et al., 2009). AgNPs, which are made from *Euphorbia prostrata* leaf extracts and have a 100% fatality rate, can also be used to suppress *S. oryzae* (L.) (Baruah & Dutta, 2009). On the other hand, a highly lethal combination of substances, such as silver-zinc mixed nanoparticles, was used to combat the damaging pest *Aphis nerii*. AgNPs shorten the cotton bollworm's life span.

In comparison to other methods, plant biomass synthesis and nanonutrients produced by microbes are straightforward and environmentally beneficial (Chaudhary, 2020). Numerous microbes and angiosperms have already been employed to synthesize nanonutrients. Metal sulphide nanonutrients including zinc sulphide, gold, and *Desulfovibrio desulfuricans* have been produced using bacteria such as *Clostridium thermoaceticum*, *D. desulfuricans*, and *Klebsiella aerogenes*. Green algae, such as *Chlorella, Chlorococcum, Scenedesmus, Cosmarium, Botryococcus*, and *Chlamydomonas*, have been utilized to synthesize a range of metallic nanoparticles (Chaudhary, 2020). Cyanobacteria, such as *Spirulina, Nostoc, Anabaena, Synechocystis, Aspergillus* spp., *Fusarium oxysporum*, and other fungus species have all been successfully used to produce metal and metal sulphide nanonutrients (Kitching et al., 2015; Panpatte et al., 2016). The ZnO-coated *Bacillus thuringiensis* NP is an efficient nanopesticide that delays the cowpea weevil's (*Callosobruchus maculatus*) larval and pupal development phase. Examples of cognitive and precise farming include the use of nanotechnology pesticides with extremely high efficacy, nanoparticle encapsulation, or nanoparticle-based DNA transfer to increase insect resistance. The current farming methods involve spraying pesticides without taking their effectiveness or persistence into account. This is why the use of pesticides with nanotechnology-based controlled release mechanisms, such as nano-encapsulation, could benefit sustainable agriculture (Chaud et al., 2021).

Two main factors make the properties of nanomaterials significantly different from other materials, namely the increase in relative surface area and quantum effects. Morphology-ratio/size, hydrophobicity, solubility-release of toxic species, surface/roughness (Gatoo et al., 2014), contamination of surface species, in synthesis/history, O_2/H_2O reactive oxygen species (ROS) generation, structure/composition, competitive receptor binding sites, and dispersion/aggregation are important properties of the nanoparticles.

18.3   ADVANCEMENT OF NANOTECHNOLOGY IN AGRICULTURE

Due to modern advancements such as nanotechnology, pest management has significantly improved along with agricultural crop production. A number of nations have also taken action in this direction to help meet the population's continued need for food and feed. The use of novel, commercially available nanocompounds that are simple to use and also meet agricultural objectives has already been made by a number of firms in the modern era. More should be done in this area since a nation such as India, which is mostly recognized for its agricultural land, needs to develop the tools and technology in order to reach the average farmers as well, so that they too may profit from this more modern technology.

Under the new nanovision, agriculture must be industrialized, standardized, and simplified into simple capacities. In the near future, the farm will be a vast biofactory that can be managed and monitored from a laptop/computer, and food will be made from specially formulated ingredients to deliver efficient calories and nutrients for the body. Agriculture's capacity to produce feedstock for industrial operations will increase as a result of nanobiotechnology. In the meantime, a new "flexible matter" nanoeconomy will make obsolete small-scale farmers who cultivate tropical agricultural items such as rubber, cocoa, coffee, and cotton by modifying the properties of industrial nanoparticles to produce less expensive, "smarter" equivalents (Manjunatha et al., 2016). In the same way that genetically modified (GM) agriculture brought about new degrees of corporate concentration throughout the food chain, proprietary nanotechnology, employed from seed to stomach, genome to gullet, would tighten agribusiness' grip on global food and farming at every stage.

Scientists have been molecularly modifying food and agriculture for two generations. The industrial food chain is further broken down and connected through agro-nano. Utilizing cutting-edge nanoscale techniques for mixing and harnessing genes, GM plants are turned into atomically modified plants. More finely made pesticides that incorporate artificial flavours and aesthetically attractive natural nutrients might kill unwanted bugs (Manjunatha et al., 2016). Visions of an automated, centralized industrial farm can now be realized using molecular sensors, molecular delivery systems, and cheap labour. However, the creation of a bioeconomy is a challenging and sophisticated process requiring the merging of numerous scientific disciplines. Nanotechnology has the ability to totally change the agriculture and food industries with new tools for the molecular treatment of diseases, fast disease detection, boosting the capacity of plants to absorb nutrients, and improve nutrient status of soil (Chaudhary et al., 2021c, d; Agri et al., 2022). Insecticides and herbicides will soon be more effective thanks to nanostructured catalysts, allowing for the use of smaller amounts. Nanotechnology will also indirectly help the environment by using alternative (renewable) energy sources, filters, or catalysts to reduce pollution and eliminate existing toxins (Bhatt et al., 2022). Controlled environment agriculture, sometimes known as "CEA," is a type of farming that is well-liked in the USA, Europe, and Japan that effectively uses modern technology for crop management (CEA) (Manjunatha et al., 2016). Plants are raised in an environment that is carefully regulated to maximize production techniques. The automated system monitors and manages the particular environments where crops are cultivated. Because of its current status, CEA technology is an excellent place to start when using nanotechnology in agriculture. Nanotechnological CEA devices with "scouting" skills could greatly improve the grower's capacity to assess the crop's viability, the ideal time to harvest it, and food security concerns such as microbial or chemical contamination. This is due to the presence of numerous monitoring and control systems.

Even while nanotechnology has not yet been fully implemented, the tiny sensors and monitoring systems that it has made feasible will have a significant impact on future precision farming approaches. One of the primary uses of nanotechnology-enabled devices will be the increased use of autonomous sensors connected to a GPS system for real-time monitoring. To monitor crop growth and soil quality, these nanosensors might be scattered across the field (Shang et al., 2019). Wireless sensors are already in use in several regions of Australia and the USA. For instance, the IT company Accenture helped install WiFi equipment at the Pickberry vineyard in Sonoma County, California. Forbes reports that Honeywell, a global technology R&D company, uses small nanosensors to monitor Minnesota's grocery stores. Store owners can identify foods that have expired with the use of this technology and get a reminder to obtain new supplies. Controlled release methods are highly valued in medicine because they can make it possible for pharmaceuticals to be absorbed more gradually, at a specific location in the body, or in reaction to an outside stimulus (Tiwari et al., 2012). Agribusiness and food firms such as Monsanto, Syngenta, and Kraft are developing and patenting these nano- and micro-formulations, which may be used in insecticides, vaccines, veterinary medicine, and nutrient-dense meals (Feistel et al., 2002). Nanocompounds have been used as an encapsulating material for the slow and controlled release of active ingredient to the targeted sites (Khati et al., 2017a, b; Kumari et al., 2020, 2021). They possess the unique property of complexation which can help in chelation of required material, for example, nano-zeolite and nanochitosan can be used in the bioformulation due to complex formation and thus entrap the nutrients and water so that it will be released slowly to the bioagent and hence help in its survival for a longer period of time (Chaudhary & Sharma, 2019; Khati et al., 2018, 2019a, b).

Value addition has been highlighted as the primary driver of increasing rural incomes at every stage along the agricultural production and consumption cycle (Anon, 2008). From the standpoint of food security, it is essential that the application of nanotechnology be expanded across all connections of the value chain of agriculture rather than being restricted to farm production levels in order to increase agricultural productivities, product quality, consumer acceptance, and resource use efficiencies. This will help reduce farm costs, raise output value, increase rural incomes, and raise the level of the natural resource base that supports agricultural production systems (Sastry et al., 2007). It is essential to see nanotechnology as an enabling technology that can function in conjunction with biotechnology and traditional technologies in order to achieve this (Salerno et al., 2008). Some of the applications of nanocompounds are listed in Table 18.1. Nanoscale devices can detect and cure an infection, vitamin deficiency, or other health issue long before symptoms become apparent at the macroscale. Smart delivery systems can be used to track the effects of pharmaceuticals, nutraceutical minerals, food supplements, bioactive substances, probiotics, pesticides, insecticides, fungicides, and vaccines (Jampilek et al., 2019). Delivery systems are crucial in agriculture for applying fertilizers and pesticides as well as for genetically modified plant enhancement. Pesticide application techniques must concentrate on improving efficacy and controlling spray drift, whereas fertilizer bioavailability issues are brought on by soil chelation, over-application, and run-offs. A practical answer to these problems is the use of controlled delivery systems for the administration of fertilizers and pesticides.

The controlled delivery strategy aims to maximize biological efficacy and minimize side effects by gradually releasing the necessary and suitable amounts of agrochemicals over time. Nanoparticles (1,000 nm) have the advantage of effective loading over microparticles due to their increased surface area, ease of attachment, and fast mass transfer. Delivery mechanisms for genetic material must get past challenges including a limited host range, cell membrane transit, and trafficking to the nucleus. A "gene gun" device, for instance, uses DNA-coated gold nanoparticles as bullets to strike plant cells and tissues in order to spread genes (Kar et al., 2010).

TABLE 18.1
Applications of Nanocompounds in Agriculture

Application	Nanoparticles	Reference
A) Pesticide Delivery		
Avermectin	Porous hollow silica (15 nm)	Li et al. (2007)
Ethiprole or phenylpyrazole	Polycaprolactone (135 nm)	Boehm et al. (2003)
Gamma cyhalothrin	Solid lipid (300 nm)	Frederiksen et al. (2003)
Tebuconazole/chlorothalonil	Polyvinyl pyridine and polyvinyl pyridine-co-styrene (100 nm)	Liu et al. (2001)
Biopesticides		
Plant origin: nanosilica for insect control *Artemisia arborescens*	Nanosilica (3–5 nm)	Barik et al. (2008)
Essential oil encapsulation	Solid lipid (200–294 nm)	Lai et al. (2006)
Microorganisms: *Lagenidium giganteum* cells in emulsion	Silica (7–14 nm)	Vander Gheynst et al. (2007)
B) Fertilizer Delivery		
NPK-controlled delivery	Nanocoating of sulphur (100 nm layer) Chitosan (78 nm)	Wilson et al. (2008)
Genetic material delivery DNA	Gold (10–15 nm) Gold (5–25 nm)	Torney et al. (2007)
C) Pesticide Sensor		
Organophosphate	Zirconium oxide (50 nm)	Torney et al. (2007)

18.4   SOME COMMON NANOCOMPOUNDS USED IN AGRICULTURE

There are a few commercially available nanocompounds on the market right now because numerous firms are working to create nanocompounds for different uses (Figure 18.2; Table 18.2). Polymeric nanoparticles are employed in the agricultural industry to administer agrochemicals slowly and precisely. Polymeric nanoparticles have a number of advantages, including increased biocompatibility and a diminished impact on organisms that shouldn't be harmed. Among the polymeric nanoparticles used in agriculture are polyethylene glycol and poly(γ-glutamic acid) (Sylwia et al., 2021). Silver NPs having antibacterial properties against a variety of phytopathogens are widely exploited. Silver nanoparticles have been found to promote plant development, according to scientists (Ahmad et al., 2020). Nanoalumino-silicate formulations are a popular choice among chemical manufacturers for pesticides (Deka et al., 2021). Titanium dioxide NPs are biocompatible nanoparticles employed as a water disinfectant (Chang et al., 2017). Carbon nanoparticles including graphene, graphene oxide, carbon dots, and fullerenes are used to improve seed germination. The use of magnetic, copper oxide, and zinc oxide nanoparticles in agriculture is

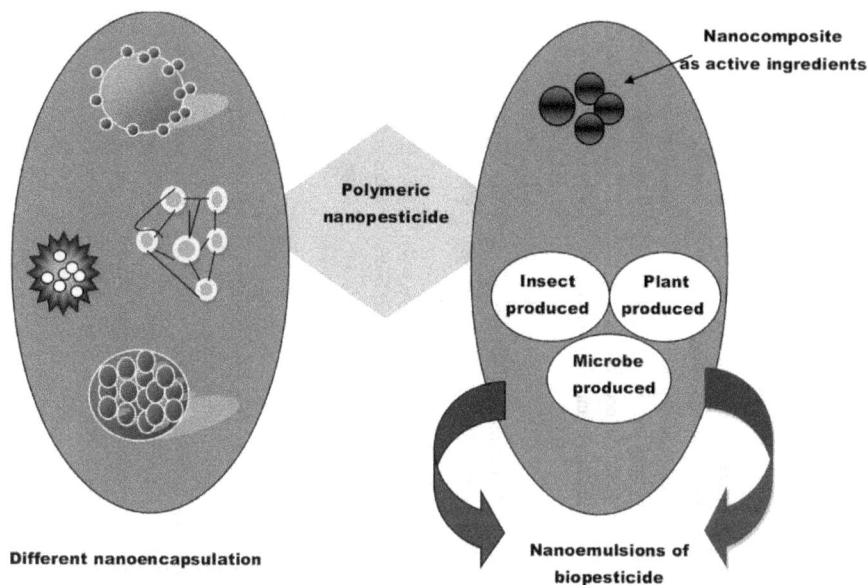

FIGURE 18.2   Nanocompounds towards sustainable agriculture.

another application for nanoparticles. Particular antibacterial nanoparticles help to stop microbial invasions. Crop productivity and soil microbial activity were also enhanced with nanophos comprising phosphate-solubilizing bacteria and nanoparticles (Chaudhary et al., 2021e). Silicon dioxide NPs also enhanced the seed germination, plant health parameters, and soil enzymes (Kukreti et al., 2020).

The typical pathogenic fungi that cause infections are *Dematophora necatrix*, *Fusarium solani*, *F. oxysporum*, and *Colletotrichum gloeosporioides*. Numerous nanoparticles, such as copper and nickel ferrite nanoparticles, are used to treat disease because they have strong antifungal capabilities. For the treatment of viral infections, such as the mosaic virus for tobacco, potatoes, and alfalfa, chitosan, zinc oxide, and silica nanoparticles are useful (Dutta et al., 2022). The quality of the seeds is one of several important elements that affect crop productivity. The abrasive seed coat of tomatoes has been observed to be penetrated by carbon nanotubes, considerably boosting germination and plant growth (Khodakovskaya et al., 2009). Similar to this, spraying maize and soybean seedlings with multiwalled carbon nanotubes increased the germination rate. The three simple steps towards the commercialization of nano-based agriproducts are mentioned through Figure 18.3.

18.5   POLICY ALTERNATIVES FOR USING NANOTECHNOLOGY TO PROMOTE SUSTAINABLE AGRICULTURAL DEVELOPMENT

    a. Creation of specialized organizations with the necessary knowledge to assess the biosafety of nanoparticles

TABLE 18.2

Commercially Available Nanocompounds in the Market

S. No.	Product Name	Company	Composition	Features	Application
1.	BEE-MICRO (Nanotechnology-based Plant Nutrition)	Nano Bee Bio Innovations	Nanozinc (1.5%), nanoiron (1.25%), nanocopper (0.25%), nanomanganese (0.5%), nanopotassium (1%), nanomagnesium (0.25%), nanoboron (0.25%), nanomolybdenum (0.05%), nanosulphur (0.25%)	Specially formulated unique combination of primary, macro- and micro-minerals in colloidal nanometre form embedded in protein hydrolysates and encapsulated by biopolymers. Fulfil the need of trace elements for the plants and improve crop health.	2 mL/L of water through foliar spraying. For NFT (Hydroponics): mix 0.75 mL/L to 1 mL/L of water. Repeat application every 15–20 days.
2.	Geolife Nano Zn Nano Technology Micro Nutrient Fertilizers	Geolife	EDTA and aminos are used to chelate a white powder formulation.	Because nano Zn is a mostly systemic formulation, unlike zinc sulphate, it is easily translocated within plants.	Foliar application 50 g/acre
3.	Krasun Nano-Gold (Flower Enhancer for cotton, vegetables, fruits, pulses, cereals, floriculture, and horticulture)	Krasun Industries, India	Nano-Gold is a micro-particle technology	It boosts the flowering stage of the plant and helps the plant to convert maximum flowers to fruits. It can be applied with all kinds of pesticides, fungicides, or nutrients. Its flower conversion helps the plant to deliver maximum yields.	5 g for 150–200 L of water at flowering stage.
4.	Katra NANO Completely water-soluble Nano Technology Fertilizer	Katra Fertilizers and Chemicals Private Limited India	Most essential secondary plant nutrients magnesium and sulphur. Magnesium sulphate 9.5%, sulphur percent by weight (12.1%), insoluble in water percent by weight 1.0%, lead (0.003%), cadmium (0.0025%), arsenic (0.01%)	Promotes chlorophyll production. Promotes the production and transfer of sugars, starches, oils, and fats. Enhances carbohydrate usage in plant. Improves nitrogen, phosphorus uptake by plant. Easy to apply with other nutrients.	Mix 20 g of powder into one pump (15 L of water) and spray at active growth stages

(Continued)

TABLE 18.2 (*Continued*)
Commercially Available Nanocompounds in the Market

S. No.	Product Name	Company	Composition	Features	Application
5.	Nano Agro Care	Kerala naturals	Nanoform of natural minerals and herbal extracts	Helps to prevent diseases like bud rot, leaf blight, stem bleeding, root wilt and coconut beetles, whitefly, moth infestation in coconut, mango, and cashew.	Foliage spraying
6.	Carbon Nano Powder, Nano Fertilizer Synergist	Nanocarbon: 90% min. Trace elements: O, S, K	Black fine powder with strong adsorption and conductivity	Encourage the development of the root system, and improve plant disease resistance. Boost crop quality, protein, fat, starch, sugar, and amino acid concentrations. Increase agricultural yields (5%–15% for field crops, 15%–30% for vegetables).	Mixing with solid fertilizers in the proportion of 3:1,000 Mixed process below 80°C
7.	Bioprime prime aavirat (growth booster with nanoparticles)	Bighaat India	Bioactive consortium with nanocomponents for active growth and overall development. It contains nutrients and active growth supplements.	Helps to increase germination, strengthens cell wall, and enhances root system in stressed conditions.	Drip and foliar application
8.	Nanomeal Flora 00-40-25+TE Nano Technology WSF+TE	Geolife Agritech India Pvt. Ltd	Utilization and mobility of phosphorus are improved by Mn, Mg, and Zn.	Helps in increasing flowering, also acts as a structural component of energy storage.	Foliar application 200–250 g/acre.
9.	LINNFIELD NANO ZNOXET	Dr. Linnfie Laboratory	Nanoparticle-based product with EDTA (12%).	Fulfilling crop nutrient requirement.	Foliar spray 0.5-1 g /L, Drip – 0.5

FIGURE 18.3   Steps in commercialization of nano-based products.

b. Creation of precise criteria based on Food Safety and Standards Authority
   (WHO standards) for the monitoring and assessment of systems utilizing
   nanoparticles
c. Proper documenting of the toxicity of nanomaterials to aquatic creatures

18.6   ECONOMIC BENEFITS POSED BY NANOCOMPOUNDS IN COMPARISON TO THEIR CONVENTIONAL COUNTERPARTS

The calculation of market predictions and prospective sales across the wide field of nanotechnology as well as for specialized green applications has come along with the introduction of a variety of green nanotechnology application areas. Early forecasts for nanotechnology usually assumed rapid development trajectories and frequently referred to the discipline as a crucial core platform technology that would significantly boost global economic growth and have a positive impact on the environment (Pokrajac et al., 2019). Despite substantial R&D investment continuing far into the second decade, the sector is still in the early phases of commercialization. The development of nanotechnology now has a more evolutionary trend than earlier predictions, and many of the early sales expectations now appear unduly optimistic.

The inclusion of total sales of items with a small contribution from nanotechnology rather than the distribution of the actual nanotechnology component of the good may have resulted in some too optimistic estimates (which indeed may be very difficult for some products that use nanomaterials and processes). There are numerous potential markets for green nanotechnologies. However, investments in research, technological development, manufacture, testing, marketing, setting standards, enacting legislation, user adoption, and monitoring must be made before these markets can be realized. The range of capital and ongoing costs, both public and private, associated with the use of green nanotechnologies must be balanced, taking into account the relative advantages of these technologies compared to traditional applications, interest rates, and the timing and distribution of various benefits and costs. The development and usage of green nanotechnology products will likely have a number of secondary effects, including effects on supply chains and spill-overs to other parties and the environment (Servin & White, 2016).

18.7   CONCLUSION

Numerous opportunities exist for nanotechnology to have an important impact on food production, agriculture, and livestock production. Applications for and benefits of nanotechnology are numerous. It is desirable to boost production through nanotechnology-enabled precision farming and to maximize output while reducing inputs through better monitoring and targeted action. Thanks to nanotechnology, plants can use water, pesticides, and fertilizers more efficiently. The development of novel products for food processing, preservation, and packaging using nanotechnology could be advantageous to farmers who grow food as well as the food industry as a whole. Nanoporous zeolites for slow-release and efficient dosage of water, fertilizers, and medications for livestock; nanocapsules for agrochemical delivery; nanocomposites for plastic film coatings used in food packaging; antimicrobial nanoemulsions for applications in decontaminating food; nanosensors/nanobiosensors for detecting pathogens; soil quality; plant health monitoring; nanobiosensors for identification of pathogens; and nanocomposites for plastic film coatings are used. However, the use of nanotechnology in the agri-food sector is receiving less attention. Additionally, present endeavours are more concerned with lessening the negative impacts that agrochemicals have on the environment and human health than they are with enhancing their properties through the use of nanotechnology for the production of food and livestock.

Experts expect the emergence of numerous nanoparticulate agroformulations with enhanced selectivity, bioavailability, and efficacy in the near future. However, as with the introduction of any new technology, a reliable risk-benefit analysis and in-depth cost accounting research are essential. This calls for the development of reliable approaches for the characterization and quantification of nanomaterials in different matrices as well as for the evaluation of their impact on the environment and on human health. It is essential to engage all stakeholders, including non-governmental organizations and consumer associations, in an open conversation in order to win consumer acceptance and support for this technology.

REFERENCES

Agri, U., Chaudhary, P., & Sharma, A. (2021). In vitro compatibility evaluation of agriusable nanochitosan on beneficial plant growth-promoting rhizobacteria and maize plant. *Natl Acad Sci Lett.* 44, 555–559.

Agri, U., Chaudhary, P., Sharma, A., & Kukreti, B. (2022). Physiological response of maize plants and its rhizospheric microbiome under the influence of potential bioinoculants and nanochitosan. *Plant Soil.* 474, 451–468.

Ahmad, S.A., Das, S.S., Khatoon, A., Ansari, M.T., Afzal, M., Hasnain, M.S., & Nayak, A.K. (2020). Bactericidal activity of silver nanoparticles: a mechanistic review. *Mat Sci Energy Technol.* 3, 756–769.

An, Y., Lian, X., Zou, Y., & Deng, D. (2014). Synthesis of porous ZnO structure for gas sensor and photocatalytic applications. *Colloids Surf A Physicochem Eng Asp.* 447, 81–87

Anon. (2008). The new face of hunger. *The Economist* (17th April 2008).

Awwad, A., & Ibrahim, Q. (2015). Optical and X-ray diffraction characterization of biosynthesis copper oxide nanoparticles using carob leaf extract. *Arab J Phys Chem*. 2, 20–24

Barik, T.K., Sahu, B., & Swain, V. (2008). Nanosilica—from medicine to pest control. *Parasitol Res*. 103(2), 253–258.

Baruah, S., & Dutta, J. (2009). Nanotechnology applications in pollution sensing and degradation in agriculture: a review. *Environ Chem Lett*. 7, 191–204.

Bhardwaj, D., Ansari, M.W., Sahoo, R.K., & Tuteja, N. (2014). Biofertilizers function as key player in sustainable agriculture by improving soil fertility, plant tolerance and crop productivity. *Microb Cell Fac*. 13(1), 1–10.

Bhatt, P., Pandey, S.C., Joshi, S., Chaudhary, P., Pathak, V.M., Huang, Y., Wu, X., Zhou, Z., & Chen, S. (2022). Nanobioremediation: a sustainable approach for the removal of toxic pollutants from the environment. *J Hazard Mater*. 427, 128033.

Boehm, A.L., Martinon, I., Zerrouk, R., Rump, E., & Fessi, H. (2003). Nanoprecipitation technique for the encapsulation of agrochemical active ingredients. *J Microencap*. 20(4), 433–441.

Chanakya, H.N., Mahapatra, D.M., Ravi, S., Chauhan, V.S., & Abitha, R. (2012). Sustainability of large-scale algal biofuel production in India. *J Indian Inst Sci*. 92(1), 63–98.

Chanakya, H.N., Mahapatra, D.M., Sarada, R., & Abitha, R. (2013). Algal biofuel production and mitigation potential in India. *Mitig Adapt Strateg Glob Chang*. 18(1), 113–136.

Chang, H.H., Cheng, T.J., Huang, C.P., & Wang, G.S. (2017). Characterization of titanium dioxide nanoparticle removal in simulated drinking water treatment processes. *Sci Total Environ*. 601, 886–894.

Chaud, M., Souto, E.B., Zielinska, A., Severino, P., Batain, F., Oliveira-Junior, J., & Alves, T. (2021). Nanopesticides in agriculture: benefits and challenge in agricultural productivity, toxicological risks to human health and environment. *Toxics*. 9(6), 131.

Chaudhary, P., Chaudhary, A., Bhatt, P., Kumar, G., Khatoon, H., Rani, A., Kumar, S., & Sharma, A. (2022a). Assessment of soil health indicators under the influence of nanocompounds and *Bacillus* spp. in field condition. *Front Environ Sci*. 9, 769871.

Chaudhary, P., Chaudhary, A., Parveen, H., Rani, A., Kumar, G., Kumar, A., & Sharma, A. (2021e). Impact of nanophos in agriculture to improve functional bacterial community and crop productivity. *BMC Plant Biol*. 21, 519.

Chaudhary, P., Khati, P., Chaudhary, A., Gangola, S., Kumar, R., & Sharma, A. (2021a). Bioinoculation using indigenous *Bacillus* spp. improves growth and yield of *Zea mays* under the influence of nanozeolite. *3 Biotech*. 11, 11.

Chaudhary, P., Khati, P., Chaudhary, A., Maithani, D., Kumar, G., & Sharma, A. (2021d). Cultivable and metagenomic approach to study the combined impact of nanogypsum and *Pseudomonas taiwanensis* on maize plant health and its rhizospheric microbiome. *PLoS One*. 16, e0250574.

Chaudhary, P., Khati, P., Gangola, S., Kumar, A., Kumar, R., & Sharma, A. (2021c). Impact of nanochitosan and *Bacillus* spp. on health, productivity and defence response in *Zea mays* under field condition. *3 Biotech*. 11, 237.

Chaudhary, P., & Sharma, A. (2019). Response of nanogypsum on the performance of plant growth promotory bacteria recovered from nanocompound infested agriculture field. *Environ Ecol*. 37, 363–372.

Chaudhary, P., Sharma, A., Chaudhary, A., Khati, P., Gangola, S., & Maithani, D. (2021b). Illumina based high throughput analysis of microbial diversity of rhizospheric soil of maize infested with nanocompounds and *Bacillus* sp. *Appl Soil Ecol*. 159, 103836.

Chaudhary, P., Singh, S., Chaudhary, A., Sharma, A., & Kumar, G. (2022b). Overview of biofertilizers in crop production and stress management for sustainable agriculture. *Front. Plant Sci*. 13, 930340.

Chaudhary, R. (2020). Green human resource management and employee green behavior: an empirical analysis. *Corp Soc Responsib Environ Manag.* 27(2), 630–641.

Chittora, D., Meena, M., Barupal, T., Swapnil, P., & Sharma, K. (2020). Cyanobacteria as a source of biofertilizers for sustainable agriculture. *Biochem Biophy Rep.* 22, 100737.

Deka, B., Babu, A., Baruah, C., & Barthakur, M. (2021). Nanopesticides: a systematic review of their prospects with special reference to tea pest management. *Front Nutr.* 8, 686131.

Dutta, P., Kumari, A., Mahanta, M., Biswas, K.K., Dudkiewicz, A., Thakuria, D., Abdelrhim, A.S., Singh, S.B., Muthukrishnan, G., Sabarinathan, K.G., Mandal, M.K., & Mazumdar, N. (2022). Advances in nanotechnology as a potential alternative for plant viral disease management. *Front Microbiol.* 13, 935193.

Feistel, L., Halle, O., Mitschker, A., Podszun, W., & Schmid, C. (2002). Production of mono-disperse crosslinked polystyrene beads, useful for producing ion exchangers, comprises a seed-feed method including a micro-encapsulation step to reduce soluble polymer content. US Patent Number 6365683–B2.

Frederiksen, H.K., Kristensen, H.G., & Pedersen, M. (2003). Solid lipid microparticle formulations of the pyrethroid gamma-cyhalothrin—incompatibility of the lipid and the pyrethroid and biological properties of the formulations. *J Control Release.* 86(2–3), 243–252.

Gajjar, P., Pettee, B., Britt, D.W., Huang, W., Johnson, W.P., & Anderson, A.J. (2009). Antimicrobial activities of commercial nanoparticles against an environmental soil microbe, *Pseudomonas putida* KT2440. *J Biol Eng.* 3, 9–10.

Gatoo, M.A., Naseem, S., Arfat, M.Y., Dar, A.M., Qasim, K., & Zubair, S. (2014). Physicochemical properties of nanomaterials: implication in associated toxic manifestations. *Biomed Res Int.* 2014, 498420.

Haddad, M.N., Hasani, A., & Mighadam, M.K. (2017). Comparison the efficiency of Aquasorb and Accepta superabsorbent polymers in improving physical, chemical, and biological properties of soil and tomato turnover under greenhouse condition. *J Water Soil.* 31, 1.

Jampilek, J., Kos, J., & Kralova, K. (2019). Potential of nanomaterial applications in dietary supplements and foods for special medical purposes. *Nanomaterials.* 9(2), 296.

Kar, M., Vijayakumar, P.S., Prasad, B.L.V., & Gupta, S.S. (2010). Synthesis and characterization of poly-L-lysine-grafted silica nanoparticles synthesized via NCA polymerization and click chemistry. *Langmuir.* 268, 5772–5781.

Karunakaran, V., & Behera, U.K. (2016). Tillage and residue management for improving productivity and resource-use efficiency in soybean (glycine max)—wheat (T*riticumaestivum*) cropping system. *Exp Agric.* 52(4), 617–634.

Khati, P., Bhatt, P., Kumar, R., & Sharma, A. (2018). Effect of nanozeolite and plant growth promoting rhizobacteria on maize. *3Biotech.* 8(141), 1–12.

Khati, P., Chaudhary, P., Gangola, S., & Sharma A. (2019a). Influence of nanozeolite on plant growth promotory bacterial isolates recovered from nanocompound infested agriculture field. *Environ Ecol.* 37 (2), 521—527

Khati, P., Gangola, S., Bhatt, P., & Sharma, A. (2017a). Nanochitosan induced growth of *Zea Mays* with soil health maintenance. *3 Biotech.* 7(81), 1–9.

Khati, P., Sharma A., Chaudhary, P., Singh, A.K., Gangola, S., & Kumar, R. (2019b). High-throughput sequencing approach to access the impact of nanozeolite treatment on species richness and evenness of soil metagenome. *Biocatal Agric Biotech.* 20, 101249.

Khati, P., Sharma A., Gangola, S., Kumar, R., Bhatt, P., & Kumar, G. (2017b). Impact of some agriusable nanocompounds on soil microbial activity: an indicator of soil health. *Clean Soil Air Water.* 45 (5), 1–7.

Khodakovskaya, M., Dervishi, E., Mahmood, M., Xu, Y., Li, Z., Watanabe, F., Biris, A.S. (2009). Carbon nanotubes are able to penetrate plant seed coat and dramatically affect seed germination and plant growth. *ACS Nano.* 3(10), 3221–3227.

Kitching, M., Ramani, M., & Marsili, E. (2015). Fungal biosynthesis of gold nanoparticles: mechanism and scale up. *Micro Biotechnol.* 8(6), 904–917.

Kukreti, B., Sharma, A., Chaudhary, P., Agri, U., & Maithani, D. (2020). Influence of nanosilicon dioxide along with bioinoculants on *Zea mays* and its rhizospheric soil. *3 Biotech.* 10, 345.

Kumari, H., Khati, P., Gangola, S., Chaudhary, P., & Sharma, A. (2021). Performance of plant growth promotory rhizobacteria on maize and soil characteristics under the influence of TiO$_2$ nanoparticles. *Pant Res J.* 19(1), 28–39.

Kumari, S., Sharma A., Chaudhary, P., & Khati, P. (2020). Management of plant vigour and soil health using two agriusable nanocompounds and plant growth promontory rhizobacteria in Fenugreek. *3 Biotech.* 10(461), 1–11.

Lai, F., Wissing, S.A., Müller, R.H., & Fadda, A.M. (2006). *Artemisia arborescens* L essential oil-loaded solid lipid nanoparticles for potential agricultural application: preparation and characterization. *AAPS PharmSciTech.* 7(1), E10–E18.

Laing, R., Gillan, V., & Devaney, E. (2017). Ivermectin–old drug, new tricks? *Trends Parasitol.* 33(6), 463–472.

Lee, L. H., Wu, T. Y., Shak, K.P.Y., Lim, S. L., Ng, K.Y., Nguyen, M.N., & Teoh, W.H. (2018). Sustainable approach to biotransform industrial sludge into organic fertilizer via vermicomposting: a mini-review. *J Chem Technol Biotechnol.* 93(4), 925–935.

Li, Z. Z., Chen, J. F., Liu, F., Liu, A. Q., Wang, Q., Sun, H. Y., & Wen, L. X. (2007). Study of UV shielding properties of novel porous hollow silica nanoparticle carriers for avermectin. *Pest Sci.* 63(3), 241–246.

Liu, Y., Yan, L., Heiden, P., & Laks, P. (2001). Use of nanoparticles for controlled release of biocides in solid wood. *J Appl Polym Sci.* 79(3), 458–465.

Lokko, Y., Heijde, M., Schebesta, K., Scholtès, P., Van Montagu, M., & Giacca, M. (2018). Biotechnology and the bioeconomy—Towards inclusive and sustainable industrial development. *New Biotechnol.* 40, 5–10.

Mahapatra, D.M., & Murthy, G.S. (2021). Long term evaluation of a pilot scale multimodal algal bioprocess for treatment of municipal wastewater. *J Clean Prod.* 311, 127690.

Mahapatra, D.M., Satapathy, K.C., & Panda, B. (2022). Biofertilizers and nanofertilizers for sustainable agriculture: phycoprospects and challenges. *Sci Total Environ.* 803, 149990.

Manjunatha, S.B., Biradar, D.P., & Aladakatti, Y.R. (2016). Nanotechnology and its applications in agriculture: a review. *J Farm Sci.* 29(1), 1–13.

Mohan, S., Singh, Y., Verna, D.K. & Hassan, S.H. (2015), Synthesis of CuO nanoparticles through green route using Citrus Limon juice and its application as nanosorbent for Cr (VI) remediation: process optimization with RSM and ANN-GA based model. *Process Safety Environ Prot.* 96, 156–166

Nicolopoulou-Stamati, P., Maipas, S., Kotampasi, C., Stamatis, P., & Hens, L. (2016). Chemical pesticides and human health: the urgent need for a new concept in agriculture. *Front Public Health.* 4, 148.

Panpatte, D.G., Jhala, Y.K., Shelat, H.N., & Vyas, R.V. (2016). Nanoparticles: the next generation technology for sustainable agriculture. In *Microbial Inoculants in Sustainable Agricultural Productivity* (pp. 289–300). Springer, New Delhi.

Pokrajac, L., Nazar, L., Chen, Z., & Mitra, S. (2019). The Waterloo Institute for Nanotechnology: Societal impact and a sustainable future. *ACS Nano.* 13, 12247–12253.

Rashed, S., Abdel Aziz, R., Moustafa, S., & Neana, S. (2016). Raising productivity and quality of sugar beet using of seed inoculation and foliar application with *Azospirillumbrasiliense* and *Bacillus megatherium*. *J Soil Sci Agric Eng.* 7(1), 89–96.

Salerno, M., Landoni, P., & Verganti, R. (2008). Funded research projects as a tool for the analysis of emerging fields: the case of nanotechnology. In *SPRU 40th Anniversary Conference–The Future of Science, Technology and Innovation Policy* (Vol. 6).

Sastry, R. K., Rao, N. H., Cahoon, R., & Tucker, T. (2007). Can nanotechnology provide the innovations for a second green revolution in Indian agriculture. In *NSF Nanoscale Science and Engineering Grantees Conference* (pp. 3–6).

Servin, A.D., & White, J.C. (2016). Nanotechnology in agriculture: next steps for understanding engineered nanoparticle exposure and risk. *Nano Impact*. 1, 9–12.

Shang, Y., Hasan, M.K., Ahammed, G.J., Li, M., Yin, H., & Zhou, J. (2019). Applications of nanotechnology in plant growth and crop protection: a review. *Molecules*. 24(14), 2558.

Sylwia, L., Krzysztof, S., Ewa, E. & Marta, D. (2015). Biocompatible polymeric nanoparticles as promising candidates for drug delivery. *Langmuir: ACS J Surf Coll*. 31, 6415–6425.

Tiwari, G., Tiwari, R., Sriwastawa, B., Bhati, L., Pandey, S., Pandey, P., & Bannerjee, S.K. (2012). Drug delivery systems: an updated review. *Int J Pharm Investig*. 2(1), 2–11.

Torney, F., Trewyn, B.G., Lin, V.S.Y., & Wang, K. (2007). Mesoporous silica nanoparticles deliver DNA and chemicals into plants. *Nat Nanotech*. 2(5), 295–300.

Tyagi, J., Chaudhary, P., Mishra, A., Khatwani, M., Dey, S., & Varma, A. (2022). Role of endophytes in abiotic stress tolerance: with special emphasis on *Serendipita indica*. *Int J Environ Res*. 16, 62.

Vander Gheynst, J.S., Scher, H.B., & Guo, H. (2007). AGRO 113-storage and delivery of aquatic microorganisms in emulsions stabilized by surface-modified silica nanoparticles. In *Abstracts of Papers of the American Chemical Society* (Vol. 234). American Chemical Society, Washington, DC.

Vermeiren, L., Devlieghere, F., & Debevere, J. (2002). Effectiveness of some recent antimicrobial packaging concepts. *Food Addit Contam*. 19(Suppl), 163–161

Wilson, B., Samanta, M.K., Santhi, K., Kumar, K.P.S., Paramakrishnan, N., & Suresh, B. (2008). Poly (n-butyl cyanoacrylate) nanoparticles coated with polysorbate 80 for the targeted delivery of rivastigmine into the brain to treat Alzheimer's disease. *Brain Res*. 1200, 159–168.

Wreford, A., Bayne, K., Edwards, P., & Renwick, A. (2019). Enabling a transformation to a bioeconomy in New Zealand. *Environ Innov Soc Transit*. 31, 184–199.

Index

For Product Safety Concerns and Information please contact our EU
representative GPSR@taylorandfrancis.com
Taylor & Francis Verlag GmbH, Kaufingerstraße 24, 80331 München, Germany

www.ingramcontent.com/pod-product-compliance
Lightning Source LLC
Chambersburg PA
CBHW060749220326
41598CB00022B/2383